Introduction to Geometry

Introduction to

GEOMETRY

second edition

H. S. M. COXETER, F. R. S.

Professor of Mathematics
University of Toronto

JOHN WILEY & SONS, INC.

New York · London · Sydney · Toronto

Preface

I am grateful to the readers of the first edition who have made suggestions for improvement. Apart from some minor corrections, the principal changes are as follows.

The equation connecting the curvatures of four mutually tangent circles, now known as the *Descartes Circle Theorem* (p. 12), is proved along the lines suggested by Mr. Beecroft on pp. 91–96 of "The Lady's and Gentleman's Diary for the year of our Lord 1842, being the second after Bissextile, designed principally for the amusement and instruction of Students in Mathematics: comprising many useful and entertaining particulars, interesting to all persons engaged in that delightful pursuit."

For *similarity* in the plane, a new treatment (pp. 73–76) was suggested by A. L. Steger when he was a sophomore at the University of Toronto. For similarity in space, a different treatment (p. 103) was suggested by Professor Maria Wonenburger. A new exercise on p. 90 introduces the useful concept of *inversive distance*. Another has been inserted on p. 127 to exhibit R. Krasnodębski's drawings of symmetrical loxodromes.

Pages 203–208 have been revised so as to clarify the treatment of *affinities* (which preserve collinearity) and *equiaffinities* (which preserve area). The new material includes some challenging exercises. For the discovery of *finite* geometries (p. 237), credit has been given to von Staudt, who anticipated Fano by 36 years.

Page 395 records the completion, in 1968, by G. Ringel and J. W. T. Youngs, of a project begun by Heawood in 1890. The result is that we now know, for every kind of surface except the sphere (or plane), the minimal number of colors that will suffice for coloring every map on the surface.

Answers are now given for practically all the exercises; a separate booklet is no longer needed. One of the prettiest answers (p. A-35) was kindly supplied by Professor P. Szász of Budapest.

H.S.M. Coxeter

Toronto, Canada
January, 1969

Preface to the first edition

For the last thirty or forty years, most Americans have somehow lost interest in geometry. The present book constitutes an attempt to revitalize this sadly neglected subject.

The four parts correspond roughly to the four years of college work. However, most of Part II can be read before Part I, and most of Part IV before Part III. The first eleven chapters (that is, Parts I and II) will provide a course for students who have some knowledge of Euclid and elementary analytic geometry but have not yet made up their minds to specialize in mathematics, or for enterprising high school teachers who wish to see what is happening just beyond their usual curriculum. Part III deals with the foundations of geometry, including projective geometry and hyperbolic non-Euclidean geometry. Part IV introduces differential geometry, combinatorial topology, and four-dimensional Euclidean geometry.

In spite of the large number of cross references, each of the twenty-two chapters is reasonably self-contained; many of them can be omitted on first reading without spoiling one's enjoyment of the rest. For instance, Chapters 1, 3, 6, 8, 13, and 17 would make a good short course. There are relevant exercises at the end of almost every section; the hardest of them are provided with hints for their solution. (Answers to some of the exercises are given at the end of the book. Answers to many of the remaining exercises are provided in a separate booklet, available from the publisher upon request.) The unifying thread that runs through the whole work is the idea of a group of transformations or, in a single word, *symmetry*.

The customary emphasis on analytic geometry is likely to give students the impression that geometry is merely a part of algebra or of analysis. It is refreshing to observe that there are some important instances (such as the Argand diagram described in Chapter 9) in which geometrical ideas are needed as essential tools in the development of these other branches of mathematics. The scope of geometry was spectacularly broadened by Klein in his *Erlanger Programm* (Erlangen program) of 1872, which stressed the fact that, besides plane and solid Euclidean geometry, there are many other geometries equally worthy of attention. For instance, many of Euclid's own propositions belong to the wider field of *affine* geometry, which is valid not only in ordinary space but also in Minkowski's space-time, so successfully exploited by Einstein in his special theory of relativity.

Geometry is useful not only in algebra, analysis, and cosmology, but also in kinematics and crystallography (where it is associated with the theory of groups), in statistics (where finite geometries help in the design of experiments), and even in botany. The subject of topology (Chapter 21) has been developed so widely that it now stands on its own feet instead of being regarded as part of geometry; but it fits into the Erlangen program, and its early stages have the added appeal of a famous unsolved problem: that of deciding whether every possible map can be colored with four colors.

The material grew out of courses of lectures delivered at summer institutes for school teachers and others at Stillwater, Oklahoma; Lunenburg, Nova Scotia; Ann Arbor, Michigan; Stanford, California; and Fredericton, New Brunswick, along with several public lectures given to the Friends of Scripta Mathematica in New York City by invitation of the late Professor Jekuthiel Ginsburg. The most popular of these separate lectures was the one on the golden section and phyllotaxis, which is embodied in Chapter 11.

Apart from the general emphasis on the idea of transformation and on the desirability of spending some time in such unusual environments as affine space and absolute space, the chief novelties are as follows: a simple treatment of the orthocenter (§ 1.6); the use of dominoes to illustrate six of the seventeen space groups of two-dimensional crystallography (§ 4.4); a construction for the invariant point of a dilative reflection (§ 5.6); a description of the general circle-preserving transformation (§ 6.7) and of the spiral similarity (§ 7.6); an "explanation" of phyllotaxis (§ 11.5); an "ordered" treatment of Sylvester's problem (§ 12.3); an economical system of axioms for affine geometry (§ 13.1); an "absolute" treatment of rotation groups (§ 15.4); an elementary treatment of the horosphere (§ 16.8) and of the extreme ternary quadratic form (§ 18.4); the correction of a prevalent error concerning the shape of the monkey saddle (§ 19.8); an application of geodesic polar coordinates to the foundations of hyperbolic trigonometry (§ 20.6); the classification of regular maps on the sphere, projective plane, torus, and Klein bottle (§ 21.3); and the suggestion of a statistical honeycomb (§ 22.5).

I offer sincere thanks to M. W. Al-Dhahir, J. J. Burckhardt, Werner Fenchel, L. M. Kelly, Peter Scherk, and F. A. Sherk for critically reading various chapters; also to H G. Forder, Martin Gardner, and C. J. Scriba for their help in proofreading, to S. H. Gould, J. E. Littlewood, and J. L. Synge for permission to quote certain passages from their published works, and to M. C. Escher, I. Kitrosser, and the Royal Society of Canada for permission to reproduce the plates.

H.S.M. Coxeter

Toronto, Canada
March, 1961

Contents

Part I

Part II

Part III

Part IV

Plates

Mathematics possesses not only truth, but supreme beauty
—a beauty cold and austere, like that of sculpture,
without appeal to any part of our weaker nature . . .
sublimely pure, and capable of a stern perfection
such as only the greatest art can show.

BERTRAND RUSSELL (1872–   )

Introduction to Geometry

Part I

1

Triangles

In this chapter we review some of the well-known propositions of elementary geometry, stressing the role of symmetry. We refer to Euclid's propositions by his own numbers, which have been used throughout the world for more than two thousand years. Since the time of F. Commandino (1509–1575), who translated the works of Archimedes, Apollonius, and Pappus, many other theorems in the same spirit have been discovered. Such results were studied in great detail during the nineteenth century. As the present tendency is to abandon them in favor of other branches of mathematics, we shall be content to mention a few that seem particularly interesting.

1.1 EUCLID

> Euclid's work will live long after all the text-books of the present day
> are superseded and forgotten. It is one of the noblest monuments of
> antiquity.
>
> Sir Thomas L. Heath (1861 -1940)*

About 300 B.C., Euclid of Alexandria wrote a treatise in thirteen books called the *Elements*. Of the author (sometimes regrettably confused with the earlier philosopher, Euclid of Megara) we know very little. Proclus (410–485 A.D.) said that he "put together the Elements, collecting many of Eudoxus's theorems, perfecting many of Theaetetus's, and also bringing to irrefragable demonstration the things which were only somewhat loosely proved by his predecessors. This man lived in the time of the first Ptolemy, [who] once asked him if there was in geometry any shorter way than that of the Elements, and he answered that there was no royal road to geometry." Heath quotes a story by Stobaeus, to the effect that someone who had begun to read geometry with Euclid asked him "What shall I get by learning these things?" Euclid called his slave and said "Give him a dime, since he must make gain out of what he learns."

* Heath **1,** p. vi. (Such references are collected at the end of the book, pp. 415–417.)

3

Of the thirteen books, the first six may be very briefly described as dealing respectively with triangles, rectangles, circles, polygons, proportion, and similarity. The next four, on the theory of numbers, include two notable achievements: IX.2 and X.9, where it is proved that there are infinitely many prime numbers, and that $\sqrt{2}$ is irrational [Hardy **2**, pp. 32–36]. Book XI is an introduction to solid geometry, XII deals with pyramids, cones, and cylinders, and XIII is on the five regular solids.

According to Proclus, Euclid "set before himself, as the end of the whole Elements, the construction of the so-called Platonic figures." This notion of Euclid's purpose is supported by the Platonic theory of a mystical correspondence between the four solids

$$\left.\begin{array}{l}\text{cube,}\\ \text{tetrahedron,}\\ \text{octahedron,}\\ \text{icosahedron}\end{array}\right\}\text{ and the four "elements"}\left\{\begin{array}{l}\text{earth,}\\ \text{fire,}\\ \text{air,}\\ \text{water}\end{array}\right.$$

[cf. Coxeter **1**, p. 18]. Evidence to the contrary is supplied by the arithmetical books VII–X, which were obviously included for their intrinsic interest rather than for any application to solid geometry.

1.2   PRIMITIVE CONCEPTS AND AXIOMS

> "When I use a word," Humpty-Dumpty said, "it means just what I choose it to mean—neither more nor less."
>
> Lewis Carroll (1832 -1898)
>
> [Dodgson **2**, Chap. 6]

In the logical development of any branch of mathematics, each definition of a concept or relation involves other concepts and relations. Therefore the only way to avoid a vicious circle is to allow certain *primitive* concepts and relations (usually as few as possible) to remain undefined [Synge **1**, pp. 32–34]. Similarly, the proof of each proposition uses other propositions, and therefore certain primitive propositions, called *postulates* or *axioms*, must remain unproved. Euclid did not specify his primitive concepts and relations, but was content to give definitions in terms of ideas that would be familiar to everybody. His five Postulates are as follows:

1.21   *A straight line may be drawn from any point to any other point.*

1.22   *A finite straight line may be extended continuously in a straight line.*

1.23   *A circle may be described with any center and any radius.*

1.24   *All right angles are equal to one another.*

1.25   *If a straight line meets two other straight lines so as to make the two interior angles on one side of it together less than two right angles, the other* *straight lines, if extended indefinitely, will meet on that side on which the angles are less than two right angles.**

It is quite natural that, after a lapse of about 2250 years, some details are now seen to be capable of improvement. (For instance, Euclid I.1 constructs an equilateral triangle by drawing two circles; but how do we know that these two circles will intersect?) The marvel is that so much of Euclid's work remains perfectly valid. In the modern treatment of his geometry [see, for instance, Coxeter **3**, pp. 161–187], it is usual to recognize the primitive concept *point* and the two primitive relations of *intermediacy* (the idea that one point may be between two others) and *congruence* (the idea that the distance between two points may be equal to the distance between two other points, or that two line segments may have the same length). There are also various versions of the axiom of *continuity,* one of which says that every convergent sequence of points has a limit.

Euclid's "principle of superposition," used in proving I.4, raises the question whether a figure can be moved without changing its internal structure. This principle is nowadays replaced by a further explicit assumption such as the axiom of "the rigidity of a triangle with a tail" (Figure 1.2*a*):

1.26   *If ABC is a triangle with D on the side BC extended, while D' is analogously related to another triangle $A'B'C'$, and if $BC = B'C'$, $CA = C'A'$, $AB = A'B'$, $BD = B'D'$, then $AD = A'D'$.*

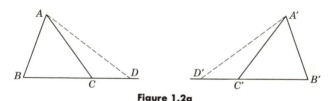

Figure 1.2a

This axiom can be used to extend the notion of congruence from line segments to more complicated figures such as angles, so that we can say precisely what we mean by the relation

$$\angle\, ABC = \angle\, A'B'C'.$$

Then we no longer need the questionable principle of superposition in order to prove Euclid I.4:

If two triangles have two sides equal to two sides respectively, and have the angles contained by the equal sides equal, they will also have their third sides equal, and their remaining angles equal respectively; in fact, they will be congruent triangles.

---

* In Chapter 15 we shall see how far we can go without using this unpleasantly complicated Fifth Postulate.

1.3 PONS ASINORUM

Minos: *It is proposed to prove I.5 by taking up the isosceles Triangle, turning it over, and then laying it down again upon itself.*

Euclid: *Surely that has too much of the Irish Bull about it, and reminds one a little too vividly of the man who walked down his own throat, to deserve a place in a strictly philosophical treatise?*

Minos: *I suppose its defenders would say that it is conceived to leave a trace of itself behind, and that the reversed Triangle is laid down upon the trace so left.*

C. L. Dodgson (1832-1898)

[Dodgson **3**, p. 48]

I.5. *The angles at the base of an isosceles triangle are equal.*

The name *pons asinorum* for this famous theorem probably arose from the bridgelike appearance of Euclid's figure (with the construction lines required in his rather complicated proof) and from the notion that anyone unable to cross this bridge must be an ass. Fortunately, a far simpler proof was supplied by Pappus of Alexandria about 340 A.D. (Figure 1.3*a*):

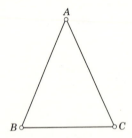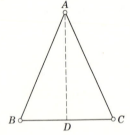

Figure 1.3a

Let *ABC* be an isosceles triangle with *AB* equal to *AC*. Let us conceive this triangle as two triangles and argue in this way. Since $AB = AC$ and $AC = AB$, the two sides *AB*, *AC* are equal to the two sides *AC*, *AB*. Also the angle *BAC* is equal to the angle *CAB*, for it is the same. Therefore all the corresponding parts (of the triangles *ABC*, *ACB*) are equal. In particular,

$$\angle ABC = \angle ACB.$$

The pedagogical difficulty of comparing the isosceles triangle *ABC* with itself is sometimes avoided by joining the apex *A* to *D*, the midpoint of the base *BC*. The median *AD* may be regarded as a *mirror* reflecting *B* into *C*. Accordingly, we say that an isosceles triangle is symmetrical by *reflection*, or that it has *bilateral symmetry*. (Of course, the idealized mirror used in geometry has no thickness and is silvered on both sides, so that it not only reflects *B* into *C* but also reflects *C* into *B*.)

Any figure, however irregular its shape may be, yields a symmetrical figure when we place it next to a mirror and waive the distinction between object and image. Such bilateral symmetry is characteristic of the external shape of most animals.

Given any point P on either side of a geometrical mirror, we can construct its reflected image P' by drawing the perpendicular from P to the mirror and extending this perpendicular line to an equal distance on the other side, so that the mirror perpendicularly bisects the line segment PP'. Working in the plane (Figure 1.3b) with a line AB for mirror, we draw two circles with centers A, B and radii AP, BP. The two points of intersection of these circles are P and its image P'.

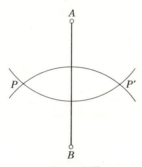

Figure 1.3b

We shall find that many geometrical proofs are shortened and made more vivid by the use of reflections. But we must remember that this procedure is merely a short cut: every such argument could have been avoided by means of a circumlocution involving congruent triangles. For instance, the above construction is valid because the triangles ABP, ABP' are congruent.

Pons asinorum has many useful consequences, such as the following five:

III.3.   *If a diameter of a circle bisects a chord which does not pass through the center, it is perpendicular to it; or, if perpendicular to it, it bisects it.*

III.20.   *In a circle the angle at the center is double the angle at the circumference, when the rays forming the angles meet the circumference in the same two points.*

III.21.   *In a circle, a chord subtends equal angles at any two points on the same one of the two arcs determined by the chord* (e.g., in Figure 1.3c, $\angle PQQ' = \angle PP'Q'$).

III.22.   *The opposite angles of any quadrangle inscribed in a circle are together equal to two right angles.*

III.32.   *If a chord of a circle be drawn from the point of contact of a tangent, the angle made by the chord with the tangent is equal to the angle subtended by the chord at a point on that part of the circumference which lies on the far side of the chord* (e.g., in Figure 1.3c, $\angle OTP' = \angle TPP'$).

We shall also have occasion to use two familiar theorems on similar triangles:

VI.2.   *If a straight line be drawn parallel to one side of a triangle, it will cut the other sides proportionately; and, if two sides of the triangle be cut proportionately, the line joining the points of section will be parallel to the remaining side.*

VI.4.   *If corresponding angles of two triangles are equal, then corresponding sides are proportional.*

Combining this last result with III.21 and 32, we deduce two significant properties of secants of a circle (Figure 1.3c):

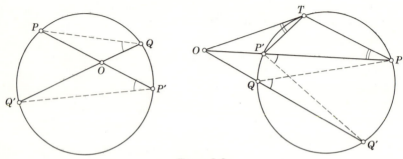

Figure 1.3c

III.35.   *If in a circle two straight lines cut each other, the rectangle contained by the segments of the one is equal to the rectangle contained by the segments of the other* (i.e., $OP \times OP' = OQ \times OQ'$).

III.36.   *If from a point outside a circle a secant and a tangent be drawn, the rectangle contained by the whole secant and the part outside the circle will be equal to the square on the tangent* (i.e., $OP \times OP' = OT^2$).

Book VI also contains an important property of area:

VI.19.   *Similar triangles are to one another in the squared ratio of their corresponding sides* (i.e., if ABC and $A'B'C'$ are similar triangles, their areas are in the ratio $AB^2 : A'B'^2$).

This result yields the following easy proof for the theorem of Pythagoras [see Heath **1**, p. 353; **2**, pp. 210, 232, 269]:

I.47.   *In a right-angled triangle, the square on the hypotenuse is equal to the sum of the squares on the two catheti.*

In the triangle ABC, right-angled at C, draw CF perpendicular to the hypotenuse AB, as in Figure 1.3d. Then we have three similar right-angled triangles ABC, ACF, CBF, with hypotenuses AB, AC, CB. By VI.19, the areas satisfy

$$\frac{ABC}{AB^2} = \frac{ACF}{AC^2} = \frac{CBF}{CB^2}.$$

Evidently, $ABC = ACF + CBF$. Therefore $AB^2 = AC^2 + CB^2$.

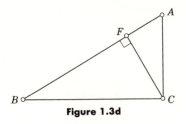

Figure 1.3d

EXERCISES

1. Using rectangular Cartesian coordinates, show that the reflection in the y-axis ($x = 0$) reverses the sign of x. What happens when we reflect in the line $x = y$?

2. Deduce I.47 from III.36 (applied to the circle with center A and radius AC).

3. Inside a square $ABDE$, take a point C so that CDE is an isosceles triangle with angles 15° at D and E. What kind of triangle is ABC?

4. Prove the Erdös-Mordell theorem: If O is any point inside a triangle ABC and P, Q, R are the feet of the perpendiculars from O upon the respective sides BC, CA, AB, then

$$OA + OB + OC \geqslant 2(OP + OQ + OR).$$

(*Hint:** Let P_1 and P_2 be the feet of the perpendiculars from R and Q upon BC. Define analogous points Q_1 and Q_2, R_1 and R_2 on the other sides. Using the similarity of the triangles PRP_1 and OBR, express P_1P in terms of RP, OR, and OB. After substituting such expressions into

$$OA + OB + OC \geqslant OA(P_1P + PP_2)/RQ + OB(Q_1Q + QQ_2)/PR$$
$$+ OC(R_1R + RR_2)/QP,$$

collect the terms involving OP, OQ, OR, respectively.)

5. Under what circumstances can the sign $\geqslant$ in Ex. 4 be replaced by $=$?

6. In the notation of Ex. 4,

$$OA \times OB \times OC \geqslant (OQ + OR)(OR + OP)(OP + OQ).$$

(A. Oppenheim, *American Mathematical Monthly,* **68** (1961), p. 230. See also L. J. Mordell, *Mathematical Gazette,* **46** (1962), pp. 213–215.)

7. Prove the Steiner-Lehmus theorem: Any triangle having two equal internal angle bisectors (each measured from a vertex to the opposite side) is isosceles. (*Hint:*† If a triangle has two different angles, the smaller angle has the longer internal bisector.)

---

* Leon Bankoff, *American Mathematical Monthly,* **65** (1958), p. 521. For other proofs see G. R. Veldkamp and H. Brabant, *Nieuw Tijdschrift voor Wiskunde,* **45** (1958), pp. 193–196; **46** (1959), p. 87.

† Court **2,** p. 72. For Lehmus's proof of 1848, see Coxeter and Greitzer **1,** p. 15.

1.4   THE MEDIANS AND THE CENTROID

> Oriental mathematics may be an interesting curiosity, but Greek mathe-
> matics is the real thing. . . . The Greeks, as Littlewood said to me once,
> are not clever schoolboys or "scholarship candidates," but "Fellows of
> another college." So Greek mathematics is "permanent," more per-
> manent even than Greek literature.   Archimedes will be remembered
> when Aeschylus is forgotten, because languages die and mathematical
> ideas do not.
>
> G. H. Hardy (1877 -1947)
>
> [Hardy **2,** p. 21]

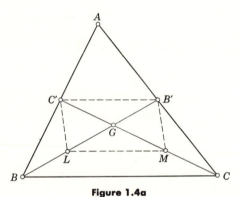

Figure 1.4a

The line joining a vertex of a triangle to the midpoint of the opposite side is called a *median*.

Let two of the three medians, say BB' and CC', meet in G (Figure 1.4a). Let L and M be the midpoints of GB and GC. By Euclid VI.2 and 4 (which were quoted on page 8), both $C'B'$ and LM are parallel to BC and half as long. Therefore $B'C'LM$ is a parallelogram. Since the diagonals of a paral-lelogram bisect each other, we have

$$B'G = GL = LB, \qquad C'G = GM = MC.$$

Thus the two medians BB', CC' trisect each other at G. In other words, this point G, which could have been defined as a point of trisection of one median, is also a point of trisection of another, and similarly of the third. We have thus proved [by the method of Court **1,** p. 58] the following theorem:

1.41   *The three medians of any triangle all pass through one point.*

This common point G of the three medians is called the *centroid* of the triangle.   Archimedes (*c.* 287–212 B.C.) obtained it as the center of gravity of a triangular plate of uniform density.

EXERCISES

1. Any triangle having two equal medians is isosceles.*

2. The sum of the medians of a triangle lies between $\frac{3}{4}$ p and p, where p is the sum of the sides. [Court **1,** pp. 60–61.]

1.5 THE INCIRCLE AND THE CIRCUMCIRCLE

Alone at nights,
I read my Bible more and Euclid less.

Robert Buchanan (1841 -1901)

(An Old Dominie's Story)

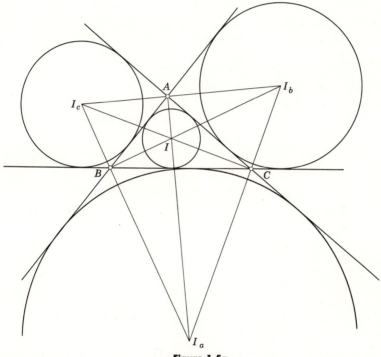

Figure 1.5a

Euclid III.3 tells us that a circle is symmetrical by reflection in any diameter (whereas an ellipse is merely symmetrical about two special diameters: the major and minor axes). It follows that the angle between two intersecting tangents is bisected by the diameter through their common point.

---

* It is to be understood that any exercise appearing in the form of a theorem is intended to be *proved.* It saves space to omit the words "Prove that" or "Show that."

By considering the loci of points equidistant from pairs of sides of a triangle ABC, we see that the internal and external bisectors of the three angles of the triangle meet by threes in four points I, I_a, I_b, I_c, as in Figure 1.5a. These points are the centers of the four circles that can be drawn to touch the three lines BC, CA, AB. One of them, the *incenter I*, being inside the triangle, is the center of the inscribed circle or *incircle* (Euclid IV.4). The other three are the *excenters* I_a, I_b, I_c: the centers of the three escribed circles or *excircles* [Court **2**, pp. 72–88]. The radii of the incircle and excircles are the *inradius r* and the *exradii* r_a, r_b, r_c.

In describing a triangle ABC, it is customary to call the sides

$$a = BC, \qquad b = CA, \qquad c = AB,$$

the semiperimeter

$$s = \tfrac{1}{2}(a + b + c),$$

the angles A, B, C, and the area Δ.

Since $A + B + C = 180°$, we have

1.51 $$\angle BIC = 90° + \tfrac{1}{2}A,$$

a result which we shall find useful in § 1.9.

Since IBC is a triangle with base a and height r, its area is $\tfrac{1}{2}ar$. Adding three such triangles we deduce

$$\Delta = \tfrac{1}{2}(a + b + c)r = sr.$$

Similarly $\Delta = \tfrac{1}{2}(b + c - a)r_a = (s - a)r_a$. Thus

1.52 $$\Delta = sr = (s - a)r_a = (s - b)r_b = (s - c)r_c.$$

From the well-known formula $\cos A = (b^2 + c^2 - a^2)/2bc$, we find also

$$\sin A = [-a^4 - b^4 - c^4 + 2b^2c^2 + 2c^2a^2 + 2a^2b^2]^{\frac{1}{2}}/2bc,$$

whence

1.53
$$\begin{aligned}
\Delta &= \tfrac{1}{2}\,bc \sin A \\
&= \tfrac{1}{4}[-a^4 - b^4 - c^4 + 2b^2c^2 + 2c^2a^2 + 2a^2b^2]^{\frac{1}{2}} \\
&= \tfrac{1}{4}[(a + b + c)(-a + b + c)(a - b + c)(a + b - c)]^{\frac{1}{2}} \\
&= [s(s - a)(s - b)(s - c)]^{\frac{1}{2}}.
\end{aligned}$$

This remarkable expression, which we shall use in § 18.4, is attributed to Heron of Alexandria (about 60 A.D.), but it was really discovered by Archimedes. (See B. L. van der Waerden, *Science Awakening*, Oxford University Press, New York, 1961, pp. 228, 277.) Combining Heron's formula with 1.52, we obtain

1.531 $$r^2 = \left(\frac{\Delta}{s}\right)^2 = \frac{(s - a)(s - b)(s - c)}{s}, \quad r_a^2 = \left(\frac{\Delta}{s - a}\right)^2 = \frac{s(s - c)(s - b)}{s - a}.$$

Another consequence of the symmetry of a circle is that the perpendicular bisectors of the three sides of a triangle all pass through the *circumcenter O*,

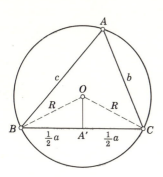

Figure 1.5b

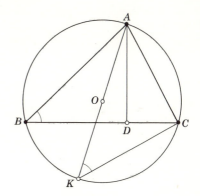

Figure 1.5c

which is the center of the circumscribed circle or *circumcircle* (Euclid IV.5). This is the only circle that can be drawn through the three vertices A, B, C. Its radius R is called the *circumradius* of the triangle. Since the "angle at the center," $\angle BOC$ (Figure 1.5b), is double the angle A, the congruent right-angled triangles OBA', OCA' each have an angle A at O, whence

$$R \sin A = BA' = \tfrac{1}{2}a,$$

1.54
$$2R = \frac{a}{\sin A} = \frac{b}{\sin B} = \frac{c}{\sin C}.$$

Draw AD perpendicular to BC, and let AK be the diameter through A of the circumcircle, as in Figure 1.5c. By Euclid III.21, the right-angled triangles ABD and AKC are similar; therefore

$$\frac{AD}{AB} = \frac{AC}{AK}, \quad AD = \frac{bc}{2R}.$$

Since $\Delta = \tfrac{1}{2}BC \times AD$, it follows that

1.55
$$\begin{aligned}
4\Delta R &= abc \\
&= s(s-b)(s-c) + s(s-c)(s-a) + s(s-a)(s-b) \\
&\qquad\qquad\qquad\qquad\qquad\qquad - (s-a)(s-b)(s-c) \\
&= \frac{\Delta^2}{s-a} + \frac{\Delta^2}{s-b} + \frac{\Delta^2}{s-c} - \frac{\Delta^2}{s} \\
&= \Delta(r_a + r_b + r_c - r).
\end{aligned}$$

Hence the five radii are connected by the formula

1.56
$$4R = r_a + r_b + r_c - r.$$

Let us now consider four circles E_1, E_2, E_3, E_4, tangent to one another at six distinct points. Each circle E_i has a *bend* ε_i, defined as the reciprocal of its radius with a suitable sign attached, namely, if all the contacts are external (as in the case of the light circles in Figure 1.5d), the bends are all positive, but if one circle surrounds the other three (as in the case of the

Figure 1.5d

heavy circles) the bend of this largest circle is taken to be negative; and a line counts as a circle of bend 0. In any case, the sum of all four bends is positive.

In a letter of November 1643 to Princess Elisabeth of Bohemia, René Descartes developed a formula relating the radii of four mutually tangent circles. In the "bend" notation it is

1.57 $$2(\varepsilon_1{}^2 + \varepsilon_2{}^2 + \varepsilon_3{}^2 + \varepsilon_4{}^2) = (\varepsilon_1 + \varepsilon_2 + \varepsilon_3 + \varepsilon_4)^2.$$

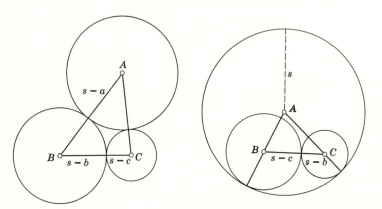

Figure 1.5e   **Figure 1.5f**

This *Descartes circle theorem* was rediscovered in 1842 by an English amateur, Philip Beecroft, who observed that the four circles E_i determine another set of four circles H_i, mutually tangent at the same six points: H_1 through the three points of contact of E_2, E_3, E_4, and so on. Let η_i denote the bend of H_i. If the centers of E_1, E_2, E_3 form a triangle ABC, H_4 is either the incircle or an excircle. In the former case (Figure 1.5e),

1.58
$$\varepsilon_1 = \frac{1}{s-a}, \quad \varepsilon_2 = \frac{1}{s-b}, \quad \varepsilon_3 = \frac{1}{s-c}, \quad \eta_4 = \mp\frac{1}{r}.$$

In the latter (Figure 1.5*f*),

$$\varepsilon_1 = -\frac{1}{s}, \quad \varepsilon_2 = \frac{1}{s-c}, \quad \varepsilon_3 = \frac{1}{s-b}, \quad \eta_4 = \pm\frac{1}{r_a}.$$

In either case, we see from 1.531 that

$$\varepsilon_2\varepsilon_3 + \varepsilon_3\varepsilon_1 + \varepsilon_1\varepsilon_2 = \left(\frac{1}{\varepsilon_1} + \frac{1}{\varepsilon_2} + \frac{1}{\varepsilon_3}\right)\varepsilon_1\varepsilon_2\varepsilon_3 = \eta_4{}^2.$$

Similarly $\eta_2\eta_3 + \eta_3\eta_1 + \eta_1\eta_2 = \varepsilon_4{}^2$, and of course we can permute the subscripts $1, 2, 3, 4$. Hence

$$(\Sigma\varepsilon_i)^2 = \varepsilon_1{}^2 + \ldots + \varepsilon_4{}^2 + 2\varepsilon_1\varepsilon_2 + \ldots + 2\varepsilon_3\varepsilon_4 = \Sigma\varepsilon_i{}^2 + \Sigma\eta_i{}^2.$$

Since this expression involves ε_i and η_i symmetrically, it is also equal to $(\Sigma\eta_i)^2$; thus

$$\varepsilon_1 + \varepsilon_2 + \varepsilon_3 + \varepsilon_4 = \eta_1 + \eta_2 + \eta_3 + \eta_4 > 0.$$

Also, since

$$
\begin{aligned}
(\varepsilon_1 + \varepsilon_2 + \varepsilon_3 - \varepsilon_4)(\varepsilon_1 + \varepsilon_2 + \varepsilon_3 + \varepsilon_4) &= (\varepsilon_1 + \varepsilon_2 + \varepsilon_3)^2 - \varepsilon_4{}^2 \\
&= \varepsilon_1{}^2 + \varepsilon_2{}^2 + \varepsilon_3{}^2 - \varepsilon_4{}^2 + 2\eta_4{}^2 \\
&= (\eta_2\eta_3 + \eta_2\eta_4 + \eta_3\eta_4) + (\eta_1\eta_3 + \ldots) + (\eta_1\eta_2 + \ldots) - (\eta_1\eta_2 + \ldots) + 2\eta_4{}^2 \\
&= 2(\eta_1\eta_4 + \eta_2\eta_4 + \eta_3\eta_4) + 2\eta_4{}^2 = 2\eta_4(\eta_1 + \eta_2 + \eta_3 + \eta_4),
\end{aligned}
$$

1.59
$$\varepsilon_1 + \varepsilon_2 + \varepsilon_3 - \varepsilon_4 = 2\eta_4.$$

Adding four such equations after squaring each side, we deduce $\Sigma\varepsilon_i{}^2 = \Sigma\eta_i{}^2$, whence

$$2\Sigma\varepsilon_i{}^2 = \Sigma\varepsilon_i{}^2 + \Sigma\eta_i{}^2 = (\Sigma\varepsilon_i)^2.$$

Thus 1.57 has been proved.

In 1936, this theorem was rediscovered again by Sir Frederick Soddy, who had received a Nobel prize in 1921 for his discovery of isotopes. He expressed the theorem in the form of a poem, *The Kiss Precise**, of which the middle verse runs as follows:

> Four circles to the kissing come,
> The smaller are the benter.
> The bend is just the inverse of
> The distance from the centre.
> Though their intrigue left Euclid dumb

---

* *Nature*, **137** (1936), p. 1021; **139** (1937), p. 62. In the next verse, Soddy announced his discovery of the analogous formula for 5 spheres in 3 dimensions. A final verse, added by Thorold Gosset (1869–1962) deals with $n+2$ spheres in n dimensions; see Coxeter, *Aequationes Mathematicae*, **1** (1968), pp. 104–121.

There's now no need for rule of thumb.
Since zero bend's a dead straight line
And concave bends have minus sign,
The sum of the squares of all four bends
Is half the square of their sum.

EXERCISES

1. Find the locus of the image of a fixed point P by reflection in a variable line through another fixed point O.

2. For the general triangle ABC, establish the identities

$$\frac{1}{r_a} + \frac{1}{r_b} + \frac{1}{r_c} = \frac{1}{r}, \quad r\, r_a r_b r_c = \Delta^2.$$

3. The lengths of the tangents from the vertex A to the incircle and to the three excircles are respectively

$$s - a, \quad s, \quad s - c, \quad s - b.$$

4. The circumcenter of an obtuse-angled triangle lies outside the triangle.

5. Where is the circumcenter of a right-angled triangle?

6. Let U, V, W be three points on the respective sides BC, CA, AB of a triangle ABC. The perpendiculars to the sides at these points are concurrent if and only if

$$AW^2 + BU^2 + CV^2 = WB^2 + UC^2 + VA^2.$$

7. A triangle is right-angled if and only if $r + 2R = s$.

8. The bends of Beecroft's eight circles satisfy

$$\varepsilon_1 + \eta_1 = \varepsilon_2 + \eta_2 = \varepsilon_3 + \eta_3 = \varepsilon_4 + \eta_4, \quad \Sigma \varepsilon_i \eta_i = 0.$$

9. For any four numbers satisfying $k + l + m + n = 0$, there is a "Beecroft configuration" having bends

$$\varepsilon_1 = k(k + l), \; \varepsilon_2 = (k + l)l, \; \varepsilon_3 = n^2 - kl, \; \varepsilon_4 = m^2 - kl,$$
$$\eta_1 = l^2 - mn, \; \eta_2 = k^2 - mn, \; \eta_3 = m(m + n), \; \eta_4 = (m + n)n.$$

(*Hint:* Express ε_3, ε_4, η_1, η_2 as rational functions of ε_1, ε_2, η_3, η_4.)

10. If three circles, externally tangent to one another, have centers forming a triangle ABC, they are all tangent to two other circles (or possibly a circle and a line) whose bends are

$$\frac{r + 4R \pm 2s}{\Delta}$$

11. Given a point P on the circumcircle of a triangle, the feet of the perpendiculars from P to the three sides all lie on a straight line. (This line is commonly called the *Simson line* of P with respect to the triangle, although it was first mentioned by W. Wallace, thirty years after Simson's death [Johnson **1**, p. 138].)

12. Given a triangle ABC and a point P in its plane (but not on a side nor on the circumcircle), let $A_1 B_1 C_1$ be the derived triangle formed by the feet of the perpendiculars from P to the sides BC, CA, AB. Let $A_2 B_2 C_2$ be derived analogously from $A_1 B_1 C_1$ (using the same P), and $A_3 B_3 C_3$ from $A_2 B_2 C_2$. Then $A_3 B_3 C_3$ is directly similar to ABC. [Casey **1**, p. 253.] (*Hint:* $\angle PBA = \angle PA_1 C_1 = \angle PC_2 B_2 = \angle PB_3 A_3$.) This result has been extended by B. M. Stewart from the third derived triangle of a triangle to the nth derived n-gon of an n-gon. (*American Mathematical Monthly* **47**, [1940], pp. 462–466).

1.6 THE EULER LINE AND THE ORTHOCENTER

Although the Greeks worked fruitfully, not only in geometry but also in the most varied fields of mathematics, nevertheless we today have gone beyond them everywhere and certainly also in geometry.

F. Klein (1849 -1925)

[Klein **2,** p. 189]

From now on, we shall have various occasions to mention the name of L. Euler (1707–1783), a Swiss who spent most of his life in Russia, making important contributions to all branches of mathematics. Some of his simplest discoveries are of such a nature that one can well imagine the ghost of Euclid saying, "Why on earth didn't I think of that?"

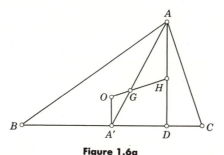

Figure 1.6a

If the circumcenter O and centroid G of a triangle coincide, each median is perpendicular to the side that it bisects, and the triangle is "isosceles three ways," that is, equilateral. Hence, if a triangle ABC is not equilateral, its circumcenter and centroid lie on a unique line OG. On this so-called *Euler line,* consider a point H such that $OH = 3OG$, that is, $GH = 2OG$ (Figure 1.6a). Since also $GA = 2A'G$, the latter half of Euclid VI.2 tells us that AH is parallel to $A'O$, which is the perpendicular bisector of BC. Thus AH is perpendicular to BC. Similarly BH is perpendicular to CA, and CH to AB.

The line through a vertex perpendicular to the opposite side is called an *altitude.* The above remarks [cf. Court **2,** p. 101] show that

The three altitudes of any triangle all pass through one point on the Euler line.

This common point H of the three altitudes is called the *orthocenter* of the triangle.

EXERCISES

1. Through each vertex of a given triangle ABC draw a line parallel to the opposite side. The perpendicular bisectors of the sides of the triangle so formed suggest an alternative proof that the three altitudes of ABC are concurrent.

2. The orthocenter of an obtuse-angled triangle lies outside the triangle.

3. Where is the orthocenter of a right-angled triangle?

4. Any triangle having two equal altitudes is isosceles.

5. Construct an isosceles triangle ABC (with base BC), given the median BB' and the altitude BE. (*Hint:* The centroid is two-thirds of the way from B to B'.) (H. Freudenthal.)

6. The altitude AD of any triangle ABC is of length

$$2R \sin B \sin C.$$

7. Find the perpendicular distance from the centroid G to the side BC.

8. If the Euler line passes through a vertex, the triangle is either right-angled or isosceles (or both).

9. If the Euler line is parallel to the side BC, the angles B and C satisfy

$$\tan B \tan C = 3.$$

1.7 THE NINE-POINT CIRCLE

> This circle is the first really exciting one to appear in any course on elementary geometry.
>
> Daniel Pedoe (1910 -   )
> [Pedoe **1**, p. 1]

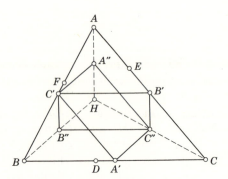

Figure 1.7a

The feet of the altitudes (that is, three points like D in Figure 1.6a) form the *orthic triangle* (or "pedal triangle") of ABC. The circumcircle of the orthic triangle is called the *nine-point circle* (or "Feuerbach circle") of the original triangle, because it contains not only the feet of the three altitudes but also six other significant points. In fact,

1.71 *The midpoints of the three sides, the midpoints of the lines joining the orthocenter to the three vertices, and the feet of the three altitudes, all lie on a circle.*

Proof [Coxeter **2,** 9.29]. Let A', B', C', A'', B'', C'' be the midpoints of BC, CA, AB, HA, HB, HC, and let D, E, F be the feet of the altitudes, as in Figure 1.7a. By Euclid VI.2 and 4 again, both $C'B'$ and $B''C''$ are parallel to BC while both $B'C''$ and $C'B''$ are parallel to AH. Since AH is perpendicular to BC, it follows that $B'C'B''C''$ is a rectangle. Similarly $C'A'C''A''$ is a rectangle. Hence $A'A''$, $B'B''$, $C'C''$ are three diameters of a circle. Since these diameters subtend right angles at D, E, F, respectively, the same circle passes through these points too.

If four points in a plane are joined in pairs by six distinct lines, they are called the *vertices* of a *complete quadrangle,* and the lines are its six *sides.* Two sides are said to be *opposite* if they have no common vertex. Any point of intersection of two opposite sides is called a *diagonal point.* There may be as many as three such points (see Figure 1.7*b*).

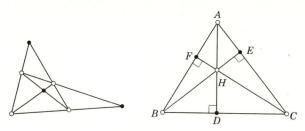

Figure 1.7b

If a triangle ABC is not right-angled, its vertices and orthocenter form a special kind of quadrangle whose opposite sides are perpendicular. In this terminology, the concurrence of the three altitudes can be expressed as follows:

1.72   *If two pairs of opposite sides of a complete quadrangle are pairs of perpendicular lines, the remaining sides are likewise perpendicular.*

Such a quadrangle $ABCH$ is called an *orthocentric quadrangle.* Its six sides

$$BC, \quad CA, \quad AB, \quad HA, \quad HB, \quad HC$$

are the sides and altitudes of the triangle ABC, and its diagonal points D, E, F are the feet of the altitudes. Among the four vertices of the quadrangle, our notation seems to give a special role to the vertex H. Clearly, however,

1.73   *Each vertex of an orthocentric quadrangle is the orthocenter of the triangle formed by the remaining three vertices.*

The four triangles (just one of which is acute-angled) all have the same orthic triangle and consequently the same nine-point circle.

It is proved in books on affine geometry [such as Coxeter **2,** 8.71] that the midpoints of the six sides of any complete quadrangle and the three diagonal points all lie on a conic. The above remarks show that, when the quadrangle is orthocentric, this "nine-point conic" reduces to a circle.

EXERCISES

1. Of the nine points described in 1.71, how many coincide when the triangle is (*a*) isosceles, (*b*) equilateral?

2. The feet of the altitudes decompose the nine-point circle into three arcs. If the triangle is scalene, the remaining six of the nine points are distributed among the three arcs as follows: One arc contains just one of the six points, another contains two, and the third contains three.

3. On the arc $A'D$ of the nine-point circle, take the point X one-third of the way from A' to D. Take points Y, Z similarly, on the arcs $B'E, C'F$. Then XYZ is an equilateral triangle.

4. The incenter and the excenters of any triangle form an orthocentric quadrangle. [Casey **1**, p. 274.]

5. In the notation of § 1.5, the Euler line of $I_a I_b I_c$ is IO.

6. The four triangles that occur in an orthocentric quadrangle have equal circumradii.

1.8   TWO EXTREMUM PROBLEMS

> *Most people have some appreciation of mathematics, just as most people can enjoy a pleasant tune; and there are probably more people really interested in mathematics than in music.*
>
> G. H. Hardy [**2**, p. 26]

> *Their interest will be stimulated if only we can eliminate the aversion toward mathematics that so many have acquired from childhood experiences.*
>
> Hans Rademacher (1892 -   )
> [Rademacher and Toeplitz **1**, p. 5]

We shall describe the problems of Fagnano and Fermat in considerable detail because of the interesting methods used in solving them. The first was proposed in 1775 by J. F. Toschi di Fagnano, who solved it by means of differential calculus. The method given here was discovered by L. Fejér while he was a student [Rademacher and Toeplitz **1**, pp. 30–32].

FAGNANO'S PROBLEM. *In a given acute-angled triangle ABC, inscribe a triangle UVW whose perimeter is as small as possible.*

Consider first an arbitrary triangle UVW with U on BC, V on CA, W on AB. Let U', U'' be the images of U by reflection in CA, AB, respectively. Then

$$UV + VW + WU = U'V + VW + WU'',$$

which is a path from U' to U'', usually a broken line with angles at V and W. Such a path from U' to U'' is minimal when it is straight, as in Figure 1.8*a*.

Hence, among all inscribed triangles with a given vertex U on BC, the one with smallest perimeter occurs when V and W lie on the straight line $U'U''$. In this way we obtain a definite triangle UVW for each choice of U on BC. The problem will be solved when we have chosen U so as to minimize $U'U''$, which is equal to the perimeter.

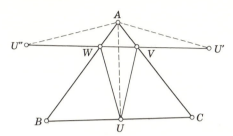

Figure 1.8a

Since AU' and AU'' are images of AU by reflection in AC and AB, they are congruent and

$$\angle U'AU'' = 2A.$$

Thus $AU'U''$ is an isosceles triangle whose angle at A is independent of the choice of U. The base $U'U''$ is minimal when the equal sides are minimal, that is, when AU is minimal. In other words, AU is the shortest distance from the given point A to the given line BC. Since the hypotenuse of a right-angled triangle is longer than either cathetus, the desired location of U is such that AU is perpendicular to BC. Thus AU is the altitude from A.

This choice of U yields a unique triangle UVW whose perimeter is smaller than that of any other inscribed triangle. Since we could equally well have begun with B or C instead of A, we see that BV and CW are the altitudes from B and C. Hence

The triangle of minimal perimeter inscribed in an acute-angled triangle ABC is the orthic triangle of ABC.

The same method can be used to prove the analogous result for spherical triangles [Steiner **2,** p. 45, No. 7].

The other problem, proposed by Pierre Fermat (1601–1665), likewise seeks to minimize the sum of three distances. The solution given here is due to J. E. Hofmann.*

FERMAT'S PROBLEM. *In a given acute-angled triangle ABC, locate a point P whose distances from A, B, C have the smallest possible sum.*

Consider first an arbitrary point P inside the triangle. Join it to A, B, C and rotate the inner triangle APB through $60°$ about B to obtain $C'P'B$, so that ABC' and PBP' are equilateral triangles, as in Figure 1.8b. Then

$$AP + BP + CP = C'P' + P'P + PC,$$

---

* Elementare Lösung einer Minimumsaufgabe, *Zeitschrift für mathematischen und naturwissenschaftlichen Unterricht,* **60** (1929), pp. 22–23.

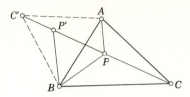

Figure 1.8b

which is a path from C' to C, usually a broken line with angles at P' and P. Such a path (joining C' to C by a sequence of three segments) is minimal when it is straight, in which case

$$\angle BPC = 180° - \angle BPP' = 120°$$

and $$\angle APB = \angle C'P'B = 180° - \angle PP'B = 120°.$$

Thus the desired point P, for which $AP + BP + CP$ is minimal, is the point from which each of the sides BC, CA, AB subtends an angle of $120°$. This "Fermat point" is most simply constructed as the second intersection of the line CC' and the circle ABC' (that is, the circumcircle of the equilateral triangle ABC').

It has been pointed out [for example by Pedoe **1**, pp. 11–12] that the triangle ABC need not be assumed to be acute-angled. The above solution is valid whenever there is no angle greater than $120°$.

Instead of the equilateral triangle ABC' on AB, we could just as well have drawn an equilateral triangle BCA' on BC, or CAB' on CA, as in Figure 1.8c. Thus the three lines AA', BB', CC' all pass through the Fermat point P, and any two of them provide an alternative construction for it. Moreover, the line segments AA', BB', CC' are all equal to $AP + BP + CP$. Hence

If equilateral triangles BCA', CAB', ABC' are drawn outwards on the sides of any triangle ABC, the line segments AA', BB', CC' are equal, concurrent, and inclined at 60° to one another.

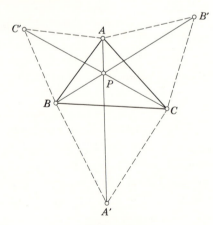

Figure 1.8c

EXERCISES

1. In Figure 1.8*a*, *UV* and *VW* make equal angles with *CA*. Deduce that the ortho-center of any triangle is the incenter of its orthic triangle. (In other words, if *ABC* is a triangular billiard table, a ball at *U*, hit in the direction *UV*, will go round the triangle *UVW* indefinitely, that is, until it is stopped by friction.)

2. How does Fagnano's problem collapse when we try to apply it to a triangle *ABC* in which the angle *A* is obtuse?

3. The circumcircles of the three equilateral triangles in Figure 1.8*c* all pass through *P*, and their centers form a fourth equilateral triangle.*

4. Three holes, at the vertices of an arbitrary triangle, are drilled through the top of a table. Through each hole a thread is passed with a weight hanging from it below the table. Above, the three threads are all tied together and then released. If the three weights are all equal, where will the knot come to rest?

5. Four villages are situated at the vertices of a square of side one mile. The inhabitants wish to connect the villages with a system of roads, but they have only enough material to make $\sqrt{3} + 1$ miles of road. How do they proceed? [Courant and Robbins **1**, p. 392.]

6. Solve Fermat's problem for a triangle *ABC* with $A > 120°$, and for a convex quadrangle *ABCD*.

7. If two points *P*, *P'*, inside a triangle *ABC*, are so situated that $\angle CBP = \angle PBP' = \angle P'BA$, $\angle ACP' = \angle P'CP = \angle PCB$, then $\angle BP'P = \angle PP'C$.

8. If four squares are placed externally (or internally) on the four sides of any parallelogram, their centers are the vertices of another square. [Yaglom **1**, pp. 96–97.]

9. Let *X*, *Y*, *Z* be the centers of squares placed externally on the sides *BC*, *CA*, *AB* of a triangle *ABC*. Then the segment *AX* is congruent and perpendicular to *YZ* (also *BY* to *ZX* and *CZ* to *XY*). (W. A. J. Luxemburg.)

10. Let *Z*, *X*, *U*, *V* be the centers of squares placed externally on the sides *AB*, *BC*, *CD*, *DA* of any simple quadrangle (or "quadrilateral") *ABCD*. Then the segment *ZU* (joining the centers of two "opposite" squares) is congruent and perpendicular to *XV*. [Forder **2**, p. 40.]

1.9 MORLEY'S THEOREM

> *Many of the proofs in mathematics are very long and intricate. Others, though not long, are very ingeniously constructed.*
>
> E. C. Titchmarsh (1899 -1963)
>
> [Titchmarsh **1**, p. 23]

One of the most surprising theorems in elementary geometry was discovered about 1899 by F. Morley (whose son Christopher wrote novels such as *Thunder on the Left*). He mentioned it to his friends, who spread it over

---

* Court [**1**, pp. 105–107]. See also *Mathesis* 1938, p. 293 (footnote, where this theorem is attributed to Napoleon); and Forder [**2**, p. 40] for some interesting generalizations.

the world in the form of mathematical gossip. At last, after ten years, a trigonometrical proof by M. Satyanarayana and an elementary proof by M. T. Naraniengar were published.*

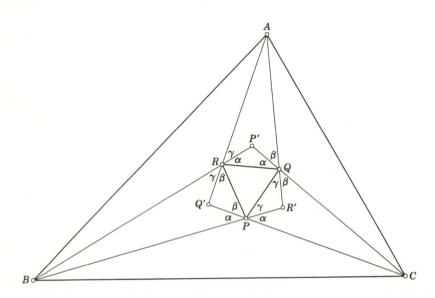

Figure 1.9a

MORLEY'S THEOREM.   *The three points of intersection of the adjacent tri-sectors of the angles of any triangle form an equilateral triangle.*

In other words, any triangle ABC yields an equilateral triangle PQR if the angles A, B, C are trisected by AQ and AR, BR and BP, CP and CQ, as in Figure 1.9a. (Much trouble is experienced if we try a direct approach, but the difficulties disappear if we work backwards, beginning with an equilateral triangle and building up a general triangle which is afterwards identified with the given triangle ABC.)

On the respective sides QR, RP, PQ of a given equilateral triangle PQR, erect isosceles triangles $P'QR$, $Q'RP$, $R'PQ$ whose base angles α, β, γ satisfy the equation and inequalities

$$\alpha + \beta + \gamma = 120°, \quad \alpha < 60°, \quad \beta < 60°, \quad \gamma < 60°.$$

* *Mathematical Questions and their Solutions from the Educational Times* (New Series), **15** (1909), pp. 23–24, 47. See also C. H. Chepmell and R. F. Davis, *Mathematical Gazette,* **11** (1923), pp. 85–86; F. Morley, *American Journal of Mathematics,* **51** (1929), pp. 465–472, H. D. Grossman, *American Mathematical Monthly,* **50** (1943), p. 552, and L. Bankoff, *Mathematics Magazine,* **35** (1962), pp. 223–224. The treatment given here is due to Raoul Bricard, *Nouvelles Annales de Mathématiques* (5), **1** (1922), pp. 254–258. A similar proof was devised independently by Bottema [**1,** p. 34].

Extend the sides of the isosceles triangles below their bases until they meet again in points A, B, C. Since $\alpha + \beta + \gamma + 60° = 180°$, we can immediately infer the measurement of some other angles, as marked in Figure 1.9a. For instance, the triangle AQR must have an angle $60° - \alpha$ at its vertex A, since its angles at Q and R are $\alpha + \beta$ and $\gamma + \alpha$.

Referring to 1.51, we see that one way to characterize the incenter I of a triangle ABC is to describe it as lying on the bisector of the angle A at such a distance that

$$\angle BIC = 90° + \tfrac{1}{2}A.$$

Applying this principle to the point P in the triangle $P'BC$, we observe that the line PP' (which is a median of both the equilateral triangle PQR and the isosceles triangle $P'QR$) bisects the angle at P'. Also the half angle at P' is $90° - \alpha$, and

$$\angle BPC = 180° - \alpha = 90° + (90° - \alpha).$$

Hence P is the incenter of the triangle $P'BC$. Likewise Q is the incenter of $Q'CA$, and R of $R'AB$. Therefore all the three small angles at C are equal; likewise at A and at B. In other words, the angles of the triangle ABC are trisected.

The three small angles at A are each $\tfrac{1}{3}A = 60° - \alpha$; similarly at B and C. Thus

$$\alpha = 60° - \tfrac{1}{3}A, \quad \beta = 60° - \tfrac{1}{3}B, \quad \gamma = 60° - \tfrac{1}{3}C.$$

By choosing these values for the base angles of our isosceles triangles, we can ensure that the above procedure yields a triangle ABC that is similar to any given triangle.

This completes the proof.

EXERCISES

1. The three lines PP', QQ', RR' (Figure 1.9a) are concurrent. In other words, the trisectors of A, B, C meet again to form another triangle $P'Q'R'$ which is perspective with the equilateral triangle PQR. (In general $P'Q'R'$ is *not* equilateral.)

2. What values of α, β, γ will make the triangle ABC (i) equilateral, (ii) right-angled isosceles? Sketch the figure in each case.

3. Let P_1 and P_2 (on CA and AB) be the images of P by reflection in CP' and BP'. Then the four points P_1, Q, R, P_2 are evenly spaced along a circle through A. In the special case when the triangle ABC is equilateral, these four points occur among the vertices of a regular enneagon (9-gon) in which A is the vertex opposite to the side QR.

2

Regular Polygons

We begin this chapter by discussing (without proofs) the possibility of constructing certain regular polygons with the instruments allowed by Euclid. We then consider all these polygons, regardless of the question of constructibility, from the standpoint of symmetry. Finally, we extend the concept of a regular polygon so as to include star polygons.

2.1 CYCLOTOMY

Euclid's postulates imply a restriction on the instruments that he allowed for making constructions, namely the restriction to ruler (or straightedge) and compasses. He constructed an equilateral triangle (I.1), a square (IV.6), a regular pentagon (IV.11), a regular hexagon (IV.15), and a regular 15-gon (IV.16). The number of sides may be doubled again and again by repeated angle bisections. It is natural to ask which other regular polygons can be constructed with Euclid's instruments. This question was completely answered by Gauss (1777–1855) at the age of nineteen [see Smith **2**, pp. 301–302]. Gauss found that a regular n-gon, say $\{n\}$, can be so constructed if the odd prime factors of n are distinct "Fermat primes"

$$F_k = 2^{2^k} + 1.$$

The only known primes of this kind are

$$F_0 = 2^1 + 1 = 3, \quad F_1 = 2^2 + 1 = 5, \quad F_2 = 2^4 + 1 = 17,$$
$$F_3 = 2^8 + 1 = 257, \quad F_4 = 2^{16} + 1 = 65537.$$

To inscribe a regular pentagon in a given circle, simpler constructions than Euclid's were given by Ptolemy and Richmond.* The former has been repeated in many textbooks. The latter is as follows (Figure 2.1a).

To inscribe a regular pentagon $P_0P_1P_2P_3P_4$ in a circle with center O: draw the radius OB perpendicular to OP_0; join P_0 to D, the midpoint of OB; bisect the angle ODP_0 to obtain N_1 on OP_0; and draw N_1P_1 perpendicular to OP_0 to obtain P_1 on the circle. Then P_0P_1 is a side of the desired pentagon.

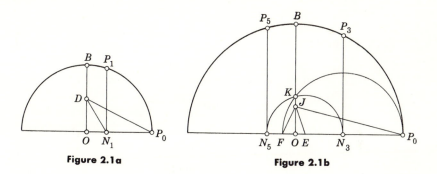

Figure 2.1a   Figure 2.1b

Richmond also gave a simple construction for the $\{17\}$ $P_0P_1 \ldots P_{16}$ (Figure 2.1b). Join P_0 to J, one quarter of the way from O to B. On the diameter through P_0 take E, F, so that $\angle OJE$ is one quarter of OJP_0 and $\angle FJE$ is 45°. Let the circle on FP_0 as diameter cut OB in K, and let the circle with center E and radius EK cut OP_0 in N_3 (between O and P_0) and N_5. Draw perpendiculars to OP_0 at these two points, to cut the original circle in P_3 and P_5. Then the arc P_3P_5 (and likewise P_1P_3) is $\frac{2}{17}$ of the circumference. (The proof involves repeated application of the principle that the roots of the equation $x^2 + 2x \cot 2C - 1 = 0$ are $\tan C$ and $-\cot C$.)

Richelot and Schwendenwein constructed the regular 257-gon in 1832. J. Hermes spent ten years on the regular 65537-gon and deposited the manuscript in a large box in the University of Göttingen, where it may still be found.

The next number of the form $F_k = 2^{2^k} + 1$ is $F_5 = 4294967297$. Fermat incorrectly assumed it to be prime. G. T. Bennett gave the following neat proof † that it is composite [Hardy and Wright **1**, p. 14]: the number

$$641 = 5^4 + 2^4 = 5 \cdot 2^7 + 1,$$

dividing both

$5^4 \cdot 2^{28} + 2^{32}$ and $5^4 \cdot 2^{28} - 1$, divides their difference, which is F_5.

---

* H. W. Richmond, *Quarterly Journal of Mathematics*, **26** (1893), pp. 296–297; see also H. E. Dudeney, *Amusements in Mathematics* (London 1917), p. 38.

† Rediscovered by P. Kanagasabapathy, *Mathematical Gazette*, **42** (1958), p. 310.

The question naturally arises whether F_k may be prime for some greater value of k. It is now known that this can happen only if F_k divides $3^{(F_k-1)/2} + 1$. Using this criterion, electronic computing machines have shown that F_k is composite for $5 \leqslant k \leqslant 16$. Therefore Hermes's construction is the last of its kind that will ever be undertaken!

EXERCISES

1. Verify the correctness of Richmond's construction for $\{5\}$ (Figure 2.1*a*).

2. Assuming Richmond's construction for $\{17\}$, how would you inscribe $\{51\}$ in the same circle?

2.2   ANGLE TRISECTION

> To trisect a given angle, we may proceed to find the sine of the angle— say *a*—then, if *x* is the sine of an angle equal to one-third of the given angle, we have $4x^3 = 3x - a$.
>
> W. W. Rouse Ball (1850 -1925)
>
> [Ball **1,** p. 327]

Gauss was almost certainly aware of the fact that his cyclotomic condition is necessary as well as sufficient, but he does not seem to have said so explicitly. The missing step was supplied by Wantzel*, who proved that, if the odd prime factors of n are *not* distinct Fermat primes, $\{n\}$ cannot be constructed with ruler and compasses. For instance, since 7 is not a Fermat prime, Euclid's instruments will not suffice for the regular heptagon $\{7\}$; and since the factors of 9 are not distinct, the same is true for the enneagon $\{9\}$.

The problem of trisecting an arbitrary angle with ruler and compasses exercised the ingenuity of professional and amateur mathematicians for two thousand years [Ball **1,** pp. 333–335]. It is, of course, easy to trisect certain particular angles, such as a right angle. But any construction for trisecting an arbitrary angle could be applied to an angle of 60°, and then we could draw a regular enneagon. In view of Wantzel's theorem, we may say that it has been known since 1837 that the classical trisection problem can never be solved.

This is probably the reason why Morley's Theorem (§1.9) was not discovered till the twentieth century: people felt uneasy about mentioning the trisectors of an angle. However, although the trisectors cannot be constructed by means of the ruler and compasses, they can be found in other ways [Cundy and Rollett **1,** pp. 208–211]. Even if these more versatile instruments had never been discovered, the theorem would still be meaningful. Most mathematicians are willing to accept the existence of things that they have not been able to construct. For instance, it was proved in 1909 that the Fermat numbers F_7 and F_8 are composite, but their smallest prime factors still remain to be computed.

EXERCISE

The number $2^n + 1$ is composite whenever n is not a power of 2.

* P. L. Wantzel, *Journal de Mathématiques pures et appliquées,* **2** (1837), pp. 366–372.

2.3 ISOMETRY

One way of describing the structure of space, preferred by both New-ton and Helmholtz, is through the notion of congruence. Congruent parts of space V, V′ are such as can be occupied by the same rigid body in two of its positions. If you move the body from one into the other position the particle of the body covering a point P of V will after-wards cover a certain point P′ of V′, and thus the result of the mo-tion is a mapping P → P′ of V upon V′. We can extend the rigid body either actually or in imagination so as to cover an arbitrarily given point P of space, and hence the congruent mapping P → P′ can be extended to the entire space.

Hermann Weyl (1885 -1955)

[Weyl **1,** p, 43]

We shall find it convenient to use the word *transformation* in the special sense of a one-to-one correspondence $P \rightarrow P'$ among all the points in the plane (or in space), that is, a rule for associating pairs of points, with the understanding that each pair has a first member P and a second member P' and that every point occurs as the first member of just one pair and also as the second member of just one pair. It may happen that the members of a pair coincide, that is, that P' coincides with P; in this case P is called an *invariant* point (or "double point") of the transformation.

In particular, an *isometry* (or "congruent transformation," or "congru-ence") is a transformation which preserves length, so that, if (P, P') and (Q, Q') are two pairs of corresponding points, we have $PQ = P'Q'$: PQ and $P'Q'$ are congruent segments. For instance, a *rotation* of the plane about P (or about a line through P perpendicular to the plane) is an isometry hav-ing P as an invariant point, but a *translation* (or "parallel displacement") has no invariant point: every point is moved.

A *reflection* is the special kind of isometry in which the invariant points consist of all the points on a line (or plane) called the *mirror*.

A still simpler kind of transformation (so simple that it may at first seem too trivial to be worth mentioning) is the *identity,* which leaves every point unchanged. The result of applying several transformations successively is called their *product.* If the product of two transformations is the identity, each is called the *inverse* of the other, and their product in the reverse order is again the identity.

2.31 *If an isometry has more than one invariant point, it must be either the identity or a reflection.*

To prove this, let A and B be two invariant points, and P any point not on the line AB (Figure 1.3*b*). The corresponding point P', satisfying

$$AP' = AP, \qquad BP' = BP,$$

must lie on the circle with center A and radius AP, and on the circle with center *B* and radius *BP*. Since *P* is not on *AB*, these circles do not touch each other but intersect in two points, one of which is *P*. Hence *P′* is either *P* itself or the image of *P* by reflection in *AB*.

2.4 SYMMETRY

When we say that a figure is "symmetrical" we mean that we can apply certain isometries, called *symmetry operations*, which leave the whole figure unchanged while permuting its parts. For example, the capital letters E and A (Figure 2.4*a*) have bilateral symmetry, the mirror being horizontal for the former, vertical for the latter. The letter N (Figure 2.4*b*) is symmetrical by a *half-turn*, or rotation through 180° (or "reflection in a point," or "central inversion"), which may be regarded as the result of reflecting horizontally and then vertically, or vice versa. The swastika (Figure 2.4*c*) is symmetrical by rotation through any number of right angles.

Figure 2.4a   Figure 2.4b

In counting the symmetry operations of a figure, it is usual to include the identity; any figure has this trivial symmetry. Thus the swastika admits four distinct symmetry operations: rotations through 1, 2, 3, or 4 right angles. The last is the identity. The first and third are inverses of each other, since their product is the identity.

This use of the word "product" suggests an algebraic symbolism in which the transformations are denoted by capital letters while 1 denotes the identity. (Instead of 1, some authors write E.) Thus if S is the counterclockwise quarter-turn, the four symmetry operations of the swastika are

$$S, \quad S^2, \quad S^3 = S^{-1} \quad \text{and} \quad S^4 = 1.$$

Since the smallest power of S that is equal to the identity is the fourth power, we say that S is of *period* 4. Similarly S², being a half-turn, is of period 2 [see Coxeter **1**, p. 39]. The only transformation of period 1 is the identity. A translation is aperiodic (that is, it has no period), but it is conveniently said to be of infinite period.

Some figures admit both reflections and rotations as symmetry operations. The letter H (Figure 2.4*d*) has a horizontal mirror (like E) and a vertical mirror (like A), as well as a center of rotational symmetry (like N) where the two mirrors intersect. Thus it has four symmetry operations: the identity 1, the horizontal reflection R_1, the vertical reflection R_2, and the half-turn $R_1R_2 = R_2R_1$.

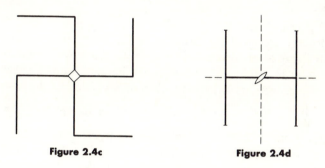

Figure 2.4c   Figure 2.4d

EXERCISES

1. Every isometry of period 2 is either a reflection or a half-turn [Bachmann **1**, pp. 2–3].

2. Express (*a*) a half-turn, (*b*) a quarter-turn, as transformations of (i) Cartesian coordinates, (ii) polar coordinates. (Take the origin to be the center of rotation.)

2.5   GROUPS

Symmetry, as wide or as narrow as you may define its meaning, is one idea by which man through the ages has tried to comprehend and create order, beauty, and perfection.

Hermann Weyl [**1**, p. 5]

A set of transformations [Birkhoff and MacLane **1**, pp. 115–118] is said to form a *group* if it contains the inverse of each and the product of any two (including the product of one with itself or with its inverse). The number of distinct transformations is called the *order* of the group. (This may be either finite or infinite.) Clearly the symmetry operations of any figure form a group. This is called the *symmetry group* of the figure. In the extreme case when the figure is completely irregular (like the numeral 6) its symmetry group is of order one, consisting of the identity alone.

The symmetry group of the letter E or A (Figure 2.4a) is the so-called *dihedral* group of order 2, generated by a single reflection and denoted by D_1. (The name is easily remembered, as the Greek origin of the word "dihedral" is almost equivalent to the Latin origin of "bilateral.") The symmetry group of the letter N (Figure 2.4b) is likewise of order 2, but in this case the generator is a half-turn and we speak of the *cyclic* group, C_2. The two groups D_1 and C_2 are abstractly identical or *isomorphic;* they are different geometrical representations of the single abstract group of order 2, defined by the relation

2.51 $$R^2 = 1$$

or $R = R^{-1}$ [Coxeter and Moser **1**, p. 1].

The symmetry group of the swastika is C_4, the cyclic group of order 4, generated by the quarter-turn S and abstractly defined by the relation $S^4 = 1$. That of the letter H (Figure 2.4d) is D_2, the dihedral group of order 4, generated by the two reflections R_1, R_2 and abstractly defined by the relations

2.52 $$R_1{}^2 = 1, \quad R_2{}^2 = 1, \quad R_1R_2 = R_2R_1.$$

Although C_4 and D_2 both have order 4, they are *not* isomorphic: they have a different structure, different "multiplication tables." To see this, it suffices to observe that C_4 contains two operations of period 4, whereas all the operations in D_2 (except the identity) are of period 2: the generators obviously, and their product also, since

$$(R_1R_2)^2 = R_1R_2R_1R_2 = R_1R_2R_2R_1 = R_1R_2{}^2R_1 = R_1R_1 = R_1{}^2 = 1.$$

This last remark illustrates what we mean by saying that 2.52 is an *abstract definition* for D_2, namely that every true relation concerning the generators R_1, R_2 is an algebraic consequence of these simple relations. An alternative abstract definition for the same group is

2.53 $$R_1{}^2 = 1, \quad R_2{}^2 = 1, \quad (R_1R_2)^2 = 1,$$

from which we can easily deduce $R_1R_2 = R_2R_1$.

The general cyclic group C_n, of order n, has the abstract definition

2.54 $$S^n = 1.$$

Its single generator S, of period n, is conveniently represented by a rotation through $360°/n$. Then S^k is a rotation through k times this angle, and the n operations in C_n are given by the values of k from 1 to n, or from 0 to $n - 1$. In particular, C_5 occurs in nature as the symmetry group of the periwinkle flower.

EXERCISE

Express a rotation through angle α about the origin as a transformation of (i) polar coordinates, (ii) Cartesian coordinates. If $f(r, \theta) = 0$ is the equation for a curve in polar coordinates, what is the equation for the transformed curve?

2.6   THE PRODUCT OF TWO REFLECTIONS

Thou in thy lake dost see
Thyself.

J. M. Legaré (1823 -1859)

(To a Lily)

In any group of transformations, the associative law

$$(RS)T = R(ST)$$

is automatically satisfied, but the commutative law

$$RS = SR$$

does not necessarily hold, and care must be taken in inverting a product, for example,

$$(RS)^{-1} = S^{-1}R^{-1},$$

not $R^{-1}S^{-1}$. (This becomes clear when we think of R and S as the operations of putting on our socks and shoes, respectively.)

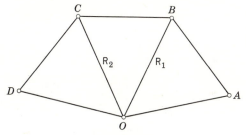

Figure 2.6a

The product of reflections in two intersecting lines (or planes) is a rotation through twice the angle between them. In fact, if *A, B, C, D,* . . . are evenly spaced on a circle with center *O*, let R_1 and R_2 be the reflections in *OB* and *OC* (Figure 2.6a). Then R_1 reflects the triangle *OAB* into *OCB*, which is reflected by R_2 to *OCD;* thus R_1R_2 is the rotation through $\angle AOC$ or $\angle BOD$, which is twice $\angle BOC$. Since a rotation is completely determined by its center and its angle, R_1R_2 is equal to the product of reflections in any two lines through *O* making the same angle as *OB* and *OC*. (The reflections in *OA* and *OB* are actually $R_1R_2R_1$ and R_1, whose product is $R_1R_2R_1^2 = R_1R_2$.) In particular, the half-turn about *O* is the product of reflections in any two perpendicular lines through *O*.

Since R_1R_2 is a counterclockwise rotation, R_2R_1 is the corresponding clockwise rotation; in fact,

$$R_2R_1 = R_2^{-1}R_1^{-1} = (R_1R_2)^{-1}.$$

This is the same as R_1R_2 if the two mirrors are at right angles, in which case R_1R_2 is a half-turn and $(R_1R_2)^2 = 1$.

EXERCISES

1. The product of quarter-turns (in the same sense) about C and B is the half-turn about the center of a square having BC for a side.

2. Let $ACPQ$ and $BARS$ be squares on the sides AC and BA of a triangle ABC. If B and C remain fixed while A varies freely, PS passes through a fixed point.

2.7 THE KALEIDOSCOPE

D_2 is a special case of the general dihedral group D_n, which is, for $n > 2$, the symmetry group of the regular n-gon, $\{n\}$. (See Figure 2.7a for the cases $n = 3, 4, 5$.) This is evidently a group of order $2n$, consisting of n rotations (through the n effectively distinct multiples of $360°/n$) and n reflections. When n is odd, each of the n mirrors joins a vertex to the midpoint of the opposite side; when n is even, $\frac{1}{2}n$ mirrors join pairs of opposite vertices and $\frac{1}{2}n$ bisect pairs of opposite sides [see Birkhoff and MacLane **1**, pp. 117–118, 135].

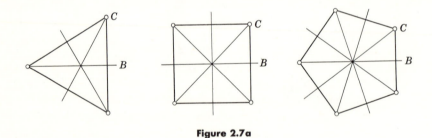

Figure 2.7a

The n rotations are just the operations of the cyclic group C_n. Thus the operations of D_n include all the operations of C_n: in technical language, C_n is a *subgroup* of D_n. The rotation through $360°/n$, which generates the subgroup, may be described as the product $S = R_1R_2$ of reflections in two adjacent mirrors (such as OB and OC in Figure 2.7a) which are inclined at $180°/n$.

Let $R_1, R_2, \ldots, R_n$ denote the n reflections in their natural order of arrangement. Then R_1R_{k+1}, being the product of reflections in two mirrors inclined at k times $180°/n$, is a rotation through k times $360°/n$:

$$R_1R_{k+1} = S^k.$$

Thus $R_{k+1} = R_1S^k$, and the n reflections may be expressed as

$$R_1, R_1S, R_1S^2, \ldots, R_1S^{n-1}.$$

In other words, D_n is generated by R_1 and S. By substituting R_1R_2 for S, we see that the same group is equally well generated by R_1 and R_2, which satisfy the relations

2.71 $R_1{}^2 = 1, \quad R_2{}^2 = 1, \quad (R_1R_2)^n = 1.$

(The first two relations come from 2.51 and the third from 2.54.) These relations can be shown to suffice for an abstract definition [see Coxeter and Moser **1**, pp. 6, 36].

A practical way to make a model of D_n is to join two ordinary mirrors by a hinge and stand them on the lines *OB*, *OC* of Figure 2.7a so that they are inclined at $180°/n$. Any object placed between the mirrors yields $2n$ visible images (including the object itself). If the object is your right hand, n of the images will look like a left hand, illustrating the principle that, since a reflection reverses sense, the product of any even number of reflections preserves sense, and the product of any odd number of reflections reverses sense.

The first published account of this instrument seems to have been by Athanasius Kircher in 1646. The name *kaleidoscope* (from καλος, beautiful; ειδος, a form; and σκοπειν, to see) was coined by Sir David Brewster, who wrote a treatise on its theory and history. He complained [Brewster **1**, p. 147] that Kircher allowed the angle between the two mirrors to be any submultiple of 360° instead of restricting it to submultiples of 180°.

The case when $n = 2$ is, of course, familiar. Standing between two perpendicular mirrors (as at a corner of a room), you see your image in each and also the image of the image, which is the way other people see you.

Having decided to use the symbol D_n for the dihedral group generated by reflections in two planes making a "dihedral" angle of $180°/n$, we naturally stretch the notation so as to allow the extreme value $n = 1$. Thus D_1 is the group of order 2 generated by a single reflection, that is, the symmetry group of the letter E or A, whereas the isomorphic group C_2, generated by a half-turn, is the symmetry group of the letter N.

According to Weyl [**1**, pp. 66, 99], it was Leonardo da Vinci who discovered that the only finite groups of isometries in the plane are

$$C_1, C_2, C_3, \ldots,$$
$$D_1, D_2, D_3, \ldots.$$

His interest in them was from the standpoint of architectural plans. Of course, the prevalent groups in architecture have always been D_1 and D_2. But the pyramids of Egypt exhibit the group D_4, and Leonardo's suggestion has been followed to some extent in modern times: the Pentagon Building in Washington has the symmetry group D_5, and the Bahai Temple near Chicago has D_9. In nature, many flowers have dihedral symmetry groups such as D_6. The symmetry group of a snowflake is usually D_6 but occasionally only D_3. [Kepler **1**, pp. 259–280.]

If you cut an apple the way most people cut an orange, the core is seen to have the symmetry group D_5. Extending the five-pointed star by straight cuts in each half, you divide the whole apple into ten pieces from each of which the core can be removed in the form of two thin flakes.

EXERCISES

1. Describe the symmetry groups of

(a) a scalene triangle,   (b) an isosceles triangle,

(c) a parabola,   (d) a parallelogram,

(e) a rhombus,   (f) a rectangle,

(g) an ellipse.

2. Use inverses and the associative law to prove algebraically the "cancellation rule" which says that the relation

$$RT = ST$$

implies R = S.

3. Show how the usual defining relations for D_3, namely 2.71 with $n = 3$, may be deduced by algebraic manipulation from the simpler relations

$$R_1{}^2 = 1, \quad R_1R_2R_1 = R_2R_1R_2.$$

4. The cyclic group C_m is a subgroup of C_n if and only if the number m is a divisor of n. In particular, if n is prime, the only subgroups of C_n are C_n itself and C_1.

2.8 STAR POLYGONS

Instead of deriving the dihedral group D_n from the regular polygon $\{n\}$, we could have derived the polygon from the group: the vertices of the polygon are just the n images of a point P_0 (the C of Figure 2.7a) on one of the two mirrors of the kaleidoscope. In fact, there is no need to use the whole group D_n: its subgroup C_n will suffice. The vertex P_k of the polygon $P_0P_1 \ldots P_{n-1}$ can be derived from the initial vertex P_0 by a rotation through k times $360°/n$.

More generally, rotations about a fixed point O through angles θ, 2θ, 3θ, . . . transform any point P_0 (distinct from O) into other points P_1, P_2, P_3, . . . on the circle with center O and radius OP_0. In general, these points become increasingly dense on the circle; but if the angle θ is commensurable with a right angle, only a finite number of them will be distinct. In particular, if $\theta = 360°/n$, where n is a positive integer greater than 2, then there will be n points P_k whose successive joins

$$P_0P_1, \ P_1P_2, \ldots, \ P_{n-1}P_0$$

are the sides of an ordinary regular n-gon.

Let us now extend this notion by allowing n to be any rational number greater than 2, say the fraction p/d (where p and d are coprime). Accordingly, we define a (generalized) *regular polygon* $\{n\}$, where $n = p/d$. Its p vertices are derived from P_0 by repeated rotations through $360°/n$, and its p sides (enclosing the center d times) are

$$P_0P_1, \ P_1P_2, \ldots, \ P_{p-1}P_0.$$

Since a ray coming out from the center without passing through a vertex will cross d of the p sides, this denominator d is called the *density* of the polygon [Coxeter **1**, pp. 93–94]. When $d = 1$, so that $n = p$, we have the ordinary regular *p*-gon, {*p*}. When *d* > 1, the sides cross one another, but the crossing points are not counted as vertices. Since *d* may be any positive integer relatively prime to *p* and less than $\frac{1}{2}p$, there is a regular polygon {*n*} for each rational number *n* > 2. In fact, it is occasionally desirable to include also the *digon* {2}, although its two sides coincide.

When *p* = 5, we have the pentagon {5} of density 1 and the *pentagram* {$\frac{5}{2}$} of density 2, which was used as a special symbol by the Babylonians and by the Pythagoreans. Similarly, the *octagram* {$\frac{8}{3}$} and the *decagram* {$\frac{10}{3}$} have density 3, while the *dodecagram* {$\frac{12}{5}$} has density 5 (Figure 2.8*a*). These particular polygons have names as well as symbols because they occur as faces of interesting polyhedra and tessellations.*

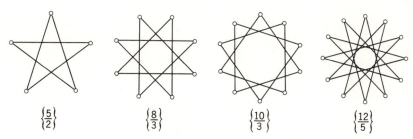

$$\left\{\frac{5}{2}\right\} \qquad \left\{\frac{8}{3}\right\} \qquad \left\{\frac{10}{3}\right\} \qquad \left\{\frac{12}{5}\right\}$$

Figure 2.8a

Polygons for which *d* > 1 are known as *star* polygons. They are frequently used in decoration. The earliest mathematical discussion of them was by Thomas Bradwardine (1290–1349), who became archbishop of Canterbury for the last month of his life. They were also studied by the great German scientist Kepler (1571–1630) [see Coxeter **1,** p. 114]. It was the Swiss mathematician L. Schläfli (1814–1895) who first used a numerical symbol such as {*p/d*}. This notation is justified by the occurrence of formulas that hold for {*n*} equally well whether *n* be an integer or a fraction. For instance, any side of {*n*} forms with the center *O* an isosceles triangle OP_0P_1 (Figure 2.8*b*) whose angle at *O* is $2\pi/n$. (As we are introducing trigonometrical ideas, it is natural to use radian measure and write 2π instead of 360°.) The base of this isosceles triangle, being a side of the polygon, is conveniently denoted by 2*l*. The other sides of the triangle are equal to the circumradius *R* of the polygon. The altitude or median from *O* is the inradius *r* of the polygon. Hence

2.81 $$R = l \csc \frac{\pi}{n}, \qquad r = l \cot \frac{\pi}{n}.$$

If *n* = *p/d*, the area of the polygon is naturally defined to be the sum of the areas of the *p* isosceles triangles, namely

---

* H. S. M. Coxeter, M. S. Longuet-Higgins, and J. C. P. Miller, Uniform polyhedra, *Philosophical Transactions of the Royal Society,* **A, 246** (1954), pp. 401–450.

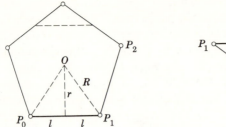

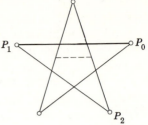

Figure 2.8b

2.82
$$plr = pl^2 \cot \frac{\pi}{n}.$$

When $d = 1$, this is simply $pl^2 \cot \pi/p$; in other cases our definition of area has the effect that every part of the interior is counted a number of times equal to the "local density" of that part; for example, the pentagonal region in the middle of the pentagram $\{\frac{5}{2}\}$ is counted twice.

The angle $P_0P_1P_2$ between two adjacent sides of $\{n\}$, being the sum of the base angles of the isosceles triangle, is the supplement of $2\pi/n$, namely

2.83
$$\left(1 - \frac{2}{n}\right)\pi.$$

The line segment joining the midpoints of two adjacent sides is called the *vertex figure* of $\{n\}$. Its length is clearly

2.84
$$2l \cos \frac{\pi}{n}$$

[Coxeter **1**, pp. 16, 94].

EXERCISES

1. If the sides of a polygon inscribed in a circle are all equal, the polygon is regular.

2. If a polygon inscribed in a circle has an odd number of vertices, and all its angles are equal, the polygon is regular. (Marcel Riesz.)

3. Find the angles of the polygons
$$\{5\}, \quad \{\tfrac{5}{2}\}, \quad \{9\}, \quad \{\tfrac{9}{2}\}, \quad \{\tfrac{9}{4}\}.$$

4. Find the radii and vertex figures of the polygons
$$\{8\}, \quad \{\tfrac{8}{3}\}, \quad \{12\}, \quad \{\tfrac{12}{5}\}.$$

5. Give polar coordinates for the kth vertex P_k of a polygon $\{n\}$ of circumradius 1 with its center at the pole, taking P_0 to be $(1, 0)$.

6. Can a square cake be cut into nine slices so that everyone gets the same amount of cake and the same amount of icing?

Isometry in the Euclidean plane

Having made some use of reflections, rotations, and translations, we naturally ask why a rotation or a translation can be achieved as a continuous displacement (or "motion") while a reflection cannot. It is also reasonable to ask whether there is any other kind of isometry that resembles a reflection in this respect. After answering these questions in terms of "sense," we shall use the information to prove a remarkable theorem (§ 3.6) and to describe the seven possible ways to repeat a pattern on an endless strip (§ 3.7).

3.1 DIRECT AND OPPOSITE ISOMETRIES

"Take care of the sense, and the sounds will take care of themselves."

Lewis Carroll

[Dodgson **1**, Chap. 9]

By several applications of Axiom 1.26, it can be proved that any point P in the plane of two congruent triangles ABC, $A'B'C'$ determines a corresponding point P' such that $AP = A'P'$, $BP = B'P'$, $CP = C'P'$. Likewise another point Q yields Q', and $PQ = P'Q'$. Hence

3.11 *Any two congruent triangles are related by a unique isometry.*

In § 1.3, we saw that Pappus's proof of *Pons asinorum* involved the comparison of two coincident triangles ABC, ACB. We see intuitively that this is a distinction of *sense*: if one is counterclockwise the other is clockwise. It is a "topological" property of the Euclidean plane that this distinction can be extended from coincident triangles to distinct triangles: any two "directed" triangles, ABC and $A'B'C'$, either agree or disagree in sense. (For a deeper investigation of this intuitive idea, see Veblen and Young [**2**, pp. 61–62] or Denk and Hofmann [**1**, p. 56].)

If ABC and $A'B'C'$ are congruent, the isometry that relates them is said to be *direct* or *opposite* according as it preserves or reverses sense, that is, according as ABC and $A'B'C'$ agree or disagree. It is easily seen that this property of the isometry is independent of the chosen triangle ABC: if the same isometry relates DEF to $D'E'F'$, where DEF agrees with ABC, then also $D'E'F'$ agrees with $A'B'C'$. Clearly, direct and opposite isometries combine like positive and negative numbers (e.g., the product of two opposite isometries is direct). Since a reflection is opposite, a rotation (which is the product of two reflections) is direct. In particular, the identity is direct. Some authors call direct and opposite isometries "displacements and reversals" or "proper and improper congruences."

Theorem 2.31 can be extended as follows:

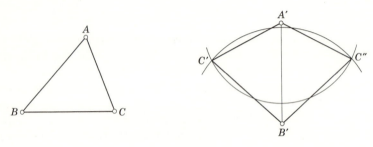

Figure 3.1a

3.12   *Two given congruent line segments (or point pairs) AB, $A'B'$ are related by just two isometries: one direct and one opposite.*

To prove this, take any point C outside the line AB, and construct C' so that the triangle $A'B'C'$ is congruent to ABC. The two possible positions of C' (marked C', C'' in Figure 3.1a) provide the two isometries. Since either can be derived from the other by reflecting in $A'B'$, one of the isometries is direct and the other opposite.

For a complete discussion we need the following theorem [Bachmann **1,** p. 3]:

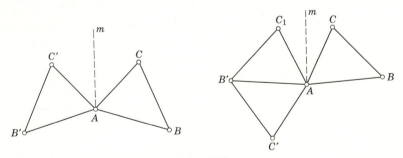

Figure 3.1b

3.13 *Every isometry of the plane is the product of at most three reflections. If there is an invariant point, "three" can be replaced by "two."*

We prove this in four stages, using 3.11. Trivially, if the triangles *ABC*, *A'B'C'* coincide, the isometry is the identity (which is the product of a reflection with itself). If *A* coincides with *A'*, and *B* with *B'*, while *C* and *C'* are distinct, the triangles are related by the reflection in *AB*. The case when only *A* coincides with *A'* can be reduced to one of the previous cases by reflecting *ABC* in *m*, the perpendicular bisector of *BB'* (see Figure 3.1*b*). Finally, the general case can be reduced to one of the first three cases by reflecting *ABC* in the perpendicular bisector of *AA'* [Coxeter **1**, p. 35].

Since a reflection reverses sense, an isometry is direct or opposite according as it is the product of an even or odd number of reflections.

Since the identity is the product of two reflections (namely of any reflection with itself), we may say simply that any isometry is the product of *two* or *three* reflections, according as it is direct or opposite. In particular,

3.14 *Any isometry with an invariant point is a rotation or a reflection according as it is direct or opposite.*

EXERCISES

1. Name two direct isometries.
2. Name one opposite isometry. Is there any other kind?
3. If *AB* and *A'B'* are related by a rotation, how can the center of rotation be constructed? (*Hint*: The perpendicular bisectors of *AA'* and *BB'* are not necessarily distinct.)
4. The product of reflections in three lines through a point is the reflection in another line through the same point [Bachmann **1**, p. 5].

3.2 TRANSLATION

> *Enoch walked with God; and he was not, for God took him.*
>
> Genesis V, 24

The particular isometries so far considered, namely reflections (which are opposite) and rotations (which are direct), have each at least one invariant point. A familiar isometry that leaves no point invariant is a *translation* [Bachmann **1**, p. 7], which may be described as the product of half-turns about two distinct points *O*, *O'* (Figure 3.2*a*). The first half-turn transforms an arbitrary point *P* into P^H, and the second transforms this into P^T, with the final result that PP^T *is parallel to OO' and twice as long*. Thus the length and direction of PP^T are constant: independent of the position of *P*. Since a translation is completely determined by its length and direction, the product of half-turns about *O* and *O'* is the same as the product of half-turns about *Q* and *Q'*, provided *QQ'* is equal and parallel to *OO'*. (This means that $OO'Q'Q$ is a parallelogram, possibly collapsing to form four collinear points, as in Figure 3.2a.) Thus, for a given translation, the center of one of the two half-turns may be arbitrarily assigned.

3.21   *The product of two translations is a translation.*

For, we may arrange the centers so that the first translation is the product of half-turns about O_1 and O_2, while the second is the product of half-turns about O_2 and O_3. When they are combined, the two half-turns about O_2 cancel, and we are left with the product of half-turns about O_1 and O_3.

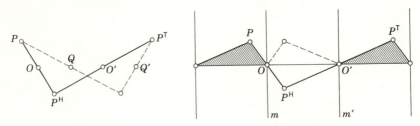

Figure 3.2a   **Figure 3.2b**

Similarly, if m and m' (Figure 3.2b) are the lines through O and O' perpendicular to OO', the half-turns about O and O' are the products of reflections in m and OO', OO' and m'. When they are combined, the two reflections in OO' cancel, and we are left with the product of reflections in m and m'. Hence

3.22   *The product of reflections in two parallel mirrors is a translation through twice the distance between the mirrors.*

If a translation T takes P to P^T and Q to Q^T, the segment QQ^T is equal and parallel to PP^T; therefore PQQ^TP^T is a parallelogram. Similarly, if another translation U takes P to Q, it also takes P^T to Q^T; therefore

$$\text{TU} = \text{UT}.$$

(In detail, if Q is P^U, Q^T is P^{UT}. But U takes P^T to P^{TU}. Therefore P^{TU} and P^{UT} coincide, for all positions of P.) In other words,

3.23   *Translations are commutative.*

The product of a half-turn H and a translation T is another half-turn; for we can express the translation as the product of two half-turns, one of which is H, say T = HH', and then we have

$$\text{HT} = \text{H}^2\text{H}' = \text{H}':$$

3.24   *The product of a half-turn and a translation is a half-turn.*

EXERCISES

1. If T is the product of half-turns about O and O', what is the product of half-turns about O' and O?

2. When a translation is expressed as the product of two reflections, to what extent can one of the two mirrors be arbitrarily assigned?

3. What is the product of rotations through opposite angles (α and $-\alpha$) about two distinct points?

4. The product of reflections in three parallel lines is the reflection in another line belonging to the same pencil of parallels.

5. Every product of three half-turns is a half-turn [Bachmann **1**, p. 7].

6. If H_1, H_2, H_3 are half-turns, $H_1H_2H_3 = H_3H_2H_1$.

7. Express the translation through distance a along the x-axis as a transformation of Cartesian coordinates. If $f(x, y) = 0$ is the equation for a curve, what is the equation for the transformed curve? Consider, for instance, the circle $x^2 + y^2 - 1 = 0$.

3.3 GLIDE REFLECTION

We are now familiar with three kinds of isometry: reflection, rotation, and translation. Another kind is the *glide reflection* (or simply "glide"), which is the product of the reflection in a line a and a translation along the same line. Picture this line as a straight path through snow; then, consecutive footprints are related by a glide. Such an isometry is determined by its *axis a* and the extent of the component translation. Since a reflection is opposite whereas a translation is direct, their product is opposite. Thus a glide reflection is an opposite isometry having no invariant point [Coxeter **1**, p. 36].

If a glide reflection G transforms an arbitrary point P into P^G (Figure 3.3a), P and P^G are equidistant from the axis a on opposite sides. Hence

The midpoint of the line segment PP^G lies on the axis for all positions of P.

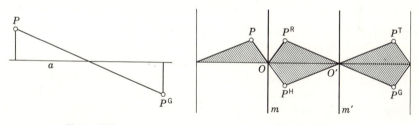

| Figure 3.3a | Figure 3.3b |

Let R_1 and T denote the component reflection and translation. They evidently commute, so that

$$G = R_1T = TR_1.$$

We have seen (Figure 3.2b) that the translation T may be expressed as the product of two half-turns or of two parallel reflections. Identifying the line a in Figure 3.3a with the line OO' in Figure 3.3b, let R, R' denote the reflections in m, m'. Then the product of the two half-turns

$$H = RR_1 = R_1R, \qquad H' = R'R_1 = R_1R'$$

is
$$T = HH' = RR_1R_1R' = RR',$$

and the glide reflection is
$$G = R_1T = R_1RR' = HR'$$
$$= TR_1 = RR'R_1 = RH'.$$

Thus a glide reflection may be expressed as the product of three reflections (two perpendicular to the third), or of a half-turn and a reflection, or of a reflection and a half-turn. Conversely, the product of any half-turn and any reflection (or vice versa) is a glide reflection, provided the center of the half-turn does not lie on the mirror. [Bachmann **1**, p. 6.]

We saw in 3.13 that any direct isometry in the plane is the product of two reflections, that is, a translation or a rotation according as the two mirrors are parallel or intersecting; also that any opposite isometry with an invariant point is a reflection. To complete the catalog of isometries, the only remaining possibility is an opposite isometry with no invariant point. If such an isometry S transforms an arbitrary point A into A', consider the half-turn H that interchanges these two points. The product HS, being an opposite isometry which leaves the point A' invariant, can only be a reflection R. Hence the given opposite isometry is the glide reflection

$$S = H^{-1}R = HR:$$

Every opposite isometry with no invariant point is a glide reflection.

In other words,

3.31   *Every product of three reflections is either a single reflection or a glide reflection.*

In particular, the product RT of any reflection and any translation is a glide reflection, degenerating to a pure reflection when the mirror for R is perpendicular to the direction of the translation T (in which case the reflections R and RT may be used as the two parallel reflections whose product is T). But since a given glide reflection G has a definite axis (the locus of midpoints of segments PP^G), its decomposition into a reflection and a translation *along the mirror* is unique (unlike its decomposition into a reflection and a half-turn, where we may either take the mirror to be any line perpendicular to the axis or equivalently take the center of the half-turn to be any point on the axis).

EXERCISES

1. If B is the midpoint of AC, what kinds of isometry will transform
(i) AB into CB,   (ii) AB into BC?

2. Every direct isometry is the product of two reflections. Every opposite isometry is the product of a reflection and a half-turn.

3. Describe the product of the reflection in OO' and the half-turn about O.

4. Describe the product of two glide reflections whose axes are perpendicular.

5. Every product of three glide reflections is a glide reflection.

6. The product of three reflections is a reflection if and only if the three mirrors are either concurrent or parallel.

7. If R_1, R_2, R_3 are three reflections, $(R_1R_2R_3)^2$ is a translation [Rademacher and Toeplitz **1**, p. 29].

8. Describe the transformation

$$(x, y) \longrightarrow (x + a, -y).$$

Justify the statement that this transforms the curve $f(x, y) = 0$ into $f(x - a, -y) = 0$.

3.4 REFLECTIONS AND HALF-TURNS

Thomsen* has developed a very beautiful theory in which geometrical properties of points O, O_1, O_2, ... and lines m, m_1, m_2, ... (understood to be all distinct) are expressed as relations among the corresponding half-turns H, H_1, H_2, ... and reflections R, R_1, R_2, ... The reader can soon convince himself that the following pairs of statements are logically equivalent:

RR_1	$= R_1R$	$\longleftrightarrow$	m and m_1 are perpendicular.
HR	$= RH$	$\longleftrightarrow$	O lies on m.
$R_1R_2R_3$	$= R_3R_2R_1$	$\longleftrightarrow$	m_1, m_2, m_3 are either concurrent or parallel.
H_1H	$= HH_2$	$\longleftrightarrow$	O is the midpoint of O_1O_2.
H_1R	$= RH_2$	$\longleftrightarrow$	m is the perpendicular bisector of O_1O_2.

EXERCISE

Interpret the relations (a) $H_1H_2H_3H_4 = 1$;　　(b) $R_1R = RR_2$.

3.5 SUMMARY OF RESULTS ON ISOMETRIES

> And thick and fast they came at last,
> And more, and more, and more.
>
> Lewis Carroll
>
> [Dodgson **2**, Chap. 4]

Some readers may have become confused with the abundance of technical terms, many of which are familiar words to which unusually precise meanings have been attached. Accordingly, let us repeat some of the definitions, stressing both their analogies and their differences.

---

* G. Thomsen, The treatment of elementary geometry by a group-calculus, *Mathematical Gazette,* **17** (1933), p. 232. Bachmann [**1**] devotes a whole book to the development of this idea.

In all the contexts that concern us here, a *transformation* is a one-to-one correspondence of the whole plane (or space) with itself. An *isometry* is a special kind of transformation, namely, the kind that preserves length. A *symmetry operation* belongs to a given figure rather than to the whole plane: it is an isometry that transforms the figure into itself.

In the plane, a *direct* (sense-preserving) isometry, being the product of two reflections, is a rotation or a translation according as it does or does not have an invariant point, that is, according as the two mirrors are intersecting or parallel. In the latter case the length of the translation is twice the distance between the mirrors; in the former, the angle of the rotation is twice the angle between the mirrors. In particular, the product of reflections in two perpendicular mirrors is a half-turn, that is, a rotation through two right angles. Moreover, the product of two half-turns is a translation.

An *opposite* (sense-reversing) isometry, being the product of three reflections, is, in general, a *glide reflection*: the product of a reflection and a translation. In the special case when the translation is the identity (i.e., a translation through zero distance), the glide reflection reduces to a single reflection, which has a whole line of invariant points, namely, all the points on the mirror.

To sum up:

3.51   *Any direct isometry is either a translation or a rotation. Any opposite isometry is either a reflection or a glide reflection.*

EXERCISES

1. If S is an opposite isometry, S^2 is a translation.

2. If R_1, R_2, R_3 are three reflections, $(R_2R_3R_1R_2R_3)^2$ is a translation along the first mirror. (*Hint*: Since $R_1R_2R_3$ and $R_2R_3R_1$ are glide reflections, their squares are commutative, by 3.23; thus

$$(R_1R_2R_3)^2(R_2R_3R_1)^2 = (R_2R_3R_1)^2(R_1R_2R_3)^2,$$

that is, R_1 and $(R_2R_3R_1R_2R_3)^2$ are commutative [cf. Bachmann **1**, p. 13].)

3.6   HJELMSLEV'S THEOREM

> . . . *a very high degree of unexpectedness, combined with inevitability and economy.*
>
> G. H. Hardy [**2**, p. 53]

We saw, in 3.12, that two congruent line segments AB, $A'B'$, are related by just two isometries: one direct and one opposite. Both isometries have the same effect on every point collinear with A and B, that is, every point on the infinite straight line AB (for instance, the midpoint of AB is transformed into the midpoint of $A'B'$). The opposite isometry is a reflection or glide reflection whose mirror or axis contains all the midpoints of segments joining pairs of corresponding points. If two of these midpoints coincide, the direct isometry is a half-turn, and they all coincide [Coxeter **3**, p. 267]. Hence

HJELMSLEV'S THEOREM. *When all the points P on one line are related by an isometry to all the points P′ on another, the midpoints of the segments PP′ are distinct and collinear or else they all coincide.*

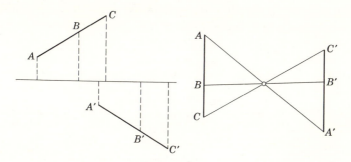

Figure 3.6a

In particular, if *A, B, C* are on one line and *A′, B′, C′* on another, with

3.61 $$AB = A'B', \qquad BC = B'C'$$

(Figure 3.6*a*), then the midpoints of *AA′, BB′, CC′* are either collinear or coincident (J. T. Hjelmslev, 1873–1950).

3.7 PATTERNS ON A STRIP

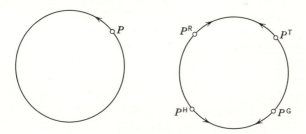

Figure 3.7a

Any kind of isometry may be used to relate two equal circles. For instance, the point *P* on the first circle of Figure 3.7*a* is transformed into P^T on the second circle by a translation, into P^R by a reflection, into P^H by a half-turn, and into P^G by a glide reflection. (Arrows have been inserted to indicate what happens to the positive sense of rotation round the first circle.) These four isometries have one important property in common: they leave invariant (as a whole) one infinite straight line, namely, the line joining the centers of the two circles. (In the fourth case this is the *only* invariant line.)

We have seen (Figure 3.2*b*) that the product of reflections in two parallel mirrors *m*, *m'* is a translation. This may be regarded as the limiting case of a rotation whose center is very far away; for the two parallel mirrors are the limiting case of two mirrors intersecting at a very small angle. Accordingly, the infinite group generated by a single translation is denoted by C_∞, and the infinite group generated by two parallel reflections is denoted by D_∞. Abstractly, C_∞ is the "free group with one generator." If T is the generating translation, the group consists of the translations

$$\ldots, \; T^{-2}, \quad T^{-1}, \quad 1, \quad T, \quad T^2, \ldots.$$

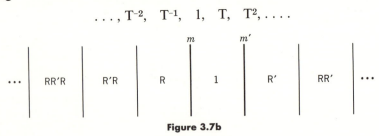

Figure 3.7b

Similarly, D_∞, generated by the reflections R, R' in parallel mirrors *m*, *m'* (Figure 3.7*b*), consists of the reflections and translations

$$\ldots, RR'R, \quad R'R, \quad R, \quad 1, \quad R', \quad RR', \quad R'RR', \ldots$$

[Coxeter **1,** p. 76]; its abstract definition is simply

$$R^2 = R'^2 = 1.$$

This group can be observed when we sit in a barber's chair between two parallel mirrors (cf. the *New Yorker,* Feb. 23, 1957, p. 39, where somehow the reflection RR'RR'R yields a demon).

    A different geometrical representation for the same abstract group D_∞ is obtained by interpreting the generators R and R' as half-turns. There is also an intermediate representation in which one of them is a reflection and the other a half-turn; but in this case their product is no longer a translation but a glide reflection.

    Continuing in this manner, we could soon obtain the complete list of the seven infinite "one-dimensional" symmetry groups: the seven essentially distinct ways to repeat a pattern on a strip or ribbon [Speiser **1,** pp. 81–82]:

Typical pattern	*Generators*	*Abstract Group*
(i)...L L L L...	1 translation ⎫	C_∞
(ii)...L Γ L Γ...	1 glide reflection ⎭	
(iii)...V V V V...	2 reflections ⎫	
(iv)...N N N N...	2 half-turns ⎬	D_∞
(v)...V Λ V Λ...	1 reflection and 1 half-turn ⎭	
(vi)...D D D D...	1 translation and 1 reflection	$C_\infty \times D_1$
(vii)...H H H H...	3 reflections	$D_\infty \times D_1$

In (iii), the two mirrors are both vertical, one in the middle of a V, reflecting it into itself, while the other reflects this V into one of its neighbors; thus one half of the V, placed between the two mirrors, yields the whole pattern. In (vi) and (vii) there is a horizontal mirror, and the symbols in the last column indicate "direct products" [Coxeter **1,** p. 42]. For all these groups, except (i) and (ii), there is some freedom in choosing the generators; for example, in (iii) or (iv) one of the two generators could be replaced by a translation.

Strictly speaking, these seven groups are not "1-dimensional" but "$1\frac{1}{2}$-dimensional;" that is, they are 2-dimensional symmetry groups involving translation in one direction. In a purely one-dimensional world there are only two infinite symmetry groups: C_∞, generated by one translation, and D_∞, generated by two reflections (in point mirrors).

EXERCISES

1. Identify the symmetry groups of the following patterns:

$$...b \; b \; b \; b... \; ,$$
$$...b \; p \; b \; p... \; ,$$
$$...b \; d \; b \; d... \; ,$$
$$...b \; q \; b \; q... \; ,$$
$$...b \; d \; p \; q \; b \; d \; p \; q... \; .$$

2. Which are the symmetry groups of (*a*) a cycloid, (*b*) a sine curve?

4

Two-dimensional crystallography

Mathematical crystallography provides one of the most important applications of elementary geometry to physics. The three-dimensional theory is complicated, but its analog in two dimensions is easy to visualize without being trivial. Patterns covering the plane arise naturally as an extension of the strip patterns considered in § 3.7. However, in spite of the restriction to two dimensions, a complete account of the enumeration of infinite symmetry groups is beyond the scope of this book.

4.1 LATTICES AND THEIR DIRICHLET REGIONS

Infinite two-dimensional groups (the symmetry groups of repeating patterns such as those commonly used on wallpaper or on tiled floors) are distinguished from infinite "one-dimensional" groups by the presence of *independent* translations, that is, translations whose directions are neither parallel nor opposite. The crystallographer E. S. Fedorov showed that there are just seventeen such two-dimensional groups of isometries. They were rediscovered in our own century by Pólya and Niggli.* The symbols by which we denote them are taken from the International Tables for X-ray Crystallography.

The simplest instance is the group **p1**, generated by two independent

* E. S. Fedorov, *Zapiski Imperatorskogo S. Peterburgskogo Mineralogicheskogo Obshchestva* (2), **28** (1891), pp. 345–390; G. Pólya and P. Niggli, *Zeitschrift für Kristallographie und Mineralogie,* **60** (1924), pp. 278–298. [See also Fricke and Klein **1,** pp. 227–233.] Fedorov's table shows that 16 of the 17 groups had been described by C. Jordan in 1869. The remaining one was recognized by L. Sohncke in 1874; but he missed three others.

Figure 4.1a

translations X, Y. Since the inverse of a translation is a translation, and the product of two translations is a translation (3.21), this group consists entirely of translations. Since $XY = YX$, these translations are simply $X^x Y^y$ for all integers x, y. Abstractly, this is the "direct product" $C_\infty \times C_\infty$, which has the single defining relation

$$XY = YX$$

[Coxeter and Moser **1**, p. 40]. Any object, such as the numeral 6 in Figure 4.1a, is transformed by the group **p1** into an infinite array of such objects, forming a pattern. Conversely, **p1** is the complete symmetry group of the pattern, provided the object has no intrinsic symmetry. If the object is a single point, the pattern is an array of points called a two-dimensional *lattice,* which may be pictured as the plan of an infinite orchard. Each lattice point is naturally associated with the symbol for the translation by which it is derived from the original point 1 (Figure 4.1b).

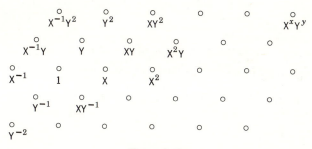

Figure 4.1b

Anyone standing in an orchard observes the alignment of trees in rows in many directions. This exhibits a characteristic property of a lattice: the line joining any two of the points contains infinitely many of them, evenly spaced, that is, a "one-dimensional lattice." In fact, the line joining the points 1 and $X^x Y^y$ contains also the points

$$X^{nx/d} Y^{ny/d} = (X^{x/d} Y^{y/d})^n$$

where d is the greatest common divisor of x and y, and n runs over all the integers. In particular, the powers of X all lie on one line, the powers of Y on another, and lines parallel to these through the remaining lattice points form a tessellation of congruent parallelograms filling the plane without interstices (Figure 4.1c). (We use the term *tessellation* for any arrangement of polygons fitting together so as to cover the whole plane without overlapping.)

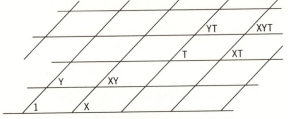

Figure 4.1c

A typical parallelogram is formed by the four points 1, X, XY, Y. The translation $T = X^x Y^y$ transforms this parallelogram into another one having the point T (instead of 1) at its "first" corner. There is thus a one-to-one correspondence between the cells or tiles of the tessellation and the transformations in the group, with the property that each transformation takes any point inside the original cell to a point similarly situated in the new cell. For this reason, the typical parallelogram is called a *fundamental region*.

The shape of the fundamental region is far from unique. Any parallelogram will serve, provided it has four lattice points for its vertices but no others on its boundary or inside [Hardy and Wright **1**, p. 28]. This is the geometrical counterpart of the algebraic statement that the group generated by X, Y is equally well generated by $X^a Y^b$, $X^c Y^d$, provided

$$ad - bc = \pm 1.$$

To express the old generators in terms of the new, we observe that

$$(X^a Y^b)^d (X^c Y^d)^{-b} = X^{ad-bc}, \quad (X^a Y^b)^{-c}(X^c Y^d)^a = Y^{ad-bc}.$$

But there is no need for the fundamental region to be a parallelogram at all; for example, we may replace each pair of opposite sides by a pair of congruent curves, as in Figure 4.1d.

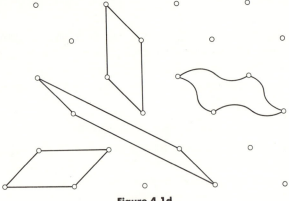

Figure 4.1d

Every possible fundamental region, whether we choose a parallelogram or any other shape, has *the same area* as the typical parallelogram of Figure 4.1c. For, in le a sufficiently large circle, the number of lattice points is equal to the number of replicas of any fundamental region (with an insignificant error due to mutilated regions at the circumference); thus every possible shape has for its area the same fraction of the area of the large circle.* It is an interesting fact that any *convex* fundamental region for the translation group is a centrally symmetrical polygon (namely, a parallelogram or a centrally symmetrical hexagon).†

Among the various possible parallelograms, we can select a standard or *reduced* parallelogram by taking the generator Y to be the shortest translation (or one of the shortest) in the group, and X to be an equal or next shortest translation in another direction. If the angle between X and Y then happens to be obtuse, we reverse the direction of Y. Thus, among all the parallelograms that can serve as a fundamental region, the reduced parallelogram has the shortest possible sides. The translations along these sides are naturally called reduced generators.

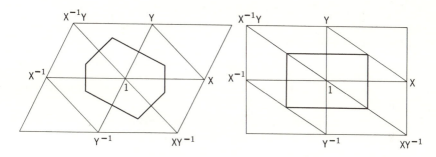

Figure 4.1e

By joining the vertices X, Y of the reduced parallelogram, and the corresponding pair of vertices of each replica, we obtain a tessellation of congruent triangles whose vertices are lattice points and whose angles are nonobtuse. Each lattice point belongs to six of the triangles; for example, the triangles surrounding the point 1 join it to pairs of adjacent points in the cycle

$$X, \quad Y, \quad X^{-1}Y, \quad X^{-1}, \quad Y^{-1}, \quad XY^{-1}$$

(Figure 4.1e). By joining the circumcenters of these six triangles, we obtain the *Dirichlet region* (or "Voronoi polygon") of the lattice: a polygon whose interior consists of all the points in the plane which are nearer to a particu-

---

* Gauss used this idea as a means of estimating π [Hilbert and Cohn-Vossen **1**, pp. 33–34].

† A. M. Macbeath. *Canadian Journal of Mathematics*, **13** (1961), p. 177.

lar lattice point (such as the point 1) than to any other lattice point.* Such regions, each surrounding a lattice point, evidently fit together to fill the whole plane; in fact, the Dirichlet region is a particular kind of fundamental region.

The lattice is symmetrical by the half-turn about the point 1 (or any other lattice point). For this half-turn interchanges the pairs of lattice points X^xY^y, $X^{-x}Y^{-y}$. (In technical language, the group **p1** has an automorphism of period 2 which replaces X and Y by their inverses.) Hence *the Dirichlet region is symmetrical by a half-turn*. Its precise shape depends on the relative lengths of the generating translations X, Y and the angle between them. If this angle is a right angle, the Dirichlet region is a rectangle (or a square), since the circumcenter of a right-angled triangle is the midpoint of the hypotenuse. In all other cases it is a hexagon (not necessarily a regular hexagon; but since it is centrally symmetrical, its pairs of opposite sides are equal and parallel).

Varying the lattice by letting the angle between the translations X and Y increase gradually to 90°, we see that two opposite sides of the hexagon shrink till they become single vertices, and then the remaining four sides form a rectangle (or square).

Reflections in the four or six sides of the Dirichlet region transform the central lattice point 1 into four or six other lattice points which we naturally call the *neighbors* of the point 1.

EXERCISES

1. Any two opposite sides of a Dirichlet region are perpendicular to the line joining their midpoints.

2. Sketch the various types of lattice that can arise if X and Y are subject to the following restrictions: they may have the same length, and the angle between them may be 90° or 60°. Indicate the Dirichlet region in each case, and state whether the symmetry group of this region is C_2, D_2, D_4, or D_6.

4.2   THE SYMMETRY GROUP OF THE GENERAL LATTICE

> *The investigation of the symmetries of a given mathematical structure has always yielded the most powerful results.*
>
> E. Artin (1898 -1962)
> [Artin **1**, p. 54]

Any given lattice is easily seen to be symmetrical by the half-turn about the midpoint of the segment joining any two lattice points [Hilbert and Cohn-Vossen **1**, p. 73]. Such midpoints form a lattice of finer mesh, whose generating translations are half as long as X and Y (see the "open" points in Figure 4.2*a*).

---

* G. L. Dirichlet, *Journal für die reine und angewandte Mathematik,* **40** (1850), pp. 216–219.

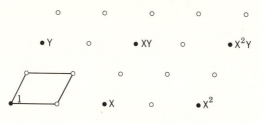

Figure 4.2a

The "general" lattice occurs when the reduced generators differ in length and the angle between them is neither 90° nor 60°. In such a case, the translations $X^x Y^y$ and the above-mentioned half-turns are its only symmetry operations. In other words, the symmetry group of the general lattice is derived from **p1** by adding an extra transformation H, which is the half-turn about the point 1. This group is denoted by **p2** [Coxeter and Moser **1,** pp. 41–42]. It is generated by the half-turn H and the translations X, Y, in terms of which the half-turn that interchanges the points 1 and $T = X^x Y^y$ is HT. (Note that T itself is the product of H and HT.) The group is equally well generated by the three half-turns HX, H, HY, or (redundantly) by these three and their product

$$HX \cdot H \cdot HY = HXY,$$

which are half-turns about the four vertices of the parallelogram shown in Figure 4.2a.

It is remarkable that any triangle or any simple quadrangle (not necessarily convex) will serve as a fundamental region for **p2**. Half-turns about the midpoints of the three or four sides may be identified with HX, H, HY (Figure 4.2b), or HX, H, HY, HXY (Figure 4.2c).

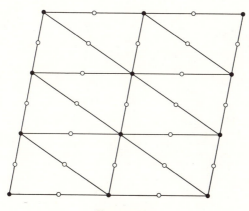

Figure 4.2b

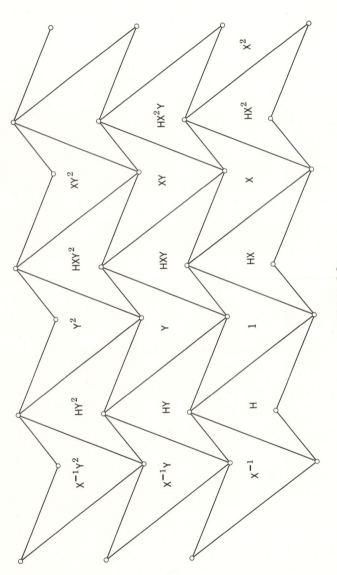

Figure 4.2c

EXERCISES

1. Why do the vertices of the quadrangles in Figure 4.2c form two superposed lattices?

2. Draw the tessellation of Dirichlet regions for a given lattice. Divide each region into two halves by means of a diagonal. The resulting tessellation is a special case of the tessellation of scalene triangles (Figure 4.2b) or of irregular quadrangles (Figure 4.2c) according as the Dirichlet region is rectangular or hexagonal.

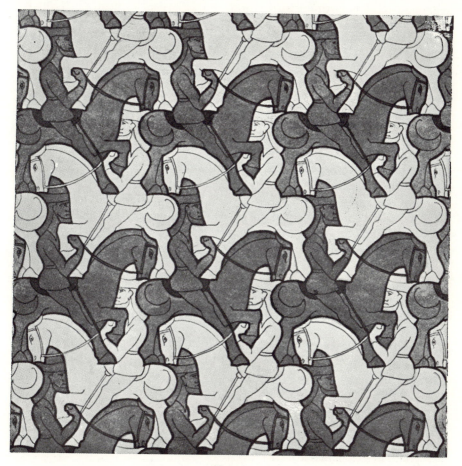

Plate I

4.3 THE ART OF M. C. ESCHER

The groups **p1** and **p2** are two of the simplest of the seventeen discrete groups of isometries involving two independent translations. Several others will be mentioned in this section and the next. Convenient generators for all of them are listed in Table I on p. 413.

The art of filling a plane with a repeating pattern reached its highest development in thirteenth-century Spain, where the Moors used all the seventeen groups in their intricate decoration of the Alhambra [Jones **1**]. Their preference for abstract patterns was due to their strict observance of the Second Commandment. The Dutch artist M. C. Escher, free from such scruples, makes an ingenious application of these groups by using animal shapes for their fundamental regions. For instance, the symmetry group of his pattern of knights on horseback (Plate I) seems at first sight to be **p1**, generated by a horizontal translation and a vertical translation. But by ignoring the distinction between the dark and light specimens we obtain the more interesting group **pg**, which is generated by two parallel glide reflections, say G and G′. We observe that the vertical translation can be expressed equally well as G^2 or G'^2. It is remarkable that the single relation

$$G^2 = G'^2$$

provides a complete abstract definition for this group [Coxeter and Moser **1**, p. 43]. Clearly, the knight and his steed (of either color) constitute a fundamental region for **pg**. But we must combine two such regions, one dark and one light, in order to obtain a fundamental region for **p1**.

Similarly, the symmetry group of Escher's pattern of beetles (Plate II) seems at first sight to be **pm**, generated by two vertical reflections and a vertical translation. But on looking more closely we see that there are both dark and light beetles, and that the colors are again interchanged by glide reflections. The complete symmetry group **cm**, whose fundamental region is the right or left half of a beetle of either color, is generated by any such vertical glide reflection along with a vertical reflection. To obtain a fundamental region for the "smaller" group **pm**, we combine the right half of a beetle of either color with the left half of an adjacent beetle of the other color.

A whole beetle (of either color) provides a fundamental region for the group **p1** (with one of its generating translations oblique) or equally well for **pg**.

EXERCISES

1. Locate the axes of two glide reflections which generate **pg** in Plates I and II.

2. Any two parallelograms whose sides are in the same two directions can together be repeated by translations to fill the plane.

4.4   SIX PATTERNS OF BRICKS

Figure 4.4*a* shows how six of the seventeen two-dimensional space groups arise as the symmetry groups of familiar patterns of rectangles, which we may think of as bricks or tiles. The generators are indicated as follows: a broken line denotes a mirror, a "lens" denotes a half-turn, a small square denotes a quarter-turn (i.e., rotation through 90°), and a "half arrow" denotes a glide reflection.

In each case, a convenient fundamental region is indicated by shading. This region is to some extent arbitrary except in the case of **pmm**, where it is entirely bounded by mirrors.

The procedure for analysing such a pattern is as follows. We observe that the symmetry group of a single brick is D_2 (of order 4), which has subgroups C_2 and D_1. If all the symmetry operations of the brick are also symmetry operations of the whole pattern, as in **cmm** and **pmm**, the fundamental region is a quarter of the brick, two of the generators are the reflections that generate D_2, and any other generator transforms the original brick into a neighboring brick. If only the subgroup C_2 or D_1 belongs to the whole pat-

Plate II

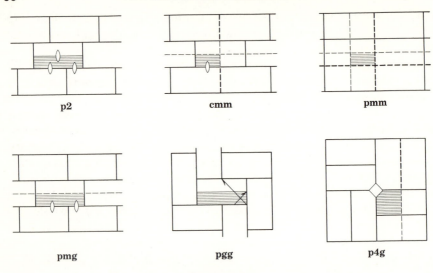

Figure 4.4a

tern (the way C_2 belongs to **p2** or **pgg**, and D_1 to **pmg** or **p4g**), the fundamental region is half a brick, and the generators are not quite so obvious.

EXERCISE

In all these patterns it is understood that a "brick" is a rectangle in which one side is twice as long as another. In each case, any brick is related to the whole pattern in the same way as any other. (In technical language, the symmetry group is *transitive* on the bricks.) Are these six the *only* transitive patterns of bricks?

4.5 THE CRYSTALLOGRAPHIC RESTRICTION

> *A mathematician, like a painter or a poet, is a maker of patterns. If his patterns are more permanent than theirs, it is because they are made with ideas.*
>
> G. H. Hardy [**2,** p. 24]

A complete account of the enumeration of the seventeen two-dimensional space-groups would occupy too much space. But it seems worthwhile to give Barlow's elegant proof * that the only possible cyclic subgroups are C_2, C_3, C_4, and C_6. In other words:

The only possible periods for a rotational symmetry operation of a lattice are 2, 3, 4, 6.

Let P be any center of rotation of period n. The remaining symmetry

* W. Barlow, *Philosophical Magazine* (6), **1** (1901), p. 17.

operations of the lattice transform P into infinitely many other centers of rotation of the same period. Let Q be one of these other centers (Figure 4.5a) at the least possible distance from P. A third center, P', is derived from P by rotation through $2\pi/n$ about Q; and a fourth, Q', is derived from Q by rotation through $2\pi/n$ about P'. Of course, the segments PQ, QP', $P'Q'$, are all equal. It may happen that P and Q' coincide; then $n = 6$. In all other cases, since Q was chosen at the least possible distance from P, we must have $PQ' \geqslant PQ$; therefore $n \leqslant 4$. (If $n = 4$, $PQP'Q'$ is a square. If $n = 5$, PQ' is obviously shorter than PQ. If $n > 6$, PQ crosses $P'Q'$, but it is no longer necessary to use Q': we already have $PP' < PQ$, which is sufficiently absurd.)

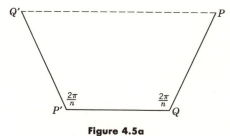

Figure 4.5a

EXERCISES

1. If S and T are rotations through $2\pi/n$ about P and Q, what is $T^{-1}ST$?

2. If a discrete group of isometries includes two rotations about distinct centers, it includes two such rotations having the same period, and therefore also a translation. If this period is greater than 2, it includes two independent translations.

4.6 REGULAR TESSELLATIONS

The mathematician's patterns, like the painter's or the poet's, must be beautiful; the ideas, like the colours or the words, must fit together in a harmonious way. Beauty is the first test: there is no permanent place in the world for ugly mathematics.

G. H. Hardy [**2,** p. 25]

It was probably Kepler (1571–1630) who first investigated the possible ways of filling the plane with equal regular polygons. We shall find it convenient to use the Schläfli symbol $\{p, q\}$ for the tessellation of regular p-gons, q surrounding each vertex [Schläfli **1,** p. 213]. The cases

$$\{6, 3\}, \quad \{4, 4\}, \quad \{3, 6\}$$

are illustrated in Figure 4.6a, where in each case the polygon drawn in heavy lines is the *vertex figure:* the q-gon whose vertices are the midpoints of the q edges at a vertex. (Since tessellations are somewhat analogous to polyhedra, it is natural to use the word *edges* for the common sides of adjacent polygons, and *faces* for the polygons themselves.)

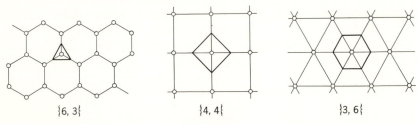

$\{6, 3\}$ $\{4, 4\}$ $\{3, 6\}$

Figure 4.6a

For a formal definition, we may say that a tessellation is *regular* if it has regular faces and a regular vertex figure at each vertex.

The tessellation $\{6, 3\}$ is often used for tiled floors in bathrooms. It can also be seen in any beehive. $\{4, 4\}$ is familiar in the form of squared paper; in terms of Cartesian coordinates, its vertices are just the points for which both x and y are integers. $\{3, 6\}$ is the dual of $\{6, 3\}$ in the following sense. The *dual* of $\{p, q\}$ is the tessellation whose edges are the perpendicular bisectors of the edges of $\{p, q\}$ (see Figure 4.6b). Thus the dual of $\{p, q\}$ is $\{q, p\}$, and vice versa; the vertices of either are the centers of the faces of the other. In particular, the dual of $\{4, 4\}$ is an equal $\{4, 4\}$.

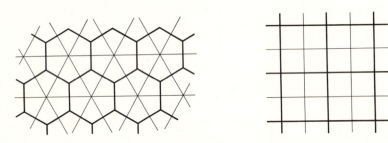

Figure 4.6b

The possible values of p and q are easily obtained by equating the angle of a p-gon, namely $(1 - 2/p)\pi$, to the value it must have if q such polygons come together at a vertex:

$$\left(1 - \frac{2}{p}\right)\pi = \frac{2\pi}{q}, \quad \frac{1}{p} + \frac{1}{q} = \frac{1}{2},$$
$$(p - 2)(q - 2) = 4.$$

The three possible ways of factorizing 4, namely

$$4 \cdot 1, \quad 2 \cdot 2, \quad 1 \cdot 4,$$

yield the three tessellations already described. However, before declaring that these are the *only* regular tessellations, we should investigate the fractional solutions of our equation; for there might conceivably be a regular "star" tessellation {*p, q*} whose face {*p*} and vertex figure {*q*} are regular polygons of the kind considered in §2.8. For instance, Figure 4.6*c* shows ten pentagons placed together at a common vertex. Although they overlap, we might expect to be able to add further pentagons so as to form a tessellation {5, $\frac{10}{3}$} (whose vertex figure is a decagram), covering the plane a number of times. But in fact this number is infinite, as we shall see.

Consider the general regular tessellation {*p, q*}, where $p = n/d$. If it covers the plane only a finite number of times, there must be a minimum distance between the centers of pairs of faces. Let *P, Q* be two such centers at this minimum distance apart. Since they are centers of rotation of period *n*, the argument used in §4.5 proves that the only possible values of *n* are 3, 4, 6. Thus $d = 1$, and these are also the only possible values of *p*. Hence *there are no regular star tessellations* [Coxeter **1**, p. 112].

It is actually possible to cover a *sphere* three times by using twelve "pentagons" whose sides are arcs of great circles [Coxeter **1**, p. 111].

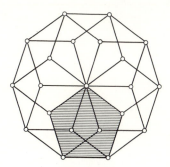

Figure 4.6c

To find the symmetry group of a regular tessellation, we treat its face the way we treated one of the bricks in §4.4. Clearly, the symmetry group of {*p, q*} is derived from the symmetry group D_p of one face by adding the reflection in a side of that face. Thus it is generated by reflections in the sides of a triangle whose angles are π/p (at the center of the face), $\pi/2$ (at the midpoint of an edge), and π/q (at a vertex). This triangle is a fundamental region, since it is transformed into neighboring triangles by the three generating reflections. Since each generator leaves invariant all the points on one side, the fundamental region is unique: it cannot be modified by addition and subtraction the way Escher modified the fundamental regions of some other groups.

The network of such triangles, filling the plane, is cut out by all the lines of symmetry of the regular tessellation. The lines of symmetry include the lines of the edges of both $\{p, q\}$ and its dual $\{q, p\}$. In the case of $\{6, 3\}$ and $\{3, 6\}$ (Figure 4.6*b*), these edge lines suffice; in the case of the two dual $\{4, 4\}$'s we need also the diagonals of the squares. In Figure 4.6*d*, alternate regions have been shaded so as to exhibit both the complete symmetry groups **p6m**, **p4m** and the "direct" subgroups **p6**, **p4** (consisting of rotations and translations) which preserve the colors and the direction of the shading [Brewster **1**, p. 94; Burnside **1**, pp. 416, 417].

Instead of deriving the network of triangles from the regular tessellation, we may conversely derive the tessellation from the network. For this purpose, we pick out a point in the network where the angles are π/p, that is, where p shaded and p white triangles come together. These $2p$ triangles combine to form a face of $\{p, q\}$.

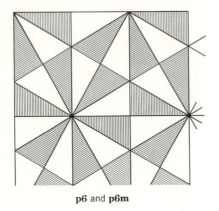

p6 and p6m   p4 and p4m

Figure 4.6d

EXERCISES

1. Justify the formal definition of "regular" on page 62. (It implies that the faces are all alike and that the vertices are all surrounded alike.)

2. Give a general argument to prove that the midpoints of the edges of a regular tessellation belong to a lattice. (*Hint*: Consider the group **p2** generated by half-turns about three such midpoints.)

3. Pick out the midpoints of the edges of $\{6, 3\}$. Verify that they belong to a lattice. Do they constitute the whole lattice?

4. Draw portions of lattices whose symmetry groups are **p2**, **pmm**, **cmm**, **p4m**, **p6m**.

4.7 SYLVESTER'S PROBLEM OF COLLINEAR POINTS

> Reductio ad absurdum, *which Euclid loved so much, is one of a mathe-matician's finest weapons. It is a far finer gambit than any chess gam-bit: a chess player may offer the sacrifice of a pawn or even a piece, but a mathematician offers the game.*
>
> G. H. Hardy [**2,** p. 34]

As we saw in § 4.1, a lattice is a discrete set of points having the property that the line joining any two of them contains not only these two but infi-nitely many. Figure 4.7a shows a *finite* "orchard" in which nine points are arranged in ten rows of three [Ball **1,** p. 105]. It was probably the investi-gation of such configurations that led Sylvester* to propose his problem of 1893:

Prove that it is not possible to arrange any finite number of real points so that a right line through every two of them shall pass through a third, unless they all lie in the same right line.

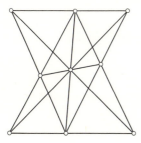

Figure 4.7a

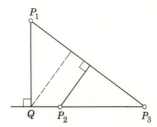

Figure 4.7b

Neither Sylvester nor any of his contemporaries were able to think of a satisfactory proof. The question was forgotten till 1933, when Karamata and Erdös revived it, and T. Gallai (*alias* Grünwald) finally succeeded, using a rather complicated argument. Sylvester's "negative" statement was rephrased "positively" by Motzkin:

If n points in the real plane are not on one straight line, then there exists a straight line containing exactly two of the points.

The following proof, which somewhat resembles Barlow's proof of the crystallographic restriction (§ 4.5), is due to L. M. Kelly.

* J. J. Sylvester, *Mathematical Questions and Solutions from the Educational Times,* **59** (1893), p. 98 (Question 11851). See also R. Steinberg, *American Mathematical Monthly,* **51** (1944), p. 170; L. M. Kelly, *ibid.,* **55** (1948), p. 28; T. Motzkin, *Transactions of the American Mathematical Society,* **70** (1951), p. 452; L. M. Kelly and W. O. J. Moser, *Canadian Journal of Mathematics,* **10** (1958), p. 213.

The n points $P_1, \ldots, P_n$ are joined by at most $\frac{1}{2} n(n - 1)$ lines P_1P_2, P_1P_3, etc. Consider the pairs P_i, P_jP_k, consisting of a point and a joining line which are not incident. Since there are at most $\frac{1}{2} n(n - 1)(n - 2)$ such pairs, there must be at least one, say P_1, P_2P_3, for which the distance P_1Q from the point to the line is the smallest such distance that occurs.

Then the line P_2P_3 contains no other point of the set. For if it contained P_4, at least two of the points P_2, P_3, P_4 would lie on one side of the perpendicular P_1Q (or possibly one of the P's would coincide with Q). Let the points be so named that these two are P_2, P_3, with P_2 nearer to Q (or coincident with Q). Then P_2, P_3P_1 (Figure 4.7b) is another pair having a smaller distance than P_1Q, which is absurd.

This completes the proof that there is always a line containing exactly two of the points. Of course, there may be more than one such line; in fact, Kelly and Moser proved that the number of such lines is at least $3n/7$.

EXERCISES

1. The above proof yields a line P_2P_3 containing only these two of the P's. The point Q actually lies *between* P_2 and P_3.

2. If n points are not all on one line, they have at least n distinct joins [Coxeter **2,** p. 31].

3. Draw a configuration of n points for which the lower limit of $3n/7$ "ordinary" joins is attained. (*Hint*: $n = 7$.)

5

Similarity in the Euclidean plane

In later chapters we shall see that Euclidean geometry is by no means the only possible geometry: other kinds are just as logical, almost as useful, and in some respects simpler. According to the famous *Erlangen program* (Klein's inaugural address at the University of Erlangen in 1872), the criterion that distinguishes one geometry from another is the group of transformations under which the propositions remain true. In the case of Euclidean geometry, we might at first expect this to be the continuous group of all isometries. But since the propositions remain valid when the scale of measurement is altered, as in a photographic enlargement, the "principal group" for Euclidean geometry [Klein **2,** p. 133] includes also "similarities" (which may change distances although of course they preserve angles). In the present chapter we classify such transformations of the Euclidean plane. In particular, "dilatations" will be seen to play a useful role in the theory of the nine-point center of a triangle. These and other "direct" similarities are treated in the standard textbooks, but "opposite" similarities (§ 5.6) seem to have been sadly neglected.

5.1 DILATATION

"If I eat one of these cakes," she thought, "it's sure to make some change in my size." . . . So she swallowed one . . . and was delighted to find that she began shrinking directly.

Lewis Carroll
[Dodgson **1,** Chap. 4]

It is convenient to extend the usual definition of *parallel* by declaring that two (infinite straight) lines are parallel if they have either no common point or two common points. (In the latter case they coincide.) This convention enables us to assert that, without any exception,

5.11   *For each point A and line r, there is just one line through A parallel to r.*

Two figures are said to be *homothetic* if they are similar and similarly placed, that is, if they are related by a dilatation (or "homothecy"), which may be defined as follows [Artin **1**, p. 54]:

A *dilatation* is a transformation which preserves (or reverses) direction: that is, it *transforms each line into a parallel line.*

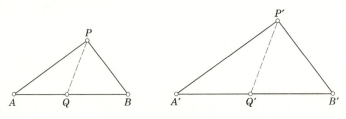

Figure 5.1a

5.12   *Two given parallel line segments AB, A'B' are related by a unique dilatation AB → A'B'.*

For, any point P not on AB is transformed into the point P' in which the line through A' parallel to AP meets the line through B' parallel to BP (Figure 5.1a); and any point Q on AB is transformed into the point Q' in which A'B' meets the line through P' parallel to PQ.

In other words, a dilatation is completely determined by its effect on any two given points [Coxeter **2**, 8.51].

Clearly, the inverse of the dilatation $AB \rightarrow A'B'$ is the dilatation $A'B' \rightarrow AB$. Also $AB \rightarrow AB$ is the identity, $AB \rightarrow BA$ is a half-turn (about the midpoint of AB), and if $ABB'A'$ is a parallelogram, $AB \rightarrow A'B'$ is a translation.

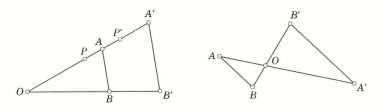

Figure 5.1b

For any dilatation which is not the identity, the two points A and B may be so chosen that A is not an invariant point and AB is not an invariant line. Such a dilatation $AB \rightarrow A'B'$ (Figure 5.1b) transforms any point P on AA' into a point P' on the parallel line through A', which is AA' itself.

Similarly, it transforms any point Q on BB' into a point Q' on BB'. If AA' and BB' are not parallel, these two invariant lines intersect in an invariant point O. Hence

5.13   *Any dilatation that is not a translation has an invariant point.*

This invariant point O is *unique*. For, a dilatation that has two invariant points O_1 and O_2 can only be the identity, which may reasonably be regarded as a translation, namely a translation through distance zero [Weyl **1**, p. 69].

Clearly, any point P is transformed into a point P' on OP. Let us write

$$OP' = \lambda OP,$$

with the convention that the number λ is positive or negative according as P and P' are on the same side of O or on opposite sides. With the help of some homothetic triangles (as in Figure 5.1b), we see that λ is a constant, that is, independent of the position of P. Moreover, any segment PQ is transformed into a segment $|\lambda|$ times as long, and oppositely directed if $\lambda < 0$. We shall use the symbol $O(\lambda)$ for the dilatation with center O and ratio λ. (Court [**2**, p. 40] prefers "(O, λ)".)

In particular, $O(1)$ is the identity and $O(-1)$ is a half-turn. Clearly, the only dilatations which are also isometries are half-turns and translations. In the case of a translation, such a symbol as $O(\lambda)$ is no longer available.

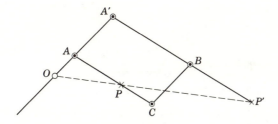

Figure 5.1c

EXERCISES

1.   What is the inverse of the dilatation $O(\lambda)$?

2.   If the product $O_1(\lambda_1)$ and $O_2(\lambda_2)$ is $O(\lambda_1\lambda_2)$, where is O?

3.   Express the dilatation $O(\lambda)$ in terms of (a) polar coordinates, (b) Cartesian coordinates.

4.   Explain the action of the *pantograph* (Figure 5.1c), an instrument invented by Christoph Scheiner about 1630 for the purpose of making a copy, reduced or enlarged, of any given figure. It is formed by four rods, hinged at the corners of a parallelogram $AA'BC$ whose angles are allowed to vary. The three collinear points O, P, P', on the respective rods AA', AC, $A'B$, remain collinear when the shape of the parallelogram is changed. The instrument is pivoted at O. When a pencil point is inserted at

P' and a tracing point at P (or vice versa), and the latter is traced over the lines of a given figure, the pencil point draws a homothetic copy. The positions of O and P are adjustable on their respective rods so as to allow various choices of the ratio $OA : OA'$. (Care must, of course, be taken to keep O and P collinear with P'.)

5. How could the pantograph be modified so as to yield a dilatation $O(\lambda)$ with λ negative?

5.2 CENTERS OF SIMILITUDE

I have often wondered why "similitude" ever got into elementary geometry. . . . I'm sure youngsters would be much more at ease with a pair of circles if they just had centers of "similarity" instead of being made to imagine that some new idea was insinuating itself.

E. H. Neville (1889-1961)

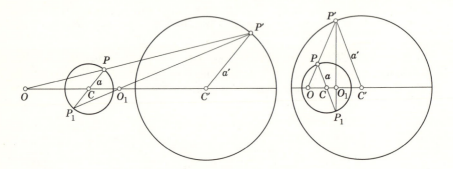

Figure 5.2a

A dilatation $O(\lambda)$, transforming C into C', transforms a circle with center C and radius r into a circle with center C' and radius $|\lambda| r$. Conversely, as we see in Figure 5.2a, if two circles have distinct centers C, C' and unequal radii a, a', they are related by two dilatations, $O(a'/a)$ and $O_1(-a'/a)$, whose centers O and O_1 divide the segment CC' externally and internally in the ratio $a : a'$ [Court **2**, p. 184]. These points O and O_1 are called the *centers of similitude* of the two circles. To construct them, we draw an arbitrary diameter PCP_1 of the first circle and a parallel radius $C'P'$ of the second (with P' on the same side of CC' as P); then O lies on PP', and O_1 on P_1P'.

If two circles are concentric or equal, they are still related by two dilatations, but there is only one center of similitude. In the case of concentric circles this is because the two dilatations have the same center. In the case of equal circles it is because one of the dilatations is a translation, which has no center. (The other is the half-turn about O_1, which is now the midpoint of CC'.)

A. Vandeghen and G. R. Veldkamp (*American Mathematical Monthly*, **71** (1964), p, 178) found that, for the triangle considered in Exercise 10 of § 1.5 (page 16), the centers of similitude of the two "Soddy circles" are the incenter and the *Gergonne point*: the point of concurrence of the lines joining the vertices to the points of contact of the respectively opposite sides with the incircle.

EXERCISES

1. If two equal circles have no common point, they have two parallel common tangents and two other common tangents through O_1 (midway between the centers). If they touch they have only three common tangents. If they intersect they have only the two parallel common tangents.

2. Any common tangent of two unequal circles passes through a center of similitude. Sketch the positions of the centers of similitude, and record the number of common tangents, in the five essentially different instances of two such circles. (Two of the five are shown in Figure 5.2a.)

3. Given two dilatations $O(\lambda)$, $O_1(\lambda_1)$, with $\lambda \neq \lambda_1$, describe the position of the unique point C on which both have the same effect.

5.3 THE NINE-POINT CENTER

Consider an arbitrary triangle ABC, with circumcenter O, centroid G, and orthocenter H. Let A', B', C' be the midpoints of the sides, and A'', B'', C'' the midpoints of the segments HA, HB, HC, as in Figure 1.7a. Clearly, both the triangles $A'B'C'$, $A''B''C''$ are homothetic to ABC, being derived from ABC by the respective dilatations $G(-\frac{1}{2})$, $H(\frac{1}{2})$. The former provides a new proof that the medians are concurrent and trisect one another.

Since $G(-\frac{1}{2})$ and $H(\frac{1}{2})$ are the two dilatations by which the nine-point circle can be derived from the circumcircle [Court **2**, p. 104], the points G, H are the centers of similitude of these two circles, and the Euler line GH contains the centers of both circles: not only the circumcenter O, as we know already, but also the *nine-point center* N. Since the values of μ for the dilatations are $\pm\frac{1}{2}$, the nine-point radius is half the circumradius, and the centers of similitude H, G divide the segment ON externally and internally in the ratio 2 : 1 (Figure 5.3a). Thus N is the midpoint of OH.

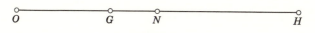

$$O \qquad\qquad G \quad N \qquad\qquad\qquad\qquad H$$

Figure 5.3a

EXERCISES

1. Using Cartesian coordinates, find the ordinates y of the centers O, G, N, H of the isosceles triangle whose vertices are $(0, 10)$, $(\pm 6, -8)$.

2. If $ABCH$ is an orthocentric quadrangle (see 1.72), the four Euler lines of the triangles BCH, CAH, ABH, ABC are concurrent.

5.4 THE INVARIANT POINT OF A SIMILARITY

> When a figure is enlarged so as to remain still of the same shape, every straight line in it remains a straight line, and every angle remains congruent to itself. All the parts of the figure are equally enlarged. When one figure is an enlarged copy of another, the two are said to be similar. The degree of enlargement necessary to make one figure equal to the other is called their ratio of similitude. The ratio of two lines in the one figure is equal to the ratio of the two corresponding lines in the other.
>
> W. K. Clifford (1845-1879)
>
> (*Mathematical Papers*, p. 631)

A *similarity* (or "similarity transformation," or "similitude") is a transformation which takes each segment AB into a segment $A'B'$ whose length is given by

$$\frac{A'B'}{AB} = \mu,$$

where μ is a constant positive number (the same for all segments) called the *ratio of magnification* (Clifford's "ratio of similitude"). It follows that any triangle is transformed into a similar triangle, and any angle into an equal (or opposite) angle. When $\mu = 1$, the similarity is an isometry. Other special cases are the dilatations $O(\pm\mu)$.

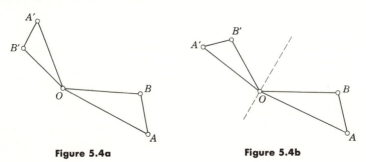

Figure 5.4a   Figure 5.4b

A less familiar instance is the *dilative rotation* (or "spiral similarity", Figure 5.4a), which is the product of a dilatation $O(\mu)$ and a rotation about O. Another is the *dilative reflection* (Figure 5.4b), which is the product of a dilatation $O(\mu)$ and the reflection in a line through O. We would not obtain anything new (in either case) if we replaced this dilatation $O(\mu)$ by $O(-\mu)$. For, since $O(-\mu) = O(-1) \cdot O(\mu)$, and $O(-1)$ is a half-turn, the product of $O(\mu)$ and a rotation through α about O is the same as the product of $O(-\mu)$ and a rotation through $\alpha + \pi$; and since $O(-1)$ is the product of two perpendicular reflections, the product of $O(\mu)$ and the reflection in a line m

through O is the same as the product of $O(-\mu)$ and the reflection in the line through O perpendicular to m. In fact, a dilative reflection has two perpendicular invariant lines (its *axes*), which are the internal and external bisectors of $\angle AOA'$ (and of $\angle BOB'$).

Clearly (cf. 3.11),

5.41 *Any two similar triangles* ABC, $A'B'C'$ *are related by a unique similarity* $ABC \rightarrow A'B'C'$, *which is direct or opposite according as the sense of* $A'B'C'$ *agrees or disagrees with that of* ABC.

In other words, a similarity is completely determined by its effect on any three given non-collinear points. For instance, the two triangles CBF, ACF used in proving Pythagoras's theorem (Figure 1.3*d*) are related by a dilative rotation, the product of the dilatation $F(AC/CB)$ and a quarter-turn; and the two triangles ABC, ACF (in the same figure) are related by a dilative reflection whose axes are the bisectors of the angle A.

Here is another way of expressing the same idea:

Any two line segments AB, $A'B'$, *are related by just two similarities: one direct and one opposite.*

For instance, the segment AB can be completed to make a square $ABCD$ on either side of the line AB, and similarly there are two ways to place a square $A'B'C'D'$ on $A'B'$. The similarity

$$ABCD \rightarrow A'B'C'D'$$

is direct or opposite according as the senses round these two squares agree or disagree.

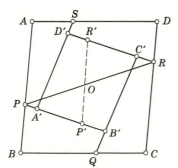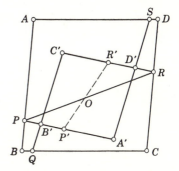

Figure 5.4c

Theorem 5.13 suggests the possibility that every similarity with $\mu \neq 1$ may have an invariant point.

If a given similarity is not a dilatation, there must be at least one line transformed into a nonparallel line. Let AB and $A'B'$ be corresponding segments on such a pair of lines, and let the given similarity (direct or opposite) be determined by similar, but not congruent, parallelograms $ABCD$ and $A'B'C'D'$ (for example, by squares, as above).

Let P, Q, R, S denote the points of intersection of the pairs of corresponding lines AB and $A'B'$, BC and $B'C'$, CD and $C'D'$, DA and $D'A'$, as in
Figure 5.4c. Suppose the given similarity transforms P (on AB) into P' (on
$A'B'$), and R (on CD) into R' (on $C'D'$). Let O be the common point of the
lines PR, $P'R'$ (which cannot be parallel, for, if they were, $PRR'P'$ would be
a parallelogram, and the segments PR, $P'R'$ would be congruent, contradicting $P'R' = \mu PR$). Since the point pairs PP' and RR' lie on parallel
lines $A'B'$ and $C'D'$,

$$\frac{OP}{OR} = \frac{OP'}{OR'}.$$

Therefore, the similarity leaves O invariant. Moreover, O is its *only* invariant point. For, if a similarity with $\mu \neq 1$ had two invariant points O_1 and
O_2, the distance O_1O_2 would be left unchanged instead of being multiplied
by μ. Hence

5.42 *Any similarity that is not an isometry has just one invariant point.*

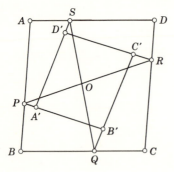 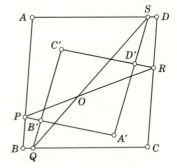

Figure 5.4d

Moreover, given two similar parallelograms $ABCD$ and $A'B'C'D'$, we can
use the method indicated in Figure 5.4d to *construct* the center of the similarity that relates them. For, having seen that O lies on the line PR, we can
apply the same reasoning (using BC and DA instead of AB and CD) to show
that O lies on the line QS. This is a different line, for, if P, Q, R, S were all
collinear we would have

$$\frac{PA}{PB} = \frac{PS}{PQ} = \frac{PA'}{PB'} \quad \text{and} \quad \frac{RC}{RD} = \frac{RQ}{RS} = \frac{RC'}{RD'},$$

making both P and R invariant. Hence, O can be constructed as the point of
intersection of the lines PR and QS.

How can the idea of continuity be used for a different proof of Theorem 5.42?

5.5 DIRECT SIMILARITY

Consider a given direct similarity whose ratio of magnification μ is not 1. Since there is an invariant point, say O, this similarity may be expressed as the product of the dilatation $O(\mu)$ and a direct *isometry* leaving O invariant. By Theorem 3.14, such an isometry is simply a rotation about O. Hence,

5.51 *Any direct similarity that is not an isometry is a dilative rotation.*

1. What is the product of two dilative rotations?

2. How can two circles be used to locate the invariant point of the direct similarity that relates two given incongruent segments on nonparallel lines? [Casey **1,** p. 186.]

5.6 OPPOSITE SIMILARITY

Consider a given opposite similarity whose ratio of magnification μ is not 1. Since there is an invariant point, say O, this similarity may be expressed as the product of the dilatation $O(\mu)$ and an opposite isometry leaving O invariant. By Theorem 3.14, such an isometry is simply the reflection in a line through O. Hence

5.61 *Any opposite similarity that is not an isometry is a dilative reflection.*

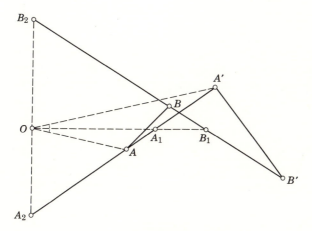

Figure 5.6a

EXERCISES

1.   If two maps of the same country on different scales are drawn on tracing paper and superposed, there is just one place that is represented by the same spot on both maps. (It is understood that one of the maps may be turned over before it is superposed on the other.) [Lachlan **1**, pp. 137, 139.]

2.   When all the points P on AB are related by a similarity to all the points P' on $A'B'$, the points dividing the segments PP' in the ratio $AB : A'B'$ (internally or externally) are distinct and collinear or else they all coincide.

3.   If S is an opposite similarity, S^2 is a dilatation.

4.   What is the product (a) of two dilative reflections? (b) of a dilative rotation and a dilative reflection?

5.   Let AB and $A'B'$ be two given segments of different lengths. Let A_1 and A_2 divide AA' internally and externally in the ratio $AB{:}A'B'$ (as in Figure 5.6a). Let B_1 and B_2 divide BB' in the same manner. Then the lines A_1B_1 and A_2B_2 are at right angles, and are the axes of the dilative reflection that transforms AB into $A'B'$. [Lachlan **1**, p. 134; Johnson **1**, p. 27.] (It has been tacitly assumed that $A_1 \neq B_1$ and $A_2 \neq B_2$. However, if A_2 and B_2 coincide, the axes are A_1B_1 and the perpendicular line through A_2.)

6.   Describe the transformation

$$(r, \theta) \rightarrow (\mu r, \theta + \alpha)$$

of polar coordinates, and the transformation

$$(x, y) \rightarrow (\mu x, - \mu y)$$

of Cartesian coordinates.

6

Circles and spheres

The present chapter shows how Euclidean geometry, in which lines and planes play a fundamental role, can be extended to *inversive* geometry, in which this role is taken over by circles and spheres. We shall see how the obvious statement, that lines and planes are circles and spheres of infinite radius, can be replaced by the sophisticated statement that lines and planes are those circles and spheres which pass through an "ideal" point, called "the point at infinity." In § 6.9 we shall briefly discuss a still more unusual geometry, called *elliptic,* which is one of the celebrated "non-Euclidean" geometries.

6.1 INVERSION IN A CIRCLE

> *Can it be that all the great scientists of the past were really playing a game, a game in which the rules are written not by man but by God? . . . When we play, we do not ask why we are playing—we just play. Play serves no moral code except that strange code which, for some unknown reason, imposes itself on the play. . . . You will search in vain through scientific literature for hints of motivation. And as for the strange moral code observed by scientists, what could be stranger than an abstract regard for truth in a world which is full of concealment, deception, and taboos? . . . In submitting to your consideration the idea that the human mind is at its best when playing, I am myself playing, and that makes me feel that what I am saying may have in it an element of truth.*
>
> J. L. Synge (1897 -   )*

All the transformations so far discussed have been similarities, which transform straight lines into straight lines and angles into equal angles. The transformation called *inversion,* which was invented by L. J. Magnus in 1831, is new in one respect but familiar in another: it transforms some

* *Hermathena,* **19** (1958), p. 40; quoted with the editor's permission.

straight lines into circles, but it still transforms angles into equal angles. Like the reflection and the half-turn, it is involutory (that is, of period 2). Like the reflection, it has infinitely many invariant points; these do not lie on a straight line but on a circle, and the center of the circle is "singular:" it has no image!

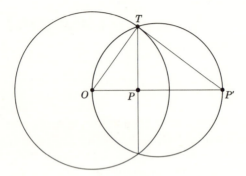

Figure 6.1a

Given a fixed circle with center O and radius k, we define the *inverse* of any point P (distinct from O) to be the point P' on the ray OP whose distance from O satisfies the equation

$$OP \times OP' = k^2.$$

It follows from this definition that the inverse of P' is P itself. Moreover, every point outside the circle of inversion is transformed into a point inside, and every point inside (except the center O) into a point outside. The circle is invariant in the strict sense that every point on it is invariant. Every line through O is invariant as a whole, but not point by point.

To construct the inverse of a given point P (other than O) inside the circle of inversion, let T be one end of the chord through P perpendicular to OP, as in Figure 6.1a. Then the tangent at T meets OP (extended) in the desired point P'. For, since the right-angled triangles OPT, OTP' are similar, and $OT = k$,

$$\frac{OP}{k} = \frac{k}{OP'}.$$

To construct the inverse of a given point P' outside the circle of inversion, let T be one of the points of intersection of this circle with the circle on OP' as diameter (Figure 6.1a). Then the desired point P is the foot of the perpendicular from T to OP'.

If $OP > \frac{1}{2}k$, the inverse of P can easily be constructed by the use of compasses alone, without a ruler. To do so, let the circle through O with center P cut the circle of inversion in Q and Q'. Then P' is the second intersection of the circles through O with centers Q and Q'. (This is easily seen by considering the similar isosceles triangles POQ, QOP'.)

There is an interesting connection between inversion and dilatation:

6.11   *The product of inversions in two concentric circles with radii k and k' is the dilatation $O(\mu)$ where $\mu = (k'/k)^2$.*

To prove this, we observe that this product transforms P into P'' (on OP) where

$$OP \times OP' = k^2, \qquad OP' \times OP'' = k'^2$$

and therefore

$$\frac{OP''}{OP} = \left(\frac{k'}{k}\right)^2.$$

EXERCISES

1.   Using compasses alone, construct the vertices of a regular hexagon.

2.   Using compasses alone, locate a point B so that the segment OB is twice as long as a given segment OA.

3.   Using compasses alone, construct the inverse of a point distant $\frac{1}{3}k$ from the center O of the circle of inversion. Describe a procedure for inverting points arbitrarily near to O.

4.   Using compasses alone, bisect a given segment.

5.   Using compasses alone, trisect a given segment. Describe a procedure for dividing a segment into any given number of equal parts.

Note. The above problems belong to the Geometry of Compasses, which was developed independently by G. Mohr in Denmark (1672) and L. Mascheroni in Italy (1797). For a concise version of the whole story, see Pedoe [**1**, pp. 23–25] or Courant and Robbins [**1**, pp. 145–151].

6.2   ORTHOGONAL CIRCLES

A circle is a happy thing to be—
Think how the joyful perpendicular
Erected at the kiss of tangency
Must meet my central point, my avatar.
And lovely as I am, yet only 3
Points are needed to determine me.

Christopher Morley (1890 -   )

Two circles are said to be *orthogonal* if they cut at right angles, that is, if they intersect in two points at either of which the radius of each is a tangent to the other (Figure 6.2*a*).

By Euclid III.36 (see p. 8) any circle through a pair of inverse points is invariant: the circle of inversion decomposes it into two arcs which invert into each other. Moreover, such a circle is orthogonal to the circle of inversion, and every circle orthogonal to the circle of inversion is invariant in this sense. Through a pair of inverse points we can draw a whole *pencil* of circles (infinitely many), and they are all orthogonal to the circle of inversion. Hence

6.21   *The inverse of a given point P is the second intersection of any two circles through P orthogonal to the circle of inversion.*

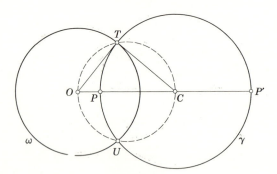

Figure 6.2a

The above remarks provide a simple solution for the problem of drawing, through a given point P, a circle (or line) orthogonal to two given circles. Let P_1, P_2 be the inverses of P in the two circles. Then the circle PP_1P_2 (or the line through these three points, if they happen to be collinear) is orthogonal to the two given circles.

If O and C are the centers of two orthogonal circles ω and γ, as in Figure 6.2a, the circle on OC as diameter passes through the points of intersection T, U. Every other point on this circle is inside one of the two orthogonal circles and outside the other. It follows that, if a and b are two perpendicular lines through O and C respectively, either a touches γ and b touches ω, or a cuts γ and b lies outside ω, or a lies outside γ and b cuts ω.

6.3   INVERSION OF LINES AND CIRCLES

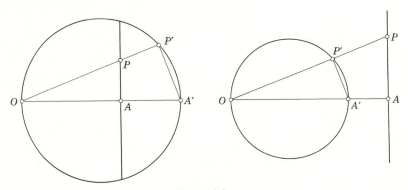

Figure 6.3a

We have seen that lines through O invert into themselves. What about other lines? Let A be the foot of the perpendicular from O to a line not through O. Let A' be the inverse of A, and P' the inverse of any other point P on the line. (See Figure 6.3a where, for simplicity, the circle of inversion has not been drawn.) Since

$$OP \times OP' = k^2 = OA \times OA',$$

the triangles OAP, $OP'A'$ are similar, and the line AP inverts into the circle on OA' as diameter, which is the locus of points P' from which OA' subtends a right angle. Thus any line not through O inverts into a circle through O, and vice versa.

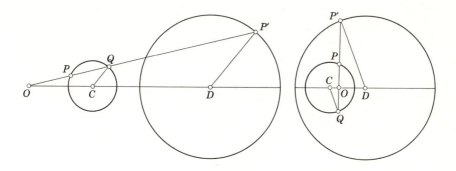

Figure 6.3b

Finally, what about a circle not through O? Let P be any point on such a circle, with center C, and let OP meet the circle again in Q. By Euclid III.35 again, the product

$$p = OP \times OQ$$

is independent of the position of P on the circle. Following Jacob Steiner (1796–1863), we call this product the *power* of O with respect to the circle. It is positive when O is outside the circle, zero when O lies on the circle, and we naturally regard it as being negative when O is inside (so that OP and OQ are measured in opposite directions). Let the dilatation $O(k^2/p)$ transform the given circle and its radius CQ into another circle (or possibly the same) and its parallel radius DP' (Figure 6.3b, cf. Figure 5.2a), so that

$$\frac{OP'}{OQ} = \frac{OD}{OC} = \frac{k^2}{p}.$$

Since $OP \times OQ = p$, we have, by multiplication,

$$OP \times OP' = k^2.$$

Thus P' is the inverse of P, and the circle with center D is the desired inverse of the given circle with center C. (The point D is usually *not* the inverse of C.)

We have thus proved that the inverse of a circle not through O is another circle of the same kind, or possibly the same circle again. The latter possibility occurs in just two cases: (1) when the given circle is orthogonal to the circle of inversion, so that $p = k^2$ and the dilatation is the identity; (2) when the given circle is the circle of inversion itself, so that $p = -k^2$ and the dilatation is a half-turn.

When p is positive (see the left half of Figure 6.3b), so that O is outside the circle with center C, this circle is orthogonal to the circle with center O and radius $\sqrt{p}$; that is, the former circle is invariant under inversion with respect to the latter. In effect, we have expressed the given inversion as the product of this new inversion, which takes P to Q, and the dilatation $O(k^2/p)$, which takes Q to P'. When p is negative (as in the right half of Figure 6.3b), P and Q are interchanged by an "anti-inversion:" the product of an inversion with radius $\sqrt{-p}$ and a half-turn [Forder **3,** p. 20].

When discussing isometries and other similarities, we distinguished between *direct* and *opposite* transformations by observing their effect on a triangle. Since we are concerned only with *sense,* the triangle could have been replaced by its circumcircle. Such a distinction can still be made for inversions (and products of inversions), which transform circles into circles. Instead of a triangle we use a circle: not an arbitrary circle but a "small" circle whose inverse is also "small," that is, a circle not surrounding O. Referring again to the left half of Figure 6.3b, we observe that P and Q describe the circle with center C in opposite senses, whereas Q and P' describe the two circles in the same sense. Thus the inverse points P and P' proceed oppositely, and

Inversion is an opposite transformation.

It follows that the product of an even number of inversions is direct. One instance is familiar: the product of inversions with respect to two concentric circles is a dilatation.

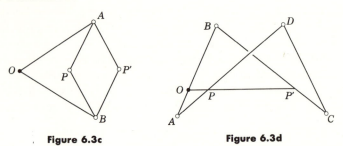

Figure 6.3c   **Figure 6.3d**

EXERCISES

1. For any two unequal circles that do not intersect, one of the two centers of similitude (§ 5.2) is the center of a circle which inverts either of the given circles into the other. For two unequal intersecting circles, both centers of similitude have this property. What happens in the case of equal intersecting circles?

2. Explain the action of *Peaucellier's cell* (Figure 6.3c), an instrument invented by A. Peaucellier in 1864 for the purpose of drawing the inverse of any given locus. It is formed by four equal rods, hinged at the corners of a rhombus $APBP'$, and two equal (longer) rods connecting two opposite corners, A and B, to a fixed pivot O. When a pencil point is inserted at P' and a tracing point at P (or vice versa), and the latter is traced over the curves of a given figure, the pencil point draws the inverse figure. In particular, if a seventh rod and another pivot are introduced so as to keep P on a circle passing through O, the locus of P' will be a straight line. This linkage gives an exact solution of the important mechanical problem of converting circular into rectilinear motion. [Lamb **2**, p. 314.]

3. Explain the action of *Hart's linkage* (Figure 6.3d), an instrument invented by H. Hart in 1874 for the same purpose as Peaucellier's cell. It requires only four rods, hinged at the corners of a "crossed parallelogram" $ABCD$ (with $AB = CD$, $BC = DA$). The three collinear points O, P, P', on the respective rods AB, AD, BC, remain collinear (on a line parallel to AC and BD) when the shape of the crossed parallelogram is changed. As before, the instrument is pivoted at O. [Lamb **2**, p. 315.]

4. With respect to a circle γ of radius r, let p be the power of an outside point O. Then the circle with center O and radius k inverts γ into a circle of radius k^2r/p.

6.4   THE INVERSIVE PLANE

Whereupon the Plumber said in tones of disgust:
"I suggest that we proceed at once to infinity."

J. L. Synge [**2**, p. 131]

We have seen that the image of a given point P by reflection in a line (Figure 1.3b) is the second intersection of any two circles through P orthogonal to the mirror, and that the inverse of P in a circle is the second intersection of any two circles through P orthogonal to the circle of inversion. Because of this analogy, inversion is sometimes called "reflection in a circle" [Blaschke **1**, p. 47], and we extend the definition of a circle so as to include a straight line as a special (or "limiting") case: a circle of infinite radius. We can then say that *any* three distinct points lie on a unique circle, and that any circle inverts into a circle.

In the same spirit, we extend the Euclidean plane by inventing an "ideal" *point at infinity O'*, which is both a common point and the common center of all straight lines, regarded as circles of infinite radius. Two circles with a common point either touch each other or intersect again. This remains obvious when one of the circles reduces to a straight line. When both of them are straight, the lines are either parallel, in which case they touch at O', or intersecting, in which case O' is their second point of intersection [Hilbert and Cohn-Vossen **1**, p. 251].

We can now assert that *every* point has an inverse. All the lines through O, being "circles" orthogonal to the circle of inversion, meet again in O', the inverse of O. When the center O is O' itself, the "circle" of inversion is straight, and the inversion reduces to a reflection.

The Euclidean plane with O' added is called the *inversive* (or "conformal") *plane*.* It gives inversion its full status as a "transformation" (§ 2.3): a one-to-one correspondence without exception.

Where two curves cross each other, their angle of intersection is naturally defined to be the angle between their tangents. In this spirit, two intersecting circles, being symmetrical by reflection in their line of centers, make equal angles at the two points of intersection. This will enable us to prove

6.41 *Any angle inverts into an equal angle (or, more strictly, an opposite angle).*

We consider first an angle at a point P which is not on the circle of inversion. Since any direction at such a point P may be described as the direction of a suitable circle through P and its inverse P', two such directions are determined by two such circles. Since these circles are self-inverse, they serve to determine the corresponding directions at P'. To show that an angle at P is still preserved when P is self-inverse, we use 6.11 to express the given inversion as the product of a dilatation and the inversion in a concentric circle that does not pass through P. Since both these transformations preserve angles, their product does likewise.

In particular, right angles invert into right angles, and

6.42 *Orthogonal circles invert into orthogonal circles (including lines as special cases).*

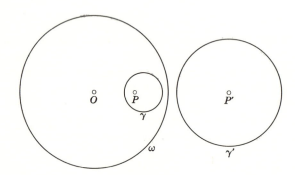

Figure 6.4a

By 6.21, inversion can be defined in terms of orthogonality. Therefore a circle and a pair of inverse points invert (in another circle) into a circle and a pair of inverse points. More precisely, if a circle γ inverts P into Q and

---

* M. Bôcher, *Bulletin of the American Mathematical Society*, **20** (1914), p. 194.

a circle ω inverts γ, P, Q into γ', P', Q', then the circle γ' inverts P' into Q'. An important special case (Figure 6.4a) arises when Q coincides with O, the center of ω, so that Q' is O', the point at infinity. Then P is the inverse of O in γ, and P' is the center of γ'. In other words, if γ inverts O into P, whereas ω inverts γ and P into γ' and P', then P' is the center of γ'.

Two circles either touch, or cut each other twice, or have no common point. In the last case (when each circle lies entirely outside the other, or else one encloses the other), we may conveniently say that the circles *miss* each other.

If two circles, α_1 and α_2, are both orthogonal to two circles β_1 and β_2, we can invert the four circles in a circle whose center is one of the points of intersection of α_1 and β_1, obtaining two orthogonal circles and two perpendicular diameters, as in the remark at the end of § 6.2. Hence, either α_1 touches α_2 and β_1 touches β_2, or α_1 cuts α_2 and β_1 misses β_2, or α_1 misses α_2 and β_1 cuts β_2.

6.5 COAXAL CIRCLES

In this section we leave the inversive plane and return to the Euclidean plane, in order to be able to speak of distances.

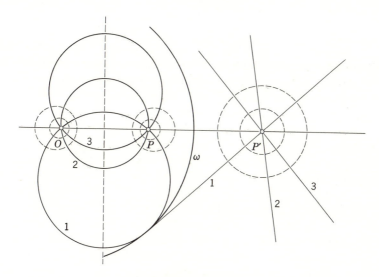

Figure 6.5a

If P and P' are inverse points in the circle ω (with center O), as in Figure 6.5a, all the lines through P' invert into all the circles through O and P: an *intersecting* (or "elliptic") *pencil of coaxal circles,* including the straight line OPP' as a degenerate case. The system of concentric circles with center P', consisting of circles orthogonal to these lines, inverts into a *nonintersecting* (or "hyperbolic") *pencil of coaxal circles* (drawn in broken lines). These circles all miss one another and are all orthogonal to the intersecting pencil. One of them degenerates to a (vertical) line, whose inverse is the circle (with center P') passing through O.

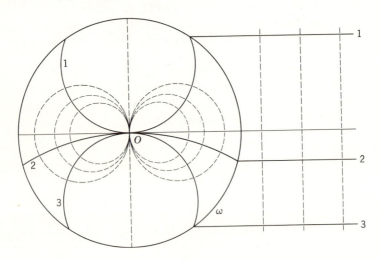

Figure 6.5b

As a kind of limiting case when O and P coincide (Figure 6.5*b*), the circles that touch a fixed line at a fixed point O constitute a *tangent* (or "parabolic") *pencil of coaxal circles*. They invert (in a circle with center O) into all the lines parallel to the fixed line. Orthogonal to these lines we have another system of the same kind, inverting into an orthogonal tangent pencil of coaxal circles. Again each member of either pencil is orthogonal to every member of the other.

Any two given circles belong to a pencil of coaxal circles of one of these three types, consisting of *all the circles orthogonal to both of any two circles orthogonal to both the given circles*. (More concisely, the coaxal circles consist of all the circles orthogonal to all the circles orthogonal to the given circles.) Two circles that cut each other belong to an intersecting pencil (and can be inverted into intersecting lines); two circles that touch each other belong to a tangent pencil (and can be inverted into parallel lines); two circles that miss each other belong to a nonintersecting pencil (by the remark at the end of § 6.4).

Each pencil contains one straight line (a circle of infinite radius) called the *radical axis* (of the pencil, or of any two of its members).* For an intersecting pencil, this is the line joining the two points common to all the circles (OP for the "unbroken" circles in Figure 6.5*a*). For a tangent pencil,

---

* Louis Gaultier, *Journal de l'École Polytechnique,* **16** (1813), p. 147.

it is the common tangent. For a nonintersecting pencil, it is the line midway between the two *limiting points* (or circles of zero radius) which are the common points of the orthogonal intersecting pencil. For each pencil there is a *line of centers*, which is the radical axis of the orthogonal pencil. Hence

6.51 *If tangents can be drawn to the circles of a coaxal pencil from a point on the radical axis, all these tangents have the same length.*

The radical axis of two given circles may be defined as the locus of points of equal power (§ 6.3) with respect to the two circles. This power can be measured as the square of a tangent except in the case when the given circles intersect in two points O, P, and we are considering a point A on the segment OP; then the power is the negative number $AO \times AP$.

It follows that, for three circles whose centers form a triangle, the three radical axes (of the circles taken in pairs) concur in a point called the *radical center*, which has the same power with respect to all three circles. If this power is positive, its square root is the length of the tangents to any of the circles, and the radical center is the center of a circle (of this radius) which is orthogonal to all the given circles. But if the power is negative, no such orthogonal circle exists.

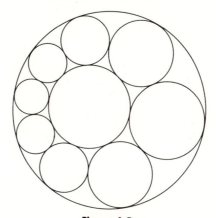

Figure 6.5c

The possibility of inverting any two nonintersecting circles into concentric circles (by taking O at either of the limiting points) provides a remarkably simple proof for Steiner's porism:* If we have two (nonconcentric) circles, one inside the other, and circles are drawn successively touching them and one another, as in Figure 6.5c, it may happen that the ring of touching circles closes, that is, that the last touches the first. Steiner's statement is that, if this happens once, it will always happen, whatever be the position of the first circle of the ring. To prove this we need only invert the original two circles into concentric circles, for which the statement is obvious.

* Forder [**3**, p. 23]. See also Coxeter, Interlocked rings of spheres, *Scripta Mathematica,* **18** (1952), pp. 113–121, or Yaglom [**2**, p. 199].

EXERCISES

1. In a pencil of coaxal circles, each member, used as a circle of inversion, interchanges the remaining members in pairs and inverts each member of the orthogonal pencil into itself.

2. The two limiting points of a nonintersecting pencil are inverses of each other in any member of the pencil.

3. If two circles have two or four common tangents, their radical axis joins the midpoints of these common tangents. If two circles have no common tangent (i.e., if one entirely surrounds the other), how can we construct their radical axis?

4. When a nonintersecting pencil of coaxal circles is inverted into a pencil of concentric circles, what happens to the limiting points?

5. In Steiner's porism, the points of contact of successive circles in the ring all lie on a circle, and this will serve to invert the two original circles into each other. Do the *centers* of the circles in the ring lie on a circle?

6. For the triangle considered in Exercise 10 of § 1.5 (page 16), the incircle is coaxal with the "two other circles" (Soddy's circles).

6.6   THE CIRCLE OF APOLLONIUS

The analogy between reflection and inversion is reinforced by the following

PROBLEM.   *To find the locus of a point P whose distances from two fixed points A, A' are in a constant ratio* $1 : \mu$, *so that*

$$A'P = \mu AP.$$

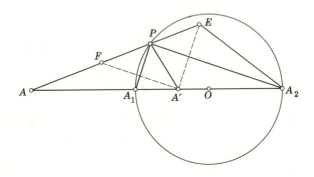

Figure 6.6a

When $\mu = 1$, the locus is evidently the perpendicular bisector of AA', that is, the line that reflects A into A'. We shall see that for other values of μ it is a circle that inverts A into A'. (Apollonius of Perga, *c.* 260–190 B.C.)

Assuming $\mu \neq 1$, let P be any point for which $A'P = \mu AP$. Let the internal and external bisectors of $\angle APA'$ meet AA' in A_1 and A_2 (as in Figure 6.6a, where $\mu = \frac{1}{2}$). Take E and F on AP so that $A'E$ is parallel to A_1P and $A'F$ is parallel to A_2P, that is, perpendicular to A_1P. Since $FP = PA' = PE$, we have

$$\frac{AA_1}{A_1A'} = \frac{AP}{PE} = \frac{AP}{PA'}, \qquad \frac{AA_2}{A'A_2} = \frac{AP}{FP} = \frac{AP}{PA'}.$$

(The former result is Euclid VI.3.) Thus A_1 and A_2 divide the segment AA' internally and externally in the ratio $1 : \mu$, and their location is independent of the position of P. Since $\angle A_1PA_2$ is a right angle, P lies on the circle with diameter A_1A_2.

Conversely, if A_1 and A_2 are defined by their property of dividing AA' in the ratio $1 : \mu$, and P is any point on the circle with diameter A_1A_2, we have

$$\frac{AP}{PE} = \frac{AA_1}{A_1A'} = \frac{1}{\mu} = \frac{AA_2}{A'A_2} = \frac{AP}{FP}.$$

Thus $FP = PE$, and P, being the midpoint of FE, is the circumcenter of the right-angled triangle EFA'. Therefore $PA' = PE$ and

$$\frac{AP}{PA'} = \frac{AP}{PE} = \frac{1}{\mu}$$

[Court **2**, p. 15].

Finally, the *circle of Apollonius* A_1A_2P inverts A into A'. For, if O is its center and k its radius, the distances $a = AO$ and $a' = A'O$ satisfy

$$\frac{a - k}{k - a'} = \frac{AA_1}{A_1A'} = \frac{AA_2}{A'A_2} = \frac{a + k}{a' + k},$$

whence

$$aa' = k^2.$$

EXERCISES

1. When μ varies while A and A' remain fixed, the circles of Apollonius form a non-intersecting pencil with A and A' for limiting points.

2. Given a line l and two points A, A' (not on l), locate points P on l for which the ratio $A'P/AP$ is maximum or minimum. (*Hint*: Consider the circle through A, A' with its center on l. The problem is due to N. S. Mendelsohn, and the hint to Richard Blum.)

3. Express k/AA' in terms of μ.

4. In the notation of Figure 5.6a (which is embodied in Figure 6.6b), the circles on A_1A_2 and B_1B_2 as diameters meet in two points O and $\bar{O}$, such that the triangles OAB and $OA'B'$ are similar, and likewise the triangles $\bar{O}AB$ and $\bar{O}A'B'$. Of the two similarities

$$OAB \rightarrow OA'B' \quad \text{and} \quad \bar{O}AB \rightarrow \bar{O}A'B',$$

one is opposite and the other direct. In fact, O is where A_1B_1 meets A_2B_2, and $\bar{O}$ lies on the four further circles $AA'P$, $BB'P$, ABT, $A'B'T$ (cf. Ex. 2 of § 5.5). [Casey **1**, p. 185.] If A' coincides with B, O lies on AB'.

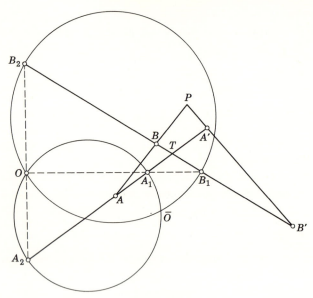

Figure 6.6b

5. Let the *inversive distance* between two nonintersecting circles be defined as the natural logarithm of the ratio of the radii (the larger to the smaller) of two concentric circles into which the given circles can be inverted. Then, if a nonintersecting pencil of coaxal circles includes α_1, α_2, α_3 (in this order), the three inversive distances satisfy

$$(\alpha_1, \alpha_2) + (\alpha_2, \alpha_3) = (\alpha_1, \alpha_3).$$

6. Two given unequal circles are related by infinitely many dilative rotations and by infinitely many dilative reflections. The locus of invariant points (in either case) is the circle having for diameter the segment joining the two centers of similitude of the given circles. (This locus is known as the *circle of similitude* of the given circles.) What is the corresponding result for two given *equal* circles?

7. The inverses, in two given circles, of a point on their circle of similitude, are images of each other by reflection in the radical axis of the two circles [Court **2**, p. 199].

6.7 CIRCLE-PRESERVING TRANSFORMATIONS

Having observed that inversion is a transformation of the whole inversive plane (including the point at infinity) into itself, taking circles into circles, we naturally ask what is the most general transformation of this kind. We distinguish two cases, according as the point at infinity is, or is not, invariant.

In the former case, not only are circles transformed into circles but also lines into lines. With the help of Euclid III.21 (see p. 7) we deduce that equality of angles is preserved, and consequently the measurement of angles is preserved, so that every triangle is transformed into a similar triangle, and the transformation is a similarity (§ 5.4).

If, on the other hand, the given transformation T takes an ordinary point

O into the point at infinity O', we consider the product J_1T, where J_1 is the inversion in the unit circle with center O. This product J_1T, leaving O' invariant, is a similarity. Let k^2 be its ratio of magnification, and J_k the inversion in the circle with center O and radius k. Since, by 6.11, J_1J_k is the dilatation $O(k^2)$, the similarity J_1T can be expressed as J_1J_kS, where S is an isometry. Thus

$$T = J_kS,$$

the product of an inversion and an isometry.

To sum up,

6.71 *Every circle-preserving transformation of the inversive plane is either a similarity or the product of an inversion and an isometry.*

It follows that every circle-preserving transformation is the product of at most four inversions (provided we regard a reflection as a special kind of inversion) [Ford **1**, p. 26]. Such a transformation is called a *homography* or an *antihomography* according as the number of inversions is even or odd. The product of two inversions (either of which could be just a reflection) is called a *rotary* or *parabolic* or *dilative* homography according as the two inverting circles are intersecting, tangent, or nonintersecting (i.e., according as the orthogonal pencil of invariant circles is nonintersecting, tangent, or intersecting). As special cases we have, respectively, a rotation, a translation, and a dilatation. The most important kind of rotary homography is the *Möbius involution,* which, being the inversive counterpart of a half-turn, is the product of inversions in two orthogonal circles (e.g., the product of the inversion in a circle and the reflection in a diameter). Any product of four inversions that cannot be reduced to a product of two is called a *loxodromic* homography [Ford **1**, p. 20].

EXERCISE

When a given circle-preserving transformation is expressed as JS (where J is an inversion and S an isometry), J and S are unique. There is an equally valid expression SJ', in which the isometry precedes the inversion. Why does this revised product involve the same S? Under what circumstances will we have $J' = J$?

6.8   INVERSION IN A SPHERE

By revolving Figures 6.1*a*, 6.2*a*, 6.3*a*, 6.3*b*, and 6.4*a* about the line of centers (OP or OA or OC), we see that the whole theory of inversion extends readily from circles in the plane to spheres in space. Given a sphere with center O and radius k, we define the inverse of any point P (distinct from O) to be the point P' on the ray OP whose distance from O satisfies

$$OP \times OP' = k^2.$$

Alternatively, P' is the second intersection of three spheres through P orthogonal to the sphere of inversion. Every sphere inverts into a sphere, provided we include, as a sphere of infinite radius, a plane, which is the inverse of a sphere through O. Thus, inversion is a transformation of *inversive* (or "conformal") *space,* which is derived from Euclidean space by postulating a *point at infinity,* which lies on all planes and lines.

Revolving the circle of Apollonius (Figure 6.6a) about the line AA', we obtain the *sphere of Apollonius,* which may be described as follows:

6.81   *Given two points A, A' and a positive number μ, let A_1 and A_2 divide AA' internally and externally in the ratio $1 : \mu$. Then the sphere on A_1A_2 as diameter is the locus of a point P whose distances from A and A' are in this ratio.*

EXERCISES

1.   If a sphere with center O inverts A into A' and B into B', the triangles OAB and $OB'A'$ are similar.

2.   In terms of $a = OA$ and $b = OB$, we have (in the notation of Ex. 1)

$$A'B' = \frac{k^2}{ab}AB.$$

3.   The "cross ratio" of any four points is preserved by any inversion:

$$\frac{AB/BD}{AC/CD} = \frac{A'B'/B'D'}{A'C'/C'D'}.$$

[Casey **1**, p. 100.]

4.   Two spheres which touch each other at O invert into parallel planes.

5.   Let α, β, γ be three spheres all touching one another. Let σ_1, σ_2, ... be a sequence of spheres touching one another successively and all touching α, β, γ. Then σ_6 touches σ_1, so that we have a ring of six spheres interlocked with the original ring of three.* (*Hint:* Invert in a sphere whose center is the point of contact of α and β.)

6.9   THE ELLIPTIC PLANE

In some unaccountable way, while he [Davidson] moved hither and thither in London, his sight moved hither and thither in a manner that corresponded, about this distant island. . . . When I said that nothing would alter the fact that the place [Antipodes Island] is eight thousand miles away, he answered that two points might be a yard away on a sheet of paper, and yet be brought together by bending the paper round.

H. G. Wells (1866-1946)

(*The Remarkable Case of Davidson's Eyes*)

Let S be the foot of the perpendicular from a point N to a plane σ, as in Figure 6.9a. A sphere (not drawn) with center N and radius NS inverts the plane σ into the sphere σ' on NS as diameter [Johnson **1**, p. 108]. We have

* Frederick Soddy, *The Hexlet, Nature,* **138** (1936), p. 958; **139** (1937), p. 77.

seen that spheres invert into spheres (or planes); therefore circles, being intersections of spheres, invert into circles (or lines). In particular, all the circles
in σ invert into circles (great or small) on the sphere σ', and all the lines in σ
invert into circles through N. Each point P in σ yields a corresponding point
P' on σ', namely, the second intersection of the line NP with σ'. Conversely,
each point P' on σ', except N, corresponds to the point P in which NP' meets
σ. The exception can be removed by making σ an inversive plane whose
point at infinity is the inverse of N.

This inversion, which puts the points of the inversive plane into one-to-
one correspondence with the points of a sphere, is known as *stereographic
projection.* It serves as one of the simplest ways to map the geographical
globe on a plane. Since angles are preserved, small islands are mapped with
the correct shape, though on various scales according to their latitude.

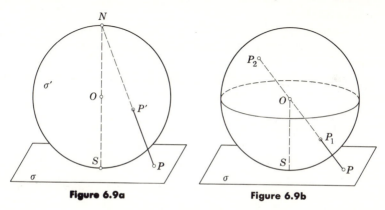

Figure 6.9a   **Figure 6.9b**

Another way is by *gnomonic* (or central) *projection,* in which the point
from which we project is not N but O, the center of the sphere, as in Figure 6.9b. Each point P in σ yields a line OP, joining it to O. This diameter
meets the sphere in two antipodal points P_1, P_2, which are both mapped on
the same point P. Each line m in σ yields a plane Om, joining it to O. This
diametral plane meets the sphere in a *great circle.* Conversely, each great
circle of the sphere, except the "equator" (whose plane is parallel to σ), corresponds to a line in σ. This time the exception can be removed by adding
to the Euclidean plane σ a *line at infinity* (representing the equator) with all
its points, called *points at infinity,* which represent pairs of antipodal points
on the equator. Thus, all the lines parallel to a given line contain the same
point at infinity, but lines in different directions have different points at infinity, all lying on the same line at infinity. (This idea is due to Kepler and
Desargues.)

When the line at infinity is treated just like any other line, the plane so
extended is called the *projective plane* or, more precisely, the *real* projective
plane [Coxeter **2**]. Two parallel lines meet in a point at infinity, and an
ordinary line meets the line at infinity in a point at infinity. Hence

6.91   *Any two lines of the projective plane meet in a point.*

Instead of taking a section of all the lines and planes through O, we could more symmetrically (though more abstractly) declare that, by definition, the points and lines of the projective plane *are* the lines and planes through O. The statement 6.91 is no longer surprising; it merely says that any two planes through O meet in a line through O.

Equivalently we could declare that, by definition, the lines of the projective plane are the great circles on a sphere, any two of which meet in a pair of antipodal points. Then the points of the projective plane are the pairs of antipodal points, abstractly identified. This abstract identification was vividly described by H. G. Wells in his short story, *The Remarkable Case of Davidson's Eyes.* (A sudden catastrophe distorted Davidson's field of vision so that he saw everything as it would have appeared from an exactly antipodal position on the earth.)

When the inversive plane is derived from the sphere by stereographic projection, distances are inevitably distorted, but the angle at which two circles intersect is preserved. In this sense, the inversive plane has a partial metric: angles are measured in the usual way, but distances are never mentioned [Graustein **1**, pp. 377, 388, 395].

On the other hand, gnomonic projection enables us, if we wish, to give the *projective* plane a *complete* metric. The distance between two points P and Q in σ (Figure 6.9a) is defined to be the angle POQ (in radian measure), and the angle between two lines m and n in σ is defined to be the angle between the planes Om and On. (This agrees with the customary measurement of distances and angles on a sphere, as used in spherical trigonometry.) We have thus obtained the *elliptic* plane* or, more precisely, the real projective plane with an elliptic metric [Coxeter **3**, Chapter VI; E. T. Bell **2**, pp. 302–311; Bachmann **1**, p. 21].

Since the points of the elliptic plane are in one-to-two correspondence with the points of the unit sphere, whose total area is 4π, it follows that the total area of the elliptic plane (according to the most natural definition of "area") is 2π. Likewise, the total length of a line (represented by a "great semicircle") is π. The simplification that results from using the elliptic plane instead of the sphere is well illustrated by the problem of computing the area of a spherical triangle ABC, whose sides are arcs of three great circles. Figure 6.9c shows these great circles, first in stereographic projection and then in gnomonic projection. The elliptic plane is decomposed, by the three lines BC, CA, AB, into four triangular regions. One of them is the given triangle Δ with angles A, B, C; the other three are marked α, β, γ in Figure 6.9c. (On the sphere, we have, of course, not only four regions but eight.) The two

---

* The name "elliptic" is possibly misleading. It does not imply any direct connection with the curve called an ellipse, but only a rather far-fetched analogy. A central conic is called an ellipse or a hyperbola according as it has no asymptote or two asymptotes. Analogously, a non-Euclidean plane is said to be elliptic or hyperbolic (Chapter 16) according as each of its lines contains no point at infinity or two points at infinity.

lines *CA, AB* decompose the plane into two *lunes* whose areas, being proportional to the supplementary angles A and $\pi - A$, are exactly $2A$ and $2(\pi - A)$. The lune with angle A is made up of the two regions Δ and α. Hence

$$\Delta + \alpha = 2A.$$

Similarly $\Delta + \beta = 2B$ and $\Delta + \gamma = 2C$. Adding these three equations and subtracting

$$\Delta + \alpha + \beta + \gamma = 2\pi,$$

we deduce Girard's "spherical excess" formula

6.92 $$\Delta = A + B + C - \pi,$$

which is equally valid for the sphere and the elliptic plane. (A. Girard, *Invention nouvelle en algèbre*, Amsterdam, 1629.)

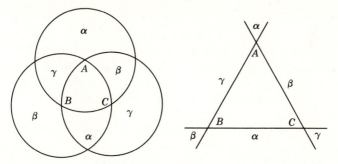

Figure 6.9c

EXERCISES

1. Two circles in the elliptic plane may have as many as four points of intersection.

2. The area of a p-gon in the elliptic plane is equal to the excess of its angle sum over the angle sum of a p-gon in the Euclidean plane.

7

Isometry and similarity
in Euclidean space

This chapter is the three-dimensional counterpart of Chapters 3 and 5. In § 7.5 we find a proof (independent of Euclid's Fifth Postulate) for the theorem (discovered by Michel Chasles in 1830) that every motion is a *twist*. In § 7.6 we see that every similarity (except the twist and the glide reflection, which are isometries) is a three-dimensional *dilative rotation*.

Most isometries are familiar in everyday life. When you walk straight forward you are undergoing a translation. When you turn a corner, it is a rotation; when you ascend a spiral staircase, a twist. The transformation that interchanges yourself and your image in an ordinary mirror is a reflection, and it is easy to see how you could combine this with a rotation or a translation to obtain a rotatory reflection or a glide reflection, respectively.

7.1   DIRECT AND OPPOSITE ISOMETRIES

A congruence is either proper, carrying a left screw into a left and a right one into a right, or it is improper or reflexive, changing a left screw into a right one and vice versa. The proper congruences are those transformations which . . . connect the positions of points of a rigid body before and after a motion.

H. Weyl [**1**, pp. 43-44]

The axioms of congruence, a sample of which was given in 1.26, can be extended in a natural manner from plane geometry to solid geometry. In space, an *isometry* (Weyl's "congruence") is still any transformation that preserves length, so that a line segment PQ is transformed into a congruent segment $P'Q'$. The most familiar examples are the *rotation* about a given line through a given angle and the *translation* in a given direction through a given distance. In the former case the axis of rotation has all its points invariant; in the latter there is no invariant point, except when the distance is zero so that the translation is the identity. A *reflection* is the special kind of isometry which has a whole plane of invariant points: the mirror. By a simple argument involving three spheres instead of two circles, we can easily prove the following analogue of Theorem 2.31:

7.11 *If an isometry has three non-collinear invariant points, it must be either the identity or a reflection.*

When two tetrahedra *ABCP, ABCP'* are images of each other by reflection in their common face, we may regard the "broken line" formed by the three edges *AB, BC, CP* as a kind of rudimentary screw, and the image formed by *AB, BC, CP'* as an oppositely oriented screw: if one is right-handed the other is left-handed. A model is easily made from two pieces of stiff wire, with right-angled bends at *B* and *C*. In this manner the idea of *sense* can be extended from two dimensions to three: we can say whether two given congruent tetrahedra agree or disagree in sense. In the former case we shall find that either tetrahedron can be *moved* (like a screw in its nut) to the position previously occupied by the other; such a motion is called a *twist*.

This distinction arises in analytic geometry when we make a coordinate transformation. If *O* is the origin and *X, Y, Z* are at unit distances along the positive coordinate axes, the sense of the tetrahedron *OXYZ* determines whether the system of axes is right-handed or left-handed. (A coordinate transformation determines an isometry transforming each point (x, y, z) into the point that has the *same* coordinates in the new system.)

Since an isometry is determined by its effect on a tetrahedron,

7.12 *Any two congruent tetrahedra ABCD, A'B'C'D' are related by a unique isometry ABCD → A'B'C'D', which is direct or opposite according as the sense of A'B'C'D' agrees or disagrees with that of ABCD.*

(Some authors, such as Weyl, say "proper or improper" instead of "direct or opposite.")

The solid analogue of Theorem 3.12 is easily seen to be:

7.13 *Two given congruent triangles are related by just two isometries: one direct and one opposite.*

As a counterpart for 3.13 we have [Coxeter **1**, p. 36]:

7.14 *Every isometry is the product of at most four reflections. If there is an invariant point, "four" can be replaced by "three."*

Since a reflection reverses sense, an isometry is direct or opposite according as it is the product of an even or odd number of reflections: 2 or 4 in the former case, 1 or 3 in the latter. In particular, a direct isometry with an invariant point is the product of just two reflections, and since the two mirrors have a common point they have a common line. Hence

7.15 *Every direct isometry with an invariant point is a rotation.*

Also, as Euler observed in 1776,

7.16 *The product of two rotations about lines through a point O is another such rotation.*

EXERCISE

The product of rotations through π about two intersecting lines that form an angle α is a rotation through 2α.

7.2 THE CENTRAL INVERSION

One of the most important opposite isometries is the *central inversion* (or "reflection in a point"), which transforms each point P into the point P' for which the midpoint of PP' is a fixed point O. This can be described as the product of reflections in any three mutually perpendicular planes through O. Taking these three mirrors to be the coordinate planes $x = 0, y = 0, z = 0$, we see that the central inversion in the origin transforms each point (x, y, z) into $(-x, -y, -z)$.

The name "central inversion," though well established in the literature of crystallography, is perhaps unfortunate: we must be careful to distinguish it from inversion in a sphere.

For most purposes the central inversion plays the same role in three dimensions as the half-turn in two. But we must remember that, since 3 is an odd number, the central inversion is an opposite isometry whereas the half-turn is direct. In space, the name *half-turn* is naturally used for the rotation through π about a line (or the "reflection in a line"), which is still direct [Lamb **1**, p. 9].

EXERCISE

What is the product of half-turns about three mutually perpendicular lines through a point?

7.3 ROTATION AND TRANSLATION

The treatment of translation in § 3.2 can be adapted to three dimensions by defining a translation as the product of two central inversions. We soon see that either the first center or the second may be arbitrarily assigned, and that the two inversions may be replaced by two half-turns about parallel axes or by two reflections in parallel mirrors.

Thus the product of two reflections is either a translation or a rotation.

The latter arises when the two mirrors intersect in a line, the axis of the rotation. In particular, the product of reflections in two perpendicular mirrors is a half-turn.

The product of reflections in two planes through a line l, being a rotation about l, is the same as the product of reflections in two other planes through l making the same dihedral angle as the given planes (in the same sense). Similarly, the product of reflections in two parallel planes, being a translation, is the same as the product of reflections in two other planes parallel to the given planes and having the same distance apart.

EXERCISE

What is the product of reflections in three parallel planes?

7.4   THE PRODUCT OF THREE REFLECTIONS

The three simplest kinds of isometry, namely rotation, translation and reflection, combine in commutative pairs to form the *twist* (or "screw displacement"), *glide reflection* and *rotatory reflection*. A twist is the product of a rotation with a translation along the direction of the axis. A glide reflection is the product of a reflection with a translation along the direction of a line lying in the mirror, that is, the product of reflections in three planes of which two are parallel while the third is perpendicular to both. A rotatory reflection is the product of a reflection with a rotation whose axis is perpendicular to the mirror. When this rotation is a half-turn, the rotatory reflection reduces to a central inversion.

Any rotatory reflection can be analysed into a central inversion and a residual rotation. For, if the rotation involved in the rotatory reflection is a rotation through θ, we may regard it as the product of a half-turn and a rotation through $\theta + \pi$ (or $\theta - \pi$). Thus a rotatory reflection can just as well be called a *rotatory inversion*: the product of a central inversion and a rotation whose axis passes through the center.

Any opposite isometry T that has an invariant point O is either a single reflection or the product of reflections in three planes through O. Its product TI with the central inversion in O, being a direct isometry with an invariant point, is simply a rotation S about a line through O. Hence the given opposite isometry is the rotatory inversion

$$T = SI^{-1} = SI:$$

7.41   *Every opposite isometry with an invariant point is a rotatory inversion.*

Since three planes that have no common point are all perpendicular to one plane α, the reflections in them (as applied to a point in α) behave like the reflections in the lines that are their sections by α. Thus we can make use of Theorem 3.31 and conclude that

7.42   *Every opposite isometry with no invariant point is a glide reflection.*

1.   What is the product of reflections in three planes through a line?

2.   Let ABC and $A'B'C'$ be two congruent triangles in distinct planes.   Consider the perpendicular bisectors of AA', BB', CC'.   If these three planes have just one common point O, the two triangles are related by a rotatory inversion with center O.   (*Hint:* If they were related by a rotation, the three planes would intersect in a line.)

3.   Every opposite isometry is expressible as the product of a reflection and a half-turn.

7.5   TWIST

The only remaining possibility is a direct isometry with no invariant point. Let S be any direct isometry (with or without an invariant point), transforming an arbitrary point A into A'.   Let R_1 be the reflection that interchanges A and A'.   Then the product R_1S is an opposite isometry leaving A' invariant. By 7.41, this is a rotatory inversion or rotatory reflection $R_2R_3R_4$, the product of a rotation R_2R_3 and a reflection R_4, the mirror for R_4 being perpendicular to the axis for R_2R_3.   Since this rotation may be expressed as the product of two reflections in various ways (§7.3), we can adjust the mirrors for R_2 and R_3 so as to make the former perpendicular to the mirror for R_1. Since both these planes remain perpendicular to the mirror for R_4, we now have

$$S = R_1R_2R_3R_4,$$

the product of the two rotations R_1R_2, R_3R_4, both of which are half-turns [Veblen and Young **2,** p. 318]:

7.51   *Every direct isometry is expressible as the product of two half-turns.*

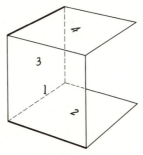

Figure 7.5a

If the isometry has an invariant point, it is a rotation, which may be expressed in various ways as the product of half-turns about two intersecting lines. When there is no invariant point, the axes of the two half-turns are either parallel, in which case the product is a translation, or *skew,* like two opposite edges of a tetrahedron. Two skew lines always lie in a pair of parallel planes, namely, the plane through each line parallel to the other.

Since a half-turn is the product of reflections in *any* two perpendicular planes through its axis, the two half-turns R_1R_2, R_3R_4 with skew axes are respectively equal to $R'_1R'_2$, $R'_3R'_4$, where the mirrors for R'_2 and R'_4 are parallel while the other two are perpendicular to them (Figure 7.5a). Hence

$$R_1R_2R_3R_4 = R'_1R'_2R'_3R'_4 = R'_1R'_3R'_2R'_4,$$

where the interchange of the middle reflections is possible since the half-turn $R'_2R'_3$ may be equally well expressed as $R'_3R'_2$. We have now fulfilled our purpose of expressing the general direct isometry as a twist: the product of the rotation $R'_1R'_3$ and the translation $R'_2R'_4$ along the axis of the rotation. (This axis meets both the skew lines at right angles, and therefore measures the shortest distance between them.) In other words,

7.52   *Every displacement is either a rotation or a translation or a twist.*

(For an alternative treatment see Thomson and Tait [**1**, § 102].)

EXERCISES

1. What kind of isometry transforms the point (x, y, z) into

(a) $(x, y, -z)$,   (b) $(-y, x, z)$,   (c) $(x, y, z + 1)$,

(d) $(-y, x, z + 1)$,   (e) $(-x, y, z + 1)$,   (f) $(-y, x, -z)$?

2. The product of half-turns about two skew lines at right angles is a twist, namely, the product of a half-turn about the line of shortest distance and a translation through twice this shortest distance. (Veblen and Young [**2**, p. 324] named this a *half twist.*)

7.6   DILATIVE ROTATION

> It can be proved by elementary methods that every Euclidean similarity other than a rigid motion has a fixed point.
>
> Hilbert and Cohn-Vossen [**1**, p. 331]

In Euclidean space, the definition of *dilatation* is exactly the same as in the plane. In fact, § 5.1 can be applied, word for word, to three dimensions, except that the special dilatation $AB \to BA$ or $O(-1)$ is not a half-turn but a central inversion (§ 7.2). Likewise, § 5.2 applies to spheres just as well as to circles: Figure 5.2a may be regarded as a plane section of two unequal spheres with their centers C, C' and their centers of similitude O, O_1. Two equal spheres are related by a translation and by a central inversion.

However, an important difference appears when we consider questions of sense. In the plane, every dilatation is direct, but in space the dilatation $O(\lambda)$ is direct or opposite according as λ is positive or negative; for example, the central inversion $O(-1)$ is opposite, as we have seen.

In space, as in the plane, two similar figures are related by a *similarity*, which in special cases may be an isometry or a dilatation. By a natural extension of the terminology we now take a *dilative rotation* to mean the product of a rotation about a line l (the *axis*) and a dilatation whose center O lies on l. The plane through O perpendicular to l is invariant, being transformed according to the two-dimensional "dilative rotation" of § 5.5. In the special case when the rotation about l is a half-turn, there are infinitely many other invariant planes, namely all the planes through l. Any such plane is transformed according to a dilative reflection.

Suppose a dilative rotation is the product of a rotation through angle α and a dilatation $O(\lambda)$ (where O lies on the axis). The following values of α and λ yield special cases which are familiar:

α	λ	Similarity
0	1	Identity
π	1	Half-turn
α	1	Rotation
π	-1	Reflection
0	-1	Central inversion
α	-1	Rotatory inversion
0	λ	Dilatation

We observe that this table includes all kinds of isometry, both direct and opposite, except the translation, twist and glide reflection (which have no invariant points). Still more surprisingly, we shall find that, with these same three exceptions, *every similarity is a dilative rotation*.

The role of similar triangles is now taken over by similar tetrahedra. Evidently

7.61 *Two given similar tetrahedra ABCD, A'B'C'D' are related by a unique similarity ABCD → A'B'C'D', which is direct or opposite according as the sense of A'B'C'D' agrees or disagrees with that of ABCD.*

In other words, a similarity is completely determined by its effect on any four given non-coplanar points, and we have the following generalization of Theorem 7.13:

7.62 *Two given similar triangles ABC, A'B'C' are related by just two similarities: one direct and one opposite.*

As a step towards proving that every similarity which is not an isometry is a dilative rotation, let us first prove

7.63 *Every similarity which is not an isometry has just one invariant point.*

Consider any given similarity S, whose ratio of magnification is $\mu \neq 1$. Let S transform an arbitrary point A into A'. If A' coincides with A, we have the desired invariant point. If not, let Q be the point that divides the segment AA' in the ratio $1:\mu$, externally or internally according as S is direct or opposite: that is, construct Q so that $QA' = \pm\mu QA$. Let D denote the direct or opposite dilatation $Q(\pm\mu^{-1})$. Then SD, having ratio of magnification 1, is a direct isometry which leaves A invariant. By Theorem 7.15, SD is a rotation about some line l through A. The plane through Q perpendicular to l is transformed into itself by both SD and D^{-1} and, therefore, also by their product S. On this invariant plane, S induces a two-dimensional similarity which, by Theorem 5.42, has an invariant point. Finally, this invariant point is unique, for if there were two distinct invariant points, the segment formed by them would be invariant instead of being multiplied by μ.

Having found the invariant point (or center) O, we can carry out a simplified version of the above procedure, with O for A. Since A' and Q both coincide with O, S is the product of a rotation about a line through O and the dilatation $O(\pm\mu)$, that is to say, S is a dilative rotation:

7.64 *Every similarity is either an isometry or a dilative rotation.*

In other words, every similarity is either a translation, a twist, a glide reflection, or a dilative rotation, provided we regard the last possibility as including all the special cases tabulated in the middle of page 102.

By Theorem 7.62, there are two dilative rotations, one direct and one opposite, which will transform a given triangle ABC into a similar (but not congruent) triangle $A'B'C'$. The ratio of magnification, $\mu \neq 1$, is given by the equation $A'B' = \mu AB$. Let A_1 and A_2 divide AA' internally and externally in the ratio $1 : \mu$. Let B_1 and B_2, C_1 and C_2 divide BB', CC' in the same manner. Consider the three spheres whose diameters are A_1A_2, B_1B_2, C_1C_2. These are "spheres of Appollonius" (Theorem 6.81); for example, the first is the locus of points whose distances from A and A' are in the ratio $1 : \mu$. Any point O for which

$$OA' = \mu OA, \qquad OB' = \mu OB, \qquad OC' = \mu OC$$

must lie on all three spheres. We have already established the existence of two such points. Hence the centers of the two dilative rotations may be constructed as the points of intersection of these three spheres.

In § 3.7 we used a translation to generate a geometric representation of the infinite cyclic group C_∞ (which is the free group with one generator). We see now that the same abstract group has a more interesting representation in which the generator is a dilative rotation. Some thirty elements of this group can be seen in the *Nautilus* shell [Thompson **2,** p. 843, Figure 418].

EXERCISES

1. How is the point (x, y, z) transformed by the general dilative rotation whose center and axis are the origin and the z-axis?

2. Find the axis and angle for the dilative rotation

$$(x, y, z) \rightarrow (\mu z, \mu x, \mu y).$$

3. How does the above classification of similarities deal with the three-dimensional "dilative reflection": the product of a dilatation $O(\lambda)$ and the reflection in a plane through O?

4. Could Theorem 7.63 be proved in the manner of the exercise at the end of §5.4 (page 73)?

7.7 SPHERE-PRESERVING TRANSFORMATIONS

The reasoning used in § 6.7 extends readily from two to three dimensions, yielding the following analog* of Theorem 6.71:

7.71 *Every sphere-preserving transformation of inversive space is either a similarity or the product of an inversion (in a sphere) and an isometry.*

EXERCISE

Every sphere-preserving transformation can be expressed as the product of r reflections and s inversions, where

$$r \leqslant 4, \quad s \leqslant 2, \quad r + s \leqslant 5.$$

---

* René Lagrange, *Produits d'inversions et métrique conforme,* Cahiers scientifiques, 23 (Gauthier-Villars, Paris, 1957), p. 7. See also Coxeter, *Annali di Matematica pura ed applicata,* **53** (1961), pp. 165–172.

Part II

8

Coordinates

In the preceding chapters, a few exercises on coordinates have been inserted for the sake of those readers who are already aquainted with analytic geometry. Other readers, having omitted such exercises, are awaiting enlightenment at the present stage. In addition to the usual rectangular Cartesian coordinates, we shall consider oblique and polar coordinates. (The polar equation for an ellipse is important because of its use in the theory of orbits.) After a brief mention of special curves we shall give an outline of Newton's application of calculus to problems of arc length and area. The section on three-dimensional space culminates in a surprising property of the doughnut-shaped *torus*.

8.1 CARTESIAN COORDINATES

> *Though the idea behind it all is childishly simple, yet the method of analytic geometry is so powerful that very ordinary boys of seventeen can use it to prove results which would have baffled the greatest of the Greek geometers—Euclid, Archimedes, and Apollonius.*

> E. T. Bell (1883-1960)
> [E. T. Bell **1,** p. 21]

Analytic geometry may be described as the representation of the points in n-dimensional space by ordered sets of n (or more) numbers called *coordinates*. For instance, any position on the earth can be specified by its latitude, longitude, and height above sea level.

The one-dimensional case is well illustrated by a thermometer. There is a certain point on the line associated with the number 0; the positive integers 1, 2, 3, . . . are evenly spaced in one direction away from 0, the negative integers $-1, -2, -3, . . .$ in the opposite direction, and the fractional numbers are interpolated in the natural manner. The *displacement* from one point x to another point x' is the positive or negative number $x' - x$.

In the two-dimensional case, the position of a point in a plane may be specified by its distances from two fixed perpendicular lines, the *axes*. This notion can be traced back over two thousand years to Archimedes of Syracuse and Apollonius of Perga, or even to the ancient Egyptians; but it was first developed systematically by two Frenchmen: Pierre Fermat (whose problem about a triangle we solved in § 1.8) and René Descartes (1596–1650). In their formulation the two distances were taken to be positive or zero. The important idea of allowing one or both to be negative was supplied by Sir Isaac Newton (1642–1727), and it was G. W. Leibniz (1646–1716) who first called them "coordinates." (The Germans write *Koordinaten,* the French *coordonnées.*)

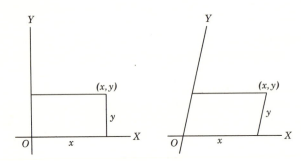

Figure 8.1a

For some purposes it is just as easy to use oblique axes, as in the second part of Figure 8.1a. Starting from the *origin O*, where the axes intersect, we reach the general point (x, y) by going a distance x along the x-axis OX and then a distance y along a line parallel to the y-axis OY. The x-axis is said to have the *equation* $y = 0$ because every point $(x, 0)$ satisfies this equation; similarly, $x = 0$ is the equation of the y-axis. On any other line through the origin, consideration of homothetic triangles shows that the ratio y/x is constant; thus any line through the origin $(0, 0)$ may be expressed as $ax + by = 0$.

To obtain the equation for any other line, we take a point (x_1, y_1) on it. In terms of new coordinates x', y', derived by translating the origin from $(0, 0)$ to (x_1, y_1), the line may be expressed as $ax' + by' = 0$. Since $x' = x - x_1$ and $y' = y - y_1$, the same line, in terms of the original coordinates, is

$$a(x - x_1) + b(y - y_1) = 0$$

or, say,

8.11 $$ax + by + c = 0.$$

Thus every line has a *linear* equation, and every linear equation determines a line. In particular, the line that makes intercepts p and q on the axes is

8.12
$$\frac{x}{p} + \frac{y}{q} = 1;$$

for, this equation is linear and is satisfied by both $(p, 0)$ and $(0, q)$. Two lines of the form 8.11 are parallel if they have the same ratio a/b (including, as one possibility, $b = 0$ for both, in which case they are parallel to the y-axis). The point of intersection of two nonparallel lines is obtained by solving the two simultaneous equations for x and y.

If $b \neq 0$, the equation 8.11 may be solved for y in the form $y = -(ax + c)/b$. More generally, points whose coordinates satisfy an equation $F(x, y) = 0$ or $y = f(x)$ can be plotted by giving convenient values to the *abscissa* x and calculating the corresponding values of the *ordinate* y. This procedure is particularly appropriate when $f(x)$ is a one-valued function of x. In other cases we may prefer to use *parametric* equations, expressing x and y as functions of a single variable (or *parameter*) t. For instance, if P_1 denotes the point (x_1, y_1), any line through P_1 has parametric equations

8.13
$$x = x_1 + Xt, \quad y = y_1 + Yt,$$

where X and Y depend on the *direction* of the line.

Sometimes, for the sake of symmetry, the single parameter t is replaced by two parameters, t_1 and t_2, related by an auxiliary equation. For instance, the general point (x, y) on the line through two given points P_1 and P_2 is given by

$$x = t_1 x_1 + t_2 x_2, \quad y = t_1 y_1 + t_2 y_2, \quad t_1 + t_2 = 1.$$

This point P, dividing the segment $P_1 P_2$ in the ratio $t_2 : t_1$, is the *centroid* (or "center of gravity") of masses t_1 at P_1 and t_2 at P_2. Positions outside the interval from P_1 (where $t_2 = 0$) to P_2 (where $t_1 = 0$) are covered by allowing t_2 or t_1 to be negative, while still satisfying $t_1 + t_2 = 1$; we may justify this by calling them "electric charges" instead of "masses."

For problems involving the distance between two points or the angle between two lines, it is often advisable to use *rectangular* axes, so that the distance from the origin to (x, y) is the square root of $x^2 + y^2$, and the distance $P_1 P_2$ is the square root of

$$(x_1 - x_2)^2 + (y_1 - y_2)^2.$$

Multiplication of the expression $l = ax + by + c$ by a suitable number enables us to *normalize* the equation $l = 0$ of the general line so that $a^2 + b^2 = 1$. Writing $l = 0$ in the form

$$(x - x_1 + 2al_1)^2 + (y - y_1 + 2bl_1)^2 = (x - x_1)^2 + (y - y_1)^2,$$

where $l_1 = ax_1 + by_1 + c$, we recognize it as the locus of points equidistant from

$$(x_1 - 2al_1, y_1 - 2bl_1) \quad \text{and} \quad (x_1, y_1);$$

in other words, the line $l = 0$ serves as a mirror which interchanges these two points by reflection. It follows that the foot of the perpendicular from P_1 to $l = 0$ is $(x_1 - al_1, y_1 - bl_1)$, and that the distance from P_1 to the line is $\pm l_1$ (provided $a^2 + b^2 = 1$). In particular, the distance from the origin to $l = 0$ is $\pm c$.

The locus of points at unit distance from the origin is the circle

$$x^2 + y^2 = 1,$$

which has the parametric equations

$$x = \cos\theta, \qquad y = \sin\theta$$

or, with $t = \tan\frac{1}{2}\theta$,

$$x = \frac{1 - t^2}{1 + t^2}, \qquad y = \frac{2t}{1 + t^2}.$$

EXERCISES

1. In terms of general Cartesian coordinates, the point (x, y) will be transformed into
   $(-x, -y)$ by the half-turn $O(-1)$ (§ 5.1),
   $(\mu x, \mu y)$ by the dilatation $O(\mu)$,
   $(x + a, y)$ by a translation along the x-axis.

2. In terms of rectangular Cartesian coordinates, the point (x, y) will be transformed into
   $(x, -y)$ by reflection in the x-axis,
   (y, x) by reflection in the line $x = y$,
   $(-y, x)$ by a quarter-turn about the origin,
   $(x + a, -y)$ by a glide reflection (in and along the x-axis),
   $(\mu x, -\mu y)$ by a dilative reflection   (§ 5.6).

3. Let M_{ij} denote the midpoint of P_iP_j. For any four points P_1, P_2, P_3, P_4, the midpoints of $M_{12}M_{34}$, $M_{13}M_{24}$, $M_{14}M_{23}$ all coincide.

8.2 POLAR COORDINATES

The deriving of short cuts from basic principles covers some of the finest achievements of the greatest mathematicians.

M. H. A. Newman (1897 -   )

(Mathematical Gazette **43** (1959), p. 170)

For problems involving directions from a fixed origin (or "pole") O, we often find it convenient to specify a point P by its *polar coordinates* (r, θ), where r is the distance OP and θ is the angle that the direction OP makes with a given *initial line* OX, which may be identified with the x-axis of rectangular Cartesian coordinates. Of course, the point (r, θ) is the same as $(r, \theta + 2n\pi)$ for any integer n. It is sometimes desirable to allow r to be negative, so that (r, θ) is the same as $(-r, \theta + \pi)$.

Given the Cartesian equation for a curve, we can deduce the polar equation for the same curve by substituting

8.21 $x = r \cos \theta, \qquad y = r \sin \theta.$

For instance, the unit circle $x^2 + y^2 = 1$ has the polar equation

$$(r \cos \theta)^2 + (r \sin \theta)^2 = 1,$$

which reduces to

$$r = 1.$$

(The positive value of r is sufficient if we allow θ to take all values from $-\pi$ to π or from 0 to 2π.) This procedure is helpful in elementary trigonometry, where students often experience some difficulty in proving (and remembering) the trigonometrical functions of obtuse and larger angles. Taking an angle XOP with $OP = 1$, we can simply *define* its cosine and sine to be the abscissa and ordinate of P.

Polar coordinates are particularly suitable for describing those isometries (§ 3.5) and similarities (§ 5.4) which have an invariant point; for this point may be used as the origin. Thus the general point (r, θ) will be transformed into

$(r, \theta + \alpha)$	by a rotation through α,
$(r, \theta + \pi)$	by a half-turn,
$(r, -\theta)$	by reflection in the initial line,
$(r, 2\alpha - \theta)$	by reflection in the line $\theta = \alpha$,
$(\mu r, \theta)$	by the dilatation $O(\mu)$,
$(\mu r, \theta + \alpha)$	by a dilative rotation with center O,
$(\mu r, 2\alpha - \theta)$	by a dilative reflection with center O and axis $\theta = \alpha$.

Likewise, inversion in the circle $r = k$ (see § 6.1) will transform (r, θ) into

$$(k^2/r, \quad \theta).$$

The Cartesian expressions for the same transformations can be deduced at once. For instance, the rotation through α about O transforms (x, y) into (x', y') where, by 8.21,

$$x' = r \cos (\theta + \alpha) = r (\cos \theta \cos \alpha - \sin \theta \sin \alpha) = x \cos \alpha - y \sin \alpha,$$
$$y' = r \sin (\theta + \alpha) = r (\cos \theta \sin \alpha + \sin \theta \cos \alpha) = x \sin \alpha + y \cos \alpha.$$

In particular, a quarter-turn transforms (x, y) into $(-y, x)$, and it follows that a necessary and sufficient condition for two points (x, y) and (x', y') to lie in perpendicular directions from the origin is

8.22 $xx' + yy' = 0.$

Such a transformation as

8.23 $x' = x \cos \alpha - y \sin \alpha,$
$y' = x \sin \alpha + y \cos \alpha$

has two distinct aspects: an "active" or *alibi* aspect, in which each point

(x, y) is moved to a new position (x', y'), and a "passive" or *alias* aspect, in which the point previously named (x, y) is renamed (x', y'). The latter aspect is sometimes used to simplify the equation of a given curve. For instance, the curve

$$ax^2 + 2hxy + by^2 = 1$$

becomes

$$a (x \cos \alpha - y \sin \alpha)^2 + 2h (x \cos \alpha - y \sin \alpha)(x \sin \alpha + y \cos \alpha)$$
$$+ b (x \sin \alpha + y \cos \alpha)^2 = 1,$$

in which the coefficient of xy is no longer $2h$ but

$$2h(\cos^2 \alpha - \sin^2 \alpha) - 2(a - b) \cos \alpha \sin \alpha = 2h \cos 2\alpha - (a - b) \sin 2\alpha.$$

Since this vanishes when $\tan 2\alpha = 2h/(a - b)$, the equation is simplified by rotating the axes through the particular angle

$$\alpha = \tfrac{1}{2} \arctan \frac{2h}{a - b}.$$

The area of a triangle OP_1P_2, where P_i has polar coordinates (r_i, θ_i), is taken to be positive if $\theta_1 < \theta_2$, negative if $\theta_1 > \theta_2$. With this convention, the area is

$$\tfrac{1}{2}r_1r_2 \sin (\theta_2 - \theta_1) = \tfrac{1}{2}r_1r_2(\sin \theta_2 \cos \theta_1 - \cos \theta_2 \sin \theta_1)$$

or, in Cartesian coordinates,

8.24
$$\tfrac{1}{2}(x_1y_2 - x_2y_1) = \tfrac{1}{2} \begin{vmatrix} x_1 & y_1 \\ x_2 & y_2 \end{vmatrix}.$$

To find the area of any triangle $P_1P_2P_3$, we choose new axes parallel to OX, OY and passing through P_3. Since the new coordinates of P_i ($i = 1$ or 2) are $(x_i - x_3, y_i - y_3)$, the area of any triangle $P_1P_2P_3$ is

8.25
$$\tfrac{1}{2} \begin{vmatrix} x_1 - x_3 & y_1 - y_3 \\ x_2 - x_3 & y_2 - y_3 \end{vmatrix} = \tfrac{1}{2} \begin{vmatrix} x_1 & y_1 & 1 \\ x_2 & y_2 & 1 \\ x_3 & y_3 & 1 \end{vmatrix}.$$

It follows that a necessary and sufficient condition for P_1, P_2, P_3 to be collinear is that this three-rowed determinant should be zero. The equation for the line P_1P_2 may be derived from this condition by writing (x, y) for (x_3, y_3).

EXERCISES

1. Use a well-known trigonometric formula to obtain an expression for the square of the distance between the points whose polar coordinates are (r_1, θ_1), (r_2, θ_2).

2. Obtain polar coordinates for the midpoint.

3. Obtain a polar equation for the line $y = x \tan \alpha$. (*Hint*: Allow r to take negative values.)

4. Use 8.22 to obtain the condition

$$aa' + bb' = 0$$

for two lines $ax + by + c = 0$ and $a'x + b'y + c' = 0$ to be perpendicular. Deduce that

$$ax + by + c = 0, \quad bx - ay + c' = 0$$

are perpendicular for all values of a, b, c, c'.

5. Use a suitable rotation of axes to simplify the equation of the curve

$$4x^2 + 24xy + 11y^2 = 5.$$

8.3   THE CIRCLE

> *A figure of universal appeal.*
>
> D. Pedoe [**1**, p. vii]

The circle with center (x', y') and radius k, being the locus of points (x, y) distant k from (x', y'), is

$$(x - x')^2 + (y - y')^2 = k^2.$$

Thus

8.31 $$x^2 + y^2 + 2gx + 2fy + c = 0$$

is a circle with center $(-g, -f)$ whenever $g^2 + f^2 > c$. If (x_1, y_1) lies on the circle, the tangent at this point P_1 is

$$x_1 x + y_1 y + g(x + x_1) + f(y + y_1) + c = 0$$

or $$(x_1 + g) x + (y_1 + f) y + (gx_1 + fy_1 + c) = 0.$$

For this line passes through P_1 and is perpendicular to the diameter

$$\frac{x + g}{x_1 + g} = \frac{y + f}{y_1 + f}.$$

The circle 8.31 is orthogonal to another circle

$$x^2 + y^2 + 2g'x + 2f'y + c' = 0$$

if, for a suitable P_1, the center of each lies on the tangent at P_1 to the other. Adding

$$(x_1 + g) g' + (y + f) f' = gx_1 + fy_1 + c$$

to the analogous relation with primed and unprimed letters interchanged, we see that the orthogonality of the two circles implies

$$2gg' + 2ff' = c + c'.$$

Conversely, any two circles that satisfy this relation are orthogonal. In particular, the circles

8.32 $x^2 + y^2 + 2gx + c = 0,$

8.33 $x^2 + y^2 + 2fy - c = 0,$

whose centers lie on the x- and y-axes respectively, are orthogonal. Keeping c constant and allowing g or f to take various values, we obtain two orthogonal pencils of coaxal circles, whose radical axes are $x = 0$ and $y = 0$ respectively. If $c = 0$, we have two orthogonal *tangent* pencils, each consisting of all the circles that touch one of the axes at the origin. If $c > 0$, the circles 8.32, for various values of g, form a *nonintersecting* pencil, including the two point circles

$$(x + g)^2 + y^2 = 0, \qquad g = \pm\sqrt{c},$$

which are the limiting points ($\mp\sqrt{c}, 0$) of the pencil. The circles 8.33, which pass through these two points, form the orthogonal *intersecting* pencil.

EXERCISES

1. The circle $x^2 + y^2 = k^2$ inverts (x, y) into

$$\left(\frac{k^2 x}{x^2 + y^2}, \frac{k^2 y}{x^2 + y^2}\right).$$

Apply this inversion to the line 8.11 and to the circle 8.31.

2. Find the locus of a point (x, y) whose distances from $(k/\mu, 0)$ and $(\mu k, 0)$ are in the ratio $1 : \mu$ (cf. § 6.6).

3. Obtain the Cartesian equation of the locus of a point the product of whose distances from $(a, 0)$ and $(-a, 0)$ is a^2. Deduce the polar equation of this "figure of eight," which is the *lemniscate of Jacob Bernoulli*.

4. Given two equal circles in contact, find the locus of the vertices of triangles for which the first is the nine-point circle (§ 1.7), the second is an excircle (§ 1.5). (*Answer:* A lemniscate.*)

5. A circle of radius b rolls without sliding on the outside of a fixed circle of radius nb. The locus of a point fixed on the circumference of the rolling circle is called an *epicycloid* (when n is an integer, an n-*cusped* epicycloid). Obtain the parametric equations

8.34 $\begin{aligned} x &= (n + 1)b \cos t - b \cos (n + 1)t, \\ y &= (n + 1)b \sin t - b \sin (n + 1)t. \end{aligned}$

Sketch the cases $n = 1$ (the *cardioid*), $n = 2$ (the *nephroid*), $n = 3$, and $n = \frac{2}{3}$. [See Robson **1**, p. 368.]

6. Shifting the origin to the cusp $(b, 0)$, obtain the polar equation

$$r = 2b (1 - \cos \theta)$$

for the cardioid (8.34 with $n = 1$). Deduce that chords through the cusp are of constant length.

---

* Richard Blum, *Canadian Mathematical Bulletin*, **1** (1958), pp. 1–3.

7. A circle of radius b rolls without sliding on the inside of a fixed circle of radius nb, where $n > 1$. Find parametric equations for the *hypocycloid* (when n is integral, the *n-cusped* hypocycloid) which is the locus of a point fixed on the circumference of the rolling circle. Sketch the cases $n = 2$ (which is surprising), $n = 3$ (the *deltoid*), and $n = 4$ (the *astroid*). Eliminate the parameter in the last two cases, obtaining, for the astroid,

$$x^{2/3} + y^{2/3} = a^{2/3} \qquad (a = 4b)$$

[Lamb **2**, pp. 297–303].

Steiner discovered that all the Simson lines for any given triangle touch a deltoid. Three of the lines, namely, those parallel to the sides of Morley's equilateral triangle (§ 1.9), are the "apsidal" tangents which the deltoid shares with the nine-point circle. Their points of contact are the vertices of the equilateral triangle XYZ described in Ex. 3 on page 20. For details, see Baker [**1**, pp. 330–349, especially p. 347].

8.4 CONICS

In addition to the straight lines, circles, planes and spheres with which every student of Euclid is familiar, the Greeks knew the properties of the curves given by cutting a cone with a plane—the ellipse, parabola and hyperbola. Kepler discovered by analysis of astronomical observations, and Newton proved mathematically on the basis of the inverse square law of gravitational attraction, that the planets move in ellipses. The geometry of ancient Greece thus became the cornerstone of modern astronomy.

J. L. Synge (1897 -   )

[Synge **2**, p. 32]

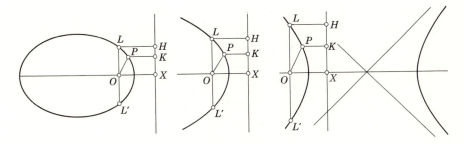

Figure 8.4a

There are several different ways to define a conic (or "conic section"). One of the most straightforward is the following (cf. § 6.6): A *conic* is the locus of a point P whose distance OP from a fixed point O is ε times its distance PK from a fixed line HX (Figure 8.4a), where ε is a positive constant.

Other definitions for a conic, proposed by Menaechmus about 340 B.C., were reconciled with this one by Pappus of Alexandria (fourth century A.D.) or possibly by Euclid [see Coolidge **1**, pp. 9–13].

The conic is called an *ellipse* if $\varepsilon < 1$, a *parabola* if $\varepsilon = 1$, a *hyperbola* if $\varepsilon > 1$. (These names are due to Apollonius.)

The point O and the line HX are called a *focus* and the corresponding *directrix*. The number ε, called the *eccentricity*, is usually denoted by e (but then, to avoid any possible misunderstanding, we should add "where e need not be the base of the natural logarithms" [Littlewood **1**, p. 43]). The chord LL' through the focus, parallel to the directrix, is called the *latus rectum;* its length is denoted by $2l$, so that

$$l = OL = \varepsilon LH.$$

In terms of polar coordinates with the initial line OX perpendicular to the directrix, we have

$$r = OP = \varepsilon PK = \varepsilon(LH - r \cos \theta)$$

8.41
$$= l - \varepsilon r \cos \theta,$$

so that

8.42
$$\frac{l}{r} = 1 + \varepsilon \cos \theta.$$

Since this equation is unchanged when we replace θ by $-\theta$, the conic is symmetrical by reflection in the initial line. When $\theta = 0$, $r = l/(1 + \varepsilon)$; and when $\theta = \pi$, $r = l/(1 - \varepsilon)$; therefore the conic meets the initial line twice except when $\varepsilon = 1$.

If $\varepsilon < 1$, 8.42 makes r finite and positive for all values of θ; therefore the ellipse is a *closed* (oval) curve. If $\varepsilon = 1$, r is still finite and positive except when $\theta = \pi$; therefore the parabola is not closed but extends to infinity in one direction. If $\varepsilon > 1$, r is positive or negative according as $\cos \theta$ is greater or less than $-1/\varepsilon$; therefore the hyperbola consists of two separate branches, given by

$$-\alpha < \theta < \alpha, \qquad \alpha < \theta < 2\pi - \alpha,$$

respectively, where $\alpha = \text{arcsec}\,(-\varepsilon)$.

Squaring 8.41, we obtain the Cartesian equation

8.43
$$x^2 + y^2 = (l - \varepsilon x)^2$$

(which indicates that a circle may be regarded as an ellipse with eccentricity zero). If $\varepsilon \neq 1$, we can divide by $1 - \varepsilon^2$ and then write a for $l/(1 - \varepsilon^2)$, obtaining

$$x^2 + \frac{y^2}{1 - \varepsilon^2} = la - 2\varepsilon ax$$

or
$$(x + \varepsilon a)^2 + \frac{a}{l} y^2 = (l + \varepsilon^2 a)a = a^2$$

or
$$\frac{(x + \varepsilon a)^2}{a^2} + \frac{y^2}{la} = 1.$$

Translating the origin to $(-\varepsilon a, 0)$, we can reduce this to

8.44
$$\frac{x^2}{a^2} \pm \frac{y^2}{b^2} = 1,$$

where
$$b^2 = |la| = |1 - \varepsilon^2| a^2,$$

so that $la = \pm b^2$, with the upper or lower sign according as $\varepsilon < 1$ or $\varepsilon > 1$. In the latter case the above definition makes a negative, but we can reverse its sign without altering the equation 8.44. A more important remark is that the equation is still unchanged when we reverse the sign of x or y. This invariance shows that the ellipse and hyperbola are symmetrical by reflection in either axis, and therefore also by the half-turn about the origin: their symmetry group is D_2, in the notation of § 2.5. For this reason the origin is called the *center*, and the ellipse and hyperbola are called *central* conics.

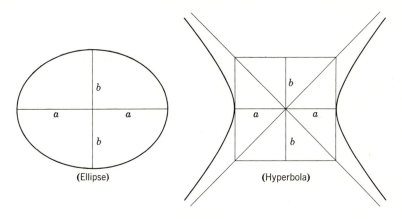

(Ellipse)   (Hyperbola)

Figure 8.4b

The geometrical significance of a and b is indicated in Figure 8.4b. For the ellipse, $2a$ and $2b$ are the *major* and *minor* axes; for the hyperbola, they are the *transverse* and *conjugate* axes. The two branches of the hyperbola

8.441
$$\frac{x^2}{a^2} - \frac{y^2}{b^2} = 1$$

lie in two opposite angular regions formed by the two lines

$$\frac{x^2}{a^2} - \frac{y^2}{b^2} = 0 \quad \text{or} \quad \left(\frac{x}{a} - \frac{y}{b}\right)\left(\frac{x}{a} + \frac{y}{b}\right) = 0.$$

These lines are called the *asymptotes* of the hyperbola. If $a = b$, they are perpendicular, and we have a *rectangular* (or "equilateral") hyperbola.

If $\varepsilon = 1$, then 8.43 reduces to

$$y^2 = 2l\left(\tfrac{1}{2}l - x\right)$$

or, by reflection in the line $x = \tfrac{1}{4}l$,

8.45 $y^2 = 2lx.$

This is the standard equation for the parabola (Figure 8.4c).

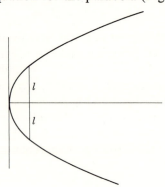

Figure 8.4c

The most convenient parametric equations are: for the ellipse

8.46 $x = a \cos t, \qquad y = b \sin t,$

for the parabola

8.47 $x = 2lt^2, \qquad y = 2lt,$

and for the hyperbola

8.48 $x = a \cosh t, \qquad y = b \sinh t,$

where

$$\cosh t = \frac{e^t + e^{-t}}{2}, \qquad \sinh t = \frac{e^t - e^{-t}}{2}.$$

(These functions will be discussed in § 8.6.)

EXERCISES

1. What kind of curve has the polar equation

$$r = \tfrac{1}{2}l \sec^2 \tfrac{1}{2}\theta ?$$

2. What kind of curve has the Cartesian equation

$$4x^2 + 24xy + 11y^2 = 5?$$

(See Ex. 5 at the end of §8.2.)

3. The sum (or difference) of the distances of a point on an ellipse (or hyperbola) from the two foci is constant.

4. Express the eccentricity of a central conic in terms of its semiaxes a and b. What is the eccentricity of a rectangular hyperbola?

5. Given points B and C, the locus of the vertex A of a triangle ABC whose Euler line is parallel to BC (as in Ex. 9 at the end of §1.6) is an ellipse whose minor axis is BC while its major axis is twice the altitude of the equilateral triangle on BC. (*Hint:* If A, B, C are (x, y) and $(\mp 1, 0)$, the circumcenter, equidistant from A and C, is $(0, \frac{1}{3}y)$.)

6. An expression such as

$$F = ax^2 + 2hxy + by^2$$

is called a *binary quadratic form*. It is said to be *definite* if $ab > h^2$, so that F has the same sign for all values of x and y except $x = y = 0$. It is said to be *positive* definite if this sign is positive. It is said to be *semidefinite* if $ab = h^2$, so that F is a times a perfect square; *positive* semidefinite if $a > 0$, so that F itself is a perfect square; *indefinite* if $ab < h^2$, so that F is positive for some values of x and y, negative for others. The equation $F = 1$ represents an ellipse if F is positive definite, a pair of parallel lines of F is positive semidefinite, and a hyperbola if F is indefinite.

7. What happens to the equation for the rectangular hyperbola $x^2 - y^2 = a^2$ when we rotate the axes through the angle $\frac{1}{4}\pi$?

8. Describe a geometrical interpretation for the parameter t in 8.46. [*Hint:* Compare $(a \cos t, b \sin t)$ with $(a \cos t, a \sin t)$.]

9. In what respect is the hyperbola 8.441 more satisfactorily represented by the equations

$$x = a \sec u, \qquad y = b \tan u$$

than by the equations 8.48?

10. The circle $r = l$ inverts the conic 8.42 into the limaçon
$$r = l(1 + \varepsilon \cos \theta).$$

Sketch this curve for various values of ε. When $\varepsilon = 1$ (so that the conic is a parabola), it is a cardioid.

11. The circle $r = a$ inverts the rectangular hyperbola $r^2 = a^2 \sec 2\theta$ into the lemniscate of Bernoulli

$$r^2 = a^2 \cos 2\theta$$

(see Ex. 3 at the end of §8.3).

8.5 TANGENT, ARC LENGTH, AND AREA

I do not know what I may appear to the world; but to myself I seem to have been only like a boy playing on the seashore, and diverting myself in now and then finding a smoother pebble or a prettier shell than ordinary, whilst the great ocean of truth lay all undiscovered before me.

Sir Isaac Newton

(Brewster's Memoirs of Newton, vol. 2, Chap. 27)

The curves with which we shall be concerned are "rectifiable," that is, there is a well-defined arc length s between any two points P and Q on such a curve. Using the temporary notation $P_0 = P$, $P_n = Q$, we subdivide the given arc PQ by $n - 1$ points $P_1, P_2, \ldots, P_{n-1}$ and consider the least upper bound

$$s = \sup \left(\sum_1^n P_{i-1} P_i \right)$$

of the lengths of the broken lines

$$PP_1 + P_1P_2 + \ldots + P_{n-1}Q,$$

for all possible subdivisions.

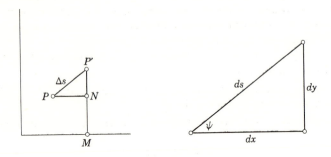

Figure 8.5a

The curve is often usefully regarded as the locus of a "moving" point. Any two points P and P' on the curve are joined by a line called a *secant*. If P is fixed while P' moves, the secant usually approaches a limiting position which is called the *tangent* at P. When we use rectangular Cartesian coordinates, we draw $P'M$ parallel to the y-axis, as in Figure 8.5a, and let N be the foot of the perpendicular from P to $P'M$. The tangent has an *angle of slope*, ψ, which may be defined as the limit of $\angle NPP'$. (The figure can be modified in an obvious manner when this angle is obtuse or negative.) To compute ψ, we consider the right-angled triangle $PP'N$ whose sides are the "increments" of x, y, s (all tending to zero):

$$\Delta x = PN, \quad \Delta y = NP', \quad \Delta s = PP'.$$

Thus

$$\cos \psi = \lim \frac{PN}{PP'} = \lim \frac{\Delta x}{\Delta s} = \frac{dx}{ds},$$

$$\sin \psi = \lim \frac{NP'}{PP'} = \lim \frac{\Delta y}{\Delta s} = \frac{dy}{ds},$$

$$\tan \psi = \lim \frac{NP'}{PN} = \lim \frac{\Delta y}{\Delta x} = \frac{dy}{dx}.$$

Since $PP'^2 = PN^2 + NP'^2$ (or since $\cos^2 \psi + \sin^2 \psi = 1$), the element of arc length ds is given by

8.51
$$ds^2 = dx^2 + dy^2,$$

and the arc length s, from (x_1, y_1) to (x_2, y_2), or from $t = t_1$ to $t = t_2$, is

$$s = \int ds = \int_{x_1}^{x_2} \left[1 + \left(\frac{dy}{dx}\right)^2 \right]^{\frac{1}{2}} dx = \int_{t_1}^{t_2} \left[\left(\frac{dx}{dt}\right)^2 + \left(\frac{dy}{dt}\right)^2 \right]^{\frac{1}{2}} dt.$$

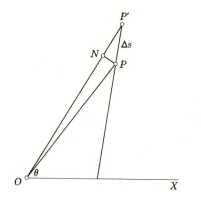
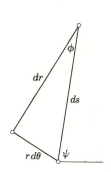

Figure 8.5b

When a curve is given by its polar equation, the direction of the tangent at P is determined either by the angle ϕ which this tangent makes with the "radius" OP or (equally well) by the angle

$$\psi = \theta + \phi$$

which it makes with the initial line OX (Figure 8.5b). As before, let P and P' be two neighboring points on the curve, so that the tangent at P is the limiting position of the secant PP'. Draw PN perpendicular to OP' [Lamb **2,** p. 254]. Then

8.52
$$\cos \phi = \lim \frac{NP'}{PP'} = \lim \frac{\Delta r}{\Delta s} = \frac{dr}{ds},$$

$$\sin \phi = \lim \frac{NP}{PP'} = \lim \frac{r \, \Delta\theta}{\Delta s} = r \frac{d\theta}{ds},$$

8.53
$$\tan \phi = \lim \frac{NP}{NP'} = \lim \frac{r \, \Delta\theta}{\Delta r} = r \frac{d\theta}{dr}.$$

Since $PP'^2 = NP'^2 + NP^2$ (or since $\cos^2 \phi + \sin^2 \phi = 1$), the element of arc length ds is now given by

8.54
$$ds^2 = dr^2 + r^2 \, d\theta^2.$$

Since the area of the thin triangle OPP' differs by a second-order infinitesimal from that of a circular sector of radius r and angle $\Delta\theta$, which is $\frac{1}{2}r^2 \Delta\theta$, the area of any closed curve surrounding the origin just once is

8.55
$$\frac{1}{2}\int_0^{2\pi} r^2 \, d\theta.$$

Such a formula can be translated into terms of Cartesian coordinates by means of the relations 8.21, which imply

8.56 $dx = dr \cos\theta - r \sin\theta \, d\theta, \quad dy = dr \sin\theta + r \cos\theta \, d\theta,$

so that $x \, dy - y \, dx = r \cos\theta \, dy - r \sin\theta \, dx = r^2(\cos^2\theta + \sin^2\theta)d\theta = r^2 \, d\theta$
and

8.57
$$\frac{1}{2}\int r^2 \, d\theta = \frac{1}{2}\int (x \, dy - y \, dx).$$

This must, of course, be interpreted as

8.58
$$\frac{1}{2}\int \left(x \frac{dy}{dt} - y \frac{dx}{dt}\right) dt,$$

where x and y are given in terms of a parameter t, and the integration is over the values of t that take us all round the curve.

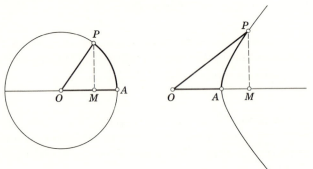

Figure 8.5c   Figure 8.5d

The same formula can be used to compute the area of the "sector" obtained by joining the origin to a given arc (Figure 8.5c or d). In polar coordinates, if the arc goes from $\theta = \theta_1$ to $\theta = \theta_2$, the area is

$$\frac{1}{2}\int_{\theta_1}^{\theta_2} r^2 \, d\theta \ .$$

When we transfer this to Cartesian coordinates, we regard the boundary of the sector as a closed "curve" consisting of the arc and the two radii. Since

$$d\frac{y}{x} = \frac{x\,dy - y\,dx}{x^2},$$

the radii (along which y/x remains constant) make a zero contribution to the integral on the right side of 8.57. Hence, if the arc goes from $t = t_1$ to $t = t_2$, the area of the sector is

8.59
$$\frac{1}{2}\int_{t_1}^{t_2} \left(x\frac{dy}{dt} - y\frac{dx}{dt}\right) dt$$

[cf. Courant **1**, p. 273].

EXERCISES

1. The line $x - (t + t')y + 2ltt' = 0$ is a secant of the parabola 8.47, meeting it in the points whose parameters are t and t'. Making t' tend to t, deduce the equation

$$x - 2ty + 2lt^2 = 0$$

for the tangent at the point whose parameter is t, so that, if (x_1, y_1) lies on the parabola 8.45, the tangent at this point is

$$y_1 y = l(x + x_1).$$

2. The line

$$\frac{x}{a}\cos\alpha + \frac{y}{b}\sin\alpha = \cos\beta$$

is a secant of the ellipse 8.46, meeting it in the points $t = \alpha \pm \beta$. Making β tend to 0, deduce the equation

$$\frac{x}{a}\cos t + \frac{y}{b}\sin t = 1$$

for the tangent at the point whose parameter is t [Robson **1**, p. 274]. Obtain analogous results for the hyperbola 8.48. Deduce that, if (x_1, y_1) lies on the central conic 8.44, the tangent at this point is

$$\frac{x_1 x}{a^2} \pm \frac{y_1 y}{b^2} = 1.$$

3. At the point t on the ellipse 8.46, the *normal*, being perpendicular to the tangent, is

$$\frac{ax}{\cos t} - \frac{by}{\sin t} = a^2 - b^2.$$

Differentiating partially with respect to t and then eliminating t, obtain the envelope of normals in the form

$$\left(\frac{ax}{a^2 - b^2}\right)^{\frac{2}{3}} + \left(\frac{by}{a^2 - b^2}\right)^{\frac{2}{3}} = 1$$

[Forder **3**, pp. 36–37; Lamb **2**, p. 350]. *Hint:*

$$\cos^3 t = \frac{ax}{a^2 - b^2}, \quad \sin^3 t = -\frac{by}{a^2 - b^2}.$$

4. Use 8.56 to reconcile 8.51 and 8.54.

8.6 HYPERBOLIC FUNCTIONS

> *The hyperbolic sine and cosine have a property in reference to the rectangular hyperbola, exactly analogous to that of the sine and cosine with reference to the circle. For this reason the former functions are called hyperbolic functions, just as the latter are called circular functions.*
>
> E. W. Hobson (1856-1933)
>
> [Hobson **1**, pp. 329-330]

As a very simple application of the formula 8.59, consider the unit circle $x^2 + y^2 = 1$ or

$$x = \cos t, \qquad y = \sin t$$

(Figure 8.5c). Since

$$\frac{dx}{dt} = -\sin t = -y \quad \text{and} \quad \frac{dy}{dt} = \cos t = x,$$

the area of the sector from $t = 0$ to any other value is

$$\tfrac{1}{2}\int_0 \left(x\frac{dy}{dt} - y\frac{dx}{dt}\right) dt = \tfrac{1}{2}\int_0 (x^2 + y^2)dt$$

$$= \tfrac{1}{2}\int_0 dt = \tfrac{1}{2}t,$$

which, of course, we knew already. More interestingly (Figure 8.5d), if the curve is the rectangular hyperbola $x^2 - y^2 = 1$ or

$$x = \cosh t, \qquad y = \sinh t,$$

so that

$$\frac{dx}{dt} = \sinh t = y \quad \text{and} \quad \frac{dy}{dt} = \cosh t = x,$$

the area of the sector is again

$$\tfrac{1}{2}\int_0 \left(x\frac{dy}{dt} - y\frac{dx}{dt}\right) dt = \tfrac{1}{2}\int_0 (x^2 - y^2)dt$$

$$= \tfrac{1}{2}\int_0 dt = \tfrac{1}{2}t.$$

Comparing the above results, we see clearly the analogy that relates the circular and hyperbolic functions. In Figures 8.5c and d, we have a sector AOP of the circle or rectangular hyperbola, respectively. In both cases $OA = 1$ and the parameter t is twice the area of the sector. In the former, $OM = \cos t$ and $PM = \sin t$. In the latter, $OM = \cosh t$ and $PM = \sinh t$.

EXERCISES

1. Find the area of the ellipse

$$x = a \cos t, \qquad y = b \sin t.$$

2. Find the area of the t sector of the general hyperbola

$$x = a \cosh t, \qquad y = b \sinh t.$$

8.7   THE EQUIANGULAR SPIRAL

In the equiangular spiral of the Nautilus or the snail-shell or Globige-
rina, the whorls continually increase in breadth, and do so in a steady
and unchanging ratio. . . . It follows that the sectors cut off by succes-
sive radii, at equal vectorial angles, are similar to one another in every
respect; and that the figure may be conceived as growing continuously
without ever changing its shape the while.

Sir D'Arcy W. Thompson (1860 -1948)

[Thompson **2,** pp. 753–754]

The circle $r = a$ may be regarded as the locus of the transform of the point
$(a, 0)$ by a *continuous rotation*, which transforms each point (r, θ) into
$(r, \theta + t)$ where t varies continuously. Similarly, the ray (or half line) $\theta = 0$
is the locus of the transform of $(a, 0)$ by a *continuous dilatation*, which yields
$(r, 0)$ for all positive values of r. By judiciously combining these two transformations, we obtain a *continuous dilative rotation*. Let μ denote the ratio of
magnification corresponding to rotation through 1 radian. Then μ^2 is the
ratio of magnification for 2 radians, μ^3 for 3 radians, . . . , μ^π for π radians,
. . . , μ^t for t radians. Thus the dilative rotation transforms the general point
(r, θ) into $(\mu^t r, \theta + t)$, where t varies continuously. The locus of the transform of $(a, 0)$ is the *equiangular spiral* (or "logarithmic spiral"), whose parametric equations

$$r = \mu^t a, \qquad \theta = t$$

may be combined into the single polar equation

8.71 $r = a\mu^\theta.$

This curve was first recognized by Descartes, who discussed it in 1638 in his letters to Mersenne.
Jacob Bernoulli (1654–1705) found it so fascinating that he arranged to have it engraved on his
tombstone (in the Münster at Basel, Switzerland) with the inscription

Eadem mutata resurgo.

These words (which E. T. Bell translates as "Though changed I shall arise the same") express a
remarkable consequence of the way the curve can be shifted along itself by a dilative rotation:
any dilatation has the same effect on it as a rotation, and vice versa. In fact, the rotation $\theta \to$
$\theta + \alpha$, changing $r = a\mu^\theta$ into $a\mu^{\theta + \alpha} = \mu^\alpha r$, is equivalent to the dilatation $O(\mu^\alpha)$. Steinhaus [**1,**
p. 97] describes this property as an optical illusion. Having drawn an equiangular spiral (Figure
8.7*a*), he remarks: "If we turn it round (in this case together with the book), it seems to grow
larger or smaller."

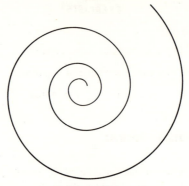

Figure 8.7a

Since

$$\frac{dr}{d\theta} = r \log \mu,$$

8.53 shows that the angle ϕ, between the position vector $\overrightarrow{OP}$ and the tangent at P, is given by

$$\cot \phi = \frac{1}{r}\frac{dr}{d\theta} = \log \mu,$$

that is, this angle is *constant*: a result which could have been foreseen from the fact that similarities preserve angles. In terms of this constant angle ϕ, which is about 80° in Figure 8.7a, we can write

$$\mu^\theta = e^{\theta \, \log \mu} = e^{\theta \, \cot \, \phi},$$

thus expressing the spiral in its classical form

8.72 $$r = ae^{\theta \, \cot \, \phi} \qquad (a \text{ and } \phi \text{ constant}).$$

By 8.52,

$$\frac{dr}{ds} = \cos \phi,$$

so that $r - s \cos \phi$ is constant. This shows that the length of the arc from $r = r_1$ to $r = r_2$ is

$$(r_2 - r_1) \sec \phi,$$

and that the length from the origin ($r = 0$) to the general point (although this involves infinitely many turns) is

$$r \sec \phi.$$

EXERCISES

1. The spiral $r = a\mu^\theta$ is homothetic to itself by means of the dilatation $O(\mu^{2\pi})$. How is it affected by inversion in the circle $r = a$?

2. Sketch the inverse of an equiangular spiral with respect to a circle whose center is on

   (i)   the spiral itself,

   (ii) the image of the spiral by the half-turn about its pole.

(Inverses of equiangular spirals are called *loxodromes*.)

8.8 THREE DIMENSIONS

> To Kästner, the Analyst, it appeared that the application of algebra to geometry would free the student from the stern discipline of Euclid. It would enable him to think for himself instead of having to watch the lips of his teacher. One of Kästner's sentences could fitly be inscribed over the door of any modern school: "The greatest satisfaction that we know is to find the truth for ourselves."
>
> J. L. Synge [**2,** p. 174]

To set up a system of Cartesian coordinates in space we use three *axial planes* meeting by pairs in three axes *OX, OY, OZ*. Starting from the origin O, we reach the general point (x, y, z) by going a distance x along the x-axis OX, then a distance y in the direction of the y-axis OY, and finally a distance z in the direction of the z-axis OZ. The three axial planes have the equations $x = 0, y = 0, z = 0$, which can be taken in pairs to determine the axes. For instance, the z-axis, consisting of all the points $(0, 0, z)$, has the two equations $x = y = 0$. Any line through the origin $(0, 0, 0)$ has parametric equations

8.81
$$x = Xt, \qquad y = Yt, \qquad z = Zt.$$

The mutual ratios of the coefficients X, Y, Z determine the direction of the line. By a translation to a new origin, we see that the parallel line through (x_1, y_1, z_1) is

8.82
$$x = x_1 + Xt, \qquad y = y_1 + Yt, \qquad z = z_1 + Zt.$$

Eliminating t, we obtain the two equations

8.83
$$\frac{x - x_1}{X} = \frac{y - y_1}{Y} = \frac{z - z_1}{Z},$$

which have to be interpreted by a special convention when $XYZ = 0$. The centroid of masses t_1 at (x_1, y_1, z_1) and t_2 at (x_2, y_2, z_2), with $t_1 + t_2 = 1$, is

$$(t_1 x_1 + t_2 x_2, \quad t_1 y_1 + t_2 y_2, \quad t_1 z_1 + t_2 z_2).$$

The origin and (x_1, y_1, z_1) are opposite vertices of the *parallelepiped* formed by the three pairs of parallel planes

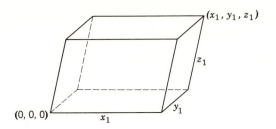

Figure 8.8a

$$x = 0, \ x = x_1; \qquad y = 0, \ y = y_1; \qquad z = 0, \ z = z_1;$$

as in Figure 8.8a. (In the tongue-twisting word "parallelepiped" we stress the syllable "ep," belonging to the Greek prefix *epi,* which occurs also in such words as "epithet" and "epicycloid.") For the rest of the present section we shall take the axes to be mutually orthogonal, so that this is a rectangular parallelepiped (or "box"). The three-dimensional extension of Pythagoras's theorem shows that the length of the diagonal

$$(0, 0, 0) \quad (x_1, y_1, z_1)$$

is the square root of $x_1^2 + y_1^2 + z_1^2$. Similarly, the distance between (x, y, z) and (x', y', z') is the square root of

$$(x - x')^2 + (y - y')^2 + (z - z')^2.$$

If the parameter t in 8.81 is adjusted so that

8.84 $$X^2 + Y^2 + Z^2 = 1,$$

it measures the distance from the origin to the general point (x, y, z) on the line. The coefficients X, Y, Z, satisfying 8.84, are called the *direction cosines* of the line, because they are the cosines of the angles which the line makes with the coordinate axes. More precisely, they are the direction cosines of one of the two rays into which the line is decomposed by the origin; the opposite ray has the direction cosines $-X, -Y, -Z$. Two rays forming an angle meet the unit sphere

$$x^2 + y^2 + z^2 = 1$$

in two points, say (X, Y, Z), (X', Y', Z'), whose coordinates are equal to the direction cosines of the rays. By "solving" the isosceles triangle which these points form with the origin, we obtain the expression

$$XX' + YY' + ZZ'$$

for the cosine of the angle between the rays. In particular, the rays (and therefore also the lines) are at right angles if

$$XX' + YY' + ZZ' = 0.$$

It follows that the plane through the origin perpendicular to the line 8.81 is

$$Xx + Yy + Zz = 0.$$

By a translation, we deduce the parallel plane through (x_1, y_1, z_1):

8.85 $$Xx + Yy + Zz = T,$$

where $T = Xx_1 + Yy_1 + Zz_1$. Still assuming

$$X^2 + Y^2 + Z^2 = X'^2 + Y'^2 + Z'^2 = 1,$$

we deduce that the two supplementary angles between the planes

$$Xx + Yy + Zz = T, \quad X'x + Y'y + Z'z = T'$$

are the angles whose cosines are

$$\pm(XX' + YY' + ZZ').$$

We see now that every plane has a linear equation, and every linear equation determines a plane. In particular, the plane that makes intercepts p, q, r on the axes is

$$\frac{x}{p} + \frac{y}{q} + \frac{z}{r} = 1.$$

Two planes of the form 8.85 are parallel if they differ only in their "constant" terms T. The line of intersection of two nonparallel planes can be reduced to the standard form 8.83 by eliminating first z and then x.

An equation connecting x, y, z (not necessarily linear) usually represents a *surface;* two such equations together represent a curve, the intersection of two surfaces. In particular, an equation

$$F(x, y) = 0,$$

involving only x and y, represents a *cylinder*, the locus of a line that passes through a variable point on the curve $F(x, y) = z = 0$ while remaining parallel to the z-axis. A *homogeneous* equation

$$f(x, y, z) = 0$$

(whose left side is merely multiplied by a power of μ when x, y, z are replaced by $\mu x, \mu y, \mu z$) represents a *cone,* the locus of a line that joins the origin to a variable point on the curve

$$f(x, y, 1) = 0, \quad z = 1.$$

Important instances are the quadric cylinders

$$\frac{x^2}{a^2} \pm \frac{y^2}{b^2} = 1 \quad \text{and} \quad y^2 = 2lx,$$

including the ordinary *cylinder of revolution* (or "right circular cylinder") $x^2 + y^2 = k^2$, and the quadric cones

$$ax^2 + by^2 + cz^2 = 0,$$

including the *cone of revolution* (or "right circular cone") $x^2 + y^2 = cz^2$.

The equation $x^2 + y^2 + z^2 = 0$, which is satisfied only by $(0, 0, 0)$, may be regarded either as a peculiar kind of cone or as a sphere of radius zero. The general sphere, having center (x', y', z') and radius k, is, of course,

$$(x - x')^2 + (y - y')^2 + (z - z')^2 = k^2.$$

We observe that this is an equation of the second degree in which the coefficients of x^2, y^2, z^2 are all equal while there are no terms in yz, zx, xy.

The sphere $x^2 + y^2 + z^2 = k^2$, whose center is the origin, inverts the point (X, Y, Z) into

$$\left(\frac{k^2 X}{X^2 + Y^2 + Z^2} , \frac{k^2 Y}{X^2 + Y^2 + Z^2} , \frac{k^2 Z}{X^2 + Y^2 + Z^2} \right).$$

The plane through this inverse point, perpendicular to the line 8.81, namely,

$$Xx + Yy + Zz = k^2,$$

is called the *polar plane* of (X, Y, Z) with respect to the sphere. If (X, Y, Z) lies in the sphere, the polar plane is simply the tangent plane.

The three-dimensional analogues of the conics are the quadric surfaces or *quadrics,* whose plane sections are conics (or occasionally pairs of lines, which may be regarded as degenerate conics). These surfaces, whose equations are of the second degree, include not only the elliptic and hyperbolic cylinders, the quadric cone, and the sphere, but also the *ellipsoid*

$$\frac{x^2}{a^2} + \frac{y^2}{b^2} + \frac{z^2}{c^2} = 1,$$

the *hyperboloid of one sheet*

8.86 $$\frac{x^2}{a^2} + \frac{y^2}{b^2} - \frac{z^2}{c^2} = 1$$

[R. J. T. Bell **1,** p. 149, Fig. 41], the *hyperboloid of two sheets*

$$\frac{x^2}{a^2} - \frac{y^2}{b^2} - \frac{z^2}{c^2} = 1$$

[Salmon **2,** p. 80, Fig. 1–4], the *elliptic paraboloid*

$$\frac{x^2}{a^2} + \frac{y^2}{b^2} = 2z,$$

and the *hyperbolic paraboloid*

8.861
$$\frac{x^2}{a^2} - \frac{y^2}{b^2} = 2z$$

[R. J. T. Bell **1**, p. 150, Fig. 42]. The nature of these surfaces can be roughly discerned by considering their sections by planes parallel to the coordinate planes. Their names were invented by G. Monge in 1805 [see Blaschke **1**, p. 131].

Important special cases are the quadrics of *revolution,* formed by revolving a conic about one of its axes. For instance, the special ellipsoid obtained by revolving an ellipse about its major or minor axis is a *prolate spheroid* or an *oblate spheroid,* respectively.

For the investigation of surfaces of revolution it is often convenient to use *cylindrical coordinates* (r, θ, z), in which the first two of the three Cartesian coordinates are replaced by polar coordinates

$$r = \sqrt{x^2 + y^2}, \qquad \theta = \arctan\frac{y}{x},$$

while z retains its usual meaning. To revolve a plane curve

$$F(x, z) = 0, \qquad y = 0$$

about the z-axis, we simply replace x by r; thus the surface of revolution is

$$F(r, z) = 0,$$

or, in Cartesian coordinates, $F(\sqrt{x^2 + y^2}, z) = 0$.

For instance, revolving the hyperbola 8.441 about its conjugate axis, we obtain the hyperboloid of revolution (of one sheet)

$$\frac{r^2}{a^2} - \frac{z^2}{b^2} = 1 \quad \text{or} \quad \frac{x^2 + y^2}{a^2} - \frac{z^2}{b^2} = 1.$$

Replacing $x^2 + y^2$ by $(x \cos \alpha + y \sin \alpha)^2 + (y \cos \alpha - x \sin \alpha)^2$, we may express this equation in the form

$$(x \cos \alpha + y \sin \alpha)^2 - (az/b)^2 = -(y \cos \alpha - x \sin \alpha)^2 + a^2$$

which shows that, for each value of α, every point on the line

$$x \cos \alpha + y \sin \alpha = az/b, \quad y \cos \alpha - x \sin \alpha = a$$

lies on the hyperboloid. Allowing α to vary from 0 to 2π, we obtain a continuous system of *generators*: lines lying entirely on the surface. Reflecting in the (x, y)-plane by reversing the sign of z, we obtain a second system of generators on the same hyperboloid. The plane

8.87
$$x \cos \alpha + y \sin \alpha = az/b,$$

through the center, touches the *asymptotic cone*

$$\frac{x^2 + y^2}{a^2} - \frac{z^2}{b^2} = 0$$

along the line

$$\frac{x}{a \cos \alpha} = \frac{y}{a \sin \alpha} = \frac{z}{b},$$

and meets the hyperboloid in two parallel lines: one in each of the planes $y \cos \alpha - x \sin \alpha = \pm a$.

Another interesting surface of revolution is the ring-shaped *torus*

$$(r - a)^2 + z^2 = b^2 \qquad (a > b),$$

which is obtained by revolving a circle of radius b about an exterior line in its plane, distant a from the center. This surface evidently contains two systems of circles: the "meridians," of radius b, and the "parallels" (in planes parallel to $z = 0$), whose radii vary between $a - b$ and $a + b$. It is less obvious that the torus contains also two "oblique" systems of circles of radius a, such that two circles of opposite systems meet twice while two distinct circles of the same system do not meet at all but are interlocked.* In fact, by expressing the equation

8.88 $$(\sqrt{x^2 + y^2} - a)^2 + z^2 = b^2$$

in the form

$$(x^2 + y^2 + z^2 - a^2 + b^2)^2 + 4(a^2 - b^2)z^2 = 4b^2(x^2 + y^2)$$

$$= 4b^2 \{ (x \cos \alpha + y \sin \alpha)^2 + (y \cos \alpha - x \sin \alpha)^2 \}$$

or

$$(x^2 + y^2 + z^2 - a^2 + b^2)^2 - 4b^2(y \cos \alpha - x \sin \alpha)^2$$

$$= 4b^2(x \cos \alpha + y \sin \alpha)^2 - 4(a^2 - b^2)z^2,$$

we see that, for each value of α, the torus contains the whole of the section of the sphere

$$x^2 + y^2 + z^2 - a^2 + b^2 = 2b(y \cos \alpha - x \sin \alpha)$$

by the plane

8.89 $$b(x \cos \alpha + y \sin \alpha) = \sqrt{a^2 - b^2}\, z.$$

Since the sphere can be expressed as

$$(x + b \sin \alpha)^2 + (y - b \cos \alpha)^2 + z^2 = a^2$$

and the plane passes through its center $(-b \sin \alpha, b \cos \alpha, 0)$, the section is a great circle, and its radius is a. Allowing α to vary from 0 to 2π, we ob-

---

* For drawings of the torus, showing all four systems of circles, see Hermann Schmidt, *Die Inversion und ihre Anwendungen* (Oldenbourg, Munich, 1950), p. 82. (The "elevation" is not as well drawn as the "plan.") See also Martin Gardner, *Scientific American*, **203** (1960), pp. 194, 196. These circles were discovered by Yvon Villarceau, *Nouvelles Annales de Mathématiques* (1), **7** (1848), pp. 345–347.

tain a continuous system of such circles, and a second system by reversing the sign of z.

The plane 8.89 meets the torus in two circles, one in each system (with α replaced by $\alpha + \pi$ in the second system). Since these two circles are sections of the two spheres

$$x^2 + y^2 + z^2 - a^2 + b^2 = \pm 2b(y \cos \alpha - x \sin \alpha),$$

their points of intersection are the two "antipodal" points

$$\left(\pm \frac{a^2 - b^2}{a} \cos \alpha, \ \pm \frac{a^2 - b^2}{a} \sin \alpha, \ \pm \frac{b}{a} \sqrt{a^2 - b^2} \ \right)$$

(with signs agreeing). Since each of these is a point of contact, 8.89 is a *bitangent* plane [R. J. T. Bell **1**, p. 267].

Comparing 8.89 with 8.87, we see that the "oblique" circles on the torus lie in the same planes (through the center) as the pairs of parallel generators of the hyperboloid of revolution

$$\frac{x^2 + y^2}{a^2 - b^2} - \frac{z^2}{b^2} = 1.$$

(This remark is due to A. W. Tucker.)

EXERCISES

1. The plane through three given points (x_i, y_i, z_i) $(i = 1, 2, 3)$ is

$$\begin{vmatrix} x_1 & y_1 & z_1 & 1 \\ x_2 & y_2 & z_2 & 1 \\ x_3 & y_3 & z_3 & 1 \\ x & y & z & 1 \end{vmatrix} = 0.$$

If the requirement of passing through a point is replaced (in one or two cases) by the requirement of being parallel to a line with direction numbers X_i, Y_i, Z_i, the corresponding row of the determinant is replaced by

$$X_i \quad Y_i \quad Z_i \quad 0.$$

2. In terms of general Cartesian coordinates, the point (x, y, z) will be transformed into

$(-x, -y, -z)$ by the central inversion $O(-1)$,
$(\mu x, \mu y, \mu z)$ by the dilatation $O(\mu)$ (§7.6),
$(x, y, z+c)$ by a translation along the z-axis.

3. In terms of rectangular coordinates, the point (x, y, z) will be transformed into

$(x, y, -z)$ by reflection in the (x, y)-plane,
(y, x, z) by reflection in the plane $x = y$,
$(-y, x, z)$ by a quarter-turn about the z-axis,
$(x, -y, z+c)$ by a glide reflection (§7.4).

4. In terms of cylindrical coordinates, the point (r, θ, z) will be transformed into

$(r, \theta + \alpha, z + c)$ by a twist,
$(\mu r, \theta + \alpha, \mu z)$ by the general dilative rotation (§7.6).

5. Obtain the condition

$$2uu' + 2vv' + 2ww' = d + d'$$

for two spheres

$$x^2 + y^2 + z^2 + 2ux + 2vy + 2wz + d = 0,$$
$$x^2 + y^2 + z^2 + 2u'x + 2v'y + 2w'z + d' = 0$$

to be orthogonal.

6. If (X, Y, Z) is outside the sphere $x^2 + y^2 + z^2 = k^2$, its polar plane contains the points of contact of all the tangent planes that pass through (X, Y, Z).

7. By expressing 8.86 in the form

$$\left(\frac{x}{a}\cos \alpha + \frac{y}{b}\sin \alpha\right)^2 + \left(\frac{y}{b}\cos \alpha - \frac{x}{a}\sin \alpha\right)^2 = 1 + \left(\frac{z}{c}\right)^2,$$

find the two systems of generators on the general hyperboloid of one sheet. Two generators of opposite systems intersect (or occasionally are parallel), but two distinct generators of the same system are skew. This observation applies also to the two systems of generators on the hyperbolic paraboloid 8.861.

9

Complex numbers

The extension of the Euclidean plane to the inversive plane (§6.4) or to the elliptic plane (§6.9) is the geometric counterpart of a familiar procedure in algebra: the extension of the concept of number. Beginning with the natural numbers such as 1 and 2, we proceed to the integers, then to the rational numbers, then to the real numbers, then to the complex numbers (and if we had time we could continue with hypercomplex numbers*). Each stage is motivated by our desire to be able to solve a certain kind of equation. Real numbers were understood remarkably well by the ancient Greeks. Complex numbers were used freely, by R. Bombelli (in his *Algebra*, Bologna, 1572) and especially by Euler, many years before they could be treated rigorously; that was how the word "imaginary" acquired its technical meaning. To put "the square root of minus one" on a firm foundation, it is convenient (though not essential) to use a geometric representation. Such an interpretation was suggested by J. Wallis (1685), formulated completely by C. Wessel (1797), rediscovered by J. R. Argand (1806), and rediscovered again by Gauss.†

The present discussion of number is not intended to be a formal development but rather to emphasize the role of geometry in the working rules. For a more complete treatment see Robinson [**1**, pp. 73–84].

9.1 RATIONAL NUMBERS

> "The northern ocean is beautiful," said the Orc, "and beautiful the delicate intricacy of the snowflake before it melts and perishes, but such beauties are as nothing to him who delights in numbers, spurning alike the wild irrationality of life and the baffling complexity of nature's laws."
>
> J. L. Synge [**2**, p. 101]

The first numbers that we consider in arithmetic are the *natural* numbers,

* See, for instance, Coxeter, Quaternions and reflections, *American Mathematical Monthly*, **53** (1946), pp. 136–146.
† See the excellent article on Complex numbers, by C. C. MacDuffee, in the *Encyclopaedia Britannica* (Chicago edition).

forming a sequence that begins with 1 and never ends. The problem of solving such an equation as

$$x + 2 = 1$$

motivates the discovery of the *integers,* which include not only the natural numbers (or "positive integers") but also zero and the negative integers. The sequence of integers

$$\ldots, -2, -1, 0, 1, 2, \ldots,$$

which has neither beginning nor end, is conveniently represented by points evenly spaced along an infinite straight line, which we may think of as the *x*-axis of ordinary analytic geometry. In this representation, addition and subtraction appear as *translations*: the transformation $x \rightarrow x + a$ shifts each point through a spaces to the right if a is positive and through $-a$ spaces to the left if a is negative; in other words, the operation of adding a is the translation that transforms 0 into a.

The problem of solving such an equation as

$$2x = 1$$

motivates the discovery of the *rational* numbers $\mu = a/b$, where a is an integer and b is a positive integer; these include not only the integers $a = a/1$, but also fractions such as 1/2 (or $\frac{1}{2}$) and $-4/3$. The rational numbers cannot be written down successively in their natural order, because between any two of them there is another, and consequently an infinity of others; for example, between a/b and c/d we find $(a + c)/(b + d)$. The corresponding points are dense on the *x*-axis, and at first sight seem to cover it completely. Multiplication and division appear as *dilatations*: the transformation $x \rightarrow \mu x$ is the dilatation $O(\mu)$, where O is the origin (which represents zero); in other words, multiplication by μ is the dilatation with center O that transforms 1 into μ. Of course, μ may be either positive (Figure 9.1a) or negative (Figure 9.1b). In particular, multiplication by -1 is the half-turn about O. (The point 1 is joined to an arbitrary point on the *y*-axis.)

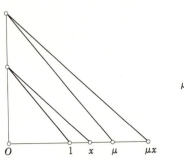

Figure 9.1a

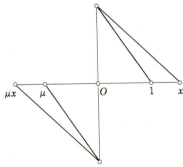

Figure 9.1b

We can derive the rational number a/b from the integer a by applying the dilatation $O(1/b)$, which transforms b into 1. (Figure 9.1c illustrates the derivation of the rational numbers $3/2$ and $-1/2$.) This construction shows clearly why we cannot allow the denominator b to be zero. There would be no harm in allowing b to be negative, but we naturally identify $a/(-b)$ with $-a/b$. In the same spirit we usually write each fraction in its "lowest terms," so that the numerator and denominator have no common factor.

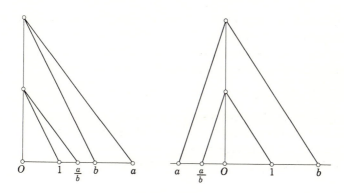

Figure 9.1c

EXERCISE

Use the method of Figure 9.1c to construct $\frac{1}{3}$.

9.2   REAL NUMBERS

> "What was that?" inquired Alice. "Reeling and Writhing, of course, to begin with," the Mock Turtle replied, "and then the different branches of Arithmetic—Ambition, Distraction, Uglification, and Derision."
>
> Lewis Carroll
> [Dodgson **1**, Chapter 9]

The problem of solving such an equation as

$$x^2 = 2$$

motivates the discovery of the *real* numbers, which include not only the rational numbers but also the *irrational* numbers (such as $\sqrt{2}$ and π), which cannot be expressed as fractions. (Actually, π cannot even be expressed as a root of an algebraic equation.) Pythagoras's proof of the irrationality of $\sqrt{2}$ was considered by Hardy [**2**, pp. 32–36] to be one of the most ancient instances of first-rate mathematics, "as fresh and significant as when it was discovered." In Cantor's treatment [Robinson **1**, p. 79], a real number is defined to be the limit of a convergent sequence of rational numbers, or (more precisely) the set of all sequences "equivalent" (in a specified sense) to a given sequence; for example, the real number π is the limit of the sequence

$$3, \quad 3.1, \quad 3.14, \quad 3.141, \quad 3.1415, \quad \ldots$$

or of the "equivalent" sequence

$$4, \quad 3.2, \quad 3.15, \quad 3.142, \quad 3.1416, \quad \ldots$$

The corresponding point on the x-axis is defined to be the limit of a convergent sequence of "rational" points.

The real numbers can be further subdivided into *algebraic* numbers (such as $\sqrt{2}$ and $\sqrt[3]{2}$), which can be expressed as roots of equations

9.21 $$a_0 x^n + a_1 x^{n-1} + \ldots + a_n = 0$$

with integers for coefficients, and *transcendental* numbers (such as π and e) which cannot be so expressed. Among the algebraic numbers, it is sometimes desirable to distinguish the *quadratic* numbers, such as $\sqrt{2}$ and $\sqrt{(2 - \sqrt{2})}$, which can be constructed with compasses.

9.3 THE ARGAND DIAGRAM

> *I met a man recently who told me that, so far from believing in the square root of minus one, he did not even believe in minus one. This is at any rate a consistent attitude.*
>
> *There are certainly many people who regard $\sqrt{2}$ as something perfectly obvious, but jib at $\sqrt{(-1)}$. This is because they think they can visualize the former as something in physical space, but not the latter. Actually $\sqrt{(-1)}$ is a much simpler concept.*
>
> E. C. Titchmarsh [**1**, p. 99]

The problem of solving such an equation as

9.31 $$x^2 + 1 = 0.$$

motivates the discovery of the *complex* numbers (so named by Gauss), which include not only the real numbers but also such "imaginary" numbers as the square root of -1. Since the real numbers occupy the whole x-axis, it is natural to try to represent the complex numbers by all the points in the (x, y)-plane, that is, to define them as ordered pairs of real numbers with suitable rules for their addition and multiplication [Synge **2**, Chapter 9]. In this so-called *Argand diagram* (invented by Caspar Wessel in 1797, a few years before Argand himself), points are added like the corresponding vectors from the origin O (which represents 0):

9.32 $(x, y) + (a, b) = (x + a, y + b)$

(Figure 9.3a). In other words, to add (a, b) we apply the translation that takes $(0, 0)$ to (a, b).

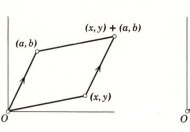 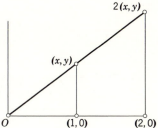

Figure 9.3a	Figure 9.3b

Multiplication by an integer still appears as a dilatation; for instance

$$2(x, y) = (x, y) + (x, y) = (2x, 2y)$$

(Figure 9.3b). In particular, multiplication by -1 is the half-turn about O. What, then, is multiplication by "the square root of -1"? This must be a transformation whose "square" is the half-turn about O. The obvious answer is a *quarter-turn* about O [Hardy **1**, p. 83].

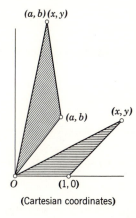 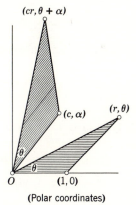

(Cartesian coordinates)	(Polar coordinates)

Figure 9.3c

Then multiplication by an arbitrary complex number should be a transformation which leaves O invariant and includes both dilatations and rotations as special cases. The obvious transformation of this kind is a *dilative rotation* (§5.4). Accordingly, the operation of multiplying the general point (x, y) by any particular point (a, b) is defined to be the dilative rotation (with center O) that transforms $(1, 0)$ into (a, b) [Klein **1**, p. 57]. To see how this works, let the points (a, b), (x, y) have polar coordinates (c, α), (r, θ), so that

$$a = c \cos \alpha, \quad b = c \sin \alpha, \quad x = r \cos \theta, \quad y = r \sin \theta.$$

Then the dilative rotation (Figure 9.3c), multiplying r by c and adding α to θ, transforms

$$r \cos \theta = x$$

and

$$r \sin \theta = y$$

into

$$\begin{aligned}
cr \cos (\theta + \alpha) &= cr (\cos \theta \cos \alpha - \sin \theta \sin \alpha) \\
&= (c \cos \alpha)(r \cos \theta) - (c \sin \alpha)(r \sin \theta) \\
&= ax - by
\end{aligned}$$

and

$$\begin{aligned}
cr \sin (\theta + \alpha) &= cr (\sin \theta \cos \alpha + \cos \theta \sin \alpha) \\
&= (c \cos \alpha)(r \sin \theta) + (c \sin \alpha)(r \cos \theta) \\
&= ay + bx.
\end{aligned}$$

Hence, finally, the rule for multiplication is (in Cartesian coordinates)

9.33 $(a, b)(x, y) = (ax - by, ay + bx)$

[Hardy **1**, p. 80].

Since $(x, y) + (a, 0) = (x + a, y)$ and $(a, 0)(x, y) = (ax, ay)$, we naturally identify the complex number $(a, 0)$ with the real number a, so that

$$a(x, y) = (ax, ay)$$

and $(x, y) = (x, 0) + (0, y) = x + y (0, 1).$

Introducing Euler's special symbol i to denote the complex number $(0, 1)$, we have

$$\begin{aligned}
(x, y) &= x + yi \\
&= x + iy
\end{aligned}$$

and $i^2 = (0, 1)(0, 1) = (-1, 0) = -1$. In this notation, the rules 9.32 and 9.33 become

$$(x + yi) + (a + bi) = (x + a) + (y + b)i,$$

$$(a + bi)(x + yi) = ax - by + (ay + bx)i,$$

which may simply be thought of as ordinary addition and multiplication, treating i as an indeterminate, followed by the insertion of -1 for i^2 [Birkhoff and MacLane **1**, pp. 95–97].

EXERCISES

1. Solve the equation $z^2 - 4z + 5 = 0$.
2. The equation $u + vi = 0$ implies both $u = 0$ and $v = 0$.
3. Express $(a + bi)^{-1}$ in the form $x + yi$.
4. Adapt Figure 9.3c to the cases (i) $b = 0$, (ii) $a^2 + b^2 = 1$.

9.4 MODULUS AND AMPLITUDE

> Abraham Demoivre *(more correctly written as de Moivre) . . . died in London on November 27, 1754. . . . Shortly before* [*his death*] *he declared that it was necessary for him to sleep some ten minutes or a quarter of an hour longer each day than the preceding one. The day after he had thus reached a total of something over twenty-three hours he slept up to the limit of twenty-four hours, and then died in his sleep.*
>
> W. W. Rouse Ball
> [Ball **2,** pp. 383 -384]

The conversion of Cartesian coordinates to polar coordinates shows that any complex number $z = x + yi$ is expressible in the form

$$r (\cos \theta + i \sin \theta).$$

The radial distance r is called the *modulus* (or "absolute value") of z, and is denoted by $|z|$. The angle θ is called the *amplitude* (or "argument") of z, and is denoted by am z. Thus

$$|x + yi| = \sqrt{x^2 + y^2}, \quad \text{am } (x + yi) = \arctan \frac{y}{x}.$$

Since multiplication by $c = a + bi$ is represented by a dilative rotation which is the product of a dilatation in the ratio $|c| : 1$ and a rotation through am c, we have

$$|cz| = |c| |z|, \quad \text{am } (cz) = \text{am } c + \text{am } z.$$

If $|c| = 1$, multiplication by c is represented by a pure rotation. Such a number c is of the form

$$\cos \alpha + i \sin \alpha,$$

where $\alpha = \text{am } c$ is the angle of the rotation. Repeating this rotation n times, we obtain de Moivre's theorem

$$(\cos \alpha + i \sin \alpha)^n = \cos n\alpha + i \sin n\alpha.$$

In particular, multiplication by

$$\omega = \cos \frac{2\pi}{n} + i \sin \frac{2\pi}{n}$$

is represented by a rotation through $2\pi/n$. Applying this rotation and its successive powers to the point 1, we obtain the vertices of a regular n-gon, $\{n\}$, inscribed in the unit circle $|z| = 1$. These n points represent the roots

$$1, \omega, \omega^2, \ldots, \omega^{n-1}$$

of the cyclotomic equation $z^n = 1$ (see §2.1). From the theory of equations we deduce the identity

$$z^n - 1 = (z - 1)(z - \omega)(z - \omega^2) \ldots (z - \omega^{n-1}).$$

In particular, the four fourth roots of 1 are

$$1, \quad i, \quad i^2 = -1, \quad i^3 = -i.$$

Since $i^4 = 1$, the higher powers yield the same cycle over and over again; for example,

$$i^5 = i, \quad i^6 = -1, \quad i^7 = -i, \quad i^8 = 1.$$

Each complex number $z = x + yi$ has a *conjugate* $\bar{z} = x - yi$, derived by reflection in the x-axis. Since

$$z\bar{z} = x^2 + y^2 = |z|^2,$$

the reciprocal of z is

$$z^{-1} = |z|^{-2}\, \bar{z}.$$

In particular, we have

$$(\cos \alpha + i \sin \alpha)^{-1} = \cos \alpha - i \sin \alpha,$$

which is de Moivre's theorem with $n = -1$.

Inversion in the circle $|z| = 1$ transforms $r(\cos \theta + i \sin \theta)$ into $r^{-1}(\cos \theta + i \sin \theta)$, that is, it transforms z into

$$\bar{z}^{-1} = |z|^{-2}z.$$

De Moivre's theorem is easily extended from integral values of n to rational values, and then to real values. In particular, a point continuously describing the circle $|z| = 1$ represents the complex number

9.41 $$\cos \theta + i \sin \theta = (\cos 1 + i \sin 1)^\theta,$$

where θ varies continuously from 0 to 2π.

Applying the same idea to the complex number

$$z = \mu(\cos 1 + i \sin 1),$$

of modulus $\mu \neq 1$, we see that its powers

$$z^\theta = \mu^\theta (\cos \theta + i \sin \theta)$$

are represented by a point describing the equiangular spiral $r = \mu^\theta$ (§ 8.6).

EXERCISE

Use de Moivre's Theorem to find
(*a*) the two square roots of $3 + 4i$;
(*b*) the three cube roots of 1, say 1, ω, ω^2;
(*c*) the six sixth roots of 1 (in terms of ω);
(*d*) the twelve twelfth roots of 1.

9.5 THE FORMULA $e^{\pi i} + 1 = 0$

> There is a famous formula—perhaps the most compact and famous of
> all formulas—developed by Euler from a discovery of de Moivre:
> $e^{i\pi} + 1 = 0$. . . . It appeals equally to the mystic, the scientist, the
> philosopher, the mathematician. For each it has its own meaning.
> Though known for over a century, de Moivre's formula came to Ben-
> jamin Peirce (1809-1880) as something of a revelation. Having dis-
> covered it one day, he turned to his students. . . . "Gentlemen," he
> said, "that is surely true, it is absolutely paradoxical; we cannot un-
> derstand it, and we don't know what it means, but we have proved it,
> and therefore we know it must be the truth."
>
> E. Kasner and J. Newman
> [**1**, pp. 103 -104]

Having accepted *real* powers such as $\mu^\theta = e^{\theta \log \mu}$, we naturally ask
whether any meaning can be attached to *complex* powers. A partial answer
is supplied by the theory of infinite series such as

$$e^x = 1 + x + \frac{x^2}{2!} + \frac{x^3}{3!} + \frac{x^4}{4!} + \frac{x^5}{5!} + \cdots,$$

$$\cos x = 1 - \frac{x^2}{2!} + \frac{x^4}{4!} - \cdots,$$

$$\sin x = x - \frac{x^3}{3!} + \frac{x^5}{5!} - \cdots.$$

The first series, which is familiar when x is real, provides a *definition* for e^x in
other cases; for instance,

$$e^i = 1 + i + \frac{i^2}{2!} + \frac{i^3}{3!} + \frac{i^4}{4!} + \frac{i^5}{5!} + \cdots$$

$$= 1 + i - \frac{1}{2!} - \frac{i}{3!} + \frac{1}{4!} + \frac{i}{5!} - \cdots$$

$$= \left(1 - \frac{1}{2!} + \frac{1}{4!} - \cdots\right) + i\left(1 - \frac{1}{3!} + \frac{1}{5!} - \cdots\right)$$

$$= \cos 1 + i \sin 1.$$

These series enable us to express 9.41 in Euler's concise form

9.51
$$\cos \theta + i \sin \theta = e^{\theta i},$$

which is a refinement of the identity

$$\theta i = \log_e (\cos \theta + i \sin \theta)$$

of R. Cotes* (1682–1716), after whose untimely death Newton said, "If Cotes had lived, we might have known something!"

Setting $\theta = \pi$ in 9.51, we obtain the "famous formula"

$$e^{\pi i} = -1,$$

which connects in such a surprising way the three important numbers

$$e = 2.71828 \ldots, \qquad \pi = 3.14159 \ldots,$$

and i.

EXERCISES

1. Evaluate $e^{\frac{1}{2}\pi i}$. Is i^i real?

2. From 9.51 deduce $\cos \theta = \dfrac{e^{\theta i} + e^{-\theta i}}{2}$, $\sin \theta = \dfrac{e^{\theta i} - e^{-\theta i}}{2i}$, the familiar formulas for $\cos(\theta + \alpha)$, $\sin(\theta + \alpha)$, and the derivatives of $\cos \theta$, $\sin \theta$.

9.6 ROOTS OF EQUATIONS

Gauss . . . was the first mathematician to use complex numbers in a really confident and scientific way.

G. H. Hardy

[Hardy and Wright **1**, p. 188]

In the field of complex numbers we can solve any quadratic equation that has real coefficients; for example, the equation 9.31 has the two roots i and $-i$. Still more remarkably, we can solve any quadratic equation with complex coefficients; indeed, not only any quadratic equation but also any equation of degree 3 or 4. In saying that we can "solve" an equation we mean here that we can find explicit expressions for the roots in terms of the coefficients. It was proved by E. Galois (who was murdered in 1832 when he was only 20 years old) that the general algebraic equation 9.21 with $n > 4$ cannot be solved in this sense [Infeld **1**]. Nevertheless, the fundamental theorem of algebra (which Gauss proved in 1799) asserts the *existence* of roots for all values of n, even when explicit expressions are not available. (For a neat proof, see Birkhoff and MacLane [**1**, pp, 101–103].) In fact, *numerical solutions* can be found, correct to any assigned number of decimal places.

EXERCISE

A ladder, 24 feet long, rests against a wall with the extra support of a cubical box of edge 7 feet, placed at the bottom of the wall with one horizontal edge against the ladder. How far up the wall does the ladder reach? (*Hint:* Take $7x$ to be the height of the top of the ladder above the top of the box. Obtain an equation whose relevant root is $7x = 9 + 4\sqrt{2}$.)

* *Harmonia mensurarum*, Cambridge, 1722, p. 28.

9.7 CONFORMAL TRANSFORMATIONS

We saw, in § 9.3, that the transformation

$$z' = z + b$$

(which adds to the complex variable z the complex constant b) is a translation, whereas

$$z' = az$$

(which multiplies z by the complex constant a) is a dilative rotation about the
point 0, including as special cases a dilatation (when a is real) and a rotation
(when $|a| = 1$). It follows that a dilative rotation about the general point
c is

$$z' - c = a(z - c)$$

or

$$z' = az + (1 - a)c.$$

Hence the general direct similarity, as described in §5.5, is the general *linear*
transformation

$$z' = az + b$$

[Ford **1**, p. 3]; and this is a translation or a dilative rotation according as
$a = 1$ or $a \neq 1$. (In the latter case, $c = b/(1 - a)$.)

Since the product of an opposite similarity and a reflection is direct, any
given opposite similarity may be expressed as the product of a given reflection and a suitable direct similarity. Using the reflection in the x-axis,
namely

$$z' = \bar{z}$$

(§ 9.4), we see that the general opposite similarity is the "conjugate" linear
transformation

$$z' = a\bar{z} + b.$$

Since the ratio of magnification is again $|a|$, this is a glide reflection (possi-
bly reducing to a pure reflection) if $|a| = 1$, and a dilative reflection other-
wise.

We saw, in § 9.4, that the transformation $z' = \bar{z}^{-1}$ is the inversion in the
unit circle $|z| = 1$. Similarly,

$$z' = \frac{k^2}{\bar{z}}$$

is the inversion in the circle $|z| = k$, of radius k. It follows that the in-
version in the general circle $|z - a| = k$ is $z' - a = k^2/(\bar{z} - \bar{a})$ or

9.71

$$z' = a + \frac{k^2}{\bar{z} - \bar{a}}.$$

By 6.71, any circle-preserving transformation that is not a similarity is the product of such an inversion and an isometry

$$z' = p\bar{z} + q \qquad \text{or} \qquad z' = pz + q,$$

where $|p| = 1$. To express this product, we replace the z on the right of 9.71 by $p\bar{z} + q$ or $pz + q$, obtaining

$$z' = a + \frac{k^2}{\bar{p}z + \bar{q} - \bar{a}} = \frac{az + b}{z + d}$$

or

$$z' = a + \frac{k^2}{\bar{p}\bar{z} + \bar{q} - \bar{a}} = \frac{a\bar{z} + b}{\bar{z} + d}$$

respectively, where b and d are certain expressions involving $k^2, p, \bar{q}, a$ and $\bar{a}$. Hence

Every circle-preserving transformation, direct or opposite, is a linear fractional transformation

9.72 $$z' = \frac{az + b}{cz + d} \quad \text{or} \quad z' = \frac{a\bar{z} + b}{c\bar{z} + d} \qquad (ad \neq bc),$$

where c may be taken to be 0 or 1 according as the transformation is, or is not, a similarity.

Conversely, *every linear fractional transformation 9.72 transforms circles into circles.* The easiest way to see this is by direct substitution in the equation $|z - u| = k$ or

$$(z - u)(\bar{z} - \bar{u}) = k^2$$

for the general circle. This is clearly transformed into an equation of the same kind. The following alternative procedure is suggested by a remark of N. S. Mendelsohn.*

If $c = 0$, the transformation is a similarity, as we have seen. If $c \neq 0$, we could arrange to have $c = 1$, as before; but we shall find it more convenient to use a different normalization, namely, to multiply all of a, b, c, d (if necessary) by such a number as to make the revised coefficients satisfy $ad - bc = 1$ [Ford 1, p. 14]. Then we have, in the notation of *continued fractions*, the identity

$$\frac{az + b}{cz + d} = ac^{-1} + \frac{1}{c +} \frac{1}{-c^{-1} +} \frac{1}{c +} \frac{1}{c^{-1}d + z},$$

which, of course, continues to hold when we replace z by $\bar{z}$ on both sides. Thus the homography

$$z_9 = \frac{az + b}{cz + d} \qquad (c \neq 0, \quad ad - bc = 1)$$

* *American Mathematical Monthly,* **51** (1944), p. 171.

may be expressed as the product of the nine simpler transformations

$$z_9 = ac^{-1} + z_8, \; z_8 = \frac{1}{z_7}, \; z_7 = c + z_6, \; z_6 = \frac{1}{z_5}, \; z_5 = -c^{-1} + z_4, \; z_4 = \frac{1}{z_3},$$

$$z_3 = c + z_2, \quad z_2 = \frac{1}{z_1}, \quad z_1 = c^{-1}d + z,$$

which are alternately translations $z' = b + z$ and Möbius involutions of the special form $z'' = 1/z$: the product of the inversion $z'' = 1/\bar{z}'$ and the reflection $z' = \bar{z}$. (The number of steps could be reduced from nine to four by using the dilative rotation $z' = c^{-2}z$; but it is interesting to observe that this more complicated transformation is itself a product of translations and "horizontal" Möbius involutions.) For the *antihomography*

$$z' = \frac{a\bar{z} + b}{c\bar{z} + d}$$

we proceed in the same way with one further reflection $z' = \bar{z}$. Since all these are circle-preserving transformations, the desired result follows.

The more powerful methods of the theory of functions of a complex variable enable us to prove [Ford **1**, pp. 3, 15] that every angle-preserving transformation of the whole inversive plane is of the form 9.72. This shows that angle-preserving transformations and circle-preserving transformations are synonymous.

EXERCISES

1. When $|a| = 1$ and $a \neq 1$, the transformation $z' = az + b$ is a rotation. Find its angle.

2. When $|a| \neq 1$, the transformation $z' = a\bar{z} + b$ is a dilative reflection. What angle does its axis make with the x-axis?

10

The five Platonic solids

We saw, in §4.6, that the Euclidean plane can be filled with squares, four at each vertex. If we try to fit squares together with only three at each vertex, we find that the figure closes as soon as we have used six squares, and we have a cube $\{4, 3\}$. Similarly, we can fill the plane with equilateral triangles, six at each vertex, and it is interesting to see what happens if we use three, four, or five instead of six. Another possibility is to use pentagons, three at each vertex, in accordance with the symbol $\{5, 3\}$.

With the possible exception of spheres, such *polyhedra* are the simplest solid figures. They provide an easy approach to the subject of topology as well as an interesting exercise in trigonometry. They can be defined and generalized in various ways [see, e.g., Hilbert and Cohn-Vossen **1,** p. 290].

10.1 PYRAMIDS, PRISMS, AND ANTIPRISMS

Although a Discourse of Solid Bodies be an uncommon and neglected Part of Geometry, yet that it is no inconsiderable or unprofitable Improvement of the Science will (no doubt) be readily granted by such, whose Genius tends as well to the Practical as Speculative Parts of it, for whom this is chiefly intended.

Abraham Sharp (1651 -1742)

(Geometry Improv'd, London, 1717, p. 65)

A *convex polygon* (such as $\{n\}$, where n is an integer) may be described as a finite region of a plane "enclosed" by a finite number of lines, in the sense that its interior lies entirely on one side of each line. Analogously, a *convex polyhedron* is a finite region of space enclosed by a finite number of planes [Coxeter **1,** p. 4]. The part of each plane that is cut off by other planes is a polygon that we call a *face*. Any common side of two faces is an *edge*.

The most familiar polyhedra are *pyramids* and *prisms*. We shall be concerned solely with "right regular" pyramids whose faces consist of a regular n-gon and n isosceles triangles, and with "right regular" prisms whose faces consist of two regular n-gons connected by n rectangles (so that there are two rectangles and one n-gon at each vertex). The height of such a prism can always be adjusted so that the rectangles become squares, and then we have an instance of a *uniform* polyhedron: all the faces are regular polygons and all the vertices are surrounded alike [Ball **1**, p. 135]. When $n = 4$, the prism is a *cube*, which is not merely uniform but *regular*: the faces are all alike, the edges are all alike, and the vertices are all alike. (The phrase "all alike" can be made precise with the aid of the theory of groups. We mean that there is a symmetry operation that will transform any face, edge, or vertex into any other face, edge, or vertex.)

The height of an n-gonal pyramid can sometimes be adjusted so that the isosceles triangles become equilateral. In fact, this can be done when $n < 6$; but six equilateral triangles fall flat into a plane instead of forming a solid angle. A triangular pyramid is called a *tetrahedron*. If three, and therefore all four, faces are equilateral, the tetrahedron is *regular*.

By slightly distorting an n-gonal prism we obtain an n-gonal *antiprism* (or "prismatoid," or "prismoid"), whose faces consist of two regular n-gons connected by $2n$ isosceles triangles. The height of such an antiprism can always be adjusted so that the isosceles triangles become equilateral, and then we have a uniform polyhedron with three triangles and an n-gon at each vertex. When $n = 3$, the antiprism is the regular *octahedron*. When $n = 5$, we can combine it with two pentagonal pyramids, one on each "base," to form the regular *icosahedron* [Coxeter **1**, p. 5]. A pair of icosahedral dice of the Ptolemaic dynasty can be seen in one of the Egyptian rooms of the British Museum in London.

We have now constructed four of the five convex regular polyhedra, namely those regarded by Plato as symbolizing the four elements: earth, fire, air, and water. The discrepancy between four elements and five solids did not upset Plato's scheme. He described the fifth as a shape that envelops the whole universe. Later it became the *quintessence* of the medieval alchemists. A model of this regular *dodecahedron* can be made by fitting together two "bowls," each consisting of a pentagon surrounded by five other pentagons. The two bowls will actually fit together because their free edges form a skew decagon like that formed by the lateral edges of a pentagonal antiprism (with isosceles lateral faces). Steinhaus [**2**, pp. 161–163] described a very neat method for building up such a model. From a sheet of cardboard, cut out two *nets* like Figure 10.1a, one for each bowl. Run a blunt knife along the five sides of the central pentagon so as to make them into hinged edges. Place one net crosswise on the other, with the scored edges outward, and bind them by running an elastic band alternately above and below the corners of the double star, holding the model flat with one hand. Removing the hand so as to allow the central pentagons to move away from each other, we see the dodecahedron rising as a perfect model (Figure 10.1b).

The most elementary properties of the five Platonic solids are collected

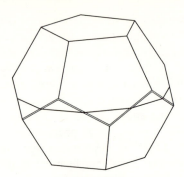

Figure 10.1a   **Figure 10.1b**

in Table II on p. 413. Each polyhedron is characterized by a Schläfli symbol $\{p, q\}$, which means that it has p-gonal faces, q at each vertex. The numbers of vertices, edges, and faces are denoted by V, E, and F. They can easily be counted in each case, but their significance will become clearer when we have expressed them as functions of p and q. We shall also obtain an expression for the *dihedral angle,* which is the angle between the planes of two adjacent faces.

EXERCISES

1. Give an alternative description of the octahedron (as a dipyramid).

2. Describe a solid having five vertices and six triangular faces.

3. Describe the following sections: (i) of a regular tetrahedron by the plane midway between two opposite edges, (ii) of a cube by the plane midway between two opposite vertices, (iii) of a dodecahedron by the plane midway between two opposite faces.

4. Six congruent rhombi, with angles 60° and 120°, will fit together to form a *rhombohedron* ("distorted cube"). From the two opposite "acute" corners of this solid, regular tetrahedra can be cut off in such a way that what remains is an octahedron. In other words, two tetrahedra and an octahedron can be fitted together to form a rhombohedron. Deduce that the tetrahedron and the octahedron have supplementary dihedral angles, and that infinitely many specimens of these two solids can be fitted together to fill the whole Euclidean space [Ball **1,** p. 147].

10.2   DRAWINGS AND MODELS

> You boil it in sawdust: you salt it in glue:
> You condense it with locusts and tape:
> Still keeping one principal object in view—
> To preserve its symmetrical shape.
>
> Lewis Carroll
>
> [Dodgson **2a,** Fit 5]

Leonardo da Vinci made skeletal models of polyhedra, using strips of wood for their edges and leaving the faces to be imagined [Pacioli **1**]. When

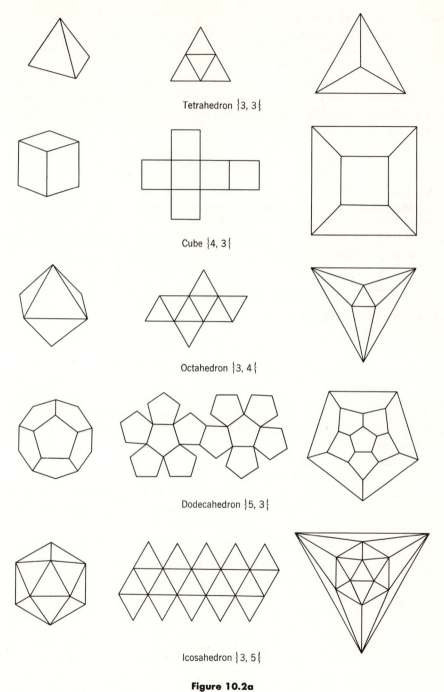

Tetrahedron {3, 3}

Cube {4, 3}

Octahedron {3, 4}

Dodecahedron {5, 3}

Icosahedron {3, 5}

Figure 10.2a

such a model is seen in perspective from a position just outside the center of one face, this face appears as a large polygon with all the remaining faces filling its interior. Such a drawing of the solid is called a *Schlegel diagram* [Hilbert and Cohn-Vossen **1**, pp. 145–146].

Figure 10.2*a* shows each of the Platonic solids in three aspects: an ordinary perspective view, a *net* which can be folded to make a cardboard model, and a Schlegel diagram. Each can be checked by observing the nature of a face and the arrangement of faces at a vertex.

EXERCISES

1. Sketch a Schlegel diagram for a pentagonal antiprism.

2. What is the smallest number of acute-angled triangles into which a given obtuse-angled triangle can be dissected? (F. W. Levi.†)

3. What is the smallest number of acute-angled triangles into which a square can be dissected? (Martin Gardner.*)

10.3 EULER'S FORMULA

Euler . . . overlooked nothing in the mathematics of his age, totally blind though he was for the last seventeen years of his life.

E. T. Bell [**2**, p. 330]

The Schlegel diagram for a polyhedron shows at a glance which vertices belong to which edges and faces. Each face appears as a region bounded by edges, except the "initial" face, which encloses all the others. To ensure a one-to-one correspondence between faces and regions we merely have to associate the initial face with the infinite exterior region.

Any polyhedron that can be represented by a Schlegel diagram is said to be *simply connected* or "Eulerian," because its numerical properties satisfy Euler's formula

$$V - E + F = 2$$

[Hilbert and Cohn-Vossen **1**, p. 290], which is valid not only for the Schlegel diagram of such a polyhedron, but for any connected "map" formed by a finite number of points and line segments decomposing a plane into nonoverlapping regions: the only restriction is that there must be at least one vertex!

A proof resembling Euler's may be expressed as follows. Any connected map can be built up, edge by edge, from the primitive map that consists of a single isolated vertex. At each stage, the new edge either joins an old vertex to a new vertex, as in Figure 10.3*a*, or joins two old vertices, as in Figure 10.3*b*. In the former case, V and E are each increased by 1 while

† *Mathematics Student,* **14** (1946).
* *Scientific American,* **202** (1960), p. 178.

F is unchanged; in the latter, *V* is unchanged while *E* and *F* are each increased by 1. In either case, the combination $V - E + F$ is unchanged. At the beginning, when there is only one vertex and one region (namely, all the rest of the plane), we have

$$V - E + F = 1 - 0 + 1 = 2.$$

This value 2 is maintained throughout the whole construction. Thus, Euler's formula holds for every plane map. In particular, it holds for every Schlegel diagram, and so for every simply connected polyhedron. (For another proof, due to von Staudt, see Rademacher and Toeplitz [**1**, pp. 75–76].)

Figure 10.3a

Figure 10.3b

In the case of the regular polyhedron $\{p, q\}$, the numerical properties satisfy the further relations

10.31 $$qV = 2E = pF.$$

In fact, if we count the *q* edges at each of the *V* vertices, we have counted every edge twice: once from each end. A similar situation arises if we count the *p* sides of each of the *F* faces, since every edge belongs to two faces.

We now have enough information to deduce expressions for *V*, *E*, *F* as functions of *p* and *q*. In fact,

$$\frac{V}{\frac{1}{q}} = \frac{E}{\frac{1}{2}} = \frac{F}{\frac{1}{p}} = \frac{V - E + F}{\frac{1}{q} - \frac{1}{2} + \frac{1}{p}} = \frac{2}{\frac{1}{q} + \frac{1}{p} - \frac{1}{2}} = \frac{4pq}{2p + 2q - pq},$$

whence

10.32

$$V = \frac{4p}{2p + 2q - pq}, \quad E = \frac{2pq}{2p + 2q - pq}, \quad F = \frac{4q}{2p + 2q - pq}.$$

Since these numbers must be positive, the possible values of *p* and *q* are restricted by the inequality $2p + 2q - pq > 0$ or

10.33 $$(p - 2)(q - 2) < 4.$$

Thus $p - 2$ and $q - 2$ are two positive integers whose product is less than 4, namely,

$$1 \cdot 1 \quad \text{or} \quad 2 \cdot 1 \quad \text{or} \quad 1 \cdot 2 \quad \text{or} \quad 3 \cdot 1 \quad \text{or} \quad 1 \cdot 3.$$

These five possibilities provide a simple proof of Euclid's assertion [Rademacher and Toeplitz **1**, pp. 84–87]:

There are just five convex regular polyhedra:

$$\{3, 3\}, \quad \{4, 3\}, \quad \{3, 4\}, \quad \{5, 3\}, \quad \{3, 5\}.$$

The inequality 10.33 is not merely a necessary condition for the existence of $\{p, q\}$ but also a sufficient condition; for in § 10.1 we saw how to construct a solid corresponding to each solution.

The same inequality arises in a more elementary manner when we construct a model of the polyhedron from its net. At a vertex we have q p-gons, each contributing an angle

$$\left(1 - \frac{2}{p}\right)\pi.$$

In order to form a solid angle, these q face angles must make a total less than 2π. Thus

$$q\left(1 - \frac{2}{p}\right)\pi < 2\pi$$

whence, as before, $(p - 2)(q - 2) < 4$.

Any maker of models soon observes that the amount by which the sum of the face angles at a vertex falls short of 2π is smaller for a complicated solid like the dodecahedron than for a simple one like the tetrahedron. Descartes proved that if this amount, say δ, is the same at every vertex, it is actually equal to $4\pi/V$ [Brückner **1**, p. 60]. In the case of $\{p, q\}$, this is an immediate consequence of the formula 10.32 for V, which yields

$$\frac{4\pi}{V} = (2p + 2q - pq)\frac{\pi}{p} = 2\pi - q\left(1 - \frac{2}{p}\right)\pi.$$

EXERCISES

1. The number of edges of $\{p, q\}$ is given by

$$E^{-1} = p^{-1} + q^{-1} - \tfrac{1}{2}.$$

2. Consider an arbitrary polyhedron having p-gonal faces for various values of p, and q faces at a vertex for various values of q. Generalize the equations 10.31 in the form

$$\Sigma q = 2E = \Sigma p,$$

where the first summation is taken over all the vertices and the last over all the faces. Deduce that every polyhedron has either a face with $p = 3$ or a vertex with $q = 3$ (or both). (*Hint:* If not, we would have $\Sigma q \geqslant 4V$ and $\Sigma p \geqslant 4F$.)

3. If the faces are all alike and the edges are all alike and the vertices are all alike, the faces are regular. Show by an example that this result for polyhedra is not valid for tessellations.

10.4   RADII AND ANGLES

A solid model of $\{p, q\}$ can evidently be built from F p-gonal pyramids of suitable altitude, placed together at their common apex, which is the center O_3 of the polyhedron.   This point O_3 is the common center of three spheres: the *circumsphere* which passes through all the vertices, the *midsphere* which touches all the edges at their midpoints, and the *insphere* which touches all the faces at their centers.   The *circumradius* $_0R$ appears as a lateral edge of any one of the pyramids (Figure 10.4a), the *midradius* $_1R$ as the altitude of a lateral face, and the *inradius* $_2R$ as the altitude of the whole pyramid.

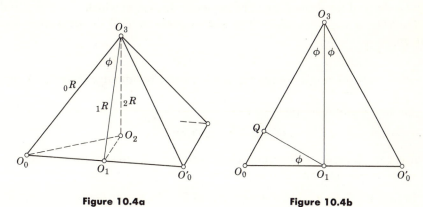

Figure 10.4a   **Figure 10.4b**

Such a p-gonal pyramid has p planes of symmetry (or "mirrors") which join its apex O_3 to the p lines of symmetry of its base.   By means of these p planes, the solid pyramid is dissected into $2p$ congruent (irregular) tetrahedra of a very special kind.   Let $O_0O_1O_2O_3$ (Figure 10.4c) be such a tetrahedron, so that O_0 is a vertex of the polyhedron, O_1 the midpoint of an edge $O_0O'_0$, O_2 the center of a face, and O_3 the center of the whole solid.   (The

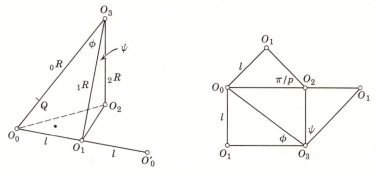

Figure 10.4c

net has been drawn to scale for the case of the cube $\{4, 3\}$, in which $O_0O_1 = O_1O_2 = O_2O_3$.) Since the plane $O_1O_2O_3$ perpendicularly bisects the edge $O_0O'_0$, O_0O_1 is perpendicular to both O_1O_2 and O_1O_3. Since $O_0O_1O_2$ is the plane of a face, the inradius O_3O_2 is perpendicular to both O_0O_2 and O_1O_2. Thus the three lines O_0O_1, O_1O_2, O_2O_3 are mutually perpendicular and the tetrahedron is "quadrirectangular": all four faces are right-angled triangles. Schläfli called such a tetrahedron an *orthoscheme* [Coxeter **1**, p. 137].

Many relations involving the radii

$$_0R = O_0O_3, \quad _1R = O_1O_3, \quad _2R = O_2O_3$$

can be derived from the four right-angled triangles, in which $O_0O_1 = l$ and $\angle O_0O_2O_1 = \pi/p$. But the whole story cannot be told till we have found the angle

$$\phi = \angle O_0O_3O_1$$

which is half the angle subtended at the center by an edge [Coxeter **1**, pp. 21–22].

Another significant angle is

$$\psi = \angle O_1O_3O_2,$$

whose complement, $\angle O_2O_1O_3$, is half the dihedral angle of the polyhedron. In other words, the dihedral angle is $\pi - 2\psi$.

In seeking these angles, it is useful to define the *vertex figure* of $\{p, q\}$: the polygon formed by the midpoints of the q edges at a vertex O_0. This is indeed a plane polygon, since its vertices lie on the circle of intersection of two spheres: the midsphere (with center O_3 and radius $_1R = O_3O_1$), and the sphere with center O_0 and radius $l = O_0O_1$. We see from 2.84 that the vertex figure of $\{p, q\}$ is a $\{q\}$ of side

$$2l \cos \frac{\pi}{p}.$$

Since its plane is perpendicular to O_3O_0, its center Q is the foot of the perpendicular from O_1 to O_3O_0 (Figure 10.4*b*), and its circumradius is

$$QO_1 = l \cos \phi.$$

By 2.81 (with $l \cos \pi/p$ for l), this circumradius is

$$l \cos \frac{\pi}{p} \csc \frac{\pi}{q}.$$

Hence

10.41 $$\cos \phi = \cos \frac{\pi}{p} \csc \frac{\pi}{q} = \cos \frac{\pi}{p} \Big/ \sin \frac{\pi}{q}$$

[Coxeter **1**, p. 21].

The right-angled triangles in Figure 10.4c now yield

$$_0R = l \csc \phi, \quad _1R = l \cot \phi,$$

$$_2R^2 = {_1R^2} - \left(l \cot \frac{\pi}{p} \right)^2, \quad \cos \psi = \frac{_2R}{_1R}.$$

In order to eliminate ϕ, it is convenient to introduce the temporary abbreviation

$$k^2 = \sin^2 \frac{\pi}{q} - \cos^2 \frac{\pi}{p} = \sin^2 \frac{\pi}{p} - \cos^2 \frac{\pi}{q},$$

so that $\sin \phi = k \csc \pi/q$. Then

$$_0R = \frac{l}{k} \sin \frac{\pi}{q}, \qquad _1R = \frac{l}{k} \cos \frac{\pi}{p},$$

10.42
$$_2R = \frac{l}{k} \cot \frac{\pi}{p} \cos \frac{\pi}{q},$$

10.43
$$\cos \psi = \cos \frac{\pi}{q} \Big/ \sin \frac{\pi}{p}.$$

This last result enables us to compute the dihedral angle

$$\pi - 2\psi = 2 \arcsin \left(\cos \frac{\pi}{q} \Big/ \sin \frac{\pi}{p} \right).$$

EXERCISES

1. Verify that $k = \sin \pi/2c$, where $c = (2 + p + q)/(10 - p - q)$ [Coxeter **4**, p. 753], and that $E = c(c + 1)$.

2. Check the values of the dihedral angle given in Table II on p. 413. (Your calculations should agree with the observation that the dihedral angles of the tetrahedron and octahedron are supplementary. See Ex. 4 at the end of § 10.1.)

3. If a polyhedron has a circumsphere and a midsphere and an insphere, and if these three spheres are concentric, then the polyhedron is regular.

10.5 RECIPROCAL POLYHEDRA

The Platonic solid $\{p, q\}$ has a *reciprocal*, which may be defined as the polyhedron enclosed by a certain set of V planes, namely, the planes of the vertex figures at the V vertices of $\{p, q\}$. Clearly, its edges bisect the edges of $\{p, q\}$ at right angles. Among these E edges, those which bisect the p sides of a face of $\{p, q\}$ all pass through a vertex of the reciprocal, and those which bisect the q edges at a vertex $\{p, q\}$ form a face of the reciprocal. Thus

The reciprocal of $\{p, q\}$ is $\{q, p\}$,

and vice versa.   There is a vertex of either for each face of the other; in fact, the centers of the faces of $\{p, q\}$ are the vertices of a smaller version of the reciprocal $\{q, p\}$ [Steinhaus **1**, pp. 72–79].

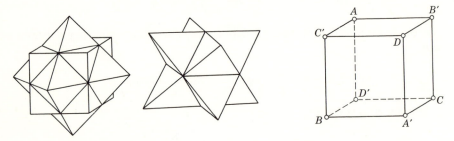

Figure 10.5a   **Figure 10.5b**

Figure 10.5a shows how the octahedron $\{3, 4\}$ arises as the reciprocal of the cube $\{4, 3\}$ (or vice versa), and how the reciprocal of a regular tetrahedron $\{3, 3\}$ is an equal tetrahedron.   The combination of two reciprocal tetrahedra, $ABCD$ and $A'B'C'D'$, occurs in nature as a crystal twin.   Pacioli [**1**, Plates XIX, XX] named it *octaedron elevatum*.   Kepler, rediscovering it a hundred years later, called it simply *stella octangula*.   The twelve edges of the two tetrahedra are the diagonals of the six faces of a cube (Figure 10.5b).

By interchanging p and q in the formula 10.32 for V (or F) we obtain the formula for F (or V).   Similarly, since Q, on O_3O_0 (Figure 10.4b), is the center of a face of $\{q, p\}$, the angular property ϕ of $\{p, q\}$ is equal to the angular property ψ of $\{q, p\}$, and therefore the expression 10.43 for the property ψ of $\{p, q\}$ could have been derived from 10.41 by the simple device of interchanging p and q.

EXERCISES

1.   A cube of edge 1, with one vertex at the origin and three edges along the Cartesian axes, has the eight vertices (x, y, z), where each of the three coordinates is either 0 or 1, independently.

2.   A cube of edge 2, with its center at the origin and its edges parallel to the Cartesian axes, has the eight vertices

$$(\pm 1, \pm 1, \pm 1).$$

3.   Where is the center of the dilatation that relates the cubes described in the two preceding exercises?

4.   Obtain coordinates for the vertices of a regular tetrahedron by selecting alternate vertices of a cube.   Find the equations of the face planes and compute the angle between two of them.

5.   Obtain coordinates for the vertices of an octahedron by locating the face centers of the cube in Ex. 2.   Find the equations of the face planes and compute the angle between two that contain a common edge.

6. The points (x, y, z) that belong to the solid octahedron are given by the inequality

$$|x| + |y| + |z| \leqslant 1.$$

7. If each edge of a regular tetrahedron is projected into an arc of a great circle on the circumsphere, at what angles do these arcs intersect? Find the equations of the planes of the six great circles.

11

The golden section and phyllotaxis

Euclid's construction for the regular pentagon depends on the division of a line segment in the ratio $\tau : 1$, where $\tau = (\sqrt{5} + 1)/2$. This algebraic number is the subject of a book published in 1509, in which Luca Pacioli describes its properties or "effects," stopping at thirteen "for the sake of our salvation." We shall see that, when it is expressed as a continued fraction [Ball **1**, pp. 55–56], all the partial quotients are 1, making it the simplest (and slowest to converge) of all infinite continued fractions. The convergents are found to be quotients of successive members of the sequence of Fibonacci numbers 1, 1, 2, 3, 5, 8, . . . , which any child can begin to write though no explicit formula was found till 1843. A subtle manifestation of this "divine proportion" in the structure of plants helps to explain the phenomenon of phyllotaxis, which is most clearly seen in the arrangement of cells on the surface of a pineapple.

11.1 EXTREME AND MEAN RATIO

> *I believe that this geometric proportion served the Creator as an idea when He introduced the continuous generation of similar objects from similar objects.*
>
> J. Kepler (1571 -1630)

As we see in Figure 2.8*b*, the vertex figure of a regular pentagon of side $2l$ is a line segment of length τl, where

11.11
$$\tau = 2 \cos \pi/5.$$

In other words, a pentagon $PQRST$ of side 1 (Figure 11.1*a*) has diagonals of length τ. Let the diagonals QS, RT intersect in U. They are said to divide each other according to the *golden section* (or "in extreme and mean ratio"). To see what this means, we observe that the homothetic isosceles triangles QTU, SRU yield

160

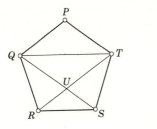

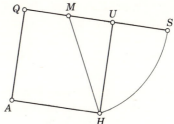

Figure 11.1a   **Figure 11.1b**

$$\frac{QU}{US} = \frac{QT}{RS} = \tau = \frac{QS}{PT} = \frac{QS}{QU}$$

(since PT and QU are opposite sides of the rhombus $PQUT$). Thus the diagonal QS is divided at U in such a way that *the ratio of the larger part to the smaller is equal to the ratio of the whole to the larger part.*

If $QU = PT = 1$, so that $QS = \tau$, then $US = \tau^{-1}$ and

$$1 + \tau^{-1} = \tau.$$

Hence τ is the positive root of the quadratic equation

11.12 $\tau^2 - \tau - 1 = 0,$

namely, $\tau = \dfrac{\sqrt{5} + 1}{2} = 1.6180339887\ldots.$

Figure 11.1*b* shows an easy way to extend a given segment QU to S so that $QS = \tau QU$. (Here M is the midpoint of the side QU of the square $AQUH$, and $MS = MH = \sqrt{5}\, MU$.) Euclid (IV.10) used such a procedure "to construct an isosceles triangle having each of the angles at the base double the remaining one" (QST or PRS in Figure 11.1*a*).

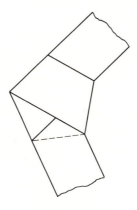

Figure 11.1c

The figure of a pentagon with diagonals can be neatly displayed by tying a simple knot in a long strip of paper and carefully pressing it flat (Figure 11.1*c*).

EXERCISES

1. Defining τ to be the ratio of the diagonal of a regular pentagon to its side, establish 11.11 by applying 1.54 to the isosceles triangle *PRS*.

2. Show how one further setting of the compasses will yield a point (in Figure 11.1*b*) dividing the given segment *QU* in the ratio $\tau : 1$.

11.2   DE DIVINA PROPORTIONE

> *Del suo secondo essentiale effecto . . . Del terço suo singulare effecto . . . Del quarto suo ineffabile effecto . . . . . . Del .10. suo supremo effecto . . . Del suo .11. excellentissimo effecto . . . Del suo .12. quasi incomprehensibile effecto . . .*
>
> Luca Pacioli (ca. 1445 -1509)
> [Pacioli **1**, pp. 6 -7]

Under the beneficent influence of the artist Piero della Francesca (ca. 1416–1492), Fra Luca Pacioli (or *Paccioli*) wrote a book about τ, called *De divina proportione,* which he illustrated with drawings of models made by his friend Leonardo da Vinci.   His enthusiasm for the subject is apparent in the above titles that he chose for various chapters.   (It is interesting to observe how closely some of his old Italian resembles English.)

"The seventh inestimable effect" is the occurrence of τ as the circumradius of a regular decagon of side 1.   (We can thus inscribe a pentagon in a given circle by first inscribing a decagon and then picking out alternate vertices.)   "The ninth effect, the best of all" is that two crossing diagonals of a regular pentagon divide one another in extreme and mean ratio.   "The twelfth almost incomprehensible effect" is the following property of the regular icosahedron $\{3, 5\}$.

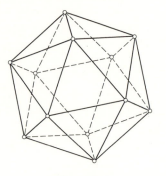

Figure 11.2a

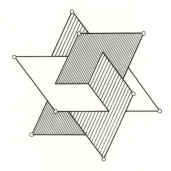

Figure 11.2b

The faces surrounding a vertex of the icosahedron belong to a pyramid whose base is a regular pentagon (similar to the vertex figure). Any two opposite edges of the icosahedron belong to a rectangle whose longer sides are diagonals of such pentagons. Since the diagonal of a pentagon is τ times its side, this rectangle is a *golden* rectangle, whose sides are in the ratio $\tau : 1$. In fact, the twelve vertices of the icosahedron (Figure 11.2*a*) are the twelve vertices of three golden rectangles in mutually perpendicular planes (Figure 11.2*b*). A model is easily made from three ordinary postcards (which are nearly golden rectangles). In the middle of each card, cut a slit parallel to the long sides and equal in length to the short sides. For a practical reason, the slit in one of the cards must be continued right to the edge. Then the cards can easily be put together so that each passes through the middle of another, in cyclic order.

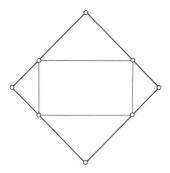

Figure 11.2c

We see from Figure 11.2*c* that a golden rectangle can be inscribed in a square so that each vertex of the rectangle divides a side of the square in the ratio $\tau : 1$. Identifying this with one of the three "equatorial" squares of a regular octahedron, we deduce that an icosahedron can be inscribed in an octahedron so that each vertex of the icosahedron divides an edge of the octahedron in the ratio $\tau : 1$.

EXERCISES

1. Using Cartesian coordinates referred to the planes of the three golden rectangles, obtain the 12 vertices of the icosahedron in the form

$$(0, \pm\tau, \pm1), \quad (\pm1, 0, \pm\tau), \quad (\pm\tau \pm1, 0).$$

2. These 12 points divide the 12 edges of the octahedron

$$(\pm\tau^2, 0, 0), \quad (0, \pm\tau^2, 0), \quad (0, 0, \pm\tau^2)$$

in the ratio $\tau : 1$.

3. Obtain coordinates for the 20 vertices of the regular dodecahedron [Coxeter **1,** p. 53].

11.3 THE GOLDEN SPIRAL

Archimedes, Leonardo, Newton—all very practical men, but with something more. A sense of wonder, perhaps . . . Very mysterious, very mysterious. And very exciting. That is the thing—excitement. You get it by being God, by constructing, and you get it by watching God, by observing things as they are.

J. L. Synge [**2,** p. 163]

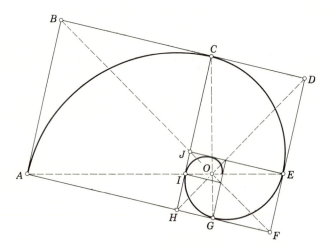

Figure 11.3a

The relation $\tau = 1 + 1/\tau$ shows that the golden rectangle $ABDF$ (Figure 11.3a) can be dissected into two pieces: a square $ABCH$ and a smaller golden rectangle. (The points B, C, D may be identified with the Q, U, S of Figure 11.1b.) From the smaller rectangle $CDFH$ we can cut off another square, leaving a still smaller rectangle, and continue the process indefinitely.

Since the rectangle $HJEF$ is homothetic to $ABDF$, the vertex J of the former lies on the diagonal BF of the latter. In fact, the lines AE, BF, CG, DH together contain all the vertices of all the rectangles. To derive $CDFH$ from $ABDF$, we may use a dilative rotation whose center is O, the point of intersection $BF \cdot DH$. This similarity, which transforms each of the points A, C, E, G, I, . . . into the next, and each of the points B, D, F, H, J, . . . into the next, is the product of a negative quarter-turn about O and the dilatation $O(\tau^{-1})$. Therefore DH is perpendicular to BF. The inverse similarity, which transforms each rectangle into the next larger one, is the product of a positive quarter-turn about O and the dilatation $O(\tau)$.

Since

$$\frac{OB}{OD} = \tau = \frac{BC}{CD},$$

OC bisects the (right) angle BOD. Thus the lines CG, AE pass through O and bisect the angles between BF, DH.

In terms of polar coordinates with pole O, the dilative rotation that transforms OE into OC transforms any point (r, θ) into $(\tau r, \theta + \frac{1}{2}\pi)$. Taking OE as the initial line, and the length OE as the unit of measurement, so that E is $(1, 0)$, we deduce the coordinates $(\tau, \frac{1}{2}\pi)$ for C, (τ^2, π) for A, $(\tau^3, \frac{3}{2}\pi)$ for the vertex opposite to A in a new square placed on AF, and so on. Similarly, G is $(\tau^{-1}, -\frac{1}{2}\pi)$, I is $(\tau^{-2}, -\pi)$ or (τ^{-2}, π), and so on. We thus obtain an infinite sequence of points . . . , I, G, E, C, A, . . . whose polar coordinates

$$r = \tau^n, \quad \theta = \tfrac{1}{2}\pi n$$

satisfy the equation $r = \tau^{2\theta/\pi}$. Hence all these points lie on the equiangular spiral

$$r = \mu^\theta,$$

where $\mu = \tau^{2/\pi}$. This true spiral [Cundy and Rollett **1**, p. 64] is closely approximated by the artificial spiral formed by circular quadrants inscribed in the successive squares, as in Figure 11.3a. (But the true spiral cuts the sides of the squares at very small angles, instead of touching them.)

EXERCISES

1. Verify the conclusion of §5.4 as applied to the similar rectangles $ABDF$ and $CDFH$ in Figure 11.3a.

2. Obtain polar coordinates for the points J, H, F, D, B. They lie on another equiangular spiral which is both congruent (by rotation through what angle?) and homothetic (by what ratio of magnification?) to the spiral $IGECA$.

11.4 THE FIBONACCI NUMBERS

> *Sint minimi 1 et 1 quos imaginaberis inaequales. Adde, fient 2. cui adde maiorem 1 fient 3. cui adde 2 fient 5. cui adde 3 fient 8. cui adde 5 fient 13. cui adde 8 fient 21.*
>
> J. Kepler [**1**, p. 270]

In 1202, Leonardo of Pisa, nicknamed Fibonacci ("son of good nature"), came across his celebrated sequence of integers f_n in connection with the breeding of rabbits.* He assumed that rabbits live forever, and that every month each pair begets a new pair which becomes productive at the age of two months. In the first month the experiment begins with a newborn pair of rabbits. In the second month there is still just one pair. In the third month there are 2; in the fourth, 3; in the fifth, 5; and so on. Let f_n denote the number of pairs of rabbits in the nth month. The first few values (and their successive ratios) may be tabulated as follows:

---

* R. C. Archibald, Golden Section, *American Mathematical Monthly*, **25** (1918), pp. 232–238.

n:	0	1	2	3	4	5	6	7	8	9	10	...
f_n:	0	1	1	2	3	5	8	13	21	34	55	...
f_{n+1}/f_n:	∞	1	2	1.5	1.6	1.6	1.625	1.6154	1.6190	1.6176	1.6182	...

Four centuries later, Kepler stated explicitly what Fibonacci must surely have noticed: that the sum of any two f's is equal to the next, so that the sequence is determined by the recursion formula

11.41
$$f_0 = 0, \quad f_1 = 1, \quad f_k + f_{k+1} = f_{k+2}.$$

Kepler observed also that the ratios $f_n : f_{n+1}$ approach $1 : \tau$ more and more closely as n increases. (See the third line of the above table.) However, another hundred years passed before R. Simson (1687–1768, who was not really the discoverer of the "Simson line") recognized f_{n+1}/f_n as being the nth convergent of the continued fraction

$$1 + \frac{1}{1+} \frac{1}{1+} \frac{1}{1+} \cdots .$$

(The "convergents" are the numbers

$$1, \quad 1 + 1 = 2, \quad 1 + \frac{1}{1+1} = \frac{3}{2}, \quad 1 + \frac{1}{1+} \frac{1}{1+1} = \frac{5}{3},$$

and so on.) To prove that $\lim (f_{n+1}/f_n) = \tau$, Simson made repeated use of the relation $\tau = 1 + 1/\tau$, as follows:

$$\tau = 1 + \frac{1}{\tau} = 1 + \frac{1}{1+} \frac{1}{\tau} = 1 + \frac{1}{1+} \frac{1}{1+} \frac{1}{\tau} = \cdots$$

Along with f_n, É. Lucas (1842–1891) considered the related numbers g_n defined by

$$g_0 = 2, \quad g_1 = 1, \quad g_k + g_{k+1} = g_{k+2}.$$

The first few values are as follows:

n:	0	1	2	3	4	5	6	7	8	9	10	...
g_n:	2	1	3	4	7	11	18	29	47	76	123	...

It is easily proved by induction that (for $n > 0$)

11.42
$$g_n = f_{n-1} + f_{n+1}.$$

Leaving this as an exercise for the reader, we proceed to establish Lucas's identities

11.43
$$f_{2n} = f_n g_n,$$

11.44
$$f_{2n+1} = f_n^2 + f_{n+1}^2,$$

which are obvious when $n = 0$ or 1. To prove them both by induction, we add the tentative relations

$$f_{2k-1} = f_{k-1}^2 + f_k^2 \quad \text{and} \quad f_{2k} = f_k g_k,$$

obtaining
$$f_{2k+1} = f_{k-1}^2 + f_k(f_{k-1} + f_{k+1}) + f_k^2$$
$$= (f_{k-1} + f_k)f_{k+1} + f_k^2$$
$$= f_k^2 + f_{k+1}^2,$$

and then add

$$f_{2k} = f_k g_k \quad \text{and} \quad f_{2k+1} = f_k^2 + f_{k+1}^2,$$

obtaining
$$f_{2k+2} = f_k^2 + f_k(f_{k-1} + f_{k+1}) + f_{k+1}^2$$
$$= f_k f_{k+1} + f_{k+1} f_{k+2}$$
$$= f_{k+1} g_{k+1}.$$

Similarly, to establish the identity

11.45
$$\tau^n = f_n \tau + f_{n-1},$$

which is obvious when $n = 1$ or 2, we add

$$\tau^k = f_k \tau + f_{k-1} \quad \text{and} \quad \tau^{k+1} = f_{k+1} \tau + f_k,$$

obtaining
$$\tau^{k+2} = f_{k+2} \tau + f_{k+1}.$$

The identity 11.45 continues to hold when n is negative, provided we define $f_{-k} = f_{-k+2} - f_{-k+1}$ (for $k > 0$), so that

$$f_{-k} = (-1)^{k+1} f_k.$$

Thus
$$\tau^{-k} = f_{-k} \tau + f_{-k-1}$$

11.46
$$= (-1)^{k+1}(f_k \tau - f_{k+1}).$$

and

11.47
$$(-\tau)^{-k} = f_{k+1} - f_k \tau$$

11.471
$$= f_{k-1} - f_k \tau^{-1}.$$

Adding 11.45 (with k for n) and 11.47, we obtain

11.48
$$g_k = \tau^k + (-\tau)^{-k}.$$

Similarly, subtracting 11.471 from 11.45 (with k for n), we obtain

11.49
$$f_k = \frac{\tau^k - (-\tau)^{-k}}{\tau + \tau^{-1}} = \frac{\tau^k - (-\tau)^{-k}}{\sqrt{5}},$$

an explicit formula which was discovered by J. P. M. Binet in 1843.

From 11.48 and 11.49 we immediately deduce

$$\tau^k = \frac{g_k + f_k \sqrt{5}}{2}, \quad (-\tau)^{-k} = \frac{g_k - f_k \sqrt{5}}{2};$$

for example,

$$\tau^3 = \sqrt{5} + 2, \qquad\qquad \tau^{-3} = \sqrt{5} - 2,$$
$$\tau^6 = 9 + 4\sqrt{5}, \qquad\qquad \tau^{-6} = 9 - 4\sqrt{5}.$$

EXERCISES

1. Express the sum of the finite series

$$1 + 1 + 2 + 3 + \ldots + f_n$$

in terms of f_{n+2}.

2. Prove Simson's identity

$$f_{n-1}\, f_{n+1} - f_n{}^2 = (-1)^n.$$

This is the basis for an amusing puzzle [Ball **1**, p. 85].*

3. Verify Lagrange's observation that the final digits of the Fibonacci numbers recur after a cycle of sixty:

$$1, 1, 2, 3, 5, 8, 3, 1, 4, \ldots, 7, 2, 9, 1, 0.$$

4. From the identity

$$(1 - t - t^2)(1 + t + 2t^2 + 3t^3 + \ldots + f_{n+1}\, t^n + \ldots) = 1,$$

which implies

$$\sum_0^\infty f_{n+1}\, t^n = \frac{1}{1 - t - t^2} = \sum_0^\infty (t + t^2)^k = \sum_0^\infty t^k (1 + t)^k = \sum_{k=0}^\infty \sum_{j=0}^k \binom{k}{j}\, t^{k+j},$$

deduce Lucas's formula for Fibonacci numbers in terms of binomial coefficients:

$$f_{n+1} = \binom{n}{0} + \binom{n-1}{1} + \binom{n-2}{2} + \ldots.$$

5. Setting $t = 0.01$ in the identity $\Sigma\, f_{n+1}\, t^n = (1 - t - t^2)^{-1}$, deduce an easy method for writing down the first 19 digits in the decimal for $\frac{10000}{9899}$.

6. Tabulate g_{n+1}/g_n and find its limit when n tends to infinity.

---

* See also Coxeter, The golden section, phyllotaxis, and Wythoff's game, *Scripta Mathematica*, **19** (1953), p. 139. More generally,

$$f_{n+h}\, f_{n+k} - f_n\, f_{n+h+k} = (-1)^n\, f_h\, f_k;$$

see A. E. Danese, *American Mathematical Monthly*, **67** (1960), p. 81.

11.5   PHYLLOTAXIS

> In the doctrine of Metamorphosis and the enunciation of the Spiral
> Theory we have . . . two remarkable generalizations which, originat-
> ing in the fertile imagination of Goethe, have passed through the chaos
> of Nature Philosophy and emerged . . . to form the groundwork of
> our present views of Plant morphology.
>
> A. H. Church (1865 -1937)
>
> [Church **1,** p. 1]

   The Fibonacci numbers have a botanical application in the phenomenon
called *phyllotaxis* (literally "leaf arrangement"). In some trees, such as the
elm and basswood, the leaves along a twig seem to occur alternately on
two opposite sides, and we speak of "$\frac{1}{2}$ phyllotaxis." In others, such as the
beech and hazel, the passage from one leaf to the next is given by a screw
displacement involving rotation through one-third of a turn, and we speak
of "$\frac{1}{3}$ phyllotaxis." Similarly, the oak and apricot exhibit $\frac{2}{5}$ phyllotaxis, the
poplar and pear $\frac{3}{8}$, the willow and almond $\frac{5}{13}$, and so on. We recognize
the fractions as quotients of alternate Fibonacci numbers, but consecutive
Fibonacci numbers could be used just as well; for example, a positive rotation through $\frac{5}{8}$ of a turn has the same effect as a negative rotation through
$\frac{3}{8}$ [Weyl **1,** p. 72].
   Another manifestation of phyllotaxis is the arrangement of the florets of
a sunflower, or of the scales of a fir cone, in spiral or helical whorls (or
"parastichies"). Such whorls are particularly evident in a pineapple (Figure 11.5a), whose more-or-less hexagonal cells are visibly arranged in rows
in various directions: 5 parallel rows sloping gently up to the right, 8 rows
sloping somewhat more steeply up to the left, and 13 rows sloping very
steeply up to the right. (Sometimes the sense is consistently reversed.) If
we regard the surface of the pineapple as a cylinder, cut it along a vertical
line (generator), and spread it out flat on a plane, we obtain a strip between
two parallel lines which represent two versions of the vertical cut. In Figure 11.5b these parallel lines are taken to be the lines $x = 0$ and $x = 1$,
and the hexagonal cells, numbered successively in the order of their distances
from the x-axis, appear as the Dirichlet regions of a lattice (§ 4.1). The lattice point marked 0 appears at the origin, and reappears at (1, 0), as if the
cylinder were rolled along the plane. The lattice point marked 1 has coordinates (τ^{-1}, h), where the height h remains to be computed; and for all
values of n the nth lattice point is (x, y) where $y = nh$ and x is the fractional part of $n\tau$ (or of $n\tau^{-1}$), namely,

$$x = n\tau - [n\tau],$$

where $[n\tau]$ is the integral part. The repetition of the strip has the effect that
the number n appears at the point

$$(n\tau + m, \, nh),$$

Figure 11.5a

for all integers n and m, and the numbers in any straight row are in arithmetic progression. When the plane is wrapped round the cylinder, such a straight line becomes a *helix*, the shape of the rail of a "spiral" staircase.

Since the ratios f_{k+1}/f_k of consecutive Fibonacci numbers converge towards τ, the number $f_k\tau$ is nearly an integer (namely f_{k+1}), that is, its fractional part is small. Therefore one of the points marked f_k is close to the y-axis (especially for large values of k), and the most nearly vertical rows are those in which the difference of the arithmetic progressions are Fibonacci numbers.

Such an arrangement can be made for any positive value of h. In Figure 11.5b the value $\frac{1}{150}$ has been chosen. We soon find by trial that this is about the right size to ensure that the neighbors of 0 are the Fibonacci numbers 5, 13, 8 and their negatives, as in a pineapple (Figure 11.5a). In other words, the most obvious rows (with the closest spacing) are those in which the arithmetic progression has difference $f_6 = 8$. The effect of increasing h would be to decrease the obtuse angle between the directions 05 and 08, so that at a certain stage this would become a right angle, and then the Dirichlet region would be a rectangle instead of a hexagon. Similarly, the effect of decreasing h would be to increase the angle between the directions of 8 and 13, so that at a certain stage this would become a right angle, and afterwards the next Fibonacci number 21 would appear as a neighbor of 0.

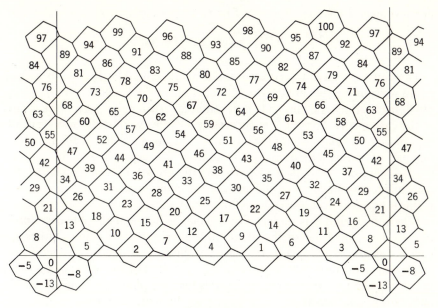

Figure 11.5b

To determine the critical values of h for which such transitions occur, we seek the condition for the points f_k and f_{k+1} to lie in perpendicular directions from 0, that is, the condition for the points

$$(f_k\tau - f_{k+1}, \; f_k h), \quad (f_{k+1}\tau - f_{k+2}, \; f_{k+1}h)$$

to lie in perpendicular directions from $(0, 0)$. By 8.22, this condition is

$$(f_k\tau - f_{k+1})(f_{k+1}\tau - f_{k+2}) + f_k f_{k+1}h^2 = 0.$$

Using 11.12, 11.41, 11.42, then 11.43, 11.44, and finally 11.46, we deduce

$$f_k f_{k+1}h^2 = (f_k^2 + f_{k+1}^2)\,\tau - f_{k+1}g_{k+1}$$
$$= f_{2k+1}\,\tau - f_{2k+2} = \tau^{-(2k+1)},$$

whence

11.51
$$h = (f_k f_{k+1})^{-\frac{1}{2}}\,\tau^{-(k+\frac{1}{2})}.$$

We conclude that the numbers of whorls in various directions (which are 5, 8, 13 in the case of the pineapple) are f_{k-1}, f_k, f_{k+1} for values of h between

$$(f_{k-1}f_k)^{-\frac{1}{2}}\,\tau^{-(k-\frac{1}{2})} \quad \text{and} \quad (f_k f_{k+1})^{-\frac{1}{2}}\,\tau^{-(k+\frac{1}{2})}.$$

As a convenient "standard" value between these critical values, we naturally choose

$$h = f_k^{-1}\,\tau^{-k}.$$

For instance, the value chosen in Figure 11.5b is approximately

$$\frac{\tau^{-6}}{f_6} = \frac{0.055727 \ldots}{8} = 0.006966 \ldots.$$

The same pattern, interpreted as an arrangement of leaves on a twig, could reasonably be called $\frac{5}{13}$ phyllotaxis or, in the general case, f_{k-1}/f_{k+1} phyllotaxis. For, as Church [1, p. 5] remarks: "Bonnet ... saw quite clearly in the case of the Apricot that successive $\frac{2}{5}$ cycles were really not vertically superposed." According to the late A. M. Turing, the gradual advance from one pair of "parastichy numbers" to another, corresponding to the continuous decrease of h, may take place during the growth of a single plant. (For convenience we have taken the girth of the cylinder as our unit of measurement; thus h will decrease if the cylinder grows faster in girth than in length.)

When the cylinder is replaced by a cone, as in the analogous description of a fir cone, the lines of the plane no longer represent helices but rather *concho-spirals* [Moseley 1]. By allowing the generators of the cone to become more and more nearly perpendicular to its axis, we approach a limiting situation in which the concho-spirals become plane equiangular spirals, cutting one another at the same angles as the helices on the cylinder, that is, at the same angles as the lines in the plane of the lattice [Church 1, p. 58]. Fibonacci numbers as large as $f_{10} = 55, f_{11} = 89, f_{12} = 144$ arise as the numbers of visible spirals in certain varieties of sunflower [see Church 1, Plates V, VII, XIII, and especially Plate VI]. However, it should be frankly admitted that in some plants the numbers do not belong to the sequence of f's but to the sequence of g's [Church 1, Plate XXV], or even to the still more anomalous sequences

$$3, 1, 4, 5, 9, \ldots; \qquad 5, 2, 7, 9, 16, \ldots$$

[Church 1, Plate IX]. Thus we must face the fact that phyllotaxis is really not a universal *law* but only a fascinatingly prevalent *tendency*.

EXERCISE

Draw a version of Figure 11.5b with $h = \frac{1}{2}$. Is this close to the value given by 11.51 with $k = 1$?

Part III

Part III

12

Ordered geometry

During the last 2000 years, the two most widely read books have undoubtedly been the Bible and the Elements. Scholars find it an interesting task to disentangle the various accounts of the Creation that are woven together in the Book of Genesis. Similarly, as Euclid collected his material from various sources, it is not surprising that we can extract from the Elements two self-contained geometries that differ in their logical foundation, their primitive concepts and axioms. They are known as *absolute* geometry and *affine* geometry. After describing them briefly in § 12.1, we shall devote the rest of this chapter to those propositions which belong to both: propositions so fundamental and "obvious" that Euclid never troubled to mention them.

12.1 THE EXTRACTION OF TWO DISTINCT GEOMETRIES FROM EUCLID

The pursuit of an idea is as exciting as the pursuit of a whale.

Henry Norris Russell (1877 -1957)

Absolute geometry, first recognized by Bolyai (1802–1860), is the part of Euclidean geometry that depends on the first four Postulates without the fifth. Thus it includes the propositions I.1–28, III.1–19, 25, 28–30; IV.4–9 (with a suitably modified definition of "square"). The study of absolute geometry is motivated by the fact that these propositions hold not only in Euclidean geometry but also in hyperbolic geometry, which we shall study in Chapter 16. In brief, absolute geometry is geometry without the assumption of a unique parallel (through a given point) to a given line.

On the other hand, in affine geometry, first recognized by Euler (1707–1783), the unique parallel plays a leading role. Euclid's third and fourth postulates become meaningless, as circles are never mentioned and angles are never measured. In fact, the only admissible isometries are half-turns and translations. The affine propositions in Euclid are those which are preserved by parallel projection from one plane to another [Yaglom **2,** p. 17]:

for example, I. 30, 33–45, and VI. 1, 2, 4, 9, 10, 24–26. The importance of affine geometry has lately been enhanced by the observation that these propositions hold not only in Euclidean geometry but also in Minkowski's geometry of time and space, which Einstein used in his special theory of relativity.

Since each of Euclid's propositions is affine or absolute or neither, we might at first imagine that the two geometries (which we shall discuss in Chapters 13 and 15, respectively) had nothing in common except Postulates I and II. However, we shall see in the present chapter that there is a quite impressive nucleus of propositions belonging properly to both. The essential idea in this nucleus is *intermediacy* (or "betweenness"), which Euclid used in his famous definition:

A line (segment) is that which lies evenly between its ends.

This suggests the possibility of regarding intermediacy as a primitive concept and using it to define a line segment as the set of all points between two given points. In the same spirit we can extend the segment to a whole infinite line. Then, if *B* lies between *A* and *C*, we can say that the three points *A*, *B*, *C* lie in *order* on their line. This relation of order can be extended from three points to four or more.

Euclid himself made no explicit use of order, except in connection with measurement: saying that one magnitude is greater or less than another. It was Pasch, in 1882, who first pointed out that a geometry of order could be developed without reference to measurement. His system of axioms was gradually improved by Peano (1889), Hilbert (1899), and Veblen (1904).

Etymologically, "geometry without measurement" looks like a contradiction in terms. But we shall find that the passage from axioms and simple theorems to "interesting" theorems resembles Euclid's work in spirit, though not in detail.

This basic geometry, the common foundation for the affine and absolute geometries, is sufficiently important to have a name. The name *descriptive* geometry, used by Bertrand Russell [**1,** p. 382], was not well chosen, because it already had a different meaning. Accordingly, we shall follow Artin [**1,** p. 73] and say *ordered* geometry.

We shall pursue this rigorous development far enough to give the reader its flavor without boring him. The whole story is a long one, adequately told by Veblen [**1**] and Forder [**1,** Chapter II, and the *Canadian Journal of Mathematics,* **19** (1967), pp. 997–1000].

It is important to remember that, in this kind of work, we must define all the concepts used (except the primitive concepts) and prove all the statements (except the axioms), however "obvious" they may seem.

EXERCISES

1. Is the ratio of two lengths along one line a concept belonging to absolute geometry or to affine geometry or to both? (*Hint:* In "one dimension," i.e., when we consider only the points on a single line, the distinction between *absolute* and *affine* disappears.)

2. Name a Euclidean theorem that belongs neither to absolute geometry nor to affine geometry.

3. The concurrence of the medians of a triangle (1.41) is a theorem belonging to both absolute geometry and affine geometry. To which geometry does the rest of § 1.4 belong?

4. Which geometry deals (a) with parallelograms? (b) with regular polygons? (c) with Fagnano's problem (§ 1.8)?

12.2   INTERMEDIACY

A discussion of order . . . has become essential to any understanding of the foundation of mathematics.

Bertrand Russell (1872 -   )

[Russell **1,** p. 199]

In Pasch's development of ordered geometry, as simplified by Veblen, the only primitive concepts are *points A, B, . . .* and the relation of *intermediacy* [*ABC*], which says that *B* is between *A* and *C*. If *B* is not between *A* and *C*, we say simply "not [*ABC*]." There are ten axioms (12.21–12.27, 12.42, 12.43, and 12.51), which we shall introduce where they are needed among the various definitions and theorems.

AXIOM 12.21   *There are at least two points.*

AXIOM 12.22   *If A and B are two distinct points, there is at least one point C for which* [*ABC*].

AXIOM 12.23   *If* [*ABC*], *then A and C are distinct:* $A \neq C$.

AXIOM 12.24   *If* [*ABC*], *then* [*CBA*] *but not* [*BCA*].

THEOREM 12.241   *If* [*ABC*] *then not* [*CAB*].
Proof.   By Axiom 12.24, [*CAB*] would imply not [*ABC*].

THEOREM 12.242   *If* [*ABC*], *then* $A \neq B \neq C$ (that is, in view of Axiom 12.23, the three points are all distinct).
Proof.   If $B = C$, the two conclusions of Axiom 12.24 are contradictory. Similarly, we cannot have $A = B$.

DEFINITIONS.   If *A* and *B* are two distinct points, the *segment AB* is the set of points *P* for which [*APB*]. We say that such a point *P* is *on* the segment. Later we shall apply the same preposition to other sets, such as "lines."

THEOREM 12.243   *Neither A nor B is on the segment AB.*
Proof.   If *A* or *B* were on the segment, we would have [*AAB*] or [*ABB*], contradicting 12.242.

THEOREM 12.244   *Segment AB = segment BA.*

Proof. By Axiom 12.24, [APB] implies [BPA].

DEFINITIONS.   The *interval* $\overline{AB}$ is the segment AB plus its *end points A* and *B*:

$$\overline{AB} = A + AB + B.$$

The *ray A/B* ("from A, away from B") is the set of points P for which [PAB]. The *line AB* is the interval $\overline{AB}$ plus the two rays A/B and B/A:

$$\text{line } AB = A/B + \overline{AB} + B/A.$$

COROLLARY 12.2441   Interval $\overline{AB}$ = interval $\overline{BA}$; line AB = line BA.

AXIOM 12.25   *If C and D are distinct points on the line AB, then A is on the line CD.*

THEOREM 12.251   *If C and D are distinct points on the line AB then*

$$\text{line } AB = \text{line } CD.$$

Proof. If A, B, C, D are not all distinct, suppose D = B. To prove that line AB = line BC, let X be any point on BC except A or B.   By 12.25, A, like X, is on BC. Therefore B is on AX, and X is on AB. Thus every point on BC is also on AB. Interchanging the roles of A and C, we see that similarly every point on AB is also on BC. Thus AB = BC. Finally, if A, B, C, D are all distinct, we have AB = BC = CD.

COROLLARY 12.2511   Two distinct points lie on just one line. Two distinct lines (if such exist) have at most one common point. (Such a common point F is called a point of intersection, and the lines are said to *meet* in F.)

COROLLARY 12.2512   Any three distinct points A, B, C on a line satisfy just one of the relations [ABC], [BCA], [CAB].

AXIOM 12.26   *If AB is a line, there is a point C not on this line.*

THEOREM 12.261   *If C is not on the line AB, then A is not on BC, nor B on CA: the three lines BC, CA, AB are distinct.*

Proof. By 12.25, if A were on BC, C would be on AB.

DEFINITIONS.   Points lying on the same line are said to be *collinear*. Three non-collinear points A, B, C determine a *triangle ABC*, which consists of these three points, called *vertices*, together with the three segments BC, CA, AB, called *sides*.

AXIOM 12.27   *If ABC is a triangle and [BCD] and [CEA], then there is, on the line DE, a point F for which [AFB].* (See Figure 12.2a.)

THEOREM 12.271   *Between two distinct points there is another point.*

Proof.   Let A and B be the two points. By 12.26, there is a point E not on the line AB. By 12.22, there is a point C for which [AEC]. By 12.251, the line AC is the same as AE. By 12.261 (applied to ABE), B is not on this line: therefore ABC is a triangle. By 12.22 again, there is a point D for which [BCD]. By 12.27, there is a point F between A and B.

THEOREM 12.272 *In the notation of Axiom* 12.27, [*DEF*].

Proof. Since *F* lies on the line *DE*, there are (by 12.2512) just five possibilities: $F = D$, $F = E$, [*EFD*], [*FDE*], [*DEF*]. Either of the first two would make *A*, *B*, *C* collinear.

If [*EFD*], we could apply 12.27 to the triangle *DCE* with [*CEA*] and [*EFD*] (Figure 12.2*b*), obtaining *X* on *AF* with [*DXC*]. Since *AF* and *CD* cannot meet more than once, we have $X = B$, so that [*DBC*]. Since [*BCD*], this contradicts 12.24.

Similarly (Figure 12.2*c*) we cannot have [*FDE*]. The only remaining possibility is [*DEF*].

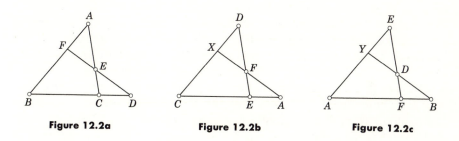

Figure 12.2a   **Figure 12.2b**   **Figure 12.2c**

This proof is typical; so let us be content to give the remaining theorems without proofs [Veblen **1**, pp. 9–15; Forder **1**, pp. 49–55].

12.273 *A line cannot meet all three sides of a triangle.* (Remember that the "sides" are not intervals, nor whole lines, but only segments.)

12.274 *If* [*ABC*] *and* [*BCD*], *then* [*ABD*].

12.275 *If* [*ABC*] *and* [*ABD*] *and* $C \neq D$, *then* [*BCD*] *or* [*BDC*], *and* [*ACD*] *or* [*ADC*].

12.276 *If* [*ABD*] *and* [*ACD*] *and* $B \neq C$, *then* [*ABC*] *or* [*ACB*].

12.277 *If* [*ABC*] *and* [*ACD*], *then* [*BCD*] *and* [*ABD*].

DEFINITION. If [*ABC*] and [*ACD*], we write [*ABCD*].

This four-point order is easily seen to have all the properties that we should expect, for example, if [*ABCD*], then [*DCBA*], but all the other orders are false.

Any point *O* on a segment *AB* decomposes the segment into two segments: *AO* and *OB*. (We are using the word *decomposes* in a technical sense [Veblen **1**, p. 21], meaning that every point on the segment *AB* except *O* itself is on just one of the two "smaller" segments.) Any point *O* on a ray from *A* decomposes the ray into a segment and a ray: *AO* and *O/A*. Any point *O* on a line decomposes the line into two "opposite" rays; if [*AOB*], the rays are *O/A* and *O/B*. The ray *O/A*, containing *B*, is sometimes more conveniently called *the ray OB*.

For any integer $n > 1$, *n* distinct collinear points decompose their line into two rays and $n - 1$ segments. The points can be named $P_1, P_2, \ldots . . P_n$ in such a way that the two rays are P_1/P_n, P_n/P_1, and the $n - 1$ segments are

$$P_1P_2, P_2P_3, \ldots, P_{n-1}P_n,$$

each containing none of the points. We say that the points are in the *order* $P_1P_2 \ldots P_n$, and write $[P_1P_2 \ldots P_n]$. Necessary and sufficient conditions for this are

$$[P_1P_2P_3], \quad [P_2P_3P_4], \quad \ldots, \quad [P_{n-2}P_{n-1}P_n].$$

Naturally, the best logical development of any subject uses the simplest or "weakest" possible set of axioms. (The worst occurs when we go to the opposite extreme and assume everything, so that there is no development at all!) In his original formulation of Axiom 12.27 [Pasch and Dehn **1**, p. 2 : "IV. Kernsatz"] Pasch made the following far stronger statement: If a line in the plane of a given triangle meets one side, it also meets another side (or else passes through a vertex). Peano's formulation, which we have adopted, excels this in two respects. The word "plane" (which we shall define in § 12.4) is not used at all, and the line DE penetrates the triangle ABC in a special manner, namely, before entering through the side CA, it comes from a point D on C/B. It might just as easily have come from a point on A/B (which is the same with C and A interchanged) or from a point on B/A or B/C (which is quite a different story). The latter possibility (with a slight change of notation) is covered by the following theorem (12.278). Axiom 12.27 is "only just strong enough"; for, although it enables us to deduce the statement 12.278 of apparently equal strength, we could not reverse the roles: if we tried instead to use 12.278 as an axiom, we would not be able to deduce 12.27 as a theorem!

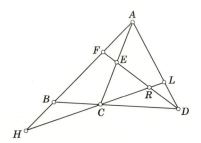

Figure 12.2d

THEOREM 12.278   *If ABC is a triangle and $[AFB]$ and $[BCD]$, then there is, on the line DF, a point E for which $[CEA]$.*

Proof.   Take H on B/F (as in Figure 12.2d) and consider the triangle DFB with $[FBH]$ and $[BCD]$. By 12.27 and 12.272, there is a point R for which $[DRF]$ and $[HCR]$. By 12.274, $[AFB]$ and $[FBH]$ imply $[AFH]$. Thus we have a triangle DAF with $[AFH]$ and $[FRD]$. By 12.27 and 12.272 again, there is a point L for which $[DLA]$ and $[HRL]$. By 12.277, $[HCR]$ and $[HRL]$

imply [*CRL*]. Thus we have a triangle *CAL* with [*ALD*] and [*LRC*]. By 12.27 a third time, there is, on the line *DR* (= *DF*), a point *E* for which [*CEA*].

EXERCISES

1. A line contains infinitely many points.

2. We have defined a segment as a set of points. At what stage in the above development can we assert that this set is never the *null* set? [Forder **1**, p. 50.]

3. In the proof of 12.272, we had to show that the relation [*FDE*] leads to a contradiction. Do this by applying 12.27 to the triangle *BFD* (instead of *EAF*).

4. Given a finite set of lines, there are infinitely many points not lying on any of the lines.

5. If *ABC* is a triangle and [*BLC*], [*CMA*], [*ANB*], then there is a point *E* for which [*AEL*] and [*MEN*]. [Forder **1**, p. 56.]

6. If *ABC* is a triangle, the three rays *B/C, A/C, A/B* have a *transversal* (that is, a line meeting them all). (K. B. Leisenring.)

7. If *ABC* is a triangle, the three rays *B/C, C/A, A/B* have no transversal.

12.3 SYLVESTER'S PROBLEM OF COLLINEAR POINTS

> *Almost any field of mathematics offered an enchanting world for discovery to Sylvester.*
>
> E. T. Bell [**1**, p. 433]

It may seem to some readers that we have been using self-evident axioms to prove trivial results. Any such feeling of irritation is likely to evaporate when it is pointed out that the machinery so far developed is sufficiently powerful to deal effectively with Sylvester's conjecture (§ 4.7), which baffled the world's mathematicians for forty years. This matter of collinearity clearly belongs to ordered geometry. Kelly's Euclidean proof involves the extraneous concept of distance: it is like using a sledge hammer to crack an almond. The really appropriate nutcracker is provided by the following argument.

THEOREM. *If n points are not all collinear, there is at least one line containing exactly two of them.*

Proof. Let $P_1, P_2, \ldots, P_n$ be the n points, so named that the first three are not collinear (Figure 12.3a). Lines joining P_1 to all the other points of the set meet the line P_2P_3 in at most $n - 1$ points (including P_2 and P_3). Let Q be any *other* point on this line. Then the line P_1Q contains P_1 but no other P_i.

Lines joining pairs of P's meet the line P_1Q in at most $\left|\dfrac{n-1}{2}\right| + 1$ points (including P_1 and Q). Let P_1A be one of the segments that arise in the decomposition of this line by all these points. (Possibly $A = Q$.) Then no joining line P_iP_j can meet the "empty" segment P_1A. By its definition, A

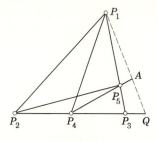

Figure 12.3a

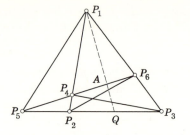

Figure 12.3b

lies on at least one joining line, say P_4P_5. If P_4 and P_5 are the only P's on this line (as in Figure 12.3*a*) our task is finished. If not, we have a joining line through A containing at least three of the P's, which we can name P_4, P_5, P_6 in such an order that the segment AP_5 contains P_4 but not P_6. (Since A decomposes the line into two opposite rays, one of which contains at least two of the three P's, this special naming is always possible. See Figure 12.3*b*.) We can now prove that the line P_1P_5 contains only these two P's.

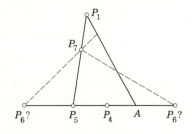

Figure 12.3c

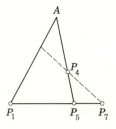

Figure 12.3d

We argue by *reductio ad absurdum*. If the line P_1P_5 contains (say) P_7, we can use 12.27 and 12.278 to deduce that the segment P_1A meets one of the joining lines, namely, P_6P_7 or P_4P_7. In fact, it meets P_6P_7 if $[P_1P_7P_5]$ (as in Figure 12.3*c*), and it meets P_4P_7 if $[P_1P_5P_7]$ or $[P_5P_1P_7]$ (as in Figure 12.3*d*). In either case our statement about the "empty" segment is contradicted.

Thus we have found, under all possible circumstances, a line (P_4P_5 or P_1P_5) containing exactly two of the P's.

EXERCISE

Justify the statement that the joining lines meet the line P_1Q in at most $\left(\dfrac{n-1}{2}\right)+1$ points. In the example shown in Figure 12.3*b*, this number (at most 11) is only 5; why? (The symbol $\left(\begin{matrix} i \\ j \end{matrix}\right)$ stands for the number of combinations of i things taken j at a time; for instance, $\left(\begin{matrix} i \\ 2 \end{matrix}\right)$ is the number of pairs, namely $\frac{1}{2}i\,(i-1)$.)

12.4 PLANES AND HYPERPLANES

If i hyperplanes in n dimensions are so placed that every n but no
n + 1 have a common point, the number of regions into which they
decompose the space is

$$\binom{i}{0} + \binom{i}{1} + \binom{i}{2} + \binom{i}{3} + \ldots + \binom{i}{n} = f(n, i).$$

Ludwig Schläfli (1814-1895)
[Schläfli **1**, p. 209]

It is remarkable that we can do so much plane geometry before defining
a plane. But now, as the Walrus said, "The time has come . . . ."

DEFINITIONS. If A, B, C are three non-collinear points, the *plane ABC*
is the set of all points collinear with pairs of points on one or two sides of
the triangle ABC. A segment, interval, ray, or line is said to be *in* a plane
if all its points are.

Axioms 12.21 to 12.27 enable us to prove all the familiar properties of
incidence in a plane, including the following two which Hilbert [**1**, p. 4] took
as axioms:

Any three non-collinear points in a plane α completely determine that
plane.

If two distinct points of a line a lie in a plane α, then every point of a
lies in α.

DEFINITIONS. An *angle* consists of a point O and two non-collinear rays
going out from O. The point O is the *vertex* and the rays are the *sides* of
the angle [Veblen **1**, p. 21; Forder **1**, p. 69]. If the sides are the rays OA
and OB, or a_1 and b_1, the angle is denoted by $\angle AOB$ or a_1b_1 (or $\angle BOA$,
or b_1a_1). The same angle a_1b_1 is determined by any points A and B on its
respective sides. If C is any point between A and B, the ray OC is said to
be *within* the angle.

From here till the statement of Axiom 12.41, we shall assume that all the
points and lines considered are *in one plane*.

A *convex region* is a set of points, any two of which can be joined by a
segment consisting entirely of points in the set, with the extra condition that
each of the points is on at least two non-collinear segments consisting entirely of points in the set. In particular, an *angular* region is the set of
all points on rays within an angle, and a *triangular* region is the set of all
points between pairs of points on distinct sides of a triangle. An angular
(or triangular) region is said to be *bounded* by the angle (or triangle).

It can be proved [Veblen **1**, p. 21] that any line containing a point of a
convex region "decomposes" it into two convex regions. In particular, a
line a decomposes a plane (in which it lies) into two *half planes*. Two points
are said to be on the *same side* of a if they are in the same half plane, on
opposite sides if they are in opposite half planes, that is, if the segment join-

ing them meets a. In the latter case we also say that a *separates* the two points. (It is unfortunate that the word "side" is used with two different meanings, both well established in the literature. However, the context will always show whether we are considering the two sides of an angle, which are rays, or the two sides of a line, which are half planes.)

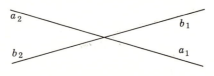

Figure 12.4a

As we remarked in § 12.2, any point O on a line a decomposes a into two rays, say a_1 and a_2. Any other line b through O is likewise decomposed by O into two rays b_1 and b_2, one in each of the half planes determined by a. Each of these rays decomposes the half plane containing it into two angular regions. Thus any two intersecting lines a and b together decompose their plane into four angular regions, bounded by the angles

$$a_1b_1, \quad b_1a_2, \quad a_2b_2, \quad b_2a_1,$$

as in Figure 12.4a. The opposite rays a_1 and a_2 are said to *separate* the rays b_1 and b_2; they likewise separate all the rays within either of the angles a_1b_1, b_1a_2 from all the rays within either of the angles a_2b_2, b_2a_1. We also say that the rays a_1 and b_1 separate all the rays between them from a_2, b_2, and from all the rays within b_1a_2, a_2b_2, or b_2a_1.

It follows from the definition of a line that two distinct points, A and B, decompose their line into three parts: the segment AB and the two rays A/B, B/A. Somewhat similarly, two nonintersecting (but coplanar) lines, a and b, decompose their plane into three regions. One of these regions lies *between* the other two, in the sense that it contains the segment AB for any A on a and B on b. Another line c is said to lie between a and b if it meets such a segment AB but does not meet a or b, and we naturally write $[acb]$.

12.401   *If ABC and $A'B'C'$ are two triads of collinear points, such that the three lines AA', BB', CC' have no intersection, and if $[ACB]$, then $[A'C'B']$.*

Analogous consideration of an angular region yields

12.402   *If ABC and $A'B'C'$ are two triads of collinear points on distinct lines, such that the three lines AA', BB', CC' have a common point O which is not between A and A', nor between B and B', nor between C and C', and if $[ACB]$, then $[A'C'B']$.*

We need one or more further axioms to determine the number of *dimensions*. If we are content to work in two dimensions we say

AXIOM 12.41   *All points are in one plane.*

If not [Forder **1**, p. 60], we say instead:

AXIOM 12.42   *If ABC is a plane, there is a point D not in this plane.*

We then define the *tetrahedron ABCD*, consisting of the four non-coplanar points A, B, C, D, called *vertices*, the six joining segments AD, BD, CD, BC, CA, AB, called *edges*, and the four triangular regions BCD, CDA, DAB, ABC, called *faces*. The *space* (or "3-space") $ABCD$ is the set of all points collinear with pairs of points in one or two faces of the tetrahedron $ABCD$.

We can now deduce the familiar properties of incidence of lines and planes [Forder **1**, pp. 61–65]. In particular, any four non-coplanar points of a space determine it, and the line joining any two points of a space lies entirely in the space. If Q is in the space $ABCD$ and P is in a face of the tetrahedron $ABCD$, then PQ meets the tetrahedron again in a point distinct from P.

If we are content to work in three dimensions, we say

AXIOM 12.43   *All points are in the same space.*

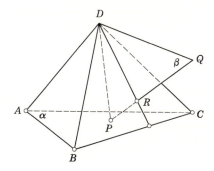

Figure 12.4b

Consequently:

THEOREM 12.431   *Two planes which meet in a point meet in another point, and so in a line.*

Proof. Let P be the common point and α one of the planes. Take A, B, C in α so that P is inside the triangle ABC. Let DPQ be a triangle in the other plane β (Figure 12.4b). If D or Q lies in α, then α and β have two common points. If not, PQ meets the tetrahedron $ABCD$ in a point R distinct from P; and DR, in β, meets the triangle ABC in a point common to α and β.

If, on the other hand, we wish to increase the number of dimensions, we replace 12.43 by

AXIOM 12.44   *If $A_0A_1A_2A_3$ is a 3-space, there is a point A_4 not in this 3-space.*

We then define the *simplex* $A_0A_1A_2A_3A_4$ which has 5 vertices A_i, 10 edges A_iA_j ($i < j$), 10 faces $A_iA_jA_k$ ($i < j < k$), and 5 *cells* $A_iA_jA_kA_l$ (which are tetrahedral regions.) The *4-space* $A_0A_1A_2A_3A_4$ is the set of points collinear with pairs of points on one or two cells of the simplex.

The possible extension to n dimensions (using mathematical induction) is now clear. The n-space $A_0A_1 \ldots A_n$ is decomposed into two convex regions (half-spaces) by an $(n-1)$-dimensional subspace such as $A_0A_1 \ldots A_{n-1}$, which is called a *hyperplane* (or "prime," or "$(n-1)$-flat").

EXERCISES

1. Any 5 coplanar points, no 3 collinear, include 4 that form a convex quadrangle.

2. A ray OC within $\angle AOB$ decomposes the angular region into two angular regions, bounded by the angles AOC and COB. [Veblen **1**, p. 24.]

3. If m distinct coplanar lines meet in a point O, they decompose their plane into $2m$ angular regions [Veblen **1**, p. 26].

4. If ABC is a triangle, the three lines BC, CA, AB decompose their plane into seven convex regions, just one of which is triangular.

5. If m coplanar lines are so placed that every 2 but no 3 have a common point, they decompose their plane into a certain number of convex regions. Call this number $f(2, m)$. Then

$$f(2, m) = f(2, m - 1) + m.$$

But $f(2, 0) = 1$. Therefore $f(2, 1) = 2$, $f(2, 2) = 4$, $f(2, 3) = 7$, and $f(2, m)$ $= 1 + m + \binom{m}{2}$.

6. If m planes in a 3-space are so placed that every 3 but no 4 have a common point, they decompose their space into (say) $f(3, m)$ convex regions. Then

$$f(3, m) = f(3, m - 1) + f(2, m - 1).$$

But $f(3, 0) = 1$. Therefore $f(3, 1) = 2$, $f(3, 2) = 4$, $f(3, 3) = 8$, $f(3, 4) = 15$, and $f(3, m) = 1 + \binom{m}{1} + \binom{m}{2} + \binom{m}{3}$. [Steiner **1**, p. 87.]

7. Obtain the analogous result for m hyperplanes in an n-space.

12.5 CONTINUITY

> *Nothing but Geometry can furnish a thread for the labyrinth of the composition of the continuum . . . and no one will arrive at a truly solid metaphysic who has not passed through that labyrinth.*
>
> G. W. Leibniz (1646-1716)
>
> [Russell **2**, pp. 108-109]

Between any two rational numbers (§ 9.1) there is another rational number, and therefore an infinity of rational numbers; but this does not mean that every real number (§ 9.2) is rational. Similarly, between any two points

(12.271) there is another point, and therefore an infinity of points; but this does not mean that the axioms in § 12.2 make the line "continuous." In fact, continuity requires at least one further axiom. There are two well-recognized approaches to this subtle subject. One, due to Cantor and Weierstrass, defines a monotonic sequence of points, with an axiom stating that *every bounded monotonic sequence has a limit* [Coxeter **2,** Axiom 10.11]. The other, due to Dedekind, obtains a general point on a line as the common origin of two opposite rays [Coxeter **3,** p. 162]. Its arithmetical counterpart is illustrated by describing $\sqrt{2}$ as the "section" between rational numbers whose squares are less than 2 and rational numbers whose squares are greater than 2. Dedekind's Axiom, though formidable in appearance, is the more readily applicable; so we shall use it here:

AXIOM 12.51 *For every partition of all the points on a line into two nonempty sets, such that no point of either lies between two points of the other, there is a point of one set which lies between every other point of that set and every point of the other set.*

This axiom is easily seen to imply several modified versions of the same statement. Instead of "the points on a *line*" we could say "the points on a *ray*" or "the points on a *segment*" or "the points on an *interval*." (In the last case, for instance, the rest of the line consists of two rays which can be added to the two sets in an obvious manner.) Another version [Forder **1,** p. 299] is:

THEOREM 12.52 *For every partition of all the rays within an angle into two nonempty sets, such that no ray of either lies between two rays of the other, there is a ray of one set which lies between every other ray of that set and every ray of the other set.*

To prove this for an angle $\angle AOB$, we consider the section of all the rays by the line AB, and apply the "segment" version of 12.51 to the segment AB.

12.6 PARALLELISM

> In the last few weeks I have begun to put down a few of my own Meditations, which are already to some extent nearly 40 years old. These I had never put in writing, so I have been compelled three or four times to go over the whole matter afresh in my head.
>
> C. F. Gauss (1777-1855)
>
> (Letter to H. K. Schumacher, May 17, 1831, as translated by Bonola [**1,** p. 67])

The idea of defining, through a given point, two rays parallel to a given line (in opposite senses), was developed independently by Gauss, Bolyai, and Lobachevsky. The following treatment is closest to that of Gauss.

THEOREM 12.61   *For any point A and any line r, not through A, there are just two rays from A, in the plane Ar, which do not meet r and which separate all the rays from A that meet r from all the other rays that do not.*

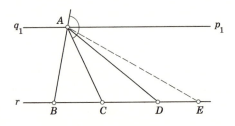

Figure 12.6a

Proof.   Taking any two distinct points B and C on r, we apply 12.52 to the angle between the rays AC and A/B (marked in Figure 12.6a). We consider the partition of all the rays within this angle into two sets according as they do or do not meet the ray C/B. Clearly, these sets are not empty, and no ray in either set lies between two in the other. We conclude that one of the sets contains a special ray p_1 which lies between every other ray of that set and every ray of the other set.

In fact, p_1 belongs to the second set. For, if it met C/B, say in D, we would have $[BCD]$. By Axiom 12.22, we could take a point E such that $[CDE]$, with the absurd conclusion that AE belongs to both sets: to the first, because E is on C/B, and to the second, because AD lies between AC and AE.

We have thus found a ray p_1, within the chosen angle, which is the "first" ray that fails to meet the ray C/B; this means that every ray within the angle between AC and p_1 does meet C/B. Interchanging the roles of B and C, we obtain another special ray q_1, on the other side of AB, which may be described (for a counterclockwise rotation) as the "last" ray that fails to meet B/C. Since the line r consists of the two rays B/C, C/B, along with the interval $\overline{BC}$, we have now found two rays p_1, q_1, which separate all the rays from A that meet r from all the other rays (from A) that do not. [Forder 1, p. 300.]

These special rays from A are said to be *parallel* to the line r in the two senses: p_1 parallel to C/B, and q_1 parallel to B/C. (Two rays are said to have the same sense if they lie on the same side of the line joining their initial points.)

For the sake of completeness, we define the rays parallel to r from a point A on r itself to be the two rays into which A decomposes r. The distinction between affine geometry and hyperbolic geometry depends on the question whether, for other positions of A, the two rays p_1, q_1 are still the two halves of one line. If they are, this line decomposes the plane into two half planes, one of which contains the whole of the line r. If not, the lines p and q (which contain the rays) decompose the plane into four angular regions

$$p_1q_1, \quad q_1p_2, \quad p_2q_2, \quad q_2p_1.$$

In this case, by 12.61, r lies entirely in the region p_1q_1.

COROLLARY 12.62 *For any point A and any line r, not through A, there is at least one line through A, in the plane Ar, which does not meet r.*

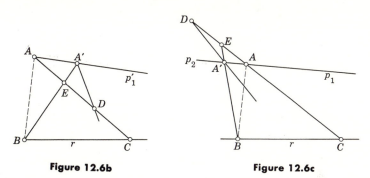

Figure 12.6b   Figure 12.6c

Another familiar property of parallelism is its "transmissibility":

THEOREM 12.63 *The parallelism of a ray and a line is maintained when the beginning of the ray is changed by the subtraction or addition of a segment.*

Proof [Gauss **1**, vol. 8, p. 203]. Let p_1 be a ray from A which is parallel to a line r through B, and let A' be any point on this ray (Figure 12.6b) or on the opposite ray p_2 (Figure 12.6c). The modified ray p'_1, beginning at A', is A'/A or A'A, respectively; it obviously does not meet r. What remains to be proved is that every ray from A', within the angle between A'B and p'_1, does meet r. Let D be any point on such a ray (Figure 12.6b) or on its opposite (Figure 12.6c). Since p_1 (from A) is parallel to r, the line AD (containing a ray within the angle between AB and p_1) meets r, say in C. The line A'B, separating A from D, meets the segment AD, say in E. By Axiom 2.27, applied to the triangle CBE with [BEA'] and [EDC] (Figure 12.6b) or to the triangle BCE with [CED] and [EA'B] (Figure 12.6c), the line A'D meets BC. Thus p'_1 is parallel to r.

This property of transmissibility enables us to say that the *line p = $\overline{AA'}$* is parallel to the line *r = BC*, provided we remember that this property is associated with a definite "sense" along each line.

Busemann [**1**, p. 139 (23.5)] has proved that it is not possible, within the framework of two-dimensional ordered geometry, to establish the "symmetry" of parallelism: that if p is parallel to r then r is parallel to p. To supply this important step we need either Axiom 12.42 [as in Coxeter **3**, pp. 165–177] or the affine axiom of parallelism (13.11) or the absolute axioms of congruence (§ 15.1).

THEOREM 12.64 *If two lines are both parallel to a third in the same sense, there is a line meeting all three.*

Proof. We have to show that, if lines p and s are both parallel to r in the same sense, then the three lines p, r, s have a transversal. In affine geometry this is obvious, so let us assume the geometry to be hyperbolic. Of the two lines parallel to r through a point A on p, one is p itself. Let q be the other, and let r be in the angular region p_1q_1, so that the rays p_1 and q_1 (from A) are parallel to r in opposite senses and s is parallel to r in the same sense as p_1. Let B and D be arbitrary points on r and s, respectively.

If D is in the region p_1q_1, the line AD is a transversal. If D is in p_1q_2, BD is a transversal. If D is in p_2q_2, both AD and BD are transversals. Finally, if D is in p_2q_1, AB is a transversal.

Hyperbolic geometry will be considered further in Chapters 15, 16, and 20.

EXERCISES

1. If p is parallel to s and [prs], then p is parallel to r. (See Figure 15.2c with s for q.)

2. Consider all the points strictly inside a given circle in the Euclidean plane. Regard all other points as nonexistent. Let chords of the circle be called lines. Then all the axioms 12.21–12.27, 12.41, and 12.51 are satisfied. Locate the two rays through a given point parallel to a given line. Note that they form an angle (as in Figure 16.2b).

13

Affine geometry

The first three sections of this chapter contain a systematic development of the foundations of affine geometry. In particular, we shall see how length may be measured along a line, though independent units are required for lines in different directions. In §§ 13.4–7 we shall investigate such topics as area, affine transformations, lattices, vectors, barycentric coordinates, and the theorems of Ceva and Menelaus. Finally, in § 13.8 and § 13.9, we shall extend these ideas from two dimensions to three.

According to Blaschke [**1**, p. 31; **2**, p. 12], the word "affine" (German *affin*) was coined by Euler. But it was only after the launching of Klein's Erlangen program (see Chapter 5) that this geometry became recognized as a self-contained discipline. Many of the propositions may seem familiar; in fact, most readers will discover that they have often been working in the affine plane without realizing that it could be so designated.

Our treatment is somewhat more geometric and less algebraic than that of Artin's *Geometric Algebra* [Artin **1**; see especially pp. 58, 63, 71]. Incidentally, we shall find that our Axiom 13.12 (which he calls DP) implies Theorem 13.122 (his D_a): this presumably means that his Axiom 4b implies 4a.

13.1 THE AXIOM OF PARALLELISM AND THE "DESARGUES" AXIOM

Mathematical language is difficult but imperishable. I do not believe that any Greek scholar of to-day can understand the idiomatic undertones of Plato's dialogues, or the jokes of Aristophanes, as thoroughly as mathematicians can understand every shade of meaning in Archimedes' works.

M. H. A. Newman
(*Mathematical Gazette* **43,** 1959, p. 167)

In this axiomatic treatment, we regard the real affine plane as a special case of the ordered plane. Accordingly, the primitive concepts are *point* and *intermediacy,* satisfying Axioms 12.21–12.27, 12.41 and 12.51. Affine geometry is derived from ordered geometry by adding the following two extra axioms:

AXIOM 13.11 *For any point A and any line r, not through A, there is at most one line through A, in the plane Ar, which does not meet r.*

AXIOM 13.12 *If A, A', B, B', C, C', O are seven distinct points, such that AA', BB', CC' are three distinct lines through O, and if the line AB is parallel to A'B', and BC to B'C', then also CA is parallel to C'A'.*

The affine axiom of parallelism (13.11) combines with 12.62 to tell us that, for any point A and any line r, there is exactly one line through A, in the plane Ar, which does not meet r. Hence the two rays from A parallel to r are always collinear, *any two lines in a plane that do not meet are parallel,* and parallelism is an *equivalence relation.* The last remark comprises three properties:

Parallelism is *reflexive.* (Each line is parallel to itself.)
Parallelism is *symmetric.* (If p is parallel to r, then r is parallel to p.)
Parallelism is *transitive.* (If p and q are parallel to r, then p is parallel to q. Euclid I. 30.)

In the manner that is characteristic of equivalence relations, every line belongs to a *pencil* of parallels whose members are all parallel to one another.

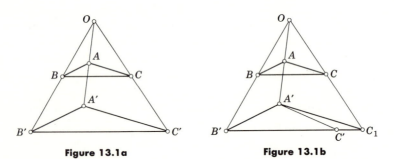

Figure 13.1a   **Figure 13.1b**

Axiom 13.12 (see Figure 13.1*a*) is probably familiar to most readers either as a corollary of Euclid VI.2 or as an affine form of Desargues's theorem. We shall see that it implies

THEOREM 13.121 *If ABC and A'B'C' are two triangles with distinct vertices, so placed that the line BC is parallel to B'C', CA to C'A', and AB to A'B', then the three lines AA', BB', CC' are either concurrent or parallel.*

Proof. If the three lines AA', BB', CC' are not all parallel, some two of them must meet. The notation being symmetrical, we may suppose that these two are AA' and BB', meeting in O, as in Figure 13.1*b*. Let OC meet $B'C'$ in C_1. By Axiom 13.12, applied to AA', BB', CC_1, the line AC is parallel to $A'C_1$ as well as to $A'C'$. By Axiom 13.11, C_1 lies on $A'C'$ as well as on $B'C'$. Since $A'B'C'$ is a triangle, C_1 coincides with C'. Thus, if AA', BB', CC' are not parallel, they are concurrent [Forder **1**, p. 158].

Roughly speaking, Axiom 13.12 is the converse of one half of Theorem 13.121. The converse of the other half is

THEOREM 13.122 *If A, A', B, B', C, C' are six distinct points on three distinct parallel lines AA', BB', CC', so placed that the line AB is parallel to $A'B'$, and BC to $B'C'$, then also CA is parallel to $C'A'$.*

Proof. Through A' draw $A'C_1$ parallel to AC, to meet $B'C'$ in C_1, as in Figure 13.1c. By 13.121, applied to the triangles ABC and $A'B'C_1$, since AA' and BB' are parallel, CC_1 is parallel to both of them, and therefore also to CC'. Hence C_1 lies on CC' as well as on $B'C'$. Since the parallel lines BB' and CC' are distinct, B' cannot lie on CC'. Therefore C_1 coincides with C', and $A'C'$ is parallel to AC.

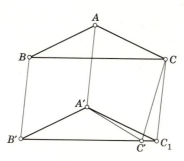

Figure 13.1c

EXERCISES

1. If a line in the plane of two parallel lines meets one of them, it meets the other also.

2. Can we always say, of three distinct parallel lines, that one lies between the other two?

13.2 DILATATIONS

> Dilatations . . . are one-to-one maps of the plane onto itself which move all points of a line into points of a parallel line.
>
> E. Artin [**1**, p. 51]

Four non-collinear points A, B, C, D are said to form a *parallelogram* $ABCD$ if the line AB is parallel to DC, and BC to AD. Its *vertices* are the four points; its *sides* are the four segments AB, BC, CD, DA, and its *diagonals* are the two segments AC, BD. Since B and D are on opposite sides of AC, the diagonals meet in a point called the *center* [Forder **1**, p. 140].

As in § 5.1, we define a *dilatation* to be a transformation which transforms each line into a parallel line. But now we must discuss more thoroughly the important theorem 5.12, which says that *two given segments, AB and A'B', on parallel lines, determine a unique dilatation* $AB \to A'B'$.

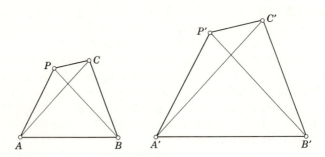

Figure 13.2a

For any point P, not on AB, we can find a corresponding point P' by drawing $A'P'$ parallel to AP, and $B'P'$ parallel to BP, as in Figure 5.1a. (The lines thus drawn through A' and B' cannot be parallel, for, if they were, AP and BP would be parallel.) Similarly, another point C yields C', as in Figure 13.2a. By 13.121, the three lines AA', BB', CC' are either concurrent or parallel. So likewise are AA', BB', PP'.

If the two parallel lines AB and $A'B'$ do not coincide, it follows that the four lines AA', BB', CC', PP' are all either concurrent or parallel. Then, by 13.12 or 13.122 (respectively), CP and $C'P'$ are parallel, so that the transformation is indeed a dilatation. If the lines AB and $A'B'$ do coincide, we can reach the same conclusion by regarding the transformation as $AC \to A'C'$ instead of $AB \to A'B'$.

We see now that a given dilatation may be specified by its effect on any given segment. The *inverse* of the dilatation $AB \to A'B'$ is the dilatation $A'B' \to AB$. The *product* of two dilatations, $AB \to A'B'$ and $A'B' \to A''B''$, is the dilatation $AB \to A''B''$. In particular, the product of a dilatation with its inverse is the *identity*, $AB \to AB$. Thus all the dilatations together form a (continuous) *group*.

The argument used in proving 5.13 shows that, for a given dilatation, the lines PP' which join pairs of corresponding points are *invariant* lines. The discussion of 5.12 shows that all these lines are either concurrent or parallel.

If the lines PP' are concurrent, their intersection O is an invariant point, and we have a *central* dilatation

$$OA \to OA'$$

(where A' lies on the line OA). The invariant point O is unique; for, if O and O_1 were two such, the dilatation would be $OO_1 \to OO_1$, which is the identity.

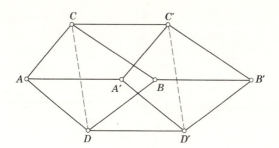

Figure 13.2b

If, on the other hand, the lines PP' are parallel, there is no invariant point, and we have a *translation* $AB \to A'B'$, where not only is AB parallel to $A'B'$ but also AA' is parallel to BB'. If these two parallel lines are distinct, $AA'B'B$ is a parallelogram. If not, we can use auxiliary parallelograms $AA'C'C$ and $C'CBB'$ (or $AA'D'D$ and $D'DBB'$) as in Figure 13.2b. Two applications of 13.122 suffice to prove that, when A, B, A' are given, B' is independent of the choice of C (or D). Hence

13.21 *Any two points A and A' determine a unique translation $A \to A'$.*

We naturally include, as a degenerate case, the identity, $A \to A$. It follows that a dilatation, other than the identity, is a translation if and only if it has no invariant point. Moreover, a given translation may be specified by its effect on any given point; in fact, the translation $A \to A'$ is the same as $B \to B'$ if $AA'B'B$ is a parallelogram, or if, for any parallelogram $AA'C'C$ based on AA', there is another parallelogram $C'CBB'$.

We next prove that dilatations are "ordered transformations:"

13.22 *The dilatation $AB \to A'B'$ transforms every point between A and B into a point between A' and B'.*

Proof. If the lines AB and $A'B'$ are distinct, the fact that $[ACB]$ implies $[A'C'B']$ follows at once from 12.401 (for a translation) or 12.402 (for a central dilatation). To obtain the analogous result for two corresponding triads on an invariant line CC', we draw six parallel lines through the six points, as in Figure 13.2c, and use the fact that $[acb]$ implies $[a'c'b']$.

To prove Theorem 3.21, which says that *the product of two translations is a translation,* we can argue thus: since translations are dilatations, the product is certainly a dilatation. If it is not a translation it has a unique invariant point O. If the first of the two given translations takes O to O', the second must take O' back to O. But the translation $O' \to O$ is the inverse of $O \to O'$. Thus the only case in which the product of two translations has an invariant point is when one of the translations is the inverse of the other. (By our convention, the product is still a translation even then.) Hence

13.23 *The product of two translations $A \to B$ and $B \to C$ is the translation $A \to C$.*

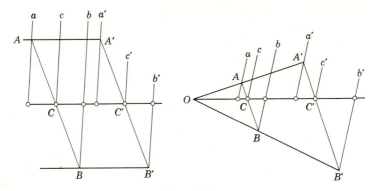

Figure 13.2c

To prove that this is a *commutative* product (as in 3.23), we consider first the easy case in which the two translations are along nonparallel lines. Completing the parallelogram $ABCD$, we observe that the translations $A \to B$ and $B \to C$ are the same as $D \to C$ and $A \to D$, respectively. Hence their product in either order is the translation $A \to C$:

$$(A \to B)(B \to C) = (A \to D)(D \to C)$$
$$= (B \to C)(A \to B).$$

To deal with the product of two translations T and X along the same line, let Y be any translation along a nonparallel line, so that X commutes with both Y and TY. Then

$$\text{TXY} = \text{TYX} = \text{XTY}$$

and therefore $$\text{TX} = \text{XT}$$

[cf. Veblen and Young **2,** p. 76].

As a special case of 5.12, we see that any two distinct points, A and B, are interchanged by a unique dilatation $AB \to BA$, or, more concisely,

$$A \leftrightarrow B,$$

which we call a *half-turn.* (Of course, $A \leftrightarrow B$ is the same as $B \leftrightarrow A$.) If C is any point outside the line AB, the half-turn transforms C into the point D in which the line through B parallel to AC meets the line through A parallel to BC (Figure 13.2d). Therefore $ADBC$ is a parallelogram, and the same half-turn can be expressed as $C \leftrightarrow D$. The invariant lines AB and CD, being the diagonals of the parallelogram, intersect in a point O, which is the invariant point of the half-turn. It follows that any segment AB has a *midpoint* which can be defined to be the invariant point of the half-turn $A \leftrightarrow B$, and we have proved that the center of a parallelogram is the midpoint of each diagonal, that is, that the two diagonals "bisect" each other. To see how the half-turn transforms an arbitrary point on AB, we merely have to join this point to C (or D) and then draw a parallel line through D (or C).

By considering their effect on an arbitrary point B, we may express any two half-turns as $A \leftrightarrow B$ and $B \leftrightarrow C$. If their product has an invariant point O, each of them must be expressible in the form $O \leftrightarrow O'$, that is, they must coincide. In every other case, there is no invariant point. Hence

13.24   *The product of two half-turns $A \leftrightarrow B$ and $B \leftrightarrow C$ is the translation $A \to C$.*

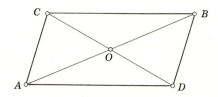

Figure 13.2d

We have seen (Figure 13.2*d*) that, if $ADBC$ is a parallelogram, the half-turn $A \leftrightarrow B$ is the same as $C \leftrightarrow D$, and the translation $A \to D$ is the same as $C \to B$. This connection between half-turns and translations remains valid when the parallelogram collapses to form a symmetrical arrangement of four collinear points, as in Figure 13.2*e*:

Figure 13.2e

13.25   *The half-turns $A \leftrightarrow B$ and $C \leftrightarrow D$ are equal if and only if the translations $A \to D$ and $C \to B$ are equal.*

In fact, the relation $(A \leftrightarrow B) = (C \leftrightarrow D)$ implies
$$(A \to D) = (A \leftrightarrow B)(B \leftrightarrow D)$$
$$= (C \leftrightarrow D)(D \leftrightarrow B) = (C \to B)$$
and, conversely, the relation $(A \to D) = (C \to B)$ implies
$$(A \leftrightarrow B) = (A \to D)(D \leftrightarrow B)$$
$$= (C \to B)(B \leftrightarrow D) = (C \leftrightarrow D).$$

In the special case when C and D coincide, we call them C' and deduce that C' is the midpoint of AB if and only if the translations $A \to C'$ and $C' \to B$

are equal. This involves the existence of parallelograms $AC'A'B'$ and $A'B'C'B$, as in Figure 13.2f. Completing the parallelogram $B'C'A'C$, we obtain a triangle ABC with A', B', C' at the midpoints of its sides. Hence

13.26 *The line joining the midpoints of two sides of a triangle is parallel to the third side, and the line through the midpoint of one side parallel to another passes through the midpoint of the third.*

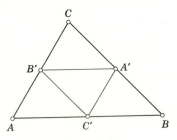

Figure 13.2f

Two figures are said to be *homothetic* if they are related by a dilatation, *congruent* if they are related by a translation or a half-turn. In particular, a directed segment AB is congruent to its "opposite" segment BA by the half-turn $A \leftrightarrow B$. Thus, in Figure 13.2f, the four small triangles

$$AC'B', \quad C'BA', \quad B'A'C, \quad A'B'C'$$

are all congruent, and each of them is homothetic to the large triangle ABC.

EXERCISES

1. Such equations as those used in proving 13.25 are easily written down if we remember that each must involve an even number of double-headed arrows (indicating half-turns). Explain this rule.

2. The translations $A \to C$ and $D \to B$ are equal if the translations $A \to D$ and $C \to B$ are equal. (This is obvious when $ADBC$ is a parallelogram, but remarkable when all the points are collinear.)

3. Setting $A = C$ in the equation

$$(A \leftrightarrow B)(B \to C) = (A \leftrightarrow C),$$

deduce that any given point C is the invariant point of a half-turn $(C \leftrightarrow B)(B \to C)$ which, by a natural extension of the symbolism, may be written as

$$C \leftrightarrow C.$$

4. If the three diagonals of a hexagon (not necessarily convex) all have the same midpoint, any two opposite sides are parallel (as in Figure 4.1e).

5. From any point A_1 on the side BC of a triangle ABC, draw A_1B_1 parallel to BA to meet CA in B_1, then B_1C_1 parallel to CB to meet AB in C_1, and then C_1A_2 parallel to AC to meet BC in A_2. If A_1 is the midpoint of BC, A_2 coincides with it. If not, continue the process, drawing A_2B_2 parallel to BA, B_2C_2 parallel to CB, and C_2A_3 parallel to AC. The path is now closed: A_3 coincides with A_1. (This is called Thom-

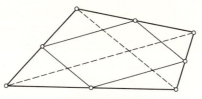

Figure 13.2g

sen's figure. See Geometrical Magic, by Nev R. Mind, *Scripta Mathematica,* **19** (1953), pp. 198–200.)

6. The midpoints of the four sides of any simple quadrangle are the vertices of a parallelogram (Figure 13.2g; cf. Figure 4.2c). This theorem was discovered by Pierre Varignon (1654–1722). It shows that the *bimedians,* which join the midpoints of opposite sides of the quadrangle, bisect each other. Thus the corollary to Hjelmslev's theorem (§ 3.6) becomes an affine theorem when we replace the hypotheses 3.61 by

$$AB = BC, \quad A'B' = B'C'.$$

7. The midpoints of the six sides of any complete quadrangle are the vertices of a centrally symmetrical hexagon (of the kind considered in Ex. 4, above).

13.3 AFFINITIES

> *"Yes, indeed," said the Unicorn, . . . "What can we measure? . . .*
> *We are experts in the theory of measurement, not its practice."*
>
> J. L. Synge [**2,** p. 51]

The results of § 13.2 may be summarized in the statement that all the translations of the affine plane form a continuous Abelian group, which is a subgroup of index 2 in the group of translations and half-turns; and the latter is a subgroup (of infinite index) in the group of dilatations [Veblen and Young **2,** pp. 79, 93].

Moreover, the group of translations is a *normal* subgroup (or "self-conjugate" subgroup)* in the group of dilatations, that is, if T is a translation while S is a dilatation, then S⁻¹TS is a translation [Artin **1,** p. 57]. To prove this, suppose if possible that the dilatation S⁻¹TS has an invariant point. Since this invariant point could have been derived from a suitable point O by applying S, we may denote it by O^S. Thus S⁻¹TS leaves O^S invariant. But S⁻¹TS transforms O^S into O^{TS}. Hence $O^{TS} = O^S$. Applying S⁻¹, we deduce $O^T = O$, which is absurd (since T has no invariant point).

If T is $A \rightarrow B$ and S is $AB \rightarrow A^S B^S$, then S⁻¹TS is $A^S \rightarrow B^S$. Accordingly, it is sometimes convenient to write T^S for S⁻¹TS [see, e.g., Coxeter **1,** p. 39] and to say that the dilatation S *transforms* the translation T into the translation T^S. (Since $A^S B^S$ is parallel to AB, T^S has the same direction as T.) In other words, a dilatation transforms the group of translations into itself in the manner of an *automorphism:* if it transforms T into T^s and another translation U into U^s, it transforms the product TU into (TU)^s = T^{s}U^s and any power of T into the same power of T^s.

It is convenient to use the italic letter *T* for the point into which the translation T transforms an arbitrarily chosen initial point (or origin) *1*. Then, if a central dilatation S has *1* as its invariant point, it not only transforms T into T^s but also transforms *T* into *T*^{s}.

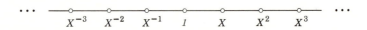

$$X^{-3} \quad X^{-2} \quad X^{-1} \quad 1 \quad X \quad X^2 \quad X^3$$

Figure 13.3a

Applying to the arbitrary point *1* all the integral powers of a given translation X, we obtain a *one-dimensional lattice* consisting of infinitely many points "evenly spaced" along a line, as in Figure 13.3a. We may regard every such point X^μ as being derived from the point *X* by a dilatation $1X \rightarrow 1X^\mu$ (which leaves the point *1* invariant). At first we take μ to be an integer; but since the same dilatation transforms each X^n into

$$(X^\mu)^n = X^{\mu n},$$

we can consistently extend the meaning of X^μ so as to allow μ to have any rational value, and finally any real value. In other words, we can interpolate new points between the points of the one-dimensional lattice and then define X^μ, for any real μ, to mean the translation $1 \rightarrow X^\mu$. The details are as follows.

For each rational number $\mu = a/b$ (where a is an integer and b is a positive integer) we derive from the point *X* a new point X^μ by means of the dilatation $1X^b \rightarrow 1X^a$. A convenient way to construct this point X^μ is to use the lattice of powers of an arbitrary translation Y along another line through the initial point *1*, drawing a line through the point *Y* parallel to the join of the points Y^b and X^a, as in Figure 13.3b (cf. Figure 9.1c).

To verify that the order of such points X^μ agrees with the order of the rational numbers μ, we take three of them and reduce their μ's to a common denominator so as to express them as $X^{a_1/b}$, $X^{a_2/b}$, $X^{a_3/b}$. If $a_1 < a_2 < a_3$, so that $[X^{a_1} X^{a_2} X^{a_3}]$, we can apply 13.22 to the dilatation $1X^b \rightarrow 1X$, with the conclusion that

$$[X^{a_1/b} \ X^{a_2/b} \ X^{a_3/b}].$$

If μ is irrational, we define X^μ to be the Dedekind section between all the rational points $X^{a/b}$ for which $a/b < \mu$ and all those for which $a/b > \mu$. More precisely, supposing for definiteness that μ is positive, we apply the "ray" version of 12.51 to two sets of points, one consisting of all the points

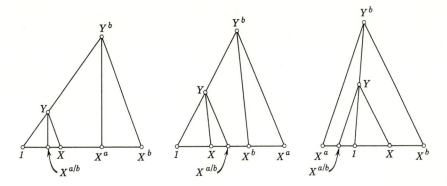

Figure 13.3b

whose exponents are positive rational numbers less than μ, and all the points between pairs of these, whereas the other set consists of the rest of the "positive" ray $1X$. (If μ is negative, we make the same kind of partition of the "negative" ray $1/X$.) Finally X^μ is, by definition, the translation $1 \to X^\mu$.

We have now interpreted the symbol X^μ for all real values of μ (including 0 and 1, which yield $X^0 = 1$ and $X^1 = X$). Conversely, *every point on the line $1X$ can be expressed in the form X^μ*.

This is obvious for any point of the interval from X^{-1} to X. Any other point T satisfies either $[1 \ X \ T]$ or $[1 \ X^{-1} \ T]$. If $[1 \ X \ T]$, the dilatation $1T \to 1X$ transforms X into a point between 1 and X, say X^λ. The inverse dilatation $1X^\lambda \to 1X$ transforms X into $X^{1/\lambda}$; therefore $T = X^{1/\lambda}$. If, on the other hand, $[1 \ X^{-1} \ T]$, we make the analogous use of $1T \to 1X^{-1}$. In either case we obtain an expression for T as a power of X.

Thus, assuming Dedekind's axiom, we have proved the "axiom of Archimedes":

13.31   *For any point T (except 1) on the line of a translation* X, *there is an integer n such that T lies between the points 1 and X^n.*

The exponent μ provides a measure of distance along the line $1X$. In fact, the segment $X^\nu X^\mu$ ($\nu < \mu$) is said to have *length $\mu - \nu$* in terms of the segment $1X$ as unit:

$$\frac{X^\nu X^\mu}{1X} = \mu - \nu.$$

Along another line $1Y$ (Figure 13.3c) we have an independent unit. Since the dilatation $1X \to 1X^\mu$ transforms the point Y into Y^μ, where the line $X^\mu Y^\mu$ is parallel to XY, we have

$$\frac{1X^\mu}{1X} = \frac{1Y^\mu}{1Y}$$

in agreement with Euclid VI.2 (see § 1.3). Thus we can define ratios of the lengths on one line, or on parallel lines, and we can compare such ratios on different lines. But affine geometry contains no machinery for comparing lengths in different directions: it is a meaningless question whether the translation Y is longer or shorter than X.

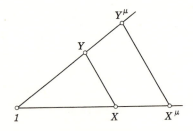

Figure 13.3c

The above definition for the length of the segment $X^\nu X^\mu$ ($\nu < \mu$) suggests the propriety of allowing the oppositely directed segment $X^\mu X^\nu$ to have the negative length $\nu - \mu$. This convention enables us to write $\mu = 1X^\mu / 1X$ for negative as well as positive values of μ, and to add lengths of collinear segments according to such formulas as

$$AB + BC = AC, \qquad BC + CA + AB = 0,$$

regardless of the order of their end points A, B, C.

Now, to set up a system of *affine coordinates* in the plane, we let (x, y) denote the point into which the origin 1 is transformed by the translation $X^x Y^y$. This simple device establishes a one-to-one correspondence between points in the plane and ordered pairs of real numbers. In particular, the point X^x is $(x, 0)$, Y^y is $(0, y)$, and the origin itself is $(0, 0)$. When x and y are integers, the points (x, y) form a *two-dimensional lattice,* as in Figure 4.1b. The remaining points (x, y) are distributed between the lattice points in the obvious manner.

In affine coordinates (as in Cartesian coordinates) a line has a linear equation. The powers of the translation $X^{-b} Y^a$ transform the origin into the points $(-\mu b, \mu a)$ whose locus is the line $ax + by = 0$. The same powers transform (x_1, y_1) into the points

$$(x_1 - \mu b, y_1 + \mu a)$$

whose locus is

$$a(x - x_1) + b(y - y_1) = 0.$$

We can thus express a line in any of the standard forms 8.11, 8.12, 8.13.

A dilatation is a special case of an *affinity,* which is any transformation (of the whole affine plane onto itself) preserving collinearity. Thus, an affinity transforms parallel lines into parallel lines, and preserves ratios of distances along parallel lines. It also preserves intermediacy (compare 13.22).

13.32 *An affinity is uniquely determined by its effect on any one triangle.*

For, if it transforms a triangle IXY into $I'X'Y'$, it transforms the point (x, y) referred to the former triangle into the point having the *same* coordinates referred to the latter. Here IXY and $I'X'Y'$ may be *any* two triangles [Veblen and Young **2**, p. 72], and we naturally speak of "the affinity $IXY \rightarrow I'X'Y'$." In particular, if ABC and ABC' are two triangles with a common side, $ABC \rightarrow ABC'$ is called a *shear* or a *strain* according as the line CC' is or is not parallel to AB. One kind of strain is sufficiently important to deserve a special name and a special symbol: the *affine reflection* $A(CC')$ or $B(CC')$, which arises when the midpoint of CC' lies on AB. In other words, any triangle ACC' determines an affine reflection $A(CC')$ whose *mirror* (or "axis") is the median through A and whose *direction* is the direction of all lines parallel to CC'.

In the language of the Erlangen program (see page 67), the principal group for affine geometry is the group of all affinities.

EXERCISES

1. The shear or strain $ABC \rightarrow ABC'$ leaves invariant every point on the line AB. What is its effect on a point P of general position?

2. Every affinity of period 2 is either a half-turn or an affine reflection.

3. If, for a given affinity, every noninvariant point lies on at least one invariant line, then the affinity is either a dilatation or a shear or a strain.

4. In terms of affine coordinates, affinities are "affine transformations"

13.33
$$x' = ax + by + l,$$
$$y' = cx + dy + m, \qquad\qquad ad \neq bc.$$

5. Describe the transformations

(i) $x' = x + 1, y' = y$;   (ii) $x' = ax, y' = ay$;

(iii) $x' = x + by, y = y$;   (iv) $x' = ax, y' = y$.

13.4 EQUIAFFINITIES

> For he, by Geometrick scale,
> Could take the size of Pots of Ale.
>
> Samuel Butler (1600 -1680)
>
> (Hudibras, I.1)

We are now ready to show how the comparison of lengths on parallel lines can be extended to yield a comparison of areas in any position [cf. Forder **1**, pp. 259–265; Coxeter **2**, pp. 125–128]. For simplicity, we restrict consideration to *polygonal* regions. (Other shapes may be included by a suitable limiting process of the kind used in integral calculus.) Clearly, any polygonal region can be dissected into a finite number of triangles.* Following H. Hadwiger and P. Glur [*Elemente der Mathematik*, **6** (1951), pp. 97–120], we declare two such regions to be *equivalent* if they can be dissected into a finite number of pieces that are congruent in pairs (by translations or by half-turns). In other words, two polygonal regions are equivalent if they can be derived from each other by dissection and rearrangement. Superposing two different dissections, we see that this kind of equivalence, which is obviously reflexive and symmetric, is also transitive; two polygons that are equivalent to the same polygon are equivalent to each other.

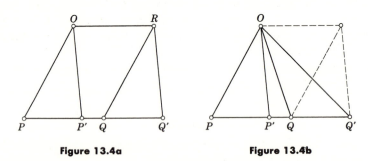

Figure 13.4a   **Figure 13.4b**

The parallelograms $OPQR$ and $OP'Q'R$ of Figure 13.4a are equivalent, since each of them consists of the trapezoid $OP'QR$ plus one of the two congruent triangles OPP', RQQ'. In some such cases, more than two pieces may be needed, but we find eventually:

13.41   *Two parallelograms are equivalent if they have one pair of opposite sides of the same length lying on the same pair of parallel lines.*

Since a parallelogram can be dissected along a diagonal to make two triangles that are congruent by a half-turn, it follows that two triangles (such as OPQ and $OP'Q'$ in Figure 13.4b) are equivalent if they have a common vertex while their sides opposite to this vertex are congruent segments on one line. In particular, if points P_0, P_1, ... , P_n are evenly spaced along a line (not through O), so that the segments P_0P_1, P_1P_2, ... are all congruent, as in Figure 13.4c, then the triangles OP_0P_1, OP_1P_2, ... are all equivalent, and we naturally say that the *area* of OP_0P_n is n times the area of OP_0P_1. By interpolation of further points on the same line, we can extend this idea to all real values of n, with the conclusion that, if Q is on the side PQ' of a triangle OPQ', as in Figure 13.4d, the *Cevian* OQ divides the area of the triangle in the same ratio that the point Q divides the side:

13.42
$$\frac{OPQ}{OPQ'} = \frac{PQ}{PQ'}.$$

* N. J. Lennes, *American Journal of Mathematics*, **33** (1911), p. 46.

We naturally regard this ratio as being negative if P lies between Q and Q', that is, if the two triangles are oppositely oriented.

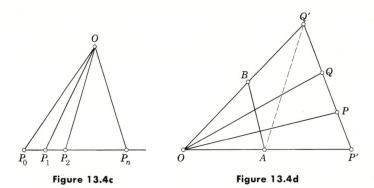

Figure 13.4c   **Figure 13.4d**

These ideas enable us to define the area of any polygon in such a way that *equivalent polygons have the same area,* and when two polygons are stuck together to make a larger polygon, the areas are added. To compute the area of a given polygon in terms of a standard triangle OAB as unit of measurement, we dissect the polygon into triangles and add the areas of the pieces, each computed as follows.

By applying a suitable translation, any given triangle can be shifted so that one vertex coincides with the vertex O of the standard triangle OAB. Accordingly, we consider a triangle OPQ. Let the line PQ meet OA in P', and OB in Q', as in Figure 13.4d. Multiplying together the three ratios

$$\frac{OPQ}{OP'Q'} = \frac{PQ}{P'Q'}, \quad \frac{OP'Q'}{OAQ'} = \frac{OP'}{OA}, \quad \frac{OAQ'}{OAB} = \frac{OQ'}{OB}$$

we obtain the desired ratio

13.43
$$\frac{OPQ}{OAB} = \frac{PQ}{P'Q'} \frac{OP'}{OA} \frac{OQ'}{OB}.$$

To obtain an analytic expression for the area of a triangle OPQ, referred to axes through the vertex O, we take the coordinates of the points

$$O, \quad A, \quad B, \quad P, \quad Q, \quad P', \quad Q'$$

to be

$$(0, 0), (1, 0), (0, 1), (x_1, y_1), (x_2, y_2), (p, 0), (0, q),$$

respectively. Since the equation

$$\frac{x}{p} + \frac{y}{q} = 1$$

for the line PQ is satisfied by (x_1, y_1) and (x_2, y_2), we have

$$\frac{1 - y_1/q}{x_1} = \frac{1}{p} = \frac{1 - y_2/q}{x_2},$$

whence

$$q = \frac{x_1 y_2 - x_2 y_1}{x_1 - x_2}.$$

Taking the product of

$$\frac{PQ}{P'Q'} = \frac{x_1 - x_2}{p}, \quad \frac{OP'}{OA} = p, \quad \frac{OQ'}{OB} = \frac{x_1 y_2 - x_2 y_1}{x_1 - x_2},$$

we obtain

13.44
$$\frac{OPQ}{OAB} = x_1 y_2 - x_2 y_1 = \begin{vmatrix} x_1 & y_1 \\ x_2 & y_2 \end{vmatrix}.$$

We deduce, as in § 8.2, that a triangle

$$(x_1, y_1) \, (x_2, y_2) \, (x_3, y_3),$$

of general position, has area PQR, where

13.45
$$\frac{PQR}{OAB} = \begin{vmatrix} x_1 & y_1 & 1 \\ x_2 & y_2 & 1 \\ x_3 & y_3 & 1 \end{vmatrix}.$$

Since the homogeneous linear transformation

$$x' = ax + by, \qquad y' = cx + dy$$

takes the triangle OAB to

$$(0, 0)(a, c)(b, d),$$

we conclude that the affinity 13.33 *preserves area* if and only if

$$ad - bc = 1.$$

An area-preserving affinity is called an *equiaffinity* (or "equiaffine collinea-tion" [Veblen and Young **2,** pp. 105–113]). The group of all equiaffinities, like the group of all dilatations, includes the group of all translations and half-turns as a normal subgroup, and is itself a normal subgroup in the group of all affinities. Equiaffinities are of many kinds. Here are some examples:

The *hyperbolic rotation* (or "Lorentz transformation")

13.46 $$x' = \mu^{-1}x, \quad y' = \mu y \qquad (\mu > 0, \quad \mu \neq 1),$$

for which $x'y' = xy$, leaves invariant each branch of the hyperbola $xy = 1$. The *crossed hyperbolic rotation*

13.47
$$x' = -\mu^{-1}x, \quad y' = -\mu y \qquad (\mu > 0, \quad \mu \neq 1)$$

interchanges the two branches. The *parabolic rotation*

13.48
$$x' = x + 1, \quad y' = 2x + y + 1,$$

for which $x'^2 - y' = x^2 - y$, leaves invariant the parabola $y = x^2$. The *elliptic rotation*

13.49
$$x' = x \cos \theta - y \sin \theta, \quad y' = x \sin \theta + y \cos \theta$$

leaves invariant the ellipse $x^2 + y^2 = 1$, and is periodic if θ is commensurable with π.

In §2.8 (page 36) we derived a regular polygon $P_0 P_1 P_2 \ldots$ from a point P_0 (other than the center) by repeated application of a rotation through $2\pi/n$. (The rotation takes P_0 to P_1, P_1 to P_2, and so on.) Although measurement of angles has no meaning in affine geometry, we can define an *affinely regular polygon* whose vertices P_j are derived from suitable point P_0 by repeated application of an equiaffinity. The polygon is said to be of type $\{n\}$ if the equiaffinity is an elliptic rotation 13.49, where $\theta = 2\pi/n$ and n is rational, so that P_j has affine coordinates

$$(\cos j\theta, \sin j\theta) \qquad\qquad (\theta = 2\pi/n).$$

Figure 13.4e shows an affinely regular pentagram ($n = 5/2$) and pentagon ($n = 5$).

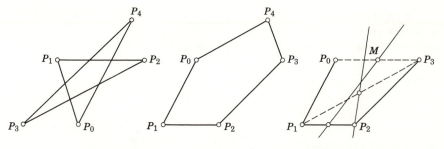

Figure 13.4e   **Figure 13.4f**

EXERCISES

1. Two triangles with a common side (such as ABC and BCD in Figure 13.2d) have the same area if and only if the line joining their remaining vertices is parallel to the common side (that is, AD parallel to BC).

2. If a pentagon has four of its diagonals parallel to four of its sides, the remaining diagonal is parallel to the remaining side.

3. When is a dilatation an equiaffinity?

4. When is a shear an equiaffinity?

5. When is a strain an equiaffinity?

6. The product of any even number of affine reflections is an equiaffinity.

7. Any translation or half-turn or shear can be expressed as the product of two affine reflections.

8. If an equiaffinity is neither a translation nor a half-turn nor a shear, it can be expressed as $P_0P_1P_2 \to P_1P_2P_3$ where P_0P_3 is parallel to P_1P_2. (See Figure 13.4*f*.)

9. Every equiaffinity can be expressed as the product of two affine reflections. (Veblen.)

10. In an affinely regular polygon $P_0P_1P_2 \ldots$, the lines P_iP_j and P_hP_k are parallel whenever $i + j = h + k$.

11. Why did we call $x^2 + y^2 = 1$ an ellipse rather than a circle (just below 13.49)?

12. What triangles and quadrangles are affinely regular?

13. Construct an affinely regular hexagon.

14. Compute the ratio P_0P_3/P_1P_2 for an affinely regular polygon of type $\{n\}$.

15. For which values of n can an affinely regular polygon of type $\{n\}$ be constructed with a parallel-ruler?

13.5 TWO-DIMENSIONAL LATTICES

> *Farey has a notice of twenty lines in the* Dictionary of National Biog-
> raphy. *. . . . His biographer does not mention the one thing in his life
> which survives.*
>
> G. H. Hardy
> [Hardy and Wright **1,** p. 37]

Our treatment of lattices in § 4.1 (as far as the description of Figure 4.1*d*) is purely affine. In fact, a lattice is the set of points whose affine coordinates are integers. Any one of the points will serve as the origin O.

Let A' be any lattice point, and A the first lattice point along the ray OA'. Following Hardy and Wright [**1,** p. 29], we call A a *visible* point, because there is no lattice point between O and A to hide A from an observer at O. In terms of affine coordinates, a necessary and sufficient condition for (x, y) to be visible is that the integers x and y be coprime, that is, that they have no common divisor greater than 1. The three visible points

$$(1, 0), \quad (1, 1), \quad (0, 1)$$

form with the origin a parallelogram. This is called a *unit cell* (or "typical parallelogram") of the lattice, because the translations transform it into infinitely many such cells filling the plane without overlapping and without interstices: it is a fundamental region for the group of translations. Thus it serves as a convenient unit for computing the area of a region.

According to Steinhaus [**2,** pp. 76–77, 260] it was G. Pick, in 1899, who discovered the following theorem:*

---

* For an extension to three dimensions, see J. E. Reeve, On the volume of lattice polyhedra, *Proceedings of the London Mathematical Society* (3), **7** (1957), pp. 378–395.

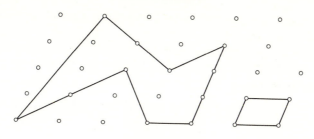

Figure 13.5a

13.51   *The area of any simple polygon whose vertices are lattice points is given by the formula*

$$\tfrac{1}{2}b + c - 1,$$

where b is the number of lattice points on the boundary while c is the number of lattice points inside.

(By a "simple" polygon we mean one whose sides do not cross one another. Figure 13.5a shows an example in which $b = 11$, $c = 3$.)

Proof.   We first observe that the expression $\tfrac{1}{2}b + c - 1$ is additive when two polygons are juxtaposed. In fact, if two polygons, involving $b_1 + c_1$ and $b_2 + c_2$ lattice points respectively, have a common side containing n ($\geqslant 0$) lattice points in addition to the two vertices at its ends, then the values of b and c for the combined polygon are

$$b = b_1 + b_2 - 2n - 2, \qquad c = c_1 + c_2 + n,$$

so that

$$\tfrac{1}{2}b + c - 1 = (\tfrac{1}{2}b_1 + c_1 - 1) + (\tfrac{1}{2}b_2 + c_2 - 1).$$

Next, the formula holds for a parallelogram having no lattice points on its sides (so that $b = 4$ and the expression reduces to $c + 1$). For, when N such parallelograms are fitted together, four at each vertex, to fill a large region, the number of lattice points involved (apart from a negligible peripheral error) is $N(c + 1)$, and this must be the same as the number of unit cells needed to fill the same region.

Splitting the parallelogram into two congruent triangles by means of a diagonal, we see that the formula holds also for a triangle having no lattice points on its sides. A triangle that does have lattice points on a side can be dealt with by joining such points to the opposite vertex so as to split the triangle into smaller triangles. This procedure may have to be repeated, but obviously only a finite number of times. Finally, as we remarked on page 204, any given polygon can be dissected into triangles; then the expressions for those pieces can be added to give the desired result.

In particular, any parallelogram for which $b = 4$ and $c = 0$ has area 1 and can serve as a unit cell. If the vertices of such a parallelogram (in counterclockwise order) are

$$(0, 0), \quad (x, y), \quad (x + x_1, y + y_1), \quad (x_1, y_1),$$

we see from 13.44 that

13.52 $$xy_1 - yx_1 = 1.$$

In other words, this is the condition for the points

13.53 $$(0, 0), \quad (x, y), \quad (x_1, y_1)$$

to form a positively oriented "empty" triangle of area $\frac{1}{2}$, which could be used just as well as $(0, 0)$ $(1, 0)$ $(0, 1)$ to generate the lattice. Thus a lattice is completely determined, apart from its position, by the area of its unit cell. Moreover, although there are infinitely many visible points in a given lattice, they all play the same role. (These properties of affine geometry are in marked contrast to Euclidean geometry, where the shape of a lattice admits unlimited variation and each lattice contains visible points at infinitely many different distances.)

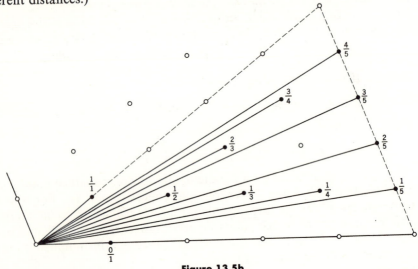

Figure 13.5b

George Pólya[*] has applied 13.52 to a useful lemma in the theory of numbers. The *Farey series* F_n of order n is the ascending sequence of fractions from 0 to 1 whose denominators do not exceed n. Thus y/x belongs to F_n if x and y are coprime and

13.54 $$0 \leqslant y \leqslant x \leqslant n.$$

For instance, F_5 is

$$\tfrac{0}{1}, \tfrac{1}{5}, \tfrac{1}{4}, \tfrac{1}{3}, \tfrac{2}{5}, \tfrac{1}{2}, \tfrac{3}{5}, \tfrac{2}{3}, \tfrac{3}{4}, \tfrac{4}{5}, \tfrac{1}{1}.$$

[*] *Acta Litterarum ac Scientiarum Regiae Universitatis Hungaricae Francisco-Josephinae*, Sectio Scientiarum Mathematicarum, **2** (1925), pp. 129–133.

The essential property of such a sequence, from which many other properties follow by simple algebra, is that 13.52 holds for any two adjacent fractions

$$\frac{y}{x} \text{ and } \frac{y_1}{x_1}.$$

To prove this, we represent each term y/x of the sequence by the point (x, y) of a lattice. For example, the terms of F_5 are the lattice points emphasized in Figure 13.5*b* (where, for convenience, the angle between the axes is obtuse). Since the fractions are in their "lowest terms," the points are visible. By 13.54, they belong to the triangle $(0, 0)$ $(n, 0)$ (n, n). A ray from the origin, rotated counterclockwise, passes through the representative points in their proper order. If y/x and y_1/x_1 are consecutive terms of the sequence, then (x, y) and (x_1, y_1) are visible points such that the triangle joining them to the origin contains no lattice point in its interior. Hence this triangle is one half of a unit cell, and 13.52 holds, as required.

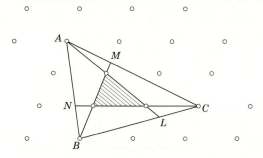

Figure 13.5c

Another result belonging to affine geometry is

13.55 *If the sides BC, CA, AB of a triangle ABC are divided at L, M, N in the respective ratios $\lambda : 1, \mu : 1, \nu : 1$, the Cevians AL, BM, CN form a triangle whose area is*

$$\frac{(\lambda\mu\nu - 1)^2}{(\lambda\mu + \lambda + 1)(\mu\nu + \mu + 1)(\nu\lambda + \nu + 1)}$$

times that of ABC.

This was discovered by Routh [**1**, p. 82; see also Dörrie **1**, pp. 41–42]. We shall give a general proof in § 13.7, but it is interesting to observe that, when $\lambda = \mu = \nu$, so that the ratio of areas is $(\lambda - 1)^3/(\lambda^3 - 1)$, the result can be deduced from 13.51. For instance, when $\lambda = \mu = \nu = 2$, so that each side is trisected [Steinhaus **2**, p. 8], the central triangle is one-seventh of the whole, and we can see this immediately by embedding the figure in a lattice, as in Figure 13.5*c*. Since the central triangle has $b = 3, c = 0$ while ABC has $b = 3, c = 3$, the ratio of areas is $\frac{1}{2}/\frac{7}{2} = \frac{1}{7}$.

EXERCISES

1. If y/x and y_1/x_1 are two consecutive terms of a Farey series, x and x_1 are coprime.

2. If y_0/x_0, y/x, y_1/x_1 are three consecutive terms of a Farey series,

$$\frac{y_0 + y_1}{x_0 + x_1} = \frac{y}{x}.$$

(C. Haros, 1802.)

3. The points A, B, C in Figure 13.5c belong to a lattice whose unit cell has seven times the area of that of the basic lattice. (For the Euclidean theory of such compound lattices, see Coxeter, Configurations and maps, *Reports of a Mathematical Colloquium* (2), **8** (1948), pp. 18–38, especially Figs. i, v, vii.)

4. Use lattices to verify 13.55 when (a) $\lambda = \mu = \nu = 3$, (b) $\lambda = \mu = \nu = \frac{3}{2}$.

5. Join the vertices A, B, C, D of a parallelogram to the midpoints of the respective sides BC, CD, DA, AB so as to form a smaller parallelogram in the middle. Its area is one-fifth that of $ABCD$. Another such parallelogram is obtained by joining A, B, C, D to the midpoints of CD, DA, AB, BC. The common part of these two small parallelograms is a centrally symmetrical octagon whose area is one-sixth that of $ABCD$ [Dörrie **1**, p. 40].

6. In the notation of 13.55, the area of the triangle LMN is

$$\frac{\lambda\mu\nu + 1}{(\lambda + 1)(\mu + 1)(\nu + 1)}$$

times that of ABC. (*Hint*: Use 13.42 to compute the relative area of CLM, etc.)

7. Of the four triangles ANM, BLN, CML, LMN, the last cannot have the smallest area unless L, M, N are the midpoints of BC, CA, AB. (H. Debrunner.*)

13.6 VECTORS AND CENTROIDS

> A vector is really the same thing as a translation, although one uses different phraseologies for vectors and translations. Instead of speaking of the translation $A \rightarrow A'$ which carries the point A into A' one speaks of the vector $\overrightarrow{AA'}$. . . . The same vector laid off from B ends in B' if the translation carrying A into A' carries B into B'.
>
> H. Weyl [**1**, p. 45]

As we saw in § 2.5, a *group* is an associative system containing an identity and, for each element, an inverse. Arithmetical instances are provided by the positive rational numbers, the positive real numbers, the complex numbers of modulus 1, and all the complex numbers except 0, combined, in each case, by ordinary multiplication. Such instances make it natural to adopt a multiplicative notation for all groups, so that the combination of S and T is ST, the inverse of S is S^{-1}, and the identity is 1. However, it is often convenient, especially in the case of Abelian (i.e., commutative)

* *Elemente der Mathematik*, **12** (1957), p. 43, Aufgabe 260.

groups, to use instead the additive notation, in which the combination of S and T is S + T, the inverse of S is −S, and the identity is 0. To see that this other notation has equally simple arithmetical instances, we merely have to consider in turn the integers, the rational numbers, the real numbers, and the complex numbers, combined, in each case, by ordinary addition.

The transition from a multiplicative group to the corresponding additive group is the foundation of the theory of logarithms [Infeld **1**, pp. 97–100].

When we go outside the domain of arithmetic, the choice between multiplication and addition is merely a matter of notation. In particular, the Abelian group of translations, which we have expressed as a multiplicative group, becomes the additive group of *vectors*.

In this notation, 13.21 asserts that any two points A and A' determine a unique vector $\overrightarrow{AA'}$ (going from A to A'), Figure 13.2b illustrates a situation in which

$$\overrightarrow{AA'} = \overrightarrow{CC'} = \overrightarrow{BB'},$$

13.23 asserts that

$$\overrightarrow{AB} + \overrightarrow{BC} = \overrightarrow{AC},$$

and 3.23 asserts that, for any two vectors **a** and **b**,

$$\mathbf{a} + \mathbf{b} = \mathbf{b} + \mathbf{a}.$$

In the same spirit the "origin" will henceforth be called O instead of 1, and the zero vector will be denoted by **0**. The integral multiples of any non-zero vector proceed from the origin to the points of a one-dimensional lattice. Two vectors **e** and **f** are said to be *independent* if neither is a (real) multiple of the other, that is, if the only numbers that satisfy the vector equation

$$x\mathbf{e} + y\mathbf{f} = \mathbf{0}$$

are $x = 0$ and $y = 0$. Two such vectors (corresponding to the translations X and Y in Figure 4.1c) provide a basis for a system of affine coordinates: they enable us to define the coordinates of any point to be the coefficients in the expression

$$x\mathbf{e} + y\mathbf{f}$$

for the *position vector* which goes from the origin to the given point. In other words, with reference to a triangle OAB, the affine coordinates of a point P are the coefficients in the expression

$$\overrightarrow{OP} = x\,\overrightarrow{OA} + y\,\overrightarrow{OB}.$$

We shall find it useful to borrow from statics the notion of the centroid (or "center of gravity") of a set of "weighted" points, that is, of points to each of which a real number is attached in a special way. For convenience, we shall call these numbers masses, although, when some of them are negative, electric charges provide a more appropriate illustration.

Let masses $t_1, \ldots, t_k$ be assigned to k distinct points $A_1, \ldots, A_k$, let O be any point (possibly coincident with one of the A's), and consider the vector

$$t_1 \overrightarrow{OA_1} + \ldots + t_k \overrightarrow{OA_k}.$$

If $t_1 + \ldots + t_k = 0$, this vector is independent of the choice of O. For, if we subtract from it the result of using O' instead, we obtain

$$t_1 (\overrightarrow{OA_1} - \overrightarrow{O'A_1}) + \ldots + t_k (\overrightarrow{OA_k} - \overrightarrow{O'A_k})$$

$$= (t_1 + \ldots + t_k) \overrightarrow{OO'} = \mathbf{0}.$$

More interestingly, if

$$t_1 + \ldots + t_k \neq 0,$$

we have

$$t_1 \overrightarrow{OA_1} + \ldots + t_k \overrightarrow{OA_k} = (t_1 + \ldots + t_k) \overrightarrow{OP},$$

where the point P is independent of the choice of O. For, if the same procedure with O' instead of O yields P' instead of P, we have, by subtraction,

$$(t_1 + \ldots + t_k) \overrightarrow{OO'} = (t_1 + \ldots + t_k)(\overrightarrow{OP} - \overrightarrow{O'P'})$$

whence

$$\overrightarrow{OP'} = \overrightarrow{OO'} + \overrightarrow{O'P'} = \overrightarrow{OP},$$

so that P' coincides with P. This point P, given by

$$\Sigma t_i \overrightarrow{OP} = \Sigma t_i \overrightarrow{OA_i},$$

is called the *centroid* (or "barycenter") of the k masses t_i at A_i.

Since, having found P, we may choose this position for O, we have

$$\Sigma t_i \overrightarrow{PA_i} = \mathbf{0}.$$

If there are only two points,

$$t_1 \overrightarrow{PA_1} = -t_2 \overrightarrow{PA_2},$$

so that P lies on the line A_1A_2 and divides the segment A_1A_2 in the ratio $t_2 : t_1$. In particular, if $t_1 = t_2$, P is the midpoint of A_1A_2.

For a triangle $A_1A_2A_3$, we have

$$(t_1 + t_2 + t_3) \overrightarrow{OP} = t_1 \overrightarrow{OA_1} + t_2 \overrightarrow{OA_2} + t_3 \overrightarrow{OA_3}$$
$$= t_1 \overrightarrow{OA_1} + (t_2 + t_3) \overrightarrow{OQ},$$

where Q is the centroid of t_2 at A_2 and t_3 at A_3. Thus, in seeking the centroid of three masses, we may replace two of them by their combined mass at their own centroid. (There is an obvious generalization to more than three masses.) In particular, when $t_1 = t_2 = t_3 (= 1$, say), Q is the midpoint of A_2A_3, and P divides A_1Q in the ratio $2:1$. Thus the "centroid" G of a triangle (§ 1.4) is the centroid of equal masses at its three vertices.

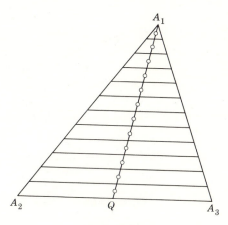

Figure 13.6a

This same point G, where the medians concur, is also the centroid of a triangular *lamina* or "plate" of uniform density. (Strictly speaking, this notion requires integral calculus.) For we may divide the triangle into thin strips parallel to the side A_2A_3, as in Figure 13.6a. The centroids of these strips evidently lie on the median A_1Q. Hence the centroid of the whole lamina lies on this median, and similarly on the others. (This argument was used by Archimedes in the third century B.C.)

EXERCISES

1. Verify in detail that
   (i) the positive rational numbers,
   (ii) the positive real numbers,
   (iii) the complex numbers of modulus 1,
   (iv) all the complex numbers except 0
form multiplicative groups; and that
   (v) the integers,
   (vi) the rational numbers,
   (vii) the real numbers,
   (viii) the complex numbers

form additive groups. Explain why the first four sets do not form additive groups, and why the last four do not form multiplicative groups.

2. If A, B, C are on one line and A', B', C' on another with

$$\frac{AB}{A'B'} = \frac{BC}{B'C'},$$

then points dividing all the segments AA', BB', CC' in the same ratio are either collinear or coincident (cf. § 3.6). (*Hint:* Consider the centroid of suitable masses at A, C, A', C'.)

3. The centroid of equal masses at the vertices of a quadrangle is the center of the Varignon parallelogram (Figure 13.2g).

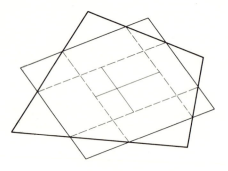

Figure 13.6b

4. The centroid of a quadrangular lamina is the center of the Wittenbauer parallelogram, whose sides join adjacent points of trisection of the sides, as in Figure 13.6b. This theorem, due to F. Wittenbauer (1857–1922) [Blaschke **2**, p. 13], was rediscovered by J. J. Welch and V. W. Foss.*

5. For what kind of quadrangle will the centroids described in the two preceding exercises coincide?

13.7 BARYCENTRIC COORDINATES

If $t_1 + t_2 \neq 0$, masses t_1 and t_2 at two fixed points A_1 and A_2 determine a unique centroid P, as in Figure 13.7a. This point is A_1 itself if $t_2 = 0$, A_2 if $t_1 = 0$. It is on the segment A_1A_2 if the t's are both positive (or both negative), on the ray A_1/A_2 if

$$t_1 > -t_2 > 0,$$

and on the ray A_2/A_1 if $\qquad t_2 > -t_1 > 0.$

* *Mathematical Gazette,* **42** (1958), p. 55; **43** (1959), p. 46.

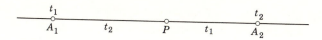

Figure 13.7a

Conversely, given a point P on the line A_1A_2, we can find numbers t_1 and t_2 such that

$$\frac{t_2}{t_1} = \frac{A_1P}{PA_2} \quad \text{or} \quad \frac{t_1}{t_2} = \frac{PA_2}{A_1P};$$

then P will be the centroid of masses t_1 and t_2 at A_1 and A_2. Since masses μt_1 and μt_2 (where $\mu \neq 0$) determine the same point as t_1 and t_2, these *barycentric coordinates* are homogeneous:

$$(t_1, t_2) = (\mu t_1, \mu t_2) \qquad (\mu \neq 0).$$

Similarly, as Möbius observed in 1827, we may set up barycentric coordinates in the plane of a *triangle of reference* $A_1A_2A_3$. If $t_1 + t_2 + t_3 \neq 0$, masses t_1, t_2, t_3 at the three vertices determine a point P (the centroid) whose coordinates are (t_1, t_2, t_3). In particular, $(1, 0, 0)$ is A_1, $(0, 1, 0)$ is A_2, $(0, 0, 1)$ is A_3, and $(0, t_2, t_3)$ is the point on A_2A_3 whose one-dimensional coordinates with respect to A_2 and A_3 are (t_2, t_3). To find coordinates for a given point P of general position, we find t_2 and t_3 from such a point Q on the line A_1P, as in Figure 13.7b, and then determine t_1 as the mass at A_1 that will balance a mass $t_2 + t_3$ at Q so as to make P the centroid. Again, as in the one-dimensional case, these coordinates are homogeneous:

$$(t_1, t_2, t_3) = (\mu t_1, \mu t_2, \mu t_3) \qquad (\mu \neq 0).$$

Joining P to A_1, A_2, A_3, we decompose $A_1A_2A_3$ into three triangles having a common vertex P. *The areas of these triangles are proportional to the barycentric coordinates of* P, as in Figure 13.7c. This fact follows at once from 13.42, since

$$\frac{t_3}{t_2} = \frac{A_2Q}{QA_3} = \frac{A_1A_2Q}{A_1QA_3} = \frac{PA_2Q}{PQA_3} = \frac{A_1A_2Q - PA_2Q}{A_1QA_3 - PQA_3} = \frac{PA_1A_2}{PA_3A_1},$$

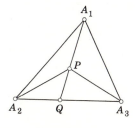

Figure 13.7b

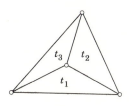

Figure 13.7c

and similarly for t_1/t_3, t_2/t_1. Positions of P outside the triangle are covered by means of our convention for the sign of the area of a directed triangle.

The inequality

$$t_1 + t_2 + t_3 \neq 0$$

enables us to normalize the coordinates so that

13.71 $$t_1 + t_2 + t_3 = 1.$$

(We merely have to divide each coordinate by the sum of all three.) These normalized barycentric coordinates are called *areal* coordinates, because they are just the areas of the triangles PA_2A_3, PA_3A_1, PA_1A_2, expressed in terms of the area of the whole triangle $A_1A_2A_3$ as unit of measurement. Areal coordinates are not homogeneous but "redundant": the position of a point is determined by two of the three, and the third is retained for the sake of symmetry. However, any expression involving them can be made homogeneous by inserting suitable powers of $t_1 + t_2 + t_3$ in appropriate places.

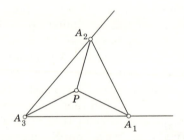

Figure 13.7d

In affine coordinates, as we have seen, a line has a linear equation. In barycentric coordinates, as we shall soon see, *a line has a linear homogeneous equation.* For this purpose we use the segments A_3A_1 and A_3A_2 as axes for affine coordinates, as in Figure 13.7d, so that the coordinates of P, A_1, A_2, A_3, which were formerly

$$(t_1, t_2, t_3), \quad (1, 0, 0), \quad (0, 1, 0), \quad (0, 0, 1),$$

are now

$$(x, y), \quad (1, 0), \quad (0, 1), \quad (0, 0).$$

By 13.44, the areas of PA_2A_3 and PA_3A_1, as fractions of the "unit" triangle $A_1A_2A_3$, are just

$$\begin{vmatrix} x & y \\ 0 & 1 \end{vmatrix} = x \quad \text{and} \quad \begin{vmatrix} 1 & 0 \\ x & y \end{vmatrix} = y.$$

By subtraction, the area of PA_1A_2 is $1 - x - y$. Hence the *areal* coordinates of P are related to the affine coordinates by the very simple formulas

$$t_1 = x, \quad t_2 = y, \quad t_3 = 1 - x - y.$$

The general line, having the affine equation 8.11, has the areal equation

$$at_1 + bt_2 + c = 0.$$

Making this homogeneous by the insertion of $t_1 + t_2 + t_3$, we deduce the barycentric equation

$$at_1 + bt_2 + c(t_1 + t_2 + t_3) = 0$$

or

$$(a + c)t_1 + (b + c)t_2 + ct_3 = 0$$

or, in a more symmetrical notation,

13.72 $$T_1t_1 + T_2t_2 + T_3t_3 = 0.$$

Thus every line has a linear homogeneous equation. In particular, the lines A_2A_3, A_3A_1, A_1A_2 have the equations

13.73 $$t_1 = 0, \quad t_2 = 0, \quad t_3 = 0.$$

The line joining two given points (r) and (s), meaning

$$(r_1, r_2, r_3) \quad \text{and} \quad (s_1, s_2, s_3),$$

has the equation

13.74
$$\begin{vmatrix} r_1 & r_2 & r_3 \\ s_1 & s_2 & s_3 \\ t_1 & t_2 & t_3 \end{vmatrix} = 0.$$

For, this equation is linear in the t's and is satisfied when the t's are replaced by the r's or the s's. Another way to obtain this result is to ask for the fixed points (r) and (s) to form with the variable point (t) a "triangle" whose area is zero. In terms of areal coordinates, with the triangle of reference as unit, the area of the triangle $(r)(s)(t)$ is, by 13.45 and 13.71,

$$\begin{vmatrix} r_1 & r_2 & 1 \\ s_1 & s_2 & 1 \\ t_1 & t_2 & 1 \end{vmatrix} = \begin{vmatrix} r_1 & r_2 & r_1 + r_2 + r_3 \\ s_1 & s_2 & s_1 + s_2 + s_3 \\ t_1 & t_2 & t_1 + t_2 + t_3 \end{vmatrix} = \begin{vmatrix} r_1 & r_2 & r_3 \\ s_1 & s_2 & s_3 \\ t_1 & t_2 & t_3 \end{vmatrix}.$$

Hence the area in general barycentric coordinates is this last determinant divided by

$$(r_1 + r_2 + r_3)(s_1 + s_2 + s_3)(t_1 + t_2 + t_3).$$

We are now ready to prove Routh's theorem 13.55 in its full generality. Identifying ABC with $A_1A_2A_3$, so that the points L, M, N are

$$(0, 1, \lambda), \quad (\mu, 0, 1), \quad (1, \nu, 0),$$

we can express the lines AL, BM, CN as

$$\lambda t_2 = t_3, \quad \mu t_3 = t_1, \quad \nu t_1 = t_2.$$

They intersect in pairs in the three points

$$(\mu, \mu\nu, 1), \quad (1, \nu, \nu\lambda), \quad (\lambda\mu, 1, \lambda),$$

forming a triangle whose area, in terms of that of the triangle of reference, is the result of dividing the determinant

$$\begin{vmatrix} \mu & \mu\nu & 1 \\ 1 & \nu & \nu\lambda \\ \lambda\mu & 1 & \lambda \end{vmatrix} = (\lambda\mu\nu - 1)^2$$

by $(\mu + \mu\nu + 1)(1 + \nu + \nu\lambda)(\lambda\mu + 1 + \lambda)$, in agreement with the statement of 13.55.

As an important special case we have

CEVA'S THEOREM. *Let the sides of a triangle ABC be divided at L, M, N in the respective ratios $\lambda : 1$, $\mu : 1$, $\nu : 1$. Then the three lines AL, BM, CN are concurrent if and only if $\lambda\mu\nu = 1$.*

The general line 13.72 meets the sides 13.73 of the triangle of reference in the points

$$(0, T_3, -T_2), \quad (-T_3, 0, T_1), \quad (T_2, -T_1, 0),$$

which divide them in the ratios

$$-\frac{T_2}{T_3}, \quad -\frac{T_3}{T_1}, \quad -\frac{T_1}{T_2},$$

whose product is -1. Conversely, any three numbers whose product is -1 can be expressed in this way for suitable values of T_1, T_2, T_3. Hence

MENELAUS'S THEOREM. *Let the sides of a triangle be divided at L, M, N in the respective ratios $\lambda : 1$, $\mu : 1$, $\nu : 1$. Then the three points L, M, N are collinear if and only if $\lambda\mu\nu = -1$.*

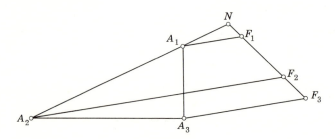

Figure 13.7e

The coefficients T_1, T_2, T_3 in the equation 13.72 for a line are sometimes called the *tangential coordinates* of the line. These homogeneous "coordinates" have a simple geometric interpretation [Salmon **1**, p. 11]: they may be regarded as *the distances from A_1, A_2, A_3 to the line,* measured in any direction (the same for all). To prove this, let A_1F_1, A_2F_2, A_3F_3 be these distances, as in Figure 13.7e. Since

$$\frac{A_1N}{NA_2} = -\frac{T_1}{T_2},$$

the homothetic triangles NA_1F_1 and NA_2F_2 yield

$$\frac{A_1F_1}{A_2F_2} = \frac{A_1N}{A_2N} = \frac{T_1}{T_2}.$$

Hence

$$\frac{A_1F_1}{T_1} = \frac{A_2F_2}{T_2},$$

and similarly each of these expressions is equal to $\dfrac{A_3F_3}{T_3}$.

Möbius's invention of homogeneous coordinates was one of the most far-reaching ideas in the history of mathematics: comparable to Leibniz's invention of differentials, which enabled him to express the equation

$$\frac{d}{dx} f(x) = f'(x)$$

in the homogeneous form

$$df(x) = f'(x)\, dx$$

(for instance, $d \sin x = \cos x\, dx$).

EXERCISES

1. Sketch the seven regions into which the lines A_2A_3, A_3A_1, A_1A_2 decompose the plane, marking each according to the signs of the three areal coordinates.

2. Verify that 13.45 yields $1 - x - y$ as the area of the triangle PA_1A_2 in Figure 13.7d.

3. In areal coordinates, the midpoint of $(s_1, s_2, s_3)(t_1, t_2, t_3)$ is

$$\left(\frac{s_1 + t_1}{2},\ \frac{s_2 + t_2}{2},\ \frac{s_3 + t_3}{2} \right).$$

4. The centroid of masses σ and τ at points whose areal coordinates are (s_1, s_2, s_3) and (t_1, t_2, t_3) is the point whose barycentric coordinates are

$$(\sigma s_1 + \tau t_1,\ \sigma s_2 + \tau t_2,\ \sigma s_3 + \tau t_3).$$

5. In barycentric coordinates, any point on the line $(s)(t)$ may be expressed in the form

$$(\sigma s_1 + \tau t_1,\ \sigma s_2 + \tau t_2,\ \sigma s_3 + \tau t_3).$$

6. Apply barycentric coordinates to Ex. 6 at the end of § 13.5. What becomes of this result when L, M, N are collinear?

7. In what way do the signs of T_1, T_2, T_3 depend on the position of the line 13.72 in relation to the triangle of reference? When T_2 and T_3 are positive, describe the cases $T_2 < T_3$, $T_2 = T_3$, $T_2 > T_3$.

13.8   AFFINE SPACE

Give me something to construct and I shall become God for the time being, pushing aside all obstacles, winning all the hard knowledge I need for the construction . . . advancing Godlike to my goal!

J. L. Synge [**2,** p. 162]

Affine geometry can be extended from two dimensions to three by using Axioms 12.42 and 12.43 instead of 12.41.   The total number of axioms is not really increased, as 13.12 now becomes a provable theorem [Forder **1,** pp. 155–157].   A line and a plane, or two planes, are said to be *parallel* if they have no common point (or if the line lies in the plane, or if the two planes coincide).   Thus any plane that meets two parallel planes meets them in parallel lines; if two planes are parallel, any line in either plane is parallel to the other plane; if two lines are parallel, any plane through either line is parallel to the other line.

The existence of parallel planes is ensured by the following theorem (cf. Axiom 13.11):

13.81   *For any point A and any plane γ, not through A, there is just one plane through A parallel to γ.*

Proof.   Let q and r be two intersecting lines in γ.   Let q' and r' be the respectively parallel lines through A.   Then the plane $q'r'$ is parallel to γ.   For otherwise, by 12.431, the two planes would meet in a line l.   Since q' and r' are parallel to γ, they cannot meet l.   Thus q' and r' are two parallels to l through A, contradicting 13.11.   This proves that $q'r'$ is parallel to γ.   Moreover, $q'r'$ is the only plane through A parallel to γ.   For, two such would meet in a line s' through A, and we could obtain a contradiction by considering their section by the plane As, where s is a line in γ not parallel to s'.

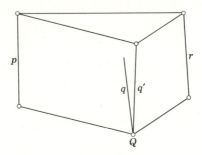

Figure 13.8a

Parallelism for lines is transitive in space as well as in a plane:

13.82   *If p and q are both parallel to r, they are parallel to each other.*

Proof [Forder **1**, p. 140].   When all three lines are in one plane, this follows at once from 13.11, so let us assume that they are not.   For any point Q on q, the planes Qp and Qr meet in a line, say q' (Figure 13.8*a*).   Any common point of q' and r would lie in both the planes Qp, pr, and therefore on their common line p; this is impossible, since p is parallel to r.   Hence q' is parallel to r.   But the only line through Q parallel to r is q.   Hence q coincides with q', and is coplanar with p.   Any common point of p and q would lie also on r.   Hence p and q are parallel.

The transitivity of parallelism provides an alternative proof for 13.81.   To establish the impossibility of a point O lying on both planes γ and $q'r'$, we imagine two lines through O, parallel to q (and q'), r (and r').   The planes γ and $q'r'$, each containing both these lines, would coincide, contradicting our assumption that A does not lie in γ.

The three face planes OBC, OCA, OAB of a tetrahedron $OABC$ form with the respectively parallel planes through A, B, C a *parallelepiped* whose faces are six parallelograms, as in Figure 13.8*b* [Forder **1**, p. 155].

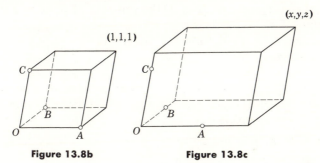

Figure 13.8b   **Figure 13.8c**

It is now easy to build up a three-dimensional theory of dilatations, translations, and vectors.   Three vectors **d**, **e**, **f** are said to be *dependent* if they are coplanar, in which case each is expressible as a linear combination of the other two.   Three vectors **e**, **f**, **g** are said to be *independent* if the only solution of the vector equation

$$x\mathbf{e} + y\mathbf{f} + z\mathbf{g} = \mathbf{0}$$

is $x = y = z = 0$.   Three such vectors provide a basis for a system of three-dimensional *affine coordinates*.   In fact, if

$$\mathbf{e} = \overrightarrow{OA}, \ \mathbf{f} = \overrightarrow{OB}, \ \mathbf{g} = \overrightarrow{OC},$$

as in Figure 13.8*c*, the general vector $\overrightarrow{OP}$ may be exhibited as a diagonal of the parallelepiped formed by drawing through P three planes parallel to OBC, OCA, OAB.   Then

$$\overrightarrow{OP} = x\mathbf{e} + y\mathbf{f} + z\mathbf{g},$$

where the terms of this sum are vectors along three edges of the parallelepiped.

In space, as in a plane, the centroid P of masses t_i at points A_i is determined by a vector $\overrightarrow{OP}$ such that

$$\Sigma t_i \, \overrightarrow{OP} = \Sigma t_i \, \overrightarrow{OA_i} \qquad (\Sigma t_i \neq 0).$$

If $\overrightarrow{OA_i} = x_i\mathbf{e} + y_i\mathbf{f} + z_i\mathbf{g}$, we deduce

$$\Sigma t_i \, \overrightarrow{OP} = \Sigma t_i x_i\mathbf{e} + \Sigma t_i y_i\mathbf{f} + \Sigma t_i z_i\mathbf{g}.$$

Hence, in terms of affine coordinates,

13.83   *The centroid of k masses t_i $(\Sigma t_i \neq 0)$ at points (x_i, y_i, z_i) $(i = 1, \ldots, k)$ is*

$$\left(\frac{\Sigma t_i x_i}{\Sigma t_i}, \quad \frac{\Sigma t_i y_i}{\Sigma t_i}, \quad \frac{\Sigma t_i z_i}{\Sigma t_i} \right).$$

In particular, if $t_1 + t_2 + t_3 = 1$, the centroid of three masses t_1, t_2, t_3 at the points

$$(1, 0, 0), \quad (0, 1, 0), \quad (0, 0, 1)$$

is (t_1, t_2, t_3).   Hence

13.84   *The affine coordinates of any point in the plane $x + y + z = 1$ are the same as its areal coordinates referred to the triangle cut out from this plane by the coordinate planes $x = 0$, $y = 0$, $z = 0$.*

It follows that there is a line

$$\frac{x}{t_1} = \frac{y}{t_2} = \frac{z}{t_3}$$

through the origin (in affine space) for each point with barycentric coordinates (t_1, t_2, t_3). On the other hand, lines lying in the plane $x + y + z = 0$ yield no corresponding points in the parallel plane $x + y + z = 1$, unless we agree to extend the affine plane by postulating a line at infinity

$$t_1 + t_2 + t_3 = 0$$

so as to form the projective plane. This possibility has already been mentioned in § 6.9; we shall explore it more systematically in Chapter 14.

EXERCISES

1.   If a line a is parallel to a plane α, and a plane through a meets α in b, then a and b are parallel lines. If another plane through a meets α in c, then b and c are parallel lines.

2.   If α, β, γ are planes intersecting in lines $\beta \cdot \gamma = a$, $\gamma \cdot \alpha = b$, $\alpha \cdot \beta = c$, and a is parallel to b, then a, b, c are all parallel.

3.   All the lines through A parallel to α are in a plane parallel to α [Forder **1**, p. 155].

4. Each of the six edges of a tetrahedron lies on a plane joining this edge to the midpoint of the opposite edge. The six planes so constructed all pass through one point: the centroid of equal masses at the four vertices.

5. Develop the theory of three-dimensional barycentric coordinates referred to a tetrahedron $A_1A_2A_3A_4$.

13.9   THREE-DIMENSIONAL LATTICES

> *The small parallelepiped built upon the three translations selected as unit translations . . . is known as the unit cell. . . . The entire crystal structure is generated through the periodic repetition, by the three unit translations, of the matter contained within the volume of the unit cell.*
>
> M. J. Buerger (1903 -   )
>
> [Buerger **1**, p. 5]

The theory of volume in affine space is more difficult than that of area in the affine plane, because of the complication introduced by M. Dehn's observation that two polyhedra of equal volume are not necessarily derivable from each other by dissection and rearrangement. A valid treatment, suggested by Mrs. Sally Ruth Struik, may be described very briefly as follows. It is found that any two tetrahedra are related by a unique *affinity ABCD →A'B'C'D'*, which transforms the whole space into itself in such a way as to preserve collinearity. In particular, a tetrahedron *ABCC'* is transformed into *ABC'C* by the *affine reflection*

$$AB(CC'),$$

which interchanges *C* and *C'* while leaving invariant every point in the plane that joins *AB* to the midpoint of *CC'*. Two tetrahedra are said to have the same *volume* if one can be transformed into the other by an *equiaffinity*: the product of an even number of affine reflections. Such a comparison is easily extended from tetrahedra to parallelepipeds, since a parallelepiped can be dissected into six tetrahedra all having the same volume.

In three dimensions, as in two, a *lattice* may be regarded as the set of points whose affine coordinates are integers. However, as it is independent of the chosen coordinate system, it is more symmetrically described as a discrete set of points whose set of position vectors is *closed under subtraction,* that is, along with any two of the vectors the set includes also their difference. Subtracting any one of the vectors from itself, we obtain the zero vector

$$\mathbf{c} - \mathbf{c} = \mathbf{0}$$

and hence also $\mathbf{0} - \mathbf{b} = -\mathbf{b}, \mathbf{a} - (-\mathbf{b}) = \mathbf{a} + \mathbf{b}, \mathbf{a} + \mathbf{a} = 2\mathbf{a}$, and so on: the set of vectors, containing the difference of any two, also contains the sum of any two, and all the integral multiples of any one. The lattice is one-, two-, or three-dimensional according to the number of independent vectors. In the three-dimensional case, a set of three independent vectors $\mathbf{e}, \mathbf{f}, \mathbf{g}$ is called a *basis* for the lattice if all the vectors are expressible in the form

13.91 $$x\mathbf{e} + y\mathbf{f} + z\mathbf{g},$$

where x, y, z are integers. If three of these vectors, say $\mathbf{r}_1, \mathbf{r}_2, \mathbf{r}_3$ form another basis for the same lattice, there must exist 18 integers

$$a_\alpha, b_\alpha, c_\alpha, A_\alpha, B_\alpha, C_\alpha \qquad (\alpha = 1, 2, 3)$$

such that

$$\mathbf{r}_\alpha = a_\alpha \mathbf{e} + b_\alpha \mathbf{f} + c_\alpha \mathbf{g}, \quad \mathbf{e} = \Sigma A_\alpha \mathbf{r}_\alpha, \quad \mathbf{f} = \Sigma B_\alpha \mathbf{r}_\alpha, \quad \mathbf{g} = \Sigma C_\alpha \mathbf{r}_\alpha$$

and therefore

$$\mathbf{r}_\alpha = a_\alpha \Sigma A_\beta \mathbf{r}_\beta + b_\alpha \Sigma B_\beta \mathbf{r}_\beta + c_\alpha \Sigma C_\beta \mathbf{r}_\beta,$$

whence

$$a_\alpha A_\beta + b_\alpha B_\beta + c_\alpha C_\beta = \begin{cases} 1 & \text{if } \alpha = \beta, \\ 0 & \text{if } \alpha \neq \beta. \end{cases}$$

Since the product of two determinants is obtained by combining the rows of one with the columns of the other, we have

$$\begin{vmatrix} a_1 & b_1 & c_1 \\ a_2 & b_2 & c_2 \\ a_3 & b_3 & c_3 \end{vmatrix} \begin{vmatrix} A_1 & A_2 & A_3 \\ B_1 & B_2 & B_3 \\ C_1 & C_2 & C_3 \end{vmatrix} = \begin{vmatrix} 1 & 0 & 0 \\ 0 & 1 & 0 \\ 0 & 0 & 1 \end{vmatrix} = 1.$$

Since the two determinants on the left are integers whose product is 1, each must be ± 1. Conversely, if $a_\alpha, b_\alpha, c_\alpha$ are given so that their determinant is ± 1, we can derive $A_\alpha, B_\alpha, C_\alpha$ by "inverting the matrix," and the given basis $\mathbf{e}, \mathbf{f}, \mathbf{g}$ yields the equally effective basis $\mathbf{r}_\alpha$. Hence

A necessary and sufficient condition for two triads of independent vectors

$$\mathbf{e}, \mathbf{f}, \mathbf{g} \quad and \quad a_\alpha \mathbf{e} + b_\alpha \mathbf{f} + c_\alpha \mathbf{g} \qquad (\alpha = 1, 2, 3)$$

to be alternative bases for the same lattice is

13.92 $$\begin{vmatrix} a_1 & b_1 & c_1 \\ a_2 & b_2 & c_2 \\ a_3 & b_3 & c_3 \end{vmatrix} = \pm 1$$

[cf. Hardy and Wright **1**, p. 28; Neville **1**, p. 5].

In other words, a lattice is derived from any one of its points by applying a *discrete group of translations:* one-, two-, or three-dimensional according as the translations are collinear, coplanar but not collinear, or not coplanar. In the one-dimensional case the generating translation is unique (except that it may be reversed), but in the other cases the two or three generators, that is, the basic vectors, may be chosen in infinitely many ways. When they have been chosen, we can use them to set up a system of affine coordinates so that, in the three-dimensional case, the vector 13.91 goes from the origin $(0, 0, 0)$ to the point (x, y, z), and the lattice consists of the points whose coordinates are integers. The eight points

$$(0, 0, 0), (1, 0, 0), (0, 1, 0), (0, 0, 1), (0, 1, 1), (1, 0, 1), (1, 1, 0), (1, 1, 1),$$

derived from the eight vectors

$$\mathbf{0}, \quad \mathbf{e}, \quad \mathbf{f}, \quad \mathbf{g}, \quad \mathbf{f} + \mathbf{g}, \quad \mathbf{g} + \mathbf{e}, \quad \mathbf{e} - \mathbf{f}, \quad \mathbf{e} + \mathbf{f} + \mathbf{g},$$

evidently form a parallelepiped, which is a *unit cell* of the lattice. By an argument analogous to that used for a two-dimensional lattice in § 4.1, *any two unit cells for the same lattice have the same volume.*

Any line joining two of the lattice points contains infinitely many of them, forming a one-dimensional sublattice of the three-dimensional lattice. In fact, the line joining $(0, 0, 0)$ and (x, y, z) contains also (nx, ny, nz) for every integer n. If x, y, z have the greatest common divisor d, the lattice point

$$(x/d, y/d, z/d)$$

lies on this same line, and the corresponding translation generates the group of the one-dimensional lattice. The lattice point (x, y, z) is *visible* if and only if the three integers x, y, z have no common divisor greater than 1.

Any triangle of lattice points determines a plane containing a two-dimensional sublattice. For, if vectors

$$\mathbf{r}_1 = x_1\mathbf{e} + y_1\mathbf{f} + z_1\mathbf{g} \quad \text{and} \quad \mathbf{r}_2 = x_2\mathbf{e} + y_2\mathbf{f} + z_2\mathbf{g}$$

have integral components, so also does $t_1\mathbf{r}_1 + t_2\mathbf{r}_2$ for any integers t_1 and t_2. The parallel plane through any other lattice point will contain a congruent sublattice. Thus we may regard all the lattice points as being distributed among an infinite sequence of parallel planes, called *rational planes* [Buerger **1**, p. 7].

Any such plane, being the join of three points whose coordinates are integers, has an equation of the form

13.93 $$Xx + Yy + Zz = N,$$

where the coefficients X, Y, Z, N are integers, so that the intercepts on the coordinate axes have the rational values $N/X, N/Y, N/Z$. (This is the reason for the name "rational" planes.) We may assume that the greatest common divisor of X, Y, Z is 1; for, any common factor of X, Y, Z would be a factor of N too, and then we could divide both sides of the equation by this number, obtaining a simpler and equally effective equation for the same plane.

Conversely, any such equation (in which the greatest common divisor of X, Y, Z is 1) represents a plane containing a two-dimensional sublattice. This is obvious when $X = 1$, since then we can assign arbitrary integral values to y, z, and solve 13.93 for x.   When X, Y, Z are all greater than 1, we consider the set of numbers

$$xX + yY + zZ,$$

where x, y, z are variable integers while X, Y, Z remain constant.   This set (like the set of lattice vectors) is an *ideal*: it contains the difference of any two of its members and (therefore) all the multiples of any one.   Let d denote its smallest positive member, and N any other member.   Then N is a multiple of d: for otherwise we could divide N by d and obtain a remainder $N - qd$, which would be a member smaller than d.   Thus every member of the set is a multiple of d.   But X, Y, Z are members.   Therefore d, being a common divisor, must be equal to 1, and the set simply consists of all the integers.   In other words, the equation 13.93 has one integral solution (and therefore infinitely many) [cf. Uspensky and Heaslet **1**, p. 54].

For each triad of integers X, Y, Z, coprime in the above sense (but not necessarily coprime in pairs), we have a sequence of parallel planes 13.93, evenly spaced, one plane for each integer N. Since every lattice point lies in one of the planes, the infinite region between any two consecutive planes is completely empty.   One of the planes, namely that for which $N = 0$, passes through the origin.   The nearest others, given by $N = \pm 1$, are appropriately called *first* rational planes [Buerger **1**, p. 9].   We shall have occasion to consider them again in § 18.3.

EXERCISES

1. How can a parallelepiped be dissected into six tetrahedra all having the same volume?

2. Identify the transformation $(x, y, z) \rightarrow (x, y, -z)$ with the affine reflection that leaves invariant the plane $z = 0$ while interchanging the points $(0, 0, \pm 1)$.

3. A lattice is transformed into itself by the central inversion that interchanges two of its points.

4. Every lattice point in a first rational plane is visible.

5. Is every rational plane through a visible point a first rational plane?

6. Find a triangle of lattice points in the first rational plane

$$6x + 10y + 15z = 1.$$

7. Obtain a formula for all the lattice points in this plane.

8. The origin is the only lattice point in the plane

$$x + \sqrt{2}y + \sqrt{3}z = 0.$$

14
Projective geometry

In affine geometry, as we have seen, parallelism plays a leading role. In projective geometry, on the other hand, there is no parallelism: every pair of coplanar lines is a pair of intersecting lines. The conflict with 12.61 is explained by the fact that the projective plane is not an "ordered" plane. The set of points on a line, like the set of lines through a point, is closed: given three, we cannot pick out one as lying "between" the other two. At first sight we might expect a geometry having no circles, no distances, no angles, no intermediacy, and no parallelism, to be somewhat meagre. But, in fact, a very beautiful and intricate collection of propositions emerges: propositions of which Euclid never dreamed, because his interest in measurement led him in a different direction. A few of these nonmetrical propositions were discovered by Pappus of Alexandria in the fourth century A.D. Others are associated with the names of two Frenchmen: the architect Girard Desargues (1591–1661) and the philosopher Blaise Pascal (1623–1662). Meanwhile, the related subject of perspective [Yaglom **2**, p. 31] had been studied by artists such as Leonardo da Vinci (1452–1519) and Albrecht Dürer (1471–1528).

Kepler's invention of points at infinity made it possible to regard the projective plane as the affine plane plus the line at infinity. A converse relationship was suggested by Poncelet's *Traité des propriétés projectives des figures* (1822) and von Staudt's *Geometrie der Lage* (1847), in which projective geometry appeared as an independent science, making it possible to regard the affine plane as the projective plane minus an arbitrary line *o*, and then to regard the Euclidean plane as the affine plane with a special rule for associating pairs of points on *o* (in "perpendicular directions") [Coxeter **2**, pp. 115, 138]. This standpoint became still clearer in 1899, when Mario Pieri placed the subject on an axiomatic foundation. Other systems of axioms, slightly different from Pieri's, have been proposed by subsequent authors. The particular system that we shall give in § 14.1 was suggested by Bachmann [**1**, pp. 76–77]. To test the consistency of a system of axioms, we apply it to a "model," in which the primitive concepts are represented by familiar concepts whose properties we are prepared to accept [Coxeter **2**, pp.

229

186–187]. In the present case a convenient model for the projective plane is provided by the affine plane plus the line at infinity (§ 6.9). We shall extend the barycentric coordinates of § 13.7 to general projective coordinates, so as to eliminate the special role of the line at infinity. The result may be regarded as a purely algebraic model in which a *point* is an ordered triad of numbers (x_1, x_2, x_3), not all zero, with the rule that $(\mu x_1, \mu x_2, \mu x_3)$ is the same point for any $\mu \neq 0$, and a *line* is a homogeneous linear equation. One advantage of this model is that the numbers x_α and μ are not necessarily real. The chosen axioms are sufficiently general to allow the coordinates to belong to any *field:* instead of real numbers we may use rational numbers, complex numbers, or even a finite field such as the residue classes modulo a prime number. Accordingly we speak of the real projective plane, the rational projective plane, the complex projective plane, or a finite projective plane.

14.1 AXIOMS FOR THE GENERAL PROJECTIVE PLANE

> *The more systematic course in the present introductory memoir . . .*
> *would have been to ignore altogether the notions of distance and*
> *metrical geometry. . . . Metrical geometry is a part of descriptive*
> *geometry, and descriptive geometry is all geometry.*
>
> Arthur Cayley *(1821 -1895)

The projective plane has already been mentioned in § 6.9. As primitive concepts we take *point, line,* and the relation of *incidence.* If a point and a line are incident, we say that the point lies *on* the line and the line passes *through* the point. The related words *join, meet* (or "intersect"), *concurrent* and *collinear* have their usual meanings. Three non-collinear points are the vertices of a *triangle* whose sides are complete lines. ("Segments" are not defined.) A *complete quadrangle,* its four vertices, its six sides, and its three diagonal points, are defined as in § 1.7. A *hexagon* $A_1 B_2 C_1 A_2 B_1 C_2$ has six vertices $A_1, B_2, \ldots, C_2$ and six sides

$$A_1 B_2, \; B_2 C_1, \; C_1 A_2, \; A_2 B_1, \; B_1 C_2, \; C_2 A_1.$$

Opposite sides are defined in the obvious manner; for example, $A_2 B_1$ is opposite to $A_1 B_2$. After these preliminary definitions, we are ready for the five axioms.

AXIOM 14.11 *Any two distinct points are incident with just one line.*

NOTATION. The line joining points A and B is denoted by AB.

---

* *Collected Mathematical Papers,* **2** (Cambridge, 1889), p. 592. Cayley, in 1859, used the word "descriptive" where today we would say "projective." His idea of the supremacy of projective geometry must now be regarded as a slight exaggeration. It is true that projective geometry includes the affine, Euclidean and non-Euclidean geometries; but it does not include the general Riemannian geometry, nor topology.

AXIOM 14.12 *Any two lines are incident with at least one point.*

THEOREM 14.121 *Any two distinct lines are incident with just one point.*

NOTATION. The point of intersection of lines a and b is denoted by $a \cdot b$; that of AB and CD by $AB \cdot CD$. The line joining $a \cdot b$ and $c \cdot d$ is denoted by $(a \cdot b)(c \cdot d)$.

AXIOM 14.13 *There exist four points of which no three are collinear.*

AXIOM 14.14 (Fano's axiom) *The three diagonal points of a complete quadrangle are never collinear.*

AXIOM 14.15 (Pappus's theorem) *If the six vertices of a hexagon lie alternately on two lines, the three points of intersection of pairs of opposite sides are collinear.*

One of the most elegant properties of projective geometry is the *principle of duality,* which asserts (in a projective plane) that every definition remains significant, and every theorem remains true, when we consistently interchange the words *point* and *line* (and consequently interchange *lie on* and *pass through, join* and *intersection, collinear* and *concurrent,* etc.). To establish this principle it will suffice to verify that *the axioms imply their own duals.* Then, given a theorem and its proof, we can immediately assert the dual theorem; for a proof of the latter could be written down mechanically by dualizing every step in the proof of the original theorem.

The dual of Axiom 14.11 is Theorem 14.121, which the reader will have no difficulty in proving (with the help of 14.12). The dual of Axiom 14.12 is one-half of 14.11. The dual of Axiom 14.13 asserts the existence of a *complete quadrilateral,* which is a set of four lines (called *sides*) intersecting in pairs in six distinct points (called *vertices*). Two vertices are said to be *opposite* if they are not joined by a side. The three joins of pairs of opposite vertices are called *diagonals.* If $PQRS$ is a quadrangle with sides

$$p = PQ, \quad q = PS, \quad r = RS, \quad s = QR, \quad w = PR, \quad u = QS,$$

as in Figure 14.1a, then $pqrs$ is a quadrilateral with vertices

$$P = p \cdot q, \quad Q = p \cdot s, \quad R = r \cdot s, \quad S = q \cdot r, \quad W = p \cdot r, \quad U = q \cdot s.$$

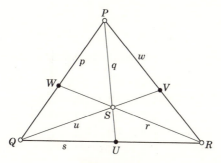

Figure 14.1a

Axiom 14.14 tells us that the three diagonal points

$$U = q \cdot s, \quad V = w \cdot u, \quad W = p \cdot r$$

are not collinear. Its dual asserts that the three diagonals of a complete quadrilateral are never concurrent. If this is false, there must exist a particular quadrilateral whose diagonals are concurrent. Let it be $pqrs$, with diagonals

$$u = QS, \quad v = WU, \quad w = PR.$$

Since these are concurrent, the point $w \cdot u = V$ must lie on v, contradicting the statement that U, V, W are not collinear.

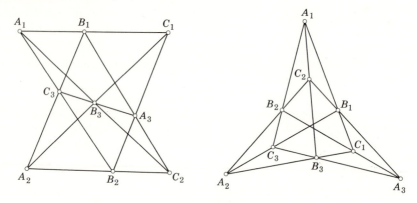

Figure 14.1b

Axiom 14.15 involves nine points and nine lines, which can be drawn in many ways (apparently different though projectively equivalent), such as the two shown in Figure 14.1b. $A_1B_2C_1A_2B_1C_2$ is a hexagon whose vertices lie alternately on the two lines $A_1B_1C_1$, $A_2B_2C_2$. The points of intersection of pairs of opposite sides are

$$A_3 = B_1C_2 \cdot B_2C_1, \quad B_3 = C_1A_2 \cdot C_2A_1, \quad C_3 = A_1B_2 \cdot A_2B_1.$$

The axiom asserts that these three points are collinear. Our notation has been devised in such a way that the three points A_i, B_j, C_k are collinear whenever

$$i + j + k \equiv 0 \quad (\text{mod } 3).*$$

Another way to express the same result is to arrange the 9 points in the form of a *matrix*

14.151
$$\begin{Vmatrix} A_1 & B_1 & C_1 \\ A_2 & B_2 & C_2 \\ A_3 & B_3 & C_3 \end{Vmatrix}.$$

* Coxeter, Self-dual configurations and regular graphs, *Bulletin of the American Mathematical Society,* **56** (1950), p. 432.

If this were a determinant that we wished to evaluate, we would proceed to multiply the elements in triads. These six "diagonal" triads, as well as the first two rows of the matrix, indicate triads of collinear points. The axiom asserts that the points in the bottom row are likewise collinear. Its inherent self-duality is seen from an analogous matrix of lines

$$\begin{Vmatrix} a_1 & b_1 & c_1 \\ a_2 & b_2 & c_2 \\ a_3 & b_3 & c_3 \end{Vmatrix}.$$

These lines can be picked out in many ways, one of which is

$$a_1 = A_3B_1C_2, \quad b_1 = A_1B_3C_2, \quad c_1 = A_2B_2C_2,$$
$$a_2 = A_2B_3C_1, \quad b_2 = A_3B_2C_1, \quad c_2 = A_1B_1C_1,$$
$$a_3 = A_1B_2C_3, \quad b_3 = A_2B_1C_3, \quad c_3 = A_3B_3C_3.$$

This completes our proof of the principle of duality.

EXERCISES

1. Every line is incident with at least three distinct points. (This statement, and the existence of a nonincident point and line, are sometimes used as axioms instead of 14.13 [Robinson **1**, p. 10; Coxeter **2**, p. 13].)

2. A set of m points and n lines is called a *configuration* (m_c, n_d) if c of the n lines pass through each of the points while d of the m points lie on each of the lines. The four numbers are not independent but satisfy $cm = dn$. The dual of (m_c, n_d) is (n_d, m_c).

In the case of a self-dual configuration, we have $m = n, c = d$, and the symbol (n_d, n_d) is conveniently abbreviated to n_d. Simple instances are the triangle 3_2, the complete quadrangle $(4_3, 6_2)$ and the complete quadrilateral $(6_2, 4_3)$. Axiom 14.14 asserts the nonexistence of the *Fano configuration* [*] 7_3. The points and lines that occur in Axiom 14.15 (Figure 14.1*b*) form the *Pappus configuration* 9_3, which may be regarded (in how many ways?) as a cycle of three triangles such as

$$A_1B_1C_2, \quad A_2B_2C_3, \quad A_3B_3C_1,$$

each inscribed in the next (cf. Figure 1.8*a*, where UVW is inscribed in ABC). The self-duality is evident.

By a suitable change of notation, Axiom 14.15 may be expressed thus: *If AB, CD, EF are concurrent, and DE, FA, BC are concurrent, then AD, BE, CF are concurrent.*

3. A particular *finite projective plane*, in which only 13 "points" and 13 "lines" exist, can be defined abstractly by calling the points P_i and the lines p_i ($i = 0, 1, \ldots, 12$) with the rule that P_i and p_j are "incident" if and only if

$$i + j \equiv 0, 1, 3 \text{ or } 9 \qquad (\text{mod } 13).$$

Construct a table to indicate the 4 points on each line and the 4 lines through each point [Veblen and Young **1**, p. 6]. Verify that all the axioms are satisfied; for example, $P_0P_1P_2P_5$ is a complete quadrangle with sides

$$P_0P_1 = p_0, \quad P_0P_2 = p_1, \quad P_1P_5 = p_8, \quad P_0P_5 = p_9, \quad P_2P_5 = p_{11}, \quad P_1P_2 = p_{12}$$

---

[*] Coxeter, *Bulletin of the American Mathematical Society,* **56** (1950), pp. 423–425.

and diagonal points $P_3 = p_0 \cdot p_{11}$, $P_4 = p_9 \cdot p_{12}$, $P_8 = p_1 \cdot p_8$. A possible matrix for Axiom 14.15 is

$$\left\| \begin{matrix} P_0 & P_2 & P_8 \\ P_3 & P_4 & P_6 \\ P_9 & P_{10} & P_5 \end{matrix} \right\|.$$

The first row may be any set of three collinear points. The second row may be any such set on a line not incident with a point in the first row. The last row is then determined; e.g., in the above instance it consists of

$$P_2P_6 \cdot P_4P_8 = P_9, \qquad P_3P_8 \cdot P_0P_6 = P_{10}, \qquad P_0P_4 \cdot P_2P_3 = P_5.$$

This differs from the general "Pappus matrix" 14.151 in that sets of collinear points occur not only in the rows and generalized diagonals but also in the columns. In other words, the 9 points form a configuration which is not merely 9_3 but $(9_4, 12_3)$. When any one of the 9 points is omitted, the remaining 8 form a self-dual configuration 8_3 which may be regarded as a pair of mutually inscribed quadrangles (such as $P_0P_9P_5P_8$ and $P_2P_3P_{10}P_6$). [Hilbert and Cohn-Vossen **1**, pp. 101–102.]

4. The geometry described in Ex. 3 is known as $PG(2, 3)$. More generally, $PG(2, p)$ is a finite plane in which each line contains $p+1$ points. Consequently, each point lies on $p+1$ lines. There are p^2+p+1 points (and the same number of lines) altogether. In other words, the whole geometry is a configuration n_d with $n = p^2+p+1$ and $d = p+1$. (Actually p is not arbitrary, e.g., although it may be any power of an odd prime, for instance, 5, 7, or 9, it cannot be 6.)* The possibility of such finite planes indicates that the projective geometry defined by Axioms 14.11 to 14.15 is not *categorical:* it is not just one geometry but many geometries, in fact, infinitely many.

5. In any finite projective geometry, Sylvester's theorem (§ 4.7) is false.

14.2 PROJECTIVE COORDINATES

Modern algebra does not seem quite so terrifying when expressed in these geometrical terms!

G. de B. Robinson (1906 -   )

[Robinson **1**, p. 94]

We saw, in § 13.7, that three real numbers t_1, t_2, t_3 will serve as barycentric coordinates for a point in the affine plane (with respect to any given triangle of reference) if and only if

$$t_1 + t_2 + t_3 \neq 0.$$

Also a linear homogeneous equation 13.72 will serve as the equation for a line if and only if the coefficients T_1, T_2, T_3 are not all equal. The remarks

---

* By not insisting on Axiom 14.14, we can develop a "geometry of characteristic 2" in which p is a power of 2. By not insisting on Axiom 14.15, we can develop a "non-Desarguesian plane." For the application to mutually orthogonal Latin squares, see Robinson **1**, p. 161, Appendix II.

just after 13.84 indicate that these artificial restrictions will be avoided when we have extended the real affine plane to the real projective plane by adding the line at infinity

14.21 $$t_1 + t_2 + t_3 = 0$$

and all its points (which are the points at infinity in various directions).

When we interpret T_1, T_2, T_3 as the distances from A_1, A_2, A_3 to the line

$$T_1 t_1 + T_2 t_2 + T_3 t_3 = 0,$$

it is obvious that a parallel line is obtained by adding the same number to all three T's. Hence the point of intersection of two parallel lines satisfies 14.21, that is, it lies on the line at infinity.

To emphasize the fact that, in projective geometry, the line at infinity no longer plays a special role, we shall abandon the barycentric coordinates (t_1, t_2, t_3) in favor of general *projective* coordinates (x_1, x_2, x_3), given by

$$t_1 = \mu_1 x_1, \quad t_2 = \mu_2 x_2, \quad t_3 = \mu_3 x_3,$$

where μ_1, μ_2, μ_3 are constants, $\mu_1 \mu_2 \mu_3 \neq 0$. Thus (x_1, x_2, x_3) is the centroid of masses $\mu_\alpha x_\alpha$ at A_α ($\alpha = 1, 2, 3$), and the line at infinity has the undistinguished equation

$$\mu_1 x_1 + \mu_2 x_2 + \mu_3 x_3 = 0.$$

The contrast between these two kinds of coordinates may also be expressed as follows. Barycentric coordinates can be referred to any given triangle; the "simplest" points

$$(1, 0, 0), \quad (0, 1, 0), \quad (0, 0, 1)$$

are the vertices, and the *unit point* $(1, 1, 1)$ is the centroid. More usefully, projective coordinates can be referred to any given quadrangle! Taking three of the four vertices to determine a system of barycentric coordinates, suppose the fourth vertex is (μ_1, μ_2, μ_3). By using these μ's for the transition to projective coordinates, we give this fourth vertex the new coordinates $(1, 1, 1)$. Just as, in affine geometry, all triangles are alike, so in projective geometry *all quadrangles are alike*.

To prove that projective coordinates provide a model (in the augmented affine plane) for the abstract projective plane described in § 14.1, we can take each of our geometric axioms and prove it analytically (i.e., algebraically).

To prove 14.11, we merely have to observe that the line joining points (y_1, y_2, y_3) and (z_1, z_2, z_3) is

14.22 $$\begin{vmatrix} y_2 & y_3 \\ z_2 & z_3 \end{vmatrix} x_1 + \begin{vmatrix} y_3 & y_1 \\ z_3 & z_1 \end{vmatrix} x_2 + \begin{vmatrix} y_1 & y_2 \\ z_1 & z_2 \end{vmatrix} x_3 = 0$$

(cf. 13.74). Similarly, for 14.12 (or rather, 14.121), the point of intersection of lines $\Sigma Y_\alpha x_\alpha = 0$ and $\Sigma Z_\alpha x_\alpha = 0$ is

$$\left(\begin{vmatrix} Y_2 & Y_3 \\ Z_2 & Z_3 \end{vmatrix}, \quad \begin{vmatrix} Y_3 & Y_1 \\ Z_3 & Z_1 \end{vmatrix}, \quad \begin{vmatrix} Y_1 & Y_2 \\ Z_1 & Z_2 \end{vmatrix} \right).$$

For 14.13, we can use the four points

14.23 $(1, 0, 0), \quad (0, 1, 0), \quad (0, 0, 1), \quad (1, 1, 1).$

The diagonal points of the quadrangle so formed are

$$(0, 1, 1), \quad (1, 0, 1), \quad (1, 1, 0).$$

If these three points lay on a line $\Sigma X_\alpha x_\alpha = 0$, we should have

14.24 $X_2 + X_3 = 0, \quad X_3 + X_1 = 0, \quad X_1 + X_2 = 0,$

whence $X_1 = X_2 = X_3 = 0$, which is absurd. This proves 14.14.

Finally, to prove 14.15 we use the coordinates 14.23 for the four points

$$A_1, \quad A_2, \quad A_3, \quad C_1.$$

On the lines $C_1 A_1$, $C_1 A_2$, $C_1 A_3$, which are

$$x_2 = x_3, \quad x_3 = x_1, \quad x_1 = x_2,$$

we take the points B_1, B_3, B_2 to be

$$(p, 1, 1), \quad (1, q, 1), \quad (1, 1, r).$$

The three lines $A_3 B_1$, $A_1 B_3$, $A_2 B_2$, being

$$x_1 = px_2, \quad x_2 = qx_3, \quad x_3 = rx_1,$$

all pass through the same point C_2 if

14.25 $pqr = 1.$

The three lines $A_3 B_3$, $A_2 B_1$, $A_1 B_2$, being

$$x_2 = qx_1, \quad x_1 = px_3, \quad x_3 = rx_2,$$

all pass through the same point C_3 if

$$qpr = 1.$$

Since this condition agrees with 14.25, the proof is complete. However, it is important to observe that the above deduction can be carried through in the more general situation where the coordinates belong not to a *field* but to an arbitrary *division ring* [Birkhoff and MacLane **1**, p. 222]. We can still speak of points and lines, but Axiom 14.15 will have to be replaced by a weaker statement if the coordinate ring includes elements p and q such that

$$pq \neq qp.$$

For instance, we might have $p = k$ and $q = j$ in a "quaternion geometry" whose coordinates are based on "units" i, j, k satisfying

$$i^2 = j^2 = k^2 = ijk = -1.$$

When the A's and B's are so chosen, 14.15 is false. We have thus established an important connection between geometry and algebra: Hilbert's discovery that, when homogeneous coordinates are used in a plane satisfying the first four axioms, *Pappus's theorem is equivalent to the commutative law for multiplication.*

EXERCISES

1. Given five points, no three collinear, we can assign the coordinates 14.23 to any four of them, and then the coordinates (x_1, x_2, x_3) of the fifth are definite (apart from the possibility of multiplying all by the same constant). If the mutual ratios of the three x's are rational, we can multiply by a "common denominator" so as to make them all integral. In this case we can derive the fifth point from the first four by a linear construction, involving a finite sequence of operations of joining two known points or taking the point of intersection of two known lines. Devise such a construction for the point $(1, 2, 3)$.

2. The four points $(1, \pm 1, \pm 1)$ form a complete quadrangle whose diagonal triangle is the triangle of reference.

3. A configuration 8_3, consisting of two mutually inscribed quadrangles, exists in the complex projective plane, but not in the real projective plane. When it does exist, its eight points appear in four pairs of "opposites" whose joins are concurrent. The complete figure is a $(9_4, 12_3)$. *Hint:* Let the two quadrangles be $P_0P_2P_4P_6$ and $P_1P_3P_5P_7$, so that the sets of three collinear points are

$$P_0P_1P_3, \ P_1P_2P_4, \ P_2P_3P_5, \ P_3P_4P_6, \ P_4P_5P_7, \ P_5P_6P_0, \ P_6P_7P_1, \ P_7P_0P_2.$$

Take $P_0P_1P_2$ as triangle of reference and let P_3, P_4, P_7 be $(1, 1, 0)$, $(0, 1, 1)$, $(1, 0, x)$. Deduce that P_5 and P_6 are $(1, 1, x + 1)$ and $(1, x + 1, x)$. Obtain an equation for x from the collinearity of $P_0P_5P_6$.

4. If p is an odd prime, a finite projective plane $PG(2, p)$ can be obtained by taking the coordinates to belong to the field $GF(p)$ which consists of the p residues (or, strictly, residue classes) modulo p [Ball **1,** pp. 60–61]. For instance, the appropriate "finite arithmetic" for $PG(2, 3)$ consists of symbols 0, 1, 2 which behave like ordinary integers except that

$$1 + 2 = 0 \quad \text{and} \quad 2 \times 2 = 1.$$

In the notation of Ex. 3 at the end of § 14.1, take $P_0P_1P_2$ to be the triangle of reference and P_5 the unit point $(1, 1, 1)$. Find coordinates for the remaining points, and equations for the lines.

Finite planes, and the analogous finite n-spaces $PG(n, p)$, were discovered by von Staudt* and rediscovered by Fano. Von Staudt took n to be 2 or 3. Fano took p to be a prime. The generalization $PG(n, p^k)$ is credited to Veblen and Bussey.

5. Taking the coordinates to belong to $GF(2)$, which consists of the two "numbers" 0 and 1 with the rule for addition

$$1 + 1 = 0,$$

we obtain a finite "geometry" in which the diagonal points of a complete quadrangle are always collinear! Our proof of 14.14 breaks down because now the equations 14.24

---

* K. G. C. von Staudt, *Beiträge zur Geometrie der Lage*, vol. I (Nürnberg, 1856), pp. 87–88; Gino Fano, *Giornale di Matematiche*, **30** (1892), pp. 114–124; Veblen and Bussey, *Transactions of the American Mathematical Society*, **7** (1906), pp. 241–259.

have not only the inadmissible solution $X_1 = X_2 = X_3 = 0$ but also the significant solution $X_1 = X_2 = X_3 = 1$, which yields the line

$$x_1 + x_2 + x_3 = 0.$$

This $PG(2, 2)$ can be described abstractly by calling its seven points P_i and its seven lines p_i ($i = 0, 1, \ldots, 6$) with the rule that P_i and p_j are incident if and only if

$$i + j \equiv 0, 1 \text{ or } 3 \quad (\text{mod } 7).$$

14.3   DESARGUES'S THEOREM

> The fundamental idea for this pure geometry came from the desire of Renaissance painters to produce a "visual" geometry. How do things really look, and how can they be presented on the plane of the drawing? For example, there will be no parallel lines, since such lines appear to the eye to converge.
>
> S. H. Gould (1909 -   )
> [Gould **1**, p. 298]

Two triangles, with their vertices named in a particular order, are said to be *perspective from a point* (or briefly, "perspective") if their three pairs of corresponding vertices are joined by concurrent lines. For instance, in Figure 14.1b, the triangles $A_1A_2A_3$ and $B_1B_3B_2$ (*sic*) are perspective from C_1. By permuting the vertices of $B_1B_3B_2$ cyclically, either forwards or backwards, we see that the same two triangles are also perspective from C_2 or C_3. In fact, one of the neatest statements of Axiom 14.15 [see Veblen and Young **1**, p. 100] is:

If two triangles are doubly perspective they are trebly perspective.

Dually, two triangles are said to be *perspective from a line* if their three pairs of corresponding sides meet in collinear points. It was observed by G. Hessenberg* that our axioms suffice for a proof of

DESARGUES'S THEOREM.   *If two triangles are perspective from a point they are perspective from a line, and conversely.*

The details are as follows. Let two triangles PQR and $P'Q'R'$ be perspective from O, as in Figure 14.3a, and let their corresponding sides meet in points

$$D = QR \cdot Q'R', \quad E = RP \cdot R'P', \quad F = PQ \cdot P'Q'.$$

We wish to prove that D, E, F are collinear. After defining four further points

$$S = PR \cdot Q'R', \quad T = PQ' \cdot OR,$$
$$U = PQ \cdot OS, \quad V = P'Q' \cdot OS,$$

we have, in general,† enough triads of collinear points to make three applications of Axiom 14.15. The "matrix" notation enables us to write simply

---

* *Mathematische Annalen*, **61** (1905), pp. 161–172.

† Pedoe **2**, pp. 35–42. See also Coxeter, *Unvergängliche Geometrie*, Birkhäuser, Basel, 1963, pp. 290–291.

$$\begin{Vmatrix} O & Q & Q' \\ P & S & R \\ D & T & U \end{Vmatrix}, \quad \begin{Vmatrix} O & P & P' \\ Q' & R' & S \\ E & V & T \end{Vmatrix}, \quad \begin{Vmatrix} P & Q' & T \\ V & U & S \\ D & E & F \end{Vmatrix}.$$

The last row of the last matrix exhibits the desired collinearity. The converse follows by the principle of duality.

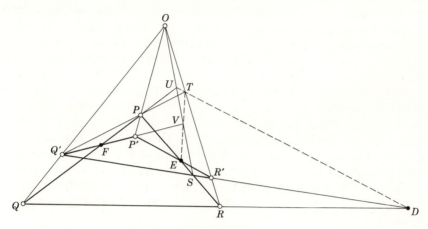

Figure 14.3a

EXERCISES

1. The triangle $(p, 1, 1)$ $(1, q, 1)$ $(1, 1, r)$ is perspective with the triangle of reference from the unit point $(1, 1, 1)$. Pairs of corresponding sides meet in the three collinear points

$$(0, q - 1, 1 - r), \quad (1 - p, 0, r - 1), \quad (p - 1, 1 - q, 0).$$

2. Desargues's theorem involves 10 points and 10 lines, forming a configuration 10_3. To obtain a symmetrical notation, consider triangles $P_{14}P_{24}P_{34}$ and $P_{15}P_{25}P_{35}$, perspective from a point P_{45} and consequently from a line $P_{23}P_{31}P_{12}$. Then three points P_{ij} are collinear if their subscripts involve just three of the numbers 1, 2, 3, 4, 5. If the remaining two of the five numbers are k and l, we may call the line p_{kl}. Then the same two triangles may be described as $p_{15} p_{25} p_{35}$ and $p_{14} p_{24} p_{34}$, perspective from the line p_{45}.

3. In the finite projective plane $PG(2, 3)$, the two triangles $P_1P_2P_7$ and $P_3P_8P_4$ are perspective from the point P_0 and from the line $P_9P_{12}P_{10}$. Identify the remaining points in Figure 14.3a. (In this special geometry, U and V both coincide with F, which is not surprising in view of the fact that Figure 14.3a involves 14 points whereas the whole plane contains only 13.)

14.4 QUADRANGULAR AND HARMONIC SETS

Desargues's theorem enables us to prove an important property of a *quadrangular set* of points, which is the section of the six sides of a complete quadrangle by any line that does not pass through a vertex:

14.41   *Each point of a quadrangular set is uniquely determined by the remaining points.*

Proof.   Let $PQRS$ be a complete quadrangle whose sides PS, QS, RS, QR, RP, PQ meet a line g (not through a vertex) in six points A, B, C, D, E, F, certain pairs of which may possibly coincide. (The first three points come from three sides all containing the same vertex S; the last three from the respectively opposite sides, which form the triangle PQR.) To show that F is uniquely determined by the remaining five points, we set up another quadrangle $P'Q'R'S'$ whose first five sides pass through A, B, C, D, E, as in Figure 14.4a. Since the two triangles PRS and $P'R'S'$ are perspective from the line g, the converse of Desargues's theorem tells us that they are also perspective from a point; thus PP' passes through the point $O = RR' \cdot SS'$. Similarly, the perspective triangles QRS and $Q'R'S'$ show that QQ' passes through this same point O. In fact, all the four lines PP' QQ', RR', SS' pass through O, so that $PQRS$ and $P'Q'R'S'$ are "persp tive quadrangles." By the direct form of Desargues's theorem, the trian$ PQR$ and $P'Q'R'$, which are perspective from the point O, are also persp tive from the line DE, which is g; that is, the sides PQ and $P'Q'$ both n g in the same point F.

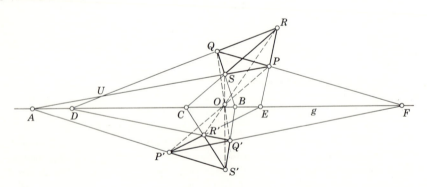

Figure 14.4a

Following Veblen and Young [**1**, p. 49] we use the symbol

$$Q(ABC, DEF)$$

to denote the statement that the six points form a quadrangular set in the above manner. This statement is evidently unchanged if we apply any permutation to ABC and the same permutation to DEF. It is also equivalent to any of

$$Q(AEF, DBC),   Q(DBF, AEC),   Q(DEC, ABF).$$

To obtain other permutations we need a new quadrangle. With the ex-

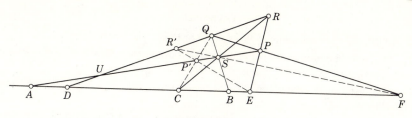

Figure 14.4b

ercise of some ingenuity we can retain two of the four old vertices, say Q and S. Defining

$$R' = QR \cdot SF, \quad P' = PS \cdot QC,$$

as in Figure 14.4*b*, we apply Axiom 14.15 to the hexagon $PRQCFS$ according to the scheme

$$\begin{Vmatrix} P & F & Q \\ C & R & S \\ R' & P' & E \end{Vmatrix},$$

with the conclusion that $R'P'$ passes through E. Now, just as the quadrangle $PQRS$ yields $Q(ABC, DEF)$, the quadrangle $P'QR'S$ yields $Q(ABF, DEC)$. In other words, the statement $Q(ABC, DEF)$ implies $Q(ABF, DEC)$, and hence also

14.42 $Q(ABC, DEF)$ *implies* $Q(DEF, ABC)$.

In the important special case $Q(ABC, ABF)$, which is abbreviated to

$$H(AB, CF),$$

we say that the four points form a *harmonic set*, or, more precisely, that F is the *harmonic conjugate* of C with respect to A and B. This means that A and B are two of the three diagonal points of a quadrangle while C and F lie respectively on the remaining sides, that is, on the sides that pass through the third diagonal point. Axiom 14.14 tells us that the harmonic conjugates C and F are distinct (except in the degenerate case when they coincide with A or B).

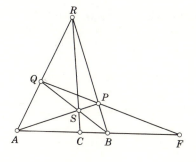 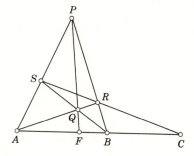

Figure 14.4c

EXERCISES

1. H(AB, CF) is equivalent to H(BA, CF) or H(AB, FC) or H(BA, FC).

2. Describe in detail a construction for the harmonic conjugate of C with respect to two given points A and B (on a line through C, as in Figure 14.4c).

3. The harmonic conjugate of $(0, 1, \lambda)$ with respect to $(0, 1, 0)$ and $(0, 0, 1)$ is $(0, 1, -\lambda)$.

4. In $PG(2, 3)$ (see Ex. 3 at the end of § 14.1), every set of four collinear points is a harmonic set in every order; e.g., H(P_0P_1, P_3P_9), H(P_0P_3, P_9P_1), H(P_0P_9, P_1P_3).

5. In Figure 6.6a, H(AA', A_1A_2). Deduce the metrical definition

$$\frac{AA_1}{A_1A'} = \frac{AA_2}{A'A_2}$$

for a harmonic set. (*Hint:* Defining E' as in Ex. 4 at the end of § 6.6, consider the quadrangle formed by P, E, E' and the point at infinity on A_1P.)

14.5 PROJECTIVITIES

A *range* is the set of all points on a line. Dually, a *pencil* is the set of all lines through a point. Ranges and pencils are instances of *one-dimensional forms*. We shall often have occasion to consider a (one-to-one) correspondence between two one-dimensional forms. The simplest possible correspondence between a range and a pencil arises when the lines of the pencil join the points of the range to another point, so that the range is a *section* of the pencil. The correspondence between two ranges that are sections of one pencil by two distinct lines is called a *perspectivity*; in such a case we write

$$X \stackrel{}{\overline{\wedge}} X' \quad \text{or} \quad X \stackrel{O}{\overline{\wedge}} X',$$

meaning that, if X and X' are corresponding points of the two ranges, their join XX' continually passes through a fixed point O, which we call the *center* of the perspectivity. There is naturally also a dual kind of perspectivity relating pencils instead of ranges.

The product of any number of perspectivities is called a *projectivity*. Two ranges (or pencils) related by a projectivity are said to be *projectively related*, and we write

$$X \stackrel{}{\overline{\wedge}} X'.$$

For instance, in the circumstances illustrated in Figure 14.5a,

$$ABCD \stackrel{O}{\overline{\wedge}} A_0B_0C_0D_0 \stackrel{O_1}{\overline{\wedge}} A'B'C'D', \qquad ABCD \stackrel{}{\overline{\wedge}} A'B'C'D'.$$

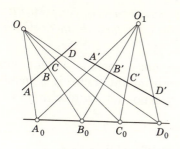

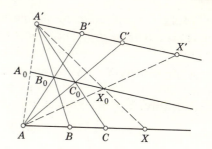

Figure 14.5a

Figure 14.5b

Analogously, we can define a projectivity relating a range to a pencil, or vice versa.

Given three distinct points A, B, C on a line, and three distinct points A', B', C' on another line, we can relate them by a pair of perspectivities in the manner of Figure 14.5b, where the *axis* (or "intermediary line") of the projectivity joins the points

$$B_0 = AB' \cdot BA', \quad C_0 = AC' \cdot CA',$$

so that

$$ABC \underset{\wedge}{\overset{A'}{\doteq}} A_0 B_0 C_0 \underset{\wedge}{\overset{A}{\doteq}} A'B'C'.$$

For each point X on AB we obtain a corresponding point X' on $A'B'$ by joining A to the point $X_0 = A'X \cdot B_0 C_0$, so that

$$ABCX \underset{\wedge}{\overset{A'}{\doteq}} A_0 B_0 C_0 X_0 \underset{\wedge}{\overset{A}{\doteq}} A'B'C'X'.$$

By Axiom 14.15, the axis $B_0 C_0$, being the "Pappus line" of the hexagon $AB'CA'BC'$, contains the point $BC' \cdot CB'$. Similarly, it contains the point of intersection of the "cross joins" of any two pairs of corresponding points. In particular, we could have derived the same point X' from a given point X by using perspectivities from B' and B (or any other pair of corresponding points) instead of A' and A.

It can be proved [Baker **1**, pp. 62–64; Robinson **1**, pp. 24–36] that the product of any number of perspectivities can be reduced to such a product of two, provided the initial and final ranges are not on the same line. In other words,

14.51 *Any projectivity relating ranges on two distinct lines is expressible as the product of two perspectivities whose centers are corresponding points (in reversed order) of the two related ranges.*

To relate two triads of distinct points ABC and $A'B'C'$ on one line, we may use an arbitrary perspectivity $ABC \underset{\wedge}{=} A_1 B_1 C_1$ to obtain a triad on another line, and then relate $A_1 B_1 C_1$ to $A'B'C'$ as in 14.51. Hence

14.52 *It is possible, by a sequence of not more than three perspectivities, to relate any three distinct collinear points to any other three distinct collinear points.*

A projectivity $X \barwedge X'$ on one line may have one or more *invariant* points (such that $X = X'$). If it has more than two invariant points, it is merely the identity, $X \barwedge X$. In fact, the above construction for a projectivity

$$ABCX \barwedge ABCX'$$

on one line involves four points on another line such that

$$ABCX \doublebarwedge A_1B_1C_1X_1 \barwedge ABCX'.$$

By 14.51, there is essentially only one projectivity $A_1B_1C_1 \barwedge ABC$. We have thus proved

THE FUNDAMENTAL THEOREM OF PROJECTIVE GEOMETRY. *A projectivity is determined when three points of one range and the corresponding three points of the other are given.*

If a projectivity relating ranges on two distinct lines has an invariant point A, this point, belonging to both ranges, must be the common point of the two lines, as in Figure 14.5c. Let B and C be any other points of the first range, B' and C' the corresponding points of the second. The fundamental theorem tells us that the perspectivity

$$ABC \stackrel{O}{\doublebarwedge} AB'C',$$

where $O = BB' \cdot CC'$, is the same as the given projectivity $ABC \barwedge AB'C'$. Hence

14.53 *A projectivity between two distinct lines is equivalent to a perspectivity if and only if their point of intersection is invariant.*

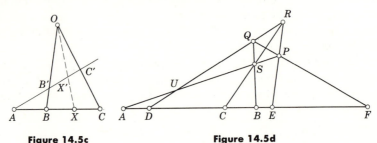

Figure 14.5c   Figure 14.5d

Returning to the notion of a projectivity between ranges on one line (i.e., a projective transformation of the line into itself), we recall that, if such a transformation is not merely the identity, it cannot have more than two invariant points. It is said to be *elliptic, parabolic,* or *hyperbolic* according as the number of invariant points is 0, 1, or 2. When coordinates are used, invariant points arise from roots of quadratic equations; thus elliptic projectivities do not occur in complex geometry, but

$$ABC \overset{}{\underset{\wedge}{\frown}} BCA$$

is elliptic in real geometry [Coxeter **2,** p. 48].

Figure 14.5d (cf. 14.4a) suggests a simple construction for a hyperbolic projectivity $ABF \overset{}{\underset{\wedge}{\frown}} ACE$ in which one of the invariant points is given:

$$ABF \overset{Q}{\underset{\wedge}{\doubleframe}} ASP \overset{R}{\underset{\wedge}{\doubleframe}} ACE.$$

Here S and P may be any two points collinear with A, and then the other two vertices of the quadrangle are

$$Q = BS \cdot FP, \quad R = CS \cdot EP.$$

The second invariant point is evidently D, on QR. When the same projectivity is expressed in the form $ADB \overset{}{\underset{\wedge}{\frown}} ADC$ (that is, when both invariant points are given), we have the analogous construction

$$ADB \overset{Q}{\underset{\wedge}{\doubleframe}} AUS \overset{R}{\underset{\wedge}{\doubleframe}} ADC,$$

where $U = AS \cdot QD$. This can still be carried out if A and D coincide (i.e., if g passes through the diagonal point $U = PS \cdot QR$ of the quadrangle), in which case we have the parabolic projectivity

$$AAB \overset{}{\underset{\wedge}{\frown}} AAC$$

[Coxeter **2,** p. 50].

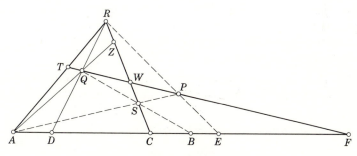

Figure 14.5e

An *involution* is a projectivity of period 2, that is, a projectivity which *interchanges* pairs of points. Figure 14.5e is derived from Figure 14.5d by adding extra points T, W, Z. We may imagine this figure to have been derived from *any* four given collinear points A, C, D, F by taking a point R outside their line, letting the joins RA, RD, RC meet an arbitrary line through F in T, Q, W, respectively, and then taking $Z = AQ \cdot RC$. Since

$$ADCF \underset{\wedge}{\overset{Q}{\doubleblank}} ZRCW \underset{\wedge}{\overset{A}{\doubleblank}} QTFW \underset{\wedge}{\overset{R}{\doubleblank}} DAFC,$$

we have

14.54
$$ADCF \underset{\wedge}{\doubleblank} DAFC.$$

But, by the fundamental theorem, there is only one projectivity $ADC \underset{\wedge}{\doubleblank} DAF$. Hence, if a projectivity interchanging A and D transforms C into F, it *interchanges C and F.* In other words,

14.55 *Any projectivity that interchanges two points is an involution.*

Applying the same set of three perspectivities to another point B, we have

$$B \underset{\wedge}{\overset{Q}{\doubleblank}} S \underset{\wedge}{\overset{A}{\doubleblank}} P \underset{\wedge}{\overset{R}{\doubleblank}} E.$$

Since $\mathbf{Q}(ABC, DEF)$, we have now proved the theorem of the quadrangular set:

14.56 *The three pairs of opposite sides of a quadrangle meet any line (not through a vertex) in three pairs of an involution.*

Combining this with 14.55, we have an alternative proof for 14.42 [Veblen and Young **1**, p. 101].

Since the involution $ACD \underset{\wedge}{\doubleblank} DFA$ is determined by its pairs AD and CF (or any other two of its pairs), it is conveniently denoted by

$$(AD)(CF)$$

or $(DA)(CF)$ or $(CF)(AD)$, etc. Thus $\mathbf{Q}(ABC, DEF)$ implies that the pair BE belongs to $(AD)(CF)$, and CF to $(AD)(BE)$, and AD to $(BE)(CF)$. The points in a pair are not necessarily distinct. When $A = D$ and $B = E$, so that $\mathbf{H}(AB, CF)$, we have the hyperbolic involution $(AA)(BB)$ which interchanges pairs of harmonic conjugates with respect to A and B. Since this same involution is expressible as $(AA)(CF)$,

14.57 *If an involution has one invariant point, it has another, and consists of the correspondence between harmonic conjugates with respect to these two points.*

It follows that there is no parabolic involution.

EXERCISES

1. Let the lines $OA, OB, \ldots, O_1A', O_1B', \ldots$ and A_0B_0 in Figure 14.5a be denoted by $a, b, \ldots, a', b', \ldots$ and o. Use the principle of duality to justify the notation

$$abcd \underset{\wedge}{\overset{o}{\doubleblank}} a'b'c'd'.$$

2. The harmonic property is invariant under a projectivity: if $\mathbf{H}(AB, CF)$ and $ABCF \underset{\wedge}{\doubleblank} A'B'C'F'$, then $\mathbf{H}(A'B', C'F')$ [Coxeter **2**, p. 23].

3. $\mathbf{H}(AB, CF)$ implies $\mathbf{H}(CF, AB)$. (*Hint:* By 14.54, $ACBF \underset{\wedge}{\doubleblank} CAFB$.)

4. Draw a quadrangle and its section, as in Figure 14.5*d*. Take an arbitrary point X on g and construct the corresponding point X' in the hyperbolic projectivity

$$ABF \underset{\wedge}{\overline{}} ACE.$$

Do the same for $ADB \underset{\wedge}{\overline{}} ADC$, and draw the modified figure that is appropriate for the parabolic projectivity $AAB \underset{\wedge}{\overline{}} AAC$.

5. Two perspectivities cannot suffice for the construction of an elliptic projectivity.

6. In the notation of Figure 14.4*b*,

$$ADCF \overset{Q}{\underset{\wedge}{\doteqdot}} AUP'P \overset{E}{\underset{\wedge}{\doteqdot}} DUR'R \overset{S}{\underset{\wedge}{\doteqdot}} DAFC.$$

7. Any projectivity may be expressed as the product of two involutions [Coxeter **2,** p. 54].

8. The projectivities on the line $x_3 = 0$ are the linear transformations

$$\mu x'_1 = c_{11}x_1 + c_{12}x_2,$$

$$\mu x'_2 = c_{21}x_1 + c_{22}x_2,$$

where $c_{11}c_{22} \neq c_{12}c_{21}$. Under what circumstances is such a projectivity (i) parabolic, (ii) an involution?

14.6 COLLINEATIONS AND CORRELATIONS

A *collineation* is a transformation (of the plane) which transforms collinear points into collinear points. Thus it transforms lines into lines, ranges into ranges, pencils into pencils, quadrangles into quadrangles, and so on. A *projective collineation* is a collineation which transforms every one-dimensional form projectively.

14.61 *Any collineation that transforms one range into a projectively related range is a projective collineation.*

Proof [Bachmann **1,** p. 85]. Let the given collineation transform the range of points X on a certain line a into a projectively related range of points X' on the corresponding line a', and let it transform the points Y on another line b into corresponding points Y' on b'. Any perspectivity relating X and Y will be transformed into a perspectivity relating X' and Y'. Hence

$$Y \overline{\overline{\wedge}} X \underset{\wedge}{\overline{}} X' \overline{\overline{\wedge}} Y',$$

so that the collineation induces a projectivity $Y \underset{\wedge}{\overline{}} Y'$ between the points of b and b', as desired.

It follows that a projective collineation is determined when two corresponding quadrangles (or quadrilaterals) are given [Coxeter **2,** p. 60].

A *perspective collineation* with center O and axis o is a collineation which leaves invariant all the lines through O and all the points on o. (By 14.61, every perspective collineation is a projective collineation.) Following Sophus Lie (1842–1899), we call a perspective collineation an *elation* or a *homology* according as the center and axis are or are not incident. A *harmonic* homology is the special case when corresponding points A and A', on a line a through O, are harmonic conjugates with respect to O and $o \cdot a$. Every projective collineation of period 2 is a harmonic homology [Coxeter **2**, p. 64].

We have seen that a collineation is a point-to-point and line-to-line transformation which preserves incidences. Somewhat analogously, a *correlation* is a point-to-line and line-to-point transformation which dualizes incidences: it transforms points A into lines a', and lines b into points B', in such a way that a' passes through B' if and only if A lies on b. Thus a correlation transforms collinear points into concurrent lines (and vice versa), ranges into pencils, quadrangles into quadrilaterals, and so on. A *projective correlation* is a correlation that transforms every one-dimensional form projectively. In a manner resembling the proof of 14.61, we can establish

14.62   *Any correlation that transforms one range into a projectively related pencil (or vice versa) is a projective correlation.*

It follows that a projective correlation is determined when a quadrangle and the corresponding quadrilateral are given [Coxeter **2**, p. 66].

A *polarity* is a projective correlation of period 2. In general, a correlation transforms a point A into a line a' and transforms this line into a new point A''. When the correlation is of period two, A'' always coincides with A and we can simplify the notation by omitting the prime ($'$). Thus a polarity relates A to a, and vice versa. Following J. D. Gergonne (1771–1859), we call a the *polar* of A, and A the *pole* of a. Clearly, the polars of all the points on a form a projectively related pencil of lines through A.

Since a polarity dualizes incidences, if A lies on b, a passes through B. In this case we say that A and B are *conjugate points*, a and b are *conjugate lines*. It may happen that A and a are incident, so that each is *self-conjugate*. We can be sure that this does not always happen, for it is easy to prove that the join of two self-conjugate points cannot be a self-conjugate line. It is slightly harder to prove that no line can contain more than two self-conjugate points [Coxeter **2**, p. 68]. The following theorem will be used in § 14.7:

14.63   *A polarity induces an involution of conjugate points on any line that is not self-conjugate.*

Proof.   On a non-self-conjugate line a, the projectivity $X \overline{\wedge} a \cdot x$ (Figure 14.6a) transforms any non-self-conjugate point B into another point $C = a \cdot b$, whose polar is AB. The same projectivity transforms C into B. Since it interchanges B and C, it must be an involution.

Dually, x and AX are paired in the involution of conjugate lines through A.

Such a triangle ABC, in which each vertex is the pole of the opposite side (so that any two vertices are conjugate points, and any two sides are conjugate lines), is said to be *self-polar*. If P is any point not on a side, its polar p does not pass through a vertex, and the polarity may be described as the unique projective correlation that transforms the quadrangle $ABCP$ into the quadrilateral $abcp$. An appropriate symbol, analogous to the symbol $(AB)(PQ)$ for an involution, is

$$(ABC)(Pp).$$

Thus any triangle ABC, any point P not on a side, and any line p not through a vertex, determine a definite polarity $(ABC)(Pp)$, in which the polar x of an arbitrary point X can be constructed by simple incidences. As a first step towards this construction we need the following lemma:*

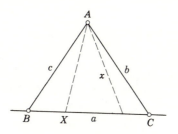

Figure 14.6a

14.64 *If the polars of the vertices of a triangle are distinct from the respectively opposite sides, they meet these sides in three collinear points.*

Proof. Let APX be a triangle whose sides PX, XA, AP meet the polars a, p, x of its vertices in points A_1, P_1, X_1, as in Figure 14.6b. The polar of $X_1 = x \cdot AP$ is, of course, $x_1 = X(a \cdot p)$. Define also the extra points $P' = a \cdot AP$, $X' = a \cdot AX$ and their polars $p' = A(a \cdot p)$, $x' = A(a \cdot x)$. By 14.54 and the polarity, we have

$$AP'PX_1 \overline{\wedge} P'AX_1P \overline{\wedge} p'ax_1p \overline{\wedge} AX'XP_1.$$

By 14.53, $AP'PX_1 \overline{\overline{\wedge}} AX'XP_1$. Since the center of this perspectivity is $P'X' \cdot PX = A_1$, the three points A_1, P_1, X_1 are collinear, as desired.

We are now ready for the construction (Figure 14.6c):

14.65 *In the polarity $(ABC)(Pp)$, the polar of a point X (not on AP, BP, or p) is the line X_1X_2 determined by*

---

* This is known as Chasles's theorem. The proof given in *The Real Projective Plane* [Coxeter **2**, p. 71] suffices for real geometry but not for the more general geometry which is developed here. Lemma 5.54 of that book is false in the finite geometry $PG(2, 3)$, which admits a quadrilateral whose three pairs of opposite vertices P_1P_2, P_3P_6, P_5P_9 are pairs of conjugate points in the polarity $P_i \to p_i$ although the four sides $P_1P_3P_9$, $P_2P_6P_9$, $P_2P_3P_5$, $P_1P_5P_6$ contain their respective poles P_0, P_7, P_8, P_{11}. (The remaining three of the thirteen points in this finite plane are the diagonal points of the quadrangle $P_0P_7P_8P_{11}$; their joins in pairs are the diagonals of the quadrilateral $p_0 p_7 p_8 p_{11}$.) See also W. G. Brown, *Canadian Mathematical Bulletin*, **3** (1960), pp. 221–223.

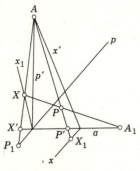

Figure 14.6b   **Figure 14.6c**

$$A_1 = a \cdot PX, \quad P_1 = p \cdot AX, \quad X_1 = AP \cdot A_1P_1,$$
$$B_2 = b \cdot PX, \quad P_2 = p \cdot BX, \quad X_2 = BP \cdot B_2P_2.$$

Proof. By 14.64, the polars a, p, x meet the lines PX, AX, AP in three collinear points, the first two of which are A_1 and P_1. Hence x passes through $X_1 = AP \cdot A_1P_1$. Similarly x passes through $X_2 = BP \cdot B_2P_2$.

In terms of coordinates, a projective collineation is a linear homogeneous transformation

14.66 $\mu x'_\alpha = \Sigma c_{\alpha\beta} x_\beta,$ $\det(c_{\alpha\beta}) \neq 0,$

where the summation is understood to be taken over the repeated index β (for each value of α). The nonvanishing of the determinant makes it possible to solve the equations for x_β in terms of x'_α so as to obtain the inverse collineation. By suitably adjusting the coefficients $c_{\alpha\beta}$, we can transform the particular quadrangle 14.23 into any given quadrangle [Coxeter **2**, p. 197].

Since the product of two correlations (e.g., a polarity and another correlation) is a collineation, any given projective correlation can be exhibited as the product of an arbitrary polarity and a suitable projective collineation. The most convenient polarity for this purpose is that in which the line

$$\Sigma X_\alpha x_\alpha = 0$$

is the polar of the point (X_1, X_2, X_3). Combining this with the general collineation 14.66, we obtain the correlation that transforms each point (y) into the line

14.661 $\Sigma(\Sigma c_{\alpha\beta} y_\beta) x_\alpha = 0,$

where again we must have $\det(c_{\alpha\beta}) \neq 0$. In fact, the correlation is associated with the *bilinear* equation

$$\Sigma\Sigma c_{\alpha\beta} x_\alpha y_\beta = 0$$

[cf. Coxeter **2**, p. 200].

The correlation is a polarity if it is the same as its *inverse*, whose equation, derived by interchanging (x) and (y), is

$$\Sigma\Sigma c_{\alpha\beta}y_{\alpha}x_{\beta} = 0, \quad \text{or} \quad \Sigma\Sigma c_{\beta\alpha}x_{\alpha}y_{\beta} = 0.$$

Thus a polarity occurs when $c_{\beta\alpha} = \lambda c_{\alpha\beta}$, where λ is the same for all α and β, so that $c_{\alpha\beta} = \lambda c_{\beta\alpha} = \lambda^2 c_{\alpha\beta}$, $\lambda^2 = 1$, $\lambda = \pm 1$. But we cannot have $\lambda = -1$, as this would make the determinant

$$\begin{vmatrix} 0 & c_{12} & -c_{31} \\ -c_{12} & 0 & c_{23} \\ c_{31} & -c_{23} & 0 \end{vmatrix} = 0.$$

Hence $\lambda = 1$, and $c_{\beta\alpha} = c_{\alpha\beta}$. In other words,

14.67 *A projective correlation is a polarity if and only if its matrix of coefficients is symmetric.*

Thus the general polarity is given by

14.68 $\qquad \Sigma\Sigma c_{\alpha\beta}x_{\alpha}y_{\beta} = 0, \qquad\qquad c_{\beta\alpha} = c_{\alpha\beta}, \quad \det(c_{\alpha\beta}) \neq 0,$

meaning that the polar of (y_1, y_2, y_3) is 14.661, or that 14.68 is the condition for points (x) and (y) to be conjugate. Setting $y_{\beta} = x_{\beta}$, we deduce the condition

$$\Sigma\Sigma c_{\alpha\beta}x_{\alpha}x_{\beta} = 0,$$

or $c_{11}x_1{}^2 + c_{22}x_2{}^2 + c_{33}x_3{}^2 + 2c_{23}x_2x_3 + 2c_{31}x_3x_1 + 2c_{12}x_1x_2 = 0$, for the point (x) to be self-conjugate. Hence

14.69 *If a polarity admits self-conjugate points, their locus is given by an equation of the second degree.*

EXERCISES

1. Given the center and axis of a perspective collineation, and one pair of corresponding points (collinear with the center), set up a construction for the transform X' of any point X [Coxeter **2,** p. 62].

2. Any two perspective triangles are related by a perspective collineation.

3. A collineation which leaves just the points of one line invariant is an elation.

4. An elation with axis o may be expressed as the product of two harmonic homologies having this same axis o [Coxeter **2,** p. 63].

5. In $PG(2, 3)$, the transformation $P_i \to P_{i+1}$ (with subscripts reduced modulo 13) is evidently a collineation of period 13. Is it a projective collineation? Consider also the transformation $P_i \to P_{3i}$.

6. What kind of collineation is
    (i) $x'_1 = x_1, \quad x'_2 = x_2, \quad x'_3 = cx_3$;
    (ii) $x'_1 = x_1 + c_1x_3, \quad x'_2 = x_2 + c_2x_3, \quad x'_3 = x_3$?

7. Use 14.64 to prove *Hesse's theorem*: If two pairs of opposite vertices of a complete quadrilateral are pairs of conjugate points (in a given polarity), then the third pair of opposite vertices is likewise a pair of conjugate points.

8. Give an analytic proof of Hesse's theorem. (*Hint*: Apply the condition 14.68 to the pairs of vertices

$$(0, 1, \pm 1), \quad (\pm 1, 0, 1), \quad (1, \pm 1, 0)$$

of the quadrilateral $x_1 \pm x_2 \pm x_3 = 0$.)

9. The bilinear equation

$$x_1 y_1 + x_2 y_2 + x_3 y_3 = 0$$

is the condition for (x) and (y) to be conjugate in the polarity $(ABC)(Pp)$, where ABC is the triangle of reference, P is $(1, 1, 1)$, and p is $x_1 + x_2 + x_3 = 0$. Are there any self-conjugate points? Consider, in particular, the case when the coordinates are residues modulo 3.

14.7 THE CONIC

The three familiar curves which we call the "conic sections" have a long history. The reputed discoverer was Menaechmus, who flourished about 350 B.C. They attracted the attention of the best of the Greek geometers down to the time of Pappus of Alexandria. . . . A vivid new interest arose in the seventeenth century. . . . It seems certain that they will always hold a place in the mathematical curriculum.

J. L. Coolidge (1873 -1954)

[Coolidge **1,** Preface]

In the projective plane there is only one kind of conic. The familiar distinction between the hyperbola, parabola, and ellipse belongs to affine geometry. To be precise, it depends on whether the line at infinity is a secant, a tangent, or a nonsecant [Coxeter **2,** p. 129].

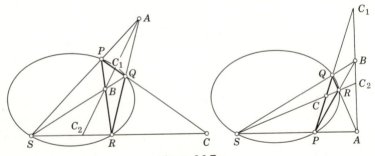

Figure 14.7a

A polarity is said to be *hyperbolic* or *elliptic* according as it does or does not admit a self-conjugate point. (In the former case it also admits a self-conjugate line: the polar of the point.) The self-conjugate point P, whose existence suffices to make a polarity hyperbolic, is by no means the only self-conjugate point: there is another on every line through P except its polar p. To prove this we use 14.63, which tells us that every such line contains an involution of conjugate points. By 14.57, this involution, having one invariant point P, has a second invariant point Q, which is, of course, another self-conjugate point of the polarity. Thus the presence of one self-conjugate point implies the presence of many (as many as the lines through a point; for example, infinitely many in real or complex geometry). Their locus is a *conic,* and their polars are its *tangents.* This simple definition, due to von Staudt, exhibits the conic as a self-dual figure: the locus of self-conjugate points and also the envelope of self-conjugate lines.

The reader must bear in mind that there are only two kinds of polarity and that there is only one kind of conic. The terminology is perhaps not very well chosen: a hyperbolic polarity has many self-conjugate points, forming a conic; an elliptic polarity has no self-conjugate points at all, but still provides a polar for each point and a pole for each line; there is no such thing as a "parabolic polarity."

A tangent justifies its name by meeting the conic only at its pole, the *point of contact.* Any other line is called a *secant* or a *nonsecant* according as it meets the conic twice or not at all, that is, according as the involution of conjugate points on it is hyperbolic or elliptic. Any two conjugate points on a secant PQ, being paired in the involution $(PP)(QQ)$, are harmonic conjugates with respect to P and Q.

Let PQR be a triangle inscribed in a conic, as in Figure 14.7a. Any line c conjugate to PQ is the polar of some point C on PQ. Let RC meet the conic again in S. Then C is one of the three diagonal points of the inscribed quadrangle $PQRS$. The other two are

$$A = PS \cdot QR, \quad B = QS \cdot RP.$$

Their join meets the sides PQ and RS in points C_1 and C_2 such that H(PQ, CC_1) and H(RS, CC_2). Since C_1 and C_2 are conjugate to C, the line AB, which contains them, is c, the polar of C. Similarly BC is the polar of A. Therefore A and B are conjugate points. These conjugate points are the intersections of c with the sides QR and RP of the given triangle. Hence

SEYDEWITZ'S THEOREM. *If a triangle is inscribed in a conic, any line conjugate to one side meets the other two sides in conjugate points.*

From this we shall have no difficulty in deducing

STEINER'S THEOREM. *Let lines x and y join a variable point on a conic to two fixed points on the same conic; then* $x \barwedge y$.

Proof. The tangents p and q, at the fixed points P and Q, intersect in D, the pole of PQ. Let c be a fixed line through D (but not through P or Q), meeting x and y in B and A, as in Figure 14.7b. By Seydewitz's theorem,

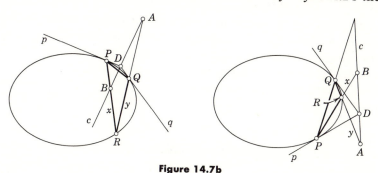

Figure 14.7b

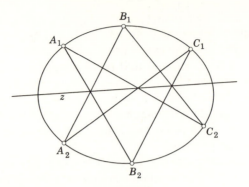

Figure 14.7c

BA is a pair of the involution of conjugate points on *c*. Hence, when the point $x \cdot y$ varies on the conic,

$$x \barwedge B \barwedge A \barwedge y.$$

The following construction for a conic through five given points, no three collinear, was discovered by Braikenridge and Maclaurin independently, about 1733 [Coxeter **2**, p. 91]. Let A_1, B_2, C_1, A_2, B_1 be the five points, as in Figure 14.7c; then the conic is the locus of the point

$$C_2 = A_1(z \cdot C_1A_2) \cdot B_1(z \cdot C_1B_2),$$

where *z* is a variable line through the point $A_1B_2 \cdot B_1A_2$. This is the converse of

PASCAL'S THEOREM.   *If a hexagon $A_1B_2C_1A_2B_1C_2$ is inscribed in a conic, the points of intersection of pairs of opposite sides, namely,*

$$B_1C_2 \cdot B_2C_1, \quad C_1A_2 \cdot C_2A_1, \quad A_1B_2 \cdot A_2B_1,$$

are collinear.

Pascal discovered his famous theorem [Coxeter **2**, p. 103] when he was only sixteen years old. More than 150 years later, it was dualized (see Figure 14.7*d*):

BRIANCHON'S THEOREM.   *If a hexagon is circumscribed about a conic, its three diagonals are concurrent.*

We saw, in § 8.4, that the familiar conics of Euclidean geometry have equations of the second degree in Cartesian coordinates. The same equations in affine coordinates remain valid in affine geometry, and yield homogeneous equations of the second degree in barycentric coordinates (§ 13.7) and in projective coordinates (§ 14.2). Thus 14.69 serves to reconcile von Staudt's definition of a conic with the classical definitions. In particular,

$$x_1x_3 = x_2{}^2$$

is a conic touching the lines $x_3 = 0$ and $x_1 = 0$ at the respective points

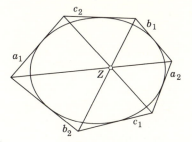

Figure 14.7d

(1, 0, 0) and (0, 0, 1). This conic can be parametrized in the form

$$x_1 : x_2 : x_3 = t^2 : t : 1,$$

which exhibits it as the locus of the point of intersection of corresponding members of the projectively related pencils of lines

$$x_1 = tx_2 \quad \text{and} \quad x_2 = tx_3.$$

If det $(c_{\alpha\beta}) = 0$, the quadratic form $\Sigma\Sigma c_{\alpha\beta}x_\alpha x_\beta$ may be expressible as the product of two linear forms $\Sigma a_\alpha x_\alpha$ and $\Sigma b_\beta x_\beta$. Accordingly, a pair of lines is sometimes regarded as a degenerate conic. In this sense, Axiom 14.15 is a special case of Pascal's theorem.

EXERCISES

1. If a quadrangle is inscribed in a conic, its diagonal points form a self-polar triangle. The tangents at the vertices of the quadrangle form a circumscribed quadrilateral whose diagonals are the sides of the same triangle [Coxeter **2**, pp. 85, 86].

2. Referring to the projectivity $x \barwedge y$ of Steiner's theorem, investigate the special positions of x and y when A or B coincides with D.

3. If a projectivity between pencils of lines x and y through P and Q has the effect $xpd \barwedge ydq$, where d is PQ, the locus of the point $x \cdot y$ is a conic through P and Q whose tangents at these points are p and q. (This construction is often used to *define* a conic; see, e.g., Robinson [**1**, p. 38].)

4. Of the conics that touch two given lines at given points, those which meet a third line (not through either of the points) do so in pairs of an involution [Coxeter **2**, p. 90].

5. If two triangles are self-polar for a given polarity, their six vertices lie on a conic or on two lines [Coxeter **2**, p. 93].

6. If two triangles have six distinct vertices, all lying on a conic, they are self-polar for some polarity [Coxeter **2**, p. 94].

7. In $PG(2, 3)$ (Ex. 3 at the end of § 14.1), the polarity $P_i \rightarrow p_i$ or $(P_4P_{10}P_{12})(P_0 p_0)$ determines a conic consisting of the four points P_0, P_7, P_8, P_{11} and the four lines p_0, p_7, p_8, p_{11}. (*Hint:* $P_0P_2P_8P_{12} \barwedge P_1P_7P_5P_4 \barwedge p_0p_2p_8p_{12}$.)

8. The equation $x_1^2 + x_2^2 - x_3^2 = 0$ represents a conic for which the triangle of reference is self-polar. Verify Pascal's theorem as applied to the inscribed hexagon

$$(0, 1, 1) \ (0, -1, 1) \ (1, 0, 1) \ (-1, 0, 1) \ (3, 4, 5) \ (4, 3, 5).$$

14.8 PROJECTIVE SPACE

> *Our Geometry is an abstract Geometry. The reasoning could be fol-*
> *lowed by a disembodied spirit who had no idea of a physical point;*
> *just as a man blind from birth could understand the Electromagnetic*
> *Theory of Light.*
>
> H. G. Forder [**1,** p. 43]

Axiom 14.12 had the effect of restricting the geometry to a single plane. If we remove this restriction, we must know exactly what we mean by a plane. First we define a *flat pencil* to be the set of lines joining a range of points (on a line) to another point. Then we define a *plane* to be the set of points on the lines of a flat pencil and the set of lines joining pairs of these points. Accordingly *we replace Axiom 14.12 by three new axioms.* The first (which may be regarded as a projective version of Pasch's axiom, 12.27) allows us to forget the role of a particular flat pencil in the definition of a plane. The second enables us to speak of more than one plane. The third (cf. 12.431) restricts the number of dimensions to three.

AXIOM 14.81   *If A, B, C are three non-collinear points, and D is a point on BC distinct from B and C, while E is a point on CA distinct from C and A, then there is a point F on AB such that D, E, F are collinear.*

AXIOM 14.82   *There is at least one point not in the plane ABC.*

AXIOM 14.83   *Any two planes meet in a line.*

We now have a different principle of duality: points, lines and planes correspond to planes, lines and points (cf. § 10.5). Two intersecting lines, *a* and *b*, determine a point $a \cdot b$ and a plane ab; these are dual concepts. Two lines that do not intersect are said to be *skew*. The theory of collineations and correlations [Coxeter **3,** pp. 63–70] is analogous to the two-dimensional case, except that the number of self-conjugate points on a line is no longer restricted to 0, 1, or 2. In fact, instead of two kinds of polarity we now have four: one "elliptic," having no self-conjugate points, two "hyperbolic," whose self-conjugate points form a quadric (nonruled or ruled), and one, the *null polarity* (or "null system"), in which every point in space is self-conjugate! The idea of defining a quadric as the locus of self-conjugate points in a three-dimensional polarity (of the second or third kind) is due to von Staudt. Another approach, using a two-dimensional polarity in an arbitrary plane ω, was devised by F. Seydewitz.* The quadric appears as the locus of the point

$$PA \cdot Qa,$$

where P and Q are fixed points (on the quadric) while A is a variable point on ω and a is its polar. This definition allows the quadric to degenerate to a cone or a pair of planes.

To sample the flavor of solid projective geometry, let us consider a few

---

* *Archiv für Mathematik und Physik,* **9** (1848), p. 158.

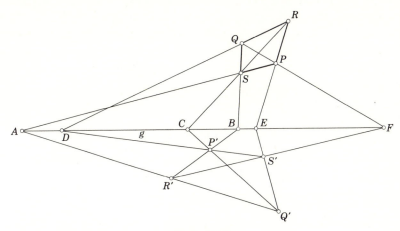

Figure 14.8a

typical theorems. Suppose a complete quadrangle $PQRS$ yields a quadrangular set $Q(ABC, DEF)$ on a line g, as in Figure 14.8a. In another plane through g, let the sides of a triangle $P'Q'R'$ pass through A, B, C, and let DP' meet EQ' in S'. Theorem 14.42 tells us that S' lies on $R'F$. This remark yields two interesting configurations: one consisting of eight lines (Figure 14.8b), and the other of two mutually inscribed tetrahedra.

GALLUCCI'S THEOREM. *If three skew lines all meet three other skew lines, any transversal to the first set of three meets any transversal to the second set.*

Proof. Let the two sets of lines be PQ', $P'Q$, RS; PQ, $P'Q'$, $R'S$. This notation agrees with Figure 14.8a, for, since PS and $Q'R'$ both pass through A, PQ' meets $R'S$, and since QS and $R'P'$ both pass through B, $P'Q$ meets $R'S$. The transversal from R to PQ' and $P'Q$ is

$$RPQ' \cdot RP'Q = REQ' \cdot RDP' = RS'.$$

The transversal from R' to PQ and $P'Q'$ is

$$R'PQ \cdot R'P'Q' = R'FQ \cdot R'FQ' = R'F.$$

Since S' lies on $R'F$, these transversals meet, as desired.

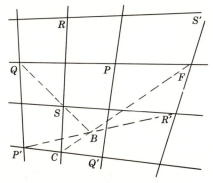

Figure 14.8b

MÖBIUS'S THEOREM. *If the four vertices of one tetrahedron lie respectively in the four face planes of another, while three vertices of the second lie in three face planes of the first, then the remaining vertex of the second lies in the remaining face plane of the first.*

Proof. Let $PQRS'$ and $P'Q'R'S$ be the two tetrahedra, with

$$P, \quad Q, \quad R, \quad S', \quad P', \quad Q', \quad S$$

in the respective planes

$$Q'R'S, P'R'S, P'Q'S, P'Q'R', QRS', PRS', PQR,$$

as in Figure 14.8a. Since $R'S'$ passes through F, on PQ, the remaining vertex R' lies in the remaining plane PQS', as desired.

Changing the notation from

$$S, \quad P, \quad Q, \quad R, \quad P', \quad Q', \quad R', \quad S'$$

to

$$S, \quad S_{14}, \quad S_{24}, \quad S_{34}, \quad S_{23}, \quad S_{13}, \quad S_{12}, \quad S_{1234},$$

we deduce the first of a remarkable "chain" of theorems due to Homersham Cox:[*]

COX'S FIRST THEOREM. *Let $\sigma_1, \sigma_2, \sigma_3, \sigma_4$ be four planes of general position through a point S. Let S_{ij} be an arbitrary point on the line $\sigma_i \cdot \sigma_j$. Let σ_{ijk} denote the plane $S_{ij}S_{ik}S_{jk}$. Then the four planes $\sigma_{234}, \sigma_{134}, \sigma_{124}, \sigma_{123}$ all pass through one point S_{1234}.*

Clearly, $\sigma_1, \sigma_2, \sigma_3, \sigma_{123}$ are the face planes of the tetrahedron $P'Q'R'S$, while $\sigma_{234}, \sigma_{134}, \sigma_{124}, \sigma_4$ are those of the inscribed-circumscribed tetrahedron $PQRS'$. Let σ_5 be a fifth plane through S. Then $S_{15}, S_{25}, S_{35}, S_{45}$ are four points in σ_5; σ_{ij5} is a plane through the line $S_{i5}S_{j5}$; and S_{ijk5} is the point $\sigma_{ij5} \cdot \sigma_{ik5} \cdot \sigma_{jk5}$. By the dual of Cox's first theorem, the four points $S_{2345}, S_{1345}, S_{1245}, S_{1235}$ all lie in one plane. Interchanging the roles of σ_4 and σ_5, we see that S_{1234} lies in this same plane $S_{2345}S_{1345}S_{1245}$, which we naturally call σ_{12345}. Hence

COX'S SECOND THEOREM. *Let $\sigma_1, \ldots, \sigma_5$ be five planes of general position through S. Then the five points $S_{2345}, S_{1345}, S_{1245}, S_{1235}, S_{1234}$ all lie in one plane σ_{12345}.*

Adding the extra digits 56 to all the subscripts in the first theorem, we deduce

COX'S THIRD THEOREM. *The six planes $\sigma_{23456}, \sigma_{13456}, \sigma_{12456}, \sigma_{12356}, \sigma_{12346}, \sigma_{12345}$ all pass through one point S_{123456}.*

The pattern is now clear: we can continue indefinitely. "Cox's $(d-3)$rd

---

[*] *Quarterly Journal of Mathematics,* **25** (1891), p. 67. See also H. W. Richmond, *Journal of the London Mathematical Society,* **16** (1941), pp. 105–112, and Coxeter, *Bulletin of the American Mathematical Society,* **56** (1950), p. 446. When we describe four planes through a point as being "of general position," we mean that their six lines of intersection are all distinct.

theorem" provides a configuration of 2^{d-1} points and 2^{d-1} planes, with d of the planes through each point and d of the points in each plane.

Our next result would be difficult to obtain without using coordinates. Since the equation of the general quadric

$$c_{11}x_1^2 + \ldots + c_{44}x_4^2 + 2c_{12}x_1x_2 + \ldots + 2c_{34}x_3x_4 = 0$$

has $4 + 6 = 10$ terms, a unique quadric $\Sigma = 0$ can be drawn through nine points of general position; for, by substituting each of the nine given sets of x's in $\Sigma = 0$, we obtain nine linear equations to solve for the mutual ratios of the ten c's. Similarly, a "pencil" (or singly infinite system) of quadrics

$$\Sigma + \mu\Sigma' = 0$$

can be drawn through eight points of general position, and a "bundle" (or doubly infinite system) of quadrics

$$\Sigma + \mu\Sigma' + \nu\Sigma'' = 0$$

can be drawn through seven points of general position. But, by solving the simultaneous quadratic equations

$$\Sigma = 0, \quad \Sigma' = 0, \quad \Sigma'' = 0$$

for the mutual ratios of the four x's, we obtain eight points of intersection for these three quadrics. Naturally these eight points lie on every quadric of the bundle. Hence

Seven points of general position determine a unique eighth point, such that every quadric through the seven passes also through the eighth.

This idea of the eighth *associated* point provides an alternative proof for Cox's first theorem (and therefore also for the theorems of Möbius and Gallucci). Let S_{1234} be defined as the common point of the three planes $\sigma_{234}, \sigma_{134}, \sigma_{124}$. (The theorem states that S_{1234} lies also on σ_{123}.) Since the plane pairs $\sigma_1\sigma_{234}, \sigma_2\sigma_{134}, \sigma_3\sigma_{124}$ form three degenerate quadrics through the eight points

$$S, S_{14}, S_{24}, S_{34}, S_{23}, S_{13}, S_{12}, S_{1234},$$

these are eight associated points. The first seven belong also to the plane pair $\sigma_4\sigma_{123}$. Since S_{1234} does not lie in σ_4, it must lie in σ_{123}, as desired.

The locus of lines meeting three given skew lines is called a *regulus*. Gallucci's theorem shows that the lines meeting three generators of the regulus (including the original three lines) form another "associated" regulus, such that every generator of either regulus meets every generator of the other. The two reguli are the two systems of generators of a *ruled quadric*.

Let a_1, b_1, c_1, d_1 be four generators of the first regulus, and a_2, b_2, c_2, d_2 four generators of the second, as in Figure 14.8c. The three lines

$$a_3 = b_1c_2 \cdot b_2c_1, \quad b_3 = c_1a_2 \cdot c_2a_1, \quad c_3 = a_1b_2 \cdot a_2b_1$$

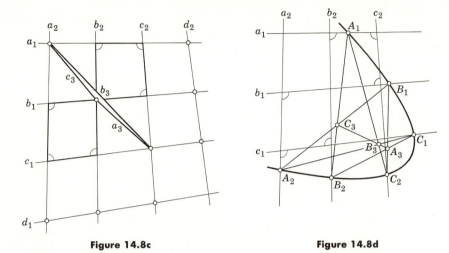

Figure 14.8c   Figure 14.8d

evidently form a triangle whose vertices are $a_1 \cdot a_2$, $b_1 \cdot b_2$, $c_1 \cdot c_2$. G. P. Dandelin, in 1824, coined the name *hexagramme mystique* for the skew hexagon $a_1b_2c_1a_2b_1c_2$. Taking the section of its sides by a plane δ of general position, he obtained a plane hexagon $A_1B_2C_1A_2B_1C_2$ whose sides A_1B_2, B_2C_1, ... lie in the planes a_1b_2, b_2c_1, ... (Figure 14.8d). The points of intersection of pairs of opposite sides, namely,

$$A_3 = B_1C_2 \cdot B_2C_1, \quad B_3 = C_1A_2 \cdot C_2A_1, \quad C_3 = A_1B_2 \cdot A_2B_1,$$

each lying in both the planes $a_3b_3c_3$ and δ, are collinear. By allowing c_2 to vary while the remaining sides of the skew hexagon remain fixed, we see from the Braikenridge-Maclaurin construction (which is the converse of Pascal's theorem, Figure 14.7c), that

The section of a ruled quadric, by a plane of general position, is a conic.

If δ, instead of being a plane of general position, is the plane d_1d_2, the vertices of the hexagon $A_1B_2C_1A_2B_1C_2$ line alternately on d_2 and d_1, as in Axiom 14.15. Thus Pappus's theorem may be regarded as a "degenerate" case of Pascal's theorem. In fact, instead of assuming Pappus's theorem and deducing Gallucci's theorem, we could have taken the latter as an axiom and deduced the former. Bachmann [**1**, p. 254] gives a particularly fine figure to illustrate this deduction.

EXERCISES

1. If a and b are two skew lines and R is a point not on either of them, $Ra \cdot Rb$ is the *only* transversal from R to the two lines.

2. Any plane through a generator of a ruled quadric contains another generator. (Such a plane is a *tangent* plane.) Any other plane section of the ruled quadric is a conic.

3. If two tetrahedra are trebly perspective they are quadruply perspective (cf. § 14.3). More precisely, if $A_1A_2A_3A_4$ is perspective with each of $B_2B_1B_4B_3$, $B_3B_4B_1B_2$,

$B_4B_3B_2B_1$, it is also perspective with $B_1B_2B_3B_4$. (*Hint*: Since A_iB_j meets A_jB_i, A_iB_i must meet A_jB_j.)

4. The four centers of perspective that were implied in Ex. 3 form a third tetrahedron which is perspective with either of the first two from each vertex of the remaining one.

5. In the finite space $PG(3, 3)$, which has 4 points on each line, there are altogether 40 points, 40 planes, and how many lines?

14.9   EUCLIDEAN SPACE

> The set of lines drawn from the artist's eye to the various points of the object . . . constitute the projection of the object and are called the Euclidean cone. Then the section of this cone made by the canvas is the desired drawing. . . . Parallel lines in the object converge in the picture to the point where the canvas is pierced by the line from the eye parallel to the given lines.
>
> S. H. Gould [**1,** p. 299]

The elementary approach to affine space is to regard it as Euclidean space without a metric; the elementary approach to projective space is to regard it as affine space plus the plane at infinity and then to ignore the special role of that plane. It is equally effective to begin with projective space and derive affine space by specializing any one plane, calling it the plane at infinity. (This is still, of course, a projective plane.) Each affine concept has its projective definition: for example, the midpoint of AB is the harmonic conjugate, with respect to A and B, of the point at infinity on AB [Coxeter **2,** p. 119]. We then derive Euclidean space by specializing one elliptic polarity in the plane at infinity, calling it the *absolute polarity*. Two lines are orthogonal if their points at infinity are conjugate in the absolute polarity; a line and a plane are orthogonal if the point at infinity on the line is the pole of the line at infinity in the plane. A sphere is the locus of the point of intersection of a line through one fixed point and the perpendicular plane through another; thus it is a special quadric according to Seydewitz's definition. Two segments with a common end are congruent if they are radii of the same sphere [Coxeter **2,** p. 146].

When we use projective coordinates (x_1, x_2, x_3, x_4), referred to an arbitrary tetrahedron

$$(1, 0, 0, 0) \quad (0, 1, 0, 0) \quad (0, 0, 1, 0) \quad (0, 0, 0, 1),$$

it is convenient to take the plane at infinity to be $x_4 = 0$. Any other equation becomes an equation in affine coordinates x_1, x_2, x_3 by the simple device of setting $x_4 = 1$. In affine terms, the tetrahedron of reference for the projective coordinates is formed by the origin and the points at infinity on the three axes. Finally, we pass from affine space to Euclidean space by declaring that two points (x_1, x_2, x_3) and (y_1, y_2, y_3) are in perpendicular directions from the origin if they satisfy the bilinear equation

$$x_1y_1 + x_2y_2 + x_3y_3 = 0,$$

that is, if the points at infinity

$$(x_1, x_2, x_3, 0) \quad \text{and} \quad (y_1, y_2, y_3, 0)$$

are conjugate in the absolute polarity.

All the theorems that we proved in § 14.8 remain valid in Euclidean space. An interesting variant of Cox's chain of theorems can be obtained by means of the following specialization. Instead of an *arbitrary* point on the line $\sigma_i \cdot \sigma_j$, we take S_{ij} to be the second intersection of this line with a fixed sphere through S. Since the sphere is a quadric through the first seven of the eight associated points

$$S, \; S_{14}, \; S_{24}, \; S_{34}, \; S_{23}, \; S_{13}, \; S_{12}, \; S_{1234},$$

it passes through S_{1234} too, and similarly through S_{1235} and the rest of the 2^{d-1} points. The 2^{d-1} planes meet the sphere in 2^{d-1} circles, which remain circles when we make an arbitrary stereographic projection, as in § 6.9. We thus obtain Clifford's chain of theorems* in the inversive (or Euclidean) plane.

CLIFFORD'S FIRST THEOREM. *Let $\sigma_1, \sigma_2, \sigma_3, \sigma_4$ be four circles of general position through a point S. Let S_{ij} be the second intersection of the circles σ_i and σ_j. Let σ_{ijk} denote the circle $S_{ij}S_{ik}S_{jk}$. Then the four circles $\sigma_{234}, \sigma_{134}, \sigma_{124}, \sigma_{123}$ all pass through one point S_{1234}.*

CLIFFORD'S SECOND THEOREM. *Let σ_5 be a fifth circle through S. Then the five points $S_{2345}, S_{1345}, S_{1245}, S_{1235}, S_{1234}$ all lie on one circle σ_{12345}.*

CLIFFORD'S THIRD THEOREM. *The six circles $\sigma_{23456}, \sigma_{13456}, \sigma_{12456}, \sigma_{12356}, \sigma_{12346}, \sigma_{12345}$ all pass through one point S_{123456}.*

And so on!

EXERCISES

1. Why is the absolute polarity elliptic?

2. Draw a careful figure for Clifford's first theorem.

3. The circumcircles of the four triangles formed by four general lines all pass through one point (cf. Ex. 2 at the end of § 5.5).

4. The circumcenters of the four triangles of Ex. 3 all lie on a circle which passes also through the point of concurrence of the four circumcircles [Forder **3,** pp. 16–22; Baker **1,** p. 328].

* W. K. Clifford, *Mathematical Papers* (London, 1882), p. 51. Apparently Clifford did not state these theorems in their full generality. Instead of circles through S he took $\sigma_1, \sigma_2, \ldots$ to be straight lines. In other words, he took S to be the point at infinity of the inversive plane. Thus his special form of the theorems could have been derived from the configuration of circles on the sphere by taking the center of the stereographic projection to be the point S on the sphere [Baker **1,** p. 133].

15

Absolute geometry

ın the present chapter we shall re-examine the material of some of the earlier chapters in the light of the axiomatic approach outlined in Chapter 12, regarding classical geometry as ordered geometry enriched with the axioms of congruence 15.11–15.15, the last of which is a restatement of 1.26. Except in §§ 15.6 and 15.8, we shall work in the domain of *absolute* geometry, that is, we shall take care not to assume any form of Euclid's fifth postulate. Accordingly, our results will be valid not only in Euclidean geometry but also in the non-Euclidean geometry of Gauss, Lobachevsky, and Bolyai.

In § 15.4 we shall give a simple account of the complete enumeration of finite groups of isometries. According to Weyl [**1**, p. 79], "This is the modern equivalent to the tabulation of the regular polyhedra by the Greeks." The relevance of these kinematical results to crystallography makes it natural, in § 15.6, to reintroduce the full machinery of Euclidean geometry. But in § 15.7 we shall return to absolute geometry for a discussion of finite groups generated by reflections. Many of the methods used remain valid also in spherical geometry.

15.1 CONGRUENCE

Every teacher certainly should know something of non-Euclidean geometry. . . . It forms one of the few parts of mathematics which . . . is talked about in wide circles, so that any teacher may be asked about it at any moment.

F. Klein [**2,** p. 135]

To give a rigorous approach to absolute geometry, we begin with ordered geometry (Chapter 12) and introduce *congruence* as a third primitive concept: an undefined equivalence relation among point pairs (or segments, or intervals). We use the notation $AB \equiv CD$ to mean "AB is congruent to CD." The following axioms are those of Pasch with some refinements due to Hilbert and R. L. Moore [see Kerékjártó **1**, pp. 90–101].

Axioms of Congruence

15.11 *If A and B are distinct points, then on any ray going out from C there is just one point D such that $AB \equiv CD$.*

15.12 *If $AB \equiv CD$ and $CD \equiv EF$, then $AB \equiv EF$.*

15.13 *$AB \equiv BA$.*

15.14 *If $[ABC]$ and $[A'B'C']$ and $AB \equiv A'B'$ and $BC \equiv B'C'$, then $AC \equiv A'C'$.*

15.15 *If ABC and A'B'C' are two triangles with $BC \equiv B'C'$, $CA \equiv C'A'$, $AB \equiv A'B'$, while D and D' are two further points such that $[BCD]$ and $[B'C'D']$ and $BD \equiv B'D'$, then $AD \equiv A'D'$.*

By two applications of 15.13, we have $AB \equiv AB$; that is, congruence is *reflexive*. From 15.11 and 15.12 we easily deduce that the relation $AB \equiv CD$ implies $CD \equiv AB$; that is, congruence is *symmetric*. Axiom 15.12 itself says that congruence is *transitive*. Hence congruence is an *equivalence* relation. This result, along with the *additive* property of 15.14, provides the basis for a theory of *length* [Forder **1**, p. 95]. Axiom 15.15 enables us to extend the relation of congruence from point pairs or segments to *angles* [Forder **1**, p. 132].

We follow Euclid in defining a *right angle* to be an angle that is congruent to its supplement; and we agree to measure angles on such a scale that the magnitude of a right angle is $\frac{1}{2}\pi$.

The statement $AB \equiv CD$ for segments is clearly equivalent to the statement $AB = CD$ for lengths, so no confusion arises from using the same symbol AB for a segment and its length. A similar remark applies to angles.

The *circle* with center O and radius r is defined as the locus of a variable point P such that $OP = r$. A point Q such that $OQ > r$ is said to be *outside* the circle. Points neither on nor outside the circle are said to be *inside*. It can be proved [Forder **1**, p. 131] that if a circle with center A has a point inside and a point outside a circle with center C, then the two circles meet in just one point on each side of the line AC. Euclid's first four postulates may now be treated as theorems, and we can prove all his propositions as far as I.26; also I.27 and 28 with the word "parallel" replaced by "nonintersecting." We can define reflection as in § 1.3, and derive its simple consequences such as *pons asinorum* (Euclid I.5) and the symmetry of a circle about its diameters (III.3; see § 1.5). But we must be careful to avoid any appeal to our usual idea about the sum of the angles of a triangle; for example, we can no longer assert that angles in the same segment of a circle are equal (Euclid III.21). Lacking such theorems as VI.2–4, which depend on the affine properties of parallelism, we have to look for some quite different way to prove the concurrence of the medians of a triangle.* On the other hand, the concurrence of the *altitudes* (of an acute-angled triangle) arises as a byproduct of Fagnano's problem, which can still be treated as in § 1.8. (Fermat's problem would require a different treatment because we can no longer assume the angles of an equilateral triangle to be $\pi/3$.)

* Bachmann **1**, pp. 74–75.

EXERCISES

1. Complete the proof that congruence is symmetric: if $AB \equiv CD$ then $CD \equiv AB$.

2. How much of § 1.5 remains valid in absolute geometry? [Kerékjártó **1**, pp. 161–163.] (See especially Exercises 1 and 3.)

3. For any simple quadrangle inscribed in a circle, the sum of two opposite angles is equal to the sum of the remaining two angles [Sommerville **1**, p. 84].

15.2 PARALLELISM

> *I have resolved to publish a work on the theory of parallels as soon as I have put the material in order. . . . The goal is not yet reached, but I have made such wonderful discoveries that I have been almost overwhelmed by them. . . . I have created a new universe from nothing.*
>
> Janos Bolyai (1802 -1860)
>
> (From a letter to his father in 1823)

Following Gauss, Bolyai, and Lobachevsky, we say that two lines are *parallel* if they "almost meet." For the precise meaning of this phrase, see § 12.6. (We use the notation p_1 for one of the two rays into which the line p is decomposed by a point that lies on it.)

The idea of the *incenter* (§ 1.5) may be extended from a triangle to the figure formed by two parallel lines and a transversal, enabling us to prove that parallelism is symmetric:

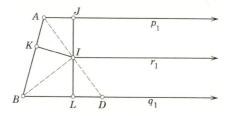

Figure 15.2a

15.21   *If p_1 is parallel to q_1, then q_1 is parallel to p_1.*

Proof. [Sommerville **1**, p. 32]. If p_1, through A, is parallel to q_1, through B, as in Figure 15.2a, the internal bisector AD of the angle at A completes a triangle ABD. Let the internal bisector of B meet AD in I. Draw perpendiculars IJ, IK, IL, to p_1, AB, q_1. Reflecting in IA and IB, we see that $IJ = IK = IL$. Let r_1 be the internal bisector of $\angle LIJ$. Reflection in the line r interchanges J and L, and therefore interchanges p and q. Since p is parallel to q, it follows that q is parallel to p in the same sense, that is, q_1 is parallel to p_1. (In the terminology of Gauss, J and L are *corresponding* points on the two parallel rays.)

We can now use the methods of ordered geometry to prove that parallelism is transitive:

15.22   *If p_1 is parallel to q_1, and q_1 is parallel to r_1, then p_1 is parallel to r_1.*

Proof [Gauss **1**, vol. 8, pp. 205–206].   We have to show that, if p_1 and r_1 are both parallel to q_1, they are parallel to each other.   We see at once that p_1 and r_1 cannot meet; for if they did, we would have two intersecting lines p and r both parallel to q in the given sense.   By Theorem 12.64, we may assume that p_1, q_1, r_1 begin from three collinear points A, B, C.   For the rest of the proof we distinguish the case in which B lies between A and C from the case in which it does not.

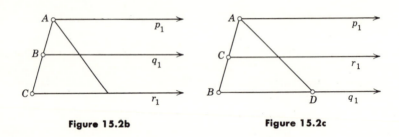

Figure 15.2b   **Figure 15.2c**

If $[ABC]$, as in Figure 15.2b, any ray from A within the angle between AC and p_1 meets q_1 (since p_1 is parallel to q_1) and then meets r_1 (since q_1 is parallel to r_1).   Therefore p_1 is parallel to r_1.

If B is not between A and C, suppose for definiteness that $[ACB]$, as in Figure 15.2c.   Any ray from A within the angle between AC and p_1 meets q_1, say in D.   Since r separates A from D, it meets the segment AD.   Therefore p_1 is parallel to r_1.

In this second part of the proof we have not used the parallelism of q_1 and r_1.   In fact,

15.23   *If a ray r_1 lies between two parallel rays, it is parallel to both.*

Having proved that parallelism is an equivalence relation, we consider the set of lines parallel to a given ray.   We naturally call this a *pencil of parallels*, since it contains a unique line through any given point [Coxeter **2**, p. 5]. Pursuing its analogy with an ordinary pencil (consisting of all the lines through a point), we may also call it a *point at infinity* or, following Hilbert, an *end*.   Instead of saying that two rays (or lines) are parallel, or that they belong to a certain pencil of parallels M, we say that they have M for a common end.   In the same spirit, the ray through A that belongs to the given pencil of parallels is denoted by AM, as if it were a segment; the same symbol AM can also be used for the whole line.

Let AM, BM be parallel rays, and ε an arbitrarily small angle. Within the angle BAM (Figure 15.2d), take a ray from A making with AM an angle less than ε. This ray cuts BM in some point C. On CM (which is C/B), take D so that $CD = CA$. The isosceles triangle CAD yields

$$\angle ADC = \angle CAD < \angle CAM < \varepsilon.$$

Hence, when BD tends to infinity, so that AD tends to the position AM, $\angle ADB$ tends to zero.

This conclusion motivates the following assertion of Bolyai [**1,** p. 207]:

15.24   *When two parallel lines are regarded as meeting at infinity, the angle of intersection must be considered as being equal to zero.*

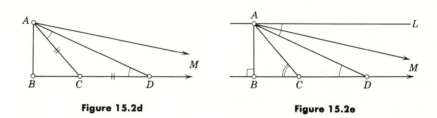

Figure 15.2d   Figure 15.2e

When AM and BM are parallel rays, we call the figure ABM an *asymptotic triangle*. Such triangles behave much like finite triangles. In particular, two of them are congruent if they agree in the finite side and one angle [Carslaw **1,** p. 49]:

15.25   *If two asymptotic triangles ABM, $A'B'M'$ have $AB = A'B'$ and $A = A'$, then also $B = B'$.*

It is a consequence of Axiom 15.11 that, if two lines have a common perpendicular, they do not intersect. The following theorem provides a kind of converse for this statement.

15.26   *If two lines are neither intersecting nor parallel, they have a common perpendicular.*

Proof. From A on the first line AL, draw AB perpendicular to the second line BM, as in Figure 15.2e. If AB is perpendicular to AL there is no more to be said. If not, suppose L is on that side of AB for which $\angle BAL$ is acute. Since the two lines are neither intersecting nor parallel, there is a smaller angle BAM such that AM is parallel to BM. If $[BCD]$ on BM, we can apply Euclid I.16 to the triangle ACD, with the conclusion that the internal angle at D is less than the external angle at C. Hence, when BD increases from 0 to ∞, so that $\angle DAL$ decreases from $\angle BAL$ to $\angle MAL$, $\angle ADB$ decreases from a right angle to zero. At the beginning of this process we have

$$\angle DAL < \angle ADB$$

(since $\angle BAL$ is acute); but at the end the inequality is reversed (since

$\angle MAL$ is positive). Hence there must be some intermediate position for which

$$\angle DAL = \angle ADB.$$

(To be precise, we can apply Dedekind's axiom 12.51 to the points on BM satisfying the two opposite inequalities.) For such a point D (Figure 15.2f) we obtain two triangles OAE, ODF by drawing EF perpendicular to BD through O, the midpoint of AD. Since these triangles are congruent, EF is perpendicular not only to BD but also to AL.

Nonintersecting lines that are not parallel are said to be *ultraparallel* (or "hyperparallel"). We are not asserting the existence of such lines, but merely showing how they must behave if they do exist.

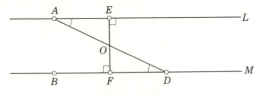

Figure 15.2f

EXERCISES

1. Prove 15.25 without referring to Carslaw **1**.

2. Give a complete proof that, if two lines have a common perpendicular, they do not intersect.

3. Example 4 on p. 16 remains valid when A is an end so that the triangle is asymptotic.

15.3   ISOMETRY

> *Beside the actual universe I can set in imagination other universes in which the laws are different.*
>
> J. L. Synge [**2**, p. 21]

The whole theory of finite groups of isometries (§§ 2.3–3.1) belongs to absolute geometry, because it is concerned with isometries having at least one invariant point. The first departure from our previous treatment (§ 3.2) is in the discussion of isometries without invariant points. We must now distinguish between a *translation,* which is the product of half-turns about two distinct points, and a *parallel displacement,* which is the product of reflections in two parallel lines.

The product of half-turns about two distinct points O, O' is a translation along a given line (called the *axis* of the translation) in a given sense through a given distance, namely, along OO' in the sense of the ray O'/O through the distance $2OO'$. Since a translation is determined by its axis and directed distance, the product of half-turns about O, O' is the same as the product of half-turns about Q, Q', provided the directed segment QQ' is congruent to OO' *on the same line* (Figure 3.2*a*). If P is on this line, the distance PP^{T} is just twice OO'. (If not, it may be greater!)

By the argument used in proving 3.21, the product of two translations with the same axis, or with intersecting axes, is a translation. (It is only in the former case that we can be sure of commutativity.) More precisely, we have

15.31 (Donkin's theorem*) *The product of three translations along the directed sides of a triangle, through twice the lengths of these sides, is the identity.*

We shall see later that the product of two translations with nonintersecting axes may be a rotation.

By the argument used in proving 3.22, if two lines have a common perpendicular, the product of reflections in them is a translation along this common perpendicular through twice the distance between them. (Such lines may be either parallel or ultraparallel according to the nature of the geometry.)

Again, as in 3.13, every isometry is the product of at most three reflections. If the isometry is direct, the number of reflections is even, namely 2. It follows from 15.26 that

15.32 *Every direct isometry (of the plane) with no invariant point is either a parallel displacement or a translation.*

It is remarkable that absolute geometry includes the whole theory of glide reflection. The only changes needed in the previous treatment (§ 3.3) are where the word "parallel" was used. (In Figure 3.3*b* we must define m, m' as being perpendicular to OO'; they are not necessarily parallel to each other.) As an immediate application of these ideas we have Hjelmslev's theorem, which is one of the best instances of a genuinely surprising result belonging to absolute geometry. The treatment in § 3.6 remains valid without changing a single word!

Likewise, the one-dimensional groups of § 3.7 belong to absolute geometry, the only change being that again the mirrors m, m' (Figure 3.7*b*) should not be said to be "parallel" but both perpendicular to the same (horizontal) line. On the other hand, the whole theory of lattices (Chapter 4) and of similarity (Chapter 5) must be abandoned.

The extension of absolute geometry from two dimensions to three presents no difficulty. In particular, much of the Euclidean theory of isometry (§ 7.1) remains valid in absolute space. It is still true that every direct isometry is the product of two half-turns, and that every opposite isometry with

---

* W. F. Donkin, On the geometrical theory of rotation, *Philosophical Magazine* (4), **1**, (1851), 187–192. Lamb [**1**, p. 6] used half-turns about the vertices A, B, C of the given triangle to construct three new triangles which, he said, "are therefore directly equal to one another, and 'symmetrically' equal to ABC." This was a mistake: all four triangles are directly congruent!

an invariant point is a rotatory inversion (possibly reducing to a reflection or to a central inversion). Moreover, the classical enumeration of the five Platonic solids (§§ 10.1–10.3) is part of absolute geometry. The few necessary changes are easily supplied; for example, the term *rectangle* must be interpreted as meaning a quadrangle whose angles are all equal (though not necessarily right angles), and a *square* is the special case when also the sides are equal.

EXERCISES

1. If *l* is a line outside the plane of a triangle *ABC*, what can be said about the three lines in which this plane meets the three planes *Al*, *Bl*, *Cl*? (If two of the three lines intersect, or are parallel, or have a common perpendicular, the same can be said of all three. This property of three lines m_1, m_2, m_3 is equivalent to $R_1R_2R_3 = R_3R_2R_1$ in the notation of § 3.4.)

2. The product of reflections in the lines *p* and *r* of Figure 15.2a is a parallel displacement which transforms *J* into *L*.

15.4   FINITE GROUPS OF ROTATIONS

> *These groups, in particular the last three, are an immensely attractive subject for geometric investigation.*
>
> H. Weyl [**1,** p. 79]

One of the simplest kinds of transformation is a *permutation* (or rearrangement) of a finite number of named objects. For instance, one way to permute the six letters *a, b, c, d, e, f* is to transpose (or interchange) *a* and *b*, to change *c* into *d*, *d* into *e*, *e* into *c*, and to leave *f* unaltered. This permutation is denoted by (*a b*)(*c d e*). The two "independent" parts, (*a b*) and (*c d e*), are called *cycles* of periods 2 and 3. A permutation that consists of just one cycle is said to be *cyclic*. Clearly, the cyclic group C_n may be represented by the powers of the generating permutation $(a_1a_2 \ldots a_n)$; for instance, the four elements of C_4 are

$$1, \quad (a\ b\ c\ d), \quad (a\ c)(b\ d), \quad (a\ d\ c\ b).$$

A cyclic permutation of period 2, such as (*a b*), is called a *transposition*. Since

$$(a_1a_2 \ldots a_n) = (a_1a_n)(a_2a_n) \ldots,$$

any permutation may be expressed as a product of transpositions. A permutation is said to be *even* or *odd* according to the parity of the number of cycles of even period; for instance, (*a c*)(*b d*) is even, but (*a b*)(*c d e*) is odd. The identity, 1, has no cycles at all, and is accordingly classified as an even permutation. It is easily proved [see Coxeter **1,** pp. 40–41] that every product of transpositions is even or odd according to the parity of the number of transpositions. It follows that the multiplication of even and odd permutations behaves like the *addition* of even and odd numbers; for example, the product of two odd permutations is even.

It follows also that every group of permutations either consists entirely of even permutations or contains equal numbers of even and odd permutations. The group of all permutations of n objects is called the *symmetric* group of order $n!$ (or of *degree* n) and is denoted by S_n. The subgroup consisting of all the even permutations is called the *alternating* group of order $\frac{1}{2} n!$ (or of degree n) and is denoted by A_n. In particular, S_2 is the same group as C_2, and A_3 the same as C_3, so we write

$$S_2 \cong C_2, \qquad A_3 \cong C_3.$$

More interestingly, $S_3 \cong D_3$ (see Figure 2.7a). For, the six elements of the dihedral group D_3, being symmetry operations of an equilateral triangle, may be regarded as permutations of the three sides of the triangle. The even permutations

$$1, \quad (a\,b\,c), \quad (a\,c\,b)$$

(which form the subgroup $A_3 \cong C_3$) are rotations, whereas the odd permutations

$$(b\,c), \quad (c\,a), \quad (a\,b)$$

are reflections in the three medians. If we regard the triangle as lying in three-dimensional (absolute) space, the rotations are about an axis through the center of the triangle, perpendicular to its plane. The reflections may then be interpreted in two alternative ways, yielding two groups which are geometrically distinct but abstractly identical or *isomorphic*: we may either reflect in three planes through the axis or rotate through half-turns about the medians themselves. In the latter representation, all the six elements of D_3 appear as rotations. We may describe this as the group of direct symmetry operations of a triangular prism. More generally, the $2n$ direct symmetry operations of an n-gonal prism form the dihedral group D_n, whereas of course the n direct symmetry operations of an n-gonal pyramid form the cyclic group C_n. The rotations of C_n all have the same axis, and D_n is derived from C_n by adding half-turns about n lines symmetrically disposed in a plane perpendicular to that axis.

We have thus found two infinite families of finite groups of rotations. Other such groups are the groups of direct symmetry operations of the five Platonic solids $\{p, q\}$. These are only three groups, not five, because any rotation that takes $\{p, q\}$ into itself also takes the reciprocal $\{q, p\}$ into itself: the octahedron has the same group of rotations as the cube, and the icosahedron the same as the dodecahedron.

The regular tetrahedron $\{3, 3\}$ is evidently symmetrical by reflection in the plane that joins any edge to the midpoint of the opposite edge. As a permutation of the four faces a, b, c, d (Figure 15.4a), this reflection is just a transposition. Thus the complete symmetry group of the tetrahedron,

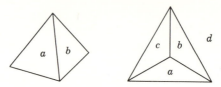

Figure 15.4a

being generated by such reflections, is isomorphic to the symmetric group S_4, which is generated by transpositions; and the rotation group, being generated by products of pairs of reflections, is isomorphic to the alternating group A_4, which is generated by products of pairs of transpositions. The 12 rotations may be counted as follows. The perpendicular from a vertex to the opposite face is the axis of a *trigonal* rotation (i.e., a rotation of period 3); the 4 vertices yield 8 such rotations. The line joining the midpoint of two opposite edges is the axis of a half-turn (or *digonal* rotation); the 3 pairs of opposite edges yield 3 such half-turns. Including the identity, we thus have $8 + 3 + 1 = 12$ rotations. As permutations, the 8 trigonal rotations are

$$(b\,c\,d), \quad (b\,d\,c), \quad (a\,c\,d), \quad (a\,d\,c), \quad (a\,b\,d), \quad (a\,d\,b), \quad (a\,b\,c), \quad (a\,c\,b)$$

and the 3 half-turns are

$$(b\,c)(a\,d), \quad (c\,a)(b\,d), \quad (a\,b)(c\,d).$$

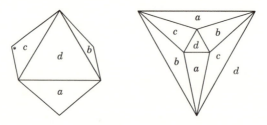

Figure 15.4b

The octahedron $\{3, 4\}$ can be derived from the tetrahedron by *truncation*: its eight faces consist of the four vertex figures of the tetrahedron and truncated versions of the four faces. Every symmetry operation of the tetrahedron is retained as a symmetry operation of the octahedron, but the octahedron also has symmetry operations that interchange the two sets of four faces. For instance, the line joining two opposite vertices is the axis of a *tetragonal* rotation (of period 4), and the line joining the midpoints of two opposite edges is the axis of a half-turn. When the four pairs of opposite faces are marked a, b, c, d, as in Figure 15.4b, such a half-turn appears as a transposition, which is one of the permutations that belong to S_4 but not

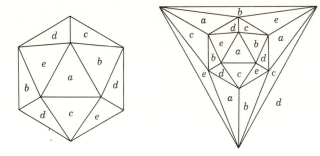

Figure 15.4c

to A_4. It follows that the rotation group of the octahedron (or of the cube) is isomorphic to the symmetric group S_4.

In Figure 15.4c, the twenty faces of the icosahedron $\{3, 5\}$ have been marked a, b, c, d, e in sets of four, in such a way that two faces marked alike have nothing in common, not even a vertex. In fact, the four a's (for instance) lie in the planes of the faces of a regular tetrahedron, and the respectively opposite faces (marked b, c, d, e) form the reciprocal tetrahedron. The twelve rotations of either tetrahedron into itself (represented by the even permutations of b, c, d, e) are also symmetry operations of the whole icosahedron. This behavior of the four a's is imitated by the b's, c's, d's and e's, so that altogether we have all the even permutations of the five letters: the rotation group of the icosahedron (or of the dodecahedron) is isomorphic to the alternating group A_5. The 60 rotations may be counted as follows: 4 pentagonal rotations about each of 6 axes, 2 trigonal rotations about each of 10 axes, 1 half-turn about each of 15 axes, and the identity [Coxeter **1**, p. 50].

We shall find that the above list exhausts the finite groups of rotations. As a first step in this direction, we observe that all the axes of rotation must pass through a fixed point. In fact, we can just as easily prove a stronger result:

15.41 *Every finite group of isometries leaves at least one point invariant.*

Proof. A finite group of isometries transforms any given point into a finite set of points, and transforms the whole set of points into itself. This, like any finite (or bounded) set of points, determines *a unique smallest sphere that contains all the points on its surface or inside*: unique because, if there were two equal smallest spheres, the points would belong to their common part, which is a "lens"; and the sphere that has the rim of the lens for a great circle is smaller than either of the two equal spheres, contradicting our supposition that these spheres are as small as possible. (The shaded area in Figure 15.4d is a section of the lens.) The group transforms this unique sphere into itself. Its surface contains some of the points, and therefore all of them. Its center is the desired invariant point.

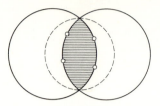

Figure 15.4d

It follows that any finite group of rotations may be regarded as operating on the surface of a sphere. In such a group G, each rotation, other than the identity, leaves just two points invariant, namely the *poles* where the axis of rotation intersects the sphere. A pole P is said to be p-gonal ($p \geqslant 2$) if it belongs to a rotation of period p. The p rotations about P, through various multiples of the angle $2\pi/p$, are those rotations of G which leave P invariant. Any other rotation of G transforms P into an "equivalent" pole, which is likewise p-gonal. Thus all the poles fall into sets of equivalent poles. All the poles in a set have the same period p, but two poles of the same period do not necessarily belong to the same set; they belong to the same set only if one is transformed into the other by a rotation that belongs to G.

Any set of equivalent p-gonal poles consists of exactly n/p poles, where n is the order of G. To prove this, take a point Q on the sphere, arbitrarily near to a pole P belonging to the set. The p rotations about P transform Q into a small p-gon round P. The other rotations of G transform this p-gon into congruent p-gons round all the other poles in the set. But the n rotations of G transform Q into just n points (including Q itself). Since these n points are distributed into p-gons round the poles, the number of poles in the set must be n/p.

The $n - 1$ rotations of G, other than the identity, consist of $p - 1$ for each p-gonal axis, that is, $\frac{1}{2}(p - 1)$ for each p-gonal pole, or

$$\tfrac{1}{2}(p - 1)n/p$$

for each set of n/p equivalent poles. Hence

$$n - 1 = \tfrac{1}{2}n \; \Sigma(p - 1)/p,$$

where the summation is over the sets of poles. This equation may be expressed as

$$2 - \frac{2}{n} = \Sigma \left(1 - \frac{1}{p}\right).$$

If $n = 1$, so that G consists of the identity alone, there are no poles, and the sum on the right has no term. In all other cases $n \geqslant 2$, and therefore

$$1 \leqslant 2 - \frac{2}{n} < 2.$$

It follows that the number of sets of poles can only be 2 or 3; for, the single term $1 - 1/p$ would be less than 1, and the sum of 4 or more terms would be

$$\geqslant 4(1 - \tfrac{1}{2}) = 2.$$

If there are 2 sets of poles, we have

$$2 - \frac{2}{n} = 1 - \frac{1}{p_1} + 1 - \frac{1}{p_2},$$

that is,

$$\frac{n}{p_1} + \frac{n}{p_2} = 2.$$

But two positive integers can have the sum 2 only if each equals 1; thus

$$p_1 = p_2 = n,$$

each of the 2 sets of poles consists of one n-gonal pole, and we have the cyclic group C_n with a pole at each end of its single axis.

Finally, in the case of 3 sets of poles we have

$$2 - \frac{2}{n} = 1 - \frac{1}{p_1} + 1 - \frac{1}{p_2} + 1 - \frac{1}{p_3},$$

whence

15.42
$$\frac{1}{p_1} + \frac{1}{p_2} + \frac{1}{p_3} = 1 + \frac{2}{n}.$$

Since this is greater than $\tfrac{1}{3} + \tfrac{1}{3} + \tfrac{1}{3} = 1$, the three periods p_i cannot all be 3 or more. Hence at least one of them is 2, say $p_3 = 2$, and we have

$$\frac{1}{p_1} + \frac{1}{p_2} = \frac{1}{2} + \frac{2}{n},$$

whence $\qquad (p_1 - 2)(p_2 - 2) = 4(1 - p_1 p_2/n) < 4$

(cf. 10.33), so that the only possibilities (with $p_1 \leqslant p_2$ for convenience) are:

$$p_1 = 2, \quad p_2 = p, \quad n = 2p; \qquad p_1 = 3, \quad p_2 = 3, \quad n = 12;$$
$$p_1 = 3, \quad p_2 = 4, \quad n = 24; \qquad p_1 = 3, \quad p_2 = 5, \quad n = 60.$$

We recognize these as the dihedral, tetrahedral, octahedral and icosahedral groups.

This completes our proof [Klein **3**, p. 129] that

15.43 *The only finite groups of rotations in three dimensions are the cyclic groups C_p ($p = 1, 2, \ldots$), the dihedral groups D_p ($p = 2, 3, \ldots$), the tetrahedral group A_4, the octahedral group S_4, and the icosahedral group A_5.*

(To avoid repetition, we have excluded D_1 which, when considered as a group of rotations, is not only abstractly but geometrically identical with C_2.)

Any solid having one of these groups for its complete symmetry group

Figure 15.4e

(such as the Archimedean *snub cube** shown in Figure 15.4e, whose group is S_4) can occur in two *enantiomorphous* varieties, *dextro* and *laevo* (i.e., right- and left-handed): mirror images that cannot be superposed by a continuous motion.

EXERCISES

1. Interpret the following permutations as rotations of the octahedron (Figure 15.4b):

$$(a\,b\,c\,d), \quad (a\,b\,c), \quad (a\,b), \quad (a\,b)(c\,d).$$

Count the rotations of each type, and check with the known order of S_4.

2. Using the symbol (p_1, p_2, p_3) for the group having three sets of poles of periods p_1, p_2, p_3, consider the possibility of stretching the notation so as to allow $(1, p, p) \cong C_p$ as well as

$$(2, 2, p) \cong D_p, \qquad (2, 3, 3) \cong A_4,$$
$$(2, 3, 4) \cong S_4, \qquad (2, 3, 5) \cong A_5.$$

15.5 FINITE GROUPS OF ISOMETRIES

Having enumerated the finite groups of rotations, we can easily solve the wider problem of enumerating the finite groups of isometries (cf. § 2.7). Since every such group leaves one point invariant, we are concerned only with isometries having fixed points. Such an isometry is a rotation or a rotatory inversion according as it is direct or opposite (7.15, 7.41).

If a finite group of isometries consists entirely of rotations, it is one of the groups G considered in § 15.4. If not, it contains such a group G as a subgroup of index 2, that is, it is a group of order $2n$ consisting of n rotations $S_1, S_2, \ldots, S_n$ and an equal number of rotatory inversions T_1,

---

* The vertices of the snub cube constitute a distribution of 24 points on a sphere for which the smallest distance between any 2 is as great as possible. This was conjectured by K. Schütte and B. L. van der Waerden (*Mathematische Annalen,* **123** (1951), pp. 108, 123) and was proved by R. M. Robinson (*ibid.,* **144** (1961), pp. 17–48). The analogous distribution of 6 or 12 points is achieved by the vertices of an octahedron or an icosahedron, respectively. For 8 points the figure is not, as we might at first expect, a cube, but a square antiprism [Fejes Tóth **1**, pp. 162–164].

$T_2, \ldots, T_n$. For, if the group consists of n rotations S_i and (say) m rotatory inversions T_i, we can multiply by T_1 so as to express the same $n + m$ isometries as S_iT_1 and T_iT_1. The n isometries S_iT_1, being rotatory inversions, are the same as T_i (suitably rearranged if necessary), and the m isometries T_iT_1, being rotations, are the same as S_i. Therefore $m = n$.

If the central inversion I belongs to the group, the n rotatory inversions are simply

$$S_iI = IS_i \quad (i = 1, 2, \ldots, n),$$

and the group is the direct product $G \times \{I\}$, where G is the subgroup consisting of the S's and $\{I\}$ denotes the group of order 2 generated by I. (As an abstract group, $\{I\}$ is, of course, the same as C_2 or D_1.)

If I does not belong, the $2n$ transformations S_i and T_iI form a group of rotations of order $2n$ which has the same multiplication table as the given group consisting of S_i and T_i. For, if $S_iT_j = T_k$,

$$S_iT_jI = T_kI,$$

and if $T_iT_j = S_k$,

$$T_iIT_jI = T_iI^2T_j = T_iT_j = S_k.$$

In other words, a group of n rotations and n rotatory inversions, not including I, is isomorphic to a rotation group G' of order $2n$ which has a subgroup G of order n. To complete our enumeration, we merely have to seek such pairs of related rotation groups. Each pair yields a "mixed" group, say $G'G$, consisting of all the rotations in the smaller group G, along with the remaining rotations in G' each multiplied by the central inversion I. Looking back at § 15.4, we see that the possible pairs are

$$C_{2n}C_n, \quad D_nC_n, \quad D_nD_{\frac{1}{2}n} \ (n \text{ even}), \quad S_4A_4.$$

Thus we can complete Table III on p. 413.

EXERCISES

1. Determine the symmetry groups of the following figures: (a) an orthoscheme $O_0O_1O_2O_3$ (Figure 10.4c) with $O_0O_1 = O_2O_3$; (b) an n-gonal antiprism (n even or odd).

2. Designate in the $G'G$ notation the direct product of the group of order 3 generated by a rotation about a vertical axis and the group of order 2 generated by the reflection in a horizontal plane.

15.6 GEOMETRICAL CRYSTALLOGRAPHY

> *The sense in which a snail's shell winds is an inheritable character*
> *founded in its genetic constitution, as is . . . the winding of the intestinal*
> *duct in the species Homo sapiens. . . . Also the deeper chemical consti-*
> *tution of our human body shows that we have a screw, a screw that is*
> *turning the same way in every one of us.* . . . A horrid manifestation of*
> *this genotypical asymmetry is a metabolic disease called phenylketo-*
> *nuria, leading to insanity, that man contracts when a small quantity of*
> *laevo-phenylalanine is added to his food, while the dextro- form has no*
> *such disastrous effects.*
>
> H. Weyl [**1,** p. 30]

The discussion of symmetry groups has been phrased in such a way as to be valid not only in Euclidean space but in absolute space. However, it seems appropriate to mention the application of these ideas to the practical science of crystallography. Accordingly, in this digression the geometry is strictly Euclidean.

Crystallographers are interested in those finite groups of isometries which arise as subgroups (and factor groups) of symmetry groups of three-dimensional lattices. By § 4.5, these are the special cases in which the only rotations that occur have periods 2, 3, 4 or 6. This crystallographic restriction reduces the rotation groups to

$$C_1, C_2, C_3, C_4, C_6, D_2, D_3, D_4, D_6, A_4, S_4,$$

the direct products to these eleven each multiplied by $\{I\}$, and the mixed groups to

$$C_2 C_1, C_4 C_2, C_6 C_3, D_2 C_2, D_3 C_3, D_4 C_4, D_6 C_6, D_4 D_2, D_6 D_3, S_4 A_4.$$

(Of course, $C_1 \times \{I\}$ is just $\{I\}$ itself.)

These 32 groups are called the *crystallographic point groups* or *"crystal classes."* Every crystal has one of them for its symmetry group, and every group except $C_6 C_3$ occurs in at least one known mineral. In the more familiar notation of Schoenflies [see, e.g., Burckhardt **1,** p. 71], the groups are respectively

$$C_1, C_2, C_3, C_4, C_6, D_2, D_3, D_4, D_6, T, O,$$
$$C_i, C_{2h}, C_{3i}, C_{4h}, C_{6h}, D_{2h}, D_{3d}, D_{4h}, D_{6h}, T_h, O_h,$$
$$C_s, S_4, C_{3h}, C_{2v}, C_{3v}, C_{4v}, C_{6v}, D_{2d}, D_{3h}, T_d.$$

To avoid possible confusion, observe that our $C_4 C_2$ and S_4 ("S" for "symmetric") are Schoenflies's S_4 and O (for "octahedral"). The 32 groups are customarily divided into seven *crystal systems,* as follows:

Triclinic:	$C_1,$	$\{I\}$.	
Monoclinic:	$C_2,$	$C_2 \times \{I\},$	$C_2 C_1.$
Orthorhombic:		$D_2,$ $D_2 \times \{I\},$	$D_2 C_2.$

---

* The DNA molecule?

Rhombohedral:	C_3,	$C_3 \times \{I\}$,		D_3,	$D_3 \times \{I\}$,	D_3C_3.	
Tetragonal:	C_4,	$C_4 \times \{I\}$,	C_4C_2,	D_4,	$D_4 \times \{I\}$,	D_4C_4,	D_4D_2.
Hexagonal:	C_6,	$C_6 \times \{I\}$,	C_6C_3,	D_6,	$D_6 \times \{I\}$,	D_6C_6,	D_6D_3.
Cubic:	A_4,	$A_4 \times \{I\}$,		S_4,	$S_4 \times \{I\}$,	S_4A_4.	

Table I (on p. 413) is a complete list of the 17 discrete groups of isometries in two dimensions involving two independent translations. The analogous groups in three dimensions are the discrete groups of isometries involving three independent translations. The enumeration of these *space groups* is the central problem of mathematical crystallography. The complete list contains $65 + 165 = 230$ groups.

The first 65 are composed entirely of *direct* isometries. Although these were enumerated as long ago as 1869 by C. Jordan [see Hilton **1**, p. 258], they are usually attributed to L. Sohncke who, in 1879, pointed out their application to crystallography. The most obvious group consists of translations alone. The remaining 64 of the 65 contain also rotations and screw displacements; 22 of them occur in 11 enantiomorphous pairs which are mirror images of each other (one containing right-handed screw displacements and the other the reflected left-handed screw displacements). This explains the phenomenon of optical activity [Sayers and Eustace **1**, pp. 238–241, 248–252]. From the standpoint of pure geometry or pure group theory, it would be more natural to ignore this distinction of sense, thus reducing the number 65 to 54, and the total of 230 to 219 [Burckhardt **1**, p. 161].

The remaining 165 groups contain not only direct but also *opposite* isometries: reflections, rotatory reflections (or rotatory inversions), and glide reflections. Their enumeration, by Fedorov in Russia (1890), Schoenflies in Germany (1891), and Barlow in England (1894), provides one of the most striking instances of independent discovery in different places using different methods. Fedorov, who obtained the 230 as $73 + 54 + 103$ instead of $65 + 165$, was probably unaware of the preliminary work of Jordan and Sohncke. It is quite certain that Schoenflies knew nothing of Fedorov, and that Barlow's work was independent of both.

EXERCISE

Determine the symmetry groups of the following figures: (a) a rectangular parallelepiped (e.g., a brick), (b) a rhombohedron; (c) a regular dodecahedron with an inscribed cube (whose 8 vertices occur among the 20 vertices of the dodecahedron).

15.7 THE POLYHEDRAL KALEIDOSCOPE

In combining three reflections . . . the effect is highly pleasing.

Sir David Brewster (1781 -1848)

[Brewster **1**, p. 93]

Table III (on p. 413) is a complete list of the finite groups of isometries. In the preceding section, we selected from this list those groups which satisfy the crystallographic restriction. Another significant way to make a selection (partly overlapping with the previous way) is to pick out those groups which are generated by reflections, namely,

$$D_n C_n \ (n \geqslant 1), \quad D_{2n} D_n \ (n \ \text{odd}), \quad D_n \times \{I\} \ (n \ \text{even}),$$
$$S_4 A_4, \qquad\qquad S_4 \times \{I\}, \qquad\qquad A_5 \times \{I\}.$$

(We have now returned to absolute geometry!)

$D_1 C_1$ (Schoenflies's C_s, which we previously denoted by $C_2 C_1$) is the group of order 2 generated by a single reflection. $D_2 C_2$ or $D_2 D_1$ (Schoenflies's C_{2v}) is the group of order 4 generated by two orthogonal reflections. The remaining groups $D_n C_n$ are the symmetry groups of the n-gonal pyramids. In other words, these are the groups D_n of § 2.7 in a different notation. (We now reserve the symbol D_n for the dihedral group of *rotations*, which is, of course, isomorphic to $D_n C_n$. Weyl [**1**, p. 80] makes the distinction by calling the rotation group D'_n and the mixed group $D'_n C_n$.)

$D_2 \times \{I\}$ is a group of order 8 (abstractly $C_2 \times C_2 \times C_2$) generated by three orthogonal reflections. The remaining groups $D_{2n} D_n$ (n odd) and $D_n \times \{I\}$ (n even) are the symmetry groups of the n-gonal prisms, or of their reciprocals, the dipyramids.

$S_4 A_4$, the symmetry group of the regular tetrahedron, is derived from the rotation group A_4 by adjoining reflections, such as the reflection in the plane $ABA'B'$ (Figure 10.5a) which joins the edge AB to the midpoint of the opposite edge CD. (The product of this reflection and the central inversion is the half-turn about the join of the midpoints of the two opposite edges CD', $C'D$ of the cube. This half-turn, which interchanges the two reciprocal tetrahedra $ABCD$, $A'B'C'D'$, is one of the twelve rotations in S_4 that do not belong to the subgroup A_4; thus it illustrates our special meaning for the "mixed" symbol $S_4 A_4$.) Since the remaining Platonic solids are centrally symmetrical, their symmetry groups are simply $S_4 \times \{I\}$ and $A_5 \times \{I\}$.

For a practical demonstration in Euclidean space, take the two hinged mirrors of § 2.7, inclined at $180°/n$, which demonstrate the group $D_n C_n$. Standing them upright on a separate horizontal mirror, we obtain the symmetry group of the n-gonal prism, i.e., the direct product of $D_n C_n$ and the group of order 2 generated by the horizontal reflection. To demonstrate the three remaining groups, remove the third mirror, and let the first two stand vertically on the table at an angle of 60°, as in the demonstration of $D_3 C_3$. Now hold the third mirror obliquely, with its horizontal edge l on the table top at right angles to one of the vertical mirrors and touching the front lower corner of the other. Gradually rotating this third mirror about its edge l from an almost horizontal position (by raising its nearer edge, opposite to l), we observe at a certain stage two faces of a regular tetrahedron $\{3, 3\}$. Each face is subdivided into six right-angled triangles, one of which is actually the exposed portion of the table top. At a later stage we see three faces of an octahedron $\{3, 4\}$; still later, four faces of an icosahedron $\{3, 5\}$. Finally, when the adjustable mirror is vertical like the others, we see a theoretically infinite number of faces of the regular tessellation $\{3, 6\}$, subdivided in the manner of Figure 4.6d. This device, employing ordinary rectangular mirrors, is a simplified version of Möbius's trihedral kaleidoscope in which the three mirrors are cut in the shape of suitable sectors of a circle [Coxeter **1**, p. 83].

When the E edges of the general Platonic solid $\{p, q\}$ are projected from its center onto a concentric sphere, they become E arcs of great circles, decomposing the surface into F regions which are "spherical p-gons." In this manner the polyhedron yields a "spherical tessellation" which closely resembles the plane tessellation of § 4.6. The symmetry group of $\{p, q\}$ is derived from the symmetry group of one face by adding the reflection in a side of that face. Thus it is generated by reflections in the sides of a spherical triangle whose angles are π/p (at the center of a face), $\pi/2$ (at the midpoint of an edge), and π/q (at a vertex). This spherical triangle is a fundamental region for the group, since it is transformed into neighboring regions by the three generating reflections.

Figure 15.7a

The network of such triangles, filling the surface of the sphere, is cut out by all the planes of symmetry of the polyhedron, namely the planes joining the center to the edges of both $\{p, q\}$ and its reciprocal $\{q, p\}$. In Figure 15.7a (where p and q are 3 and 5), alternate regions have been blackened so as to exhibit both the complete symmetry group $A_5 \times \{I\}$ and the rotational subgroup A_5, which preserves the coloring.

Instead of deriving the network of spherical triangles from the regular polyhedron, we may conversely derive the polyhedron from the network. The ten triangles in the middle of Figure 15.7a evidently combine to form a face of the blown-up dodecahedron, and the six triangles surrounding a point where the angles are 60° combine to form a face of the blown-up icosahedron.

1. Interpret the symbol $\{p, 2\}$ as a spherical tessellation ("dihedron") whose faces consist of two hemispheres, and $\{2, p\}$ as another whose faces consist of p lunes.

2. How many planes of symmetry does each Platonic solid have? Provided p and q are greater than 2, this number is always a multiple of 3, namely $3c$ in the notation of Ex. 1 at the end of § 10.4.

3. Dividing 4π by the area of the fundamental region, obtain a formula for the order of the symmetry group of $\{p, q\}$. Reconcile this with the formula for E (the number of edges) in 10.32.

15.8   DISCRETE GROUPS GENERATED BY INVERSIONS

In the present section we make one more digression into Euclidean space, so as to be able to talk about inversion. (The absolute theory of inversion presents difficulties that would take us too far afield. [See Sommerville **1,** Chapter VIII.])

Figure 15.7*a*, being an orthogonal projection, represents 10 of the 15 great circles by ellipses. (The difficult task of drawing it was undertaken by J. F. Petrie about 1932). An easier, and perhaps more significant, way to represent such figures is by stereographic projection (§ 6.9), so that the great circles remain circles (or lines) [Burnside **1,** pp. 406–407]. The reader can readily do this for himself, with the aid of the following simple instructions.

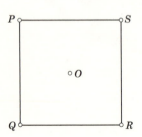
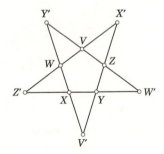

Figure 15.8a

Figure 15.8*a* shows a square *PQRS* with center *O*, and a regular pentagon *VWXYZ* with its sides extended to form a pentagram *V'X'Z'W'Y'*. With radius *PQ* and centers *P, Q, R, S*, draw four circles. These, along with two lines through *O* parallel to the sides of the square, represent 6 great circles, one in each of the 6 planes of symmetry of the tetrahedron $\{3, 3\}$, which are the planes joining pairs of opposite edges of a cube. Adding the circumcircle and diagonals of the square, we have altogether 9 great circles, one in each of the 9 planes of symmetry of the cube $\{4, 3\}$, which include 3 planes parallel to its faces.

With radius VX' ($=VX$) and centers V, W, X, Y, Z, draw five circles. With radius VW' ($=V'W'$) and centers V', W', X', Y', Z', draw five more circles. These ten circles, along with the five lines VV', WW', XX', YY', ZZ', represent 15 great circles (Figure 15.7a), one in each of the 15 planes of symmetry of the icosahedron $\{3, 5\}$ or of the dodecahedron $\{5, 3\}$. (These planes join pairs of opposite edges of either solid.)

To justify these statements we merely have to examine the curvilinear triangles* and observe that each has angles π/p, π/q, $\pi/2$.

Since stereographic projection is an inversion (Figure 6.9a), and since an inversion transforms a reflection into an inversion, the figures so constructed are, in fact, representations of the abstract groups S_4, $S_4 \times C_2$, and $A_5 \times C_2$ as groups generated by inversions. In other words, they are configurations of circles so arranged that the whole figure is symmetrical by inversion in each circle. (Of course, any straight lines that occur are to be regarded as circles of infinite radius. As we saw in § 6.4, inversion in such a "circle" is simply reflection in the line.) Any one of the regions into which the plane is decomposed will serve as a fundamental region, and the generators of the group may be taken to be the inversions in its sides.

For a group generated by just one inversion, we may invert the circle into a straight line so as to obtain the group D_1 of order 2, generated by a single reflection (§ 2.5). The groups generated by inversions in two intersecting circles are essentially the same as the groups D_n of order $2n$, generated by reflections in two intersecting lines (§ 2.7). If the circles of two generating inversions are in contact, they can be inverted into parallel lines, and we have the limiting case D_∞ (Figure 3.7b). Two nonintersecting circles can be inverted into concentric circles. Inversions in them generate an infinite sequence of concentric circles whose radii are in geometric progression. Abstractly, the group is again D_∞, but the center is a "point of accumulation" (§ 7.6). So is the point of contact in the case of the group generated by inversions in two touching circles. A group is said to be *discrete* if it has no points of accumulation. Thus, in describing discrete groups generated by inversions, we may insist that every two of the generating circles intersect properly, and do not touch.

For a discrete group generated by three inversions, the fundamental region is a curvilinear triangle whose angles are submultiples of π: say π/p_1, π/p_2, π/p_3. For instance, two radii of a circle, forming an angle π/p, cut out a sector which may be regarded as a "triangle" with angles π/p, $\pi/2$, $\pi/2$; this is a fundamental region for the group $D_p \times D_1$ of order $4p$, generated by reflections in the radii and inversion in the circle. In this case

---

* For the effect of projecting in a different direction, see Coxeter, *American Mathematical Monthly*, **45** (1938), pp. 523–525, Figs. 4 and 5.

15.81
$$\frac{1}{p_1} + \frac{1}{p_2} + \frac{1}{p_3} > 1,$$

so that the angle sum of the triangle is greater than π: an obvious consequence of the fact that the sector is derived from a spherical triangle (see
§ 6.9) by stereographic projection, which preserves angles. Every solution
of the inequality 15.81 (cf. 15.42) is a triangle that can be drawn with great
circles on a sphere. We thus obtain again the symmetry groups

$$S_4, \quad S_4 \times C_2, \quad A_5 \times C_2$$

of the Platonic solids.

When $1/p_1 + 1/p_2 + 1/p_3 = 1$, so that the angle sum is exactly π, we
have the infinite "Euclidean" groups **p6m**, **p4m**, **p31m** (see Table I and
Figure 4.6*d*). We could transform all the straight lines into circles by means
of an arbitrary inversion; but then, since the pattern is infinitely extended,
the center of inversion would be a point of accumulation.

When $1/p_1 + 1/p_2 + 1/p_3 < 1$, so that the angle sum of the fundamental
region is less than π, we may still take two of the three sides to be straight, but
now their point of intersection A is outside the circle q to which the third side
belongs, with the result that there is a circle Ω orthogonal to all three (Figure 15.8*b*); the tangents from A to q are radii of Ω.

Since Ω is invariant for each of the generating inversions, it is invariant
for the whole group. The circle q decomposes the interior of Ω into two

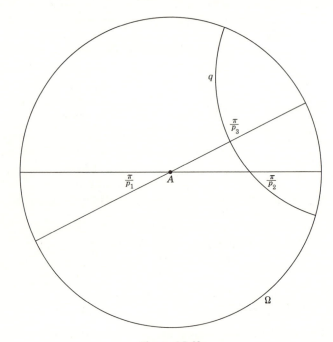

Figure 15.8b

unequal regions and inverts each of these regions into the other.   Therefore the number of triangles is the same in both regions.   But the larger region includes a replica of the smaller.   Hence, by Bolzano's definition of an infinite set (namely, a set that has the same power as a proper subset), the number of triangles is infinite; that is, *the group is infinite.*

Figure 15.8c

The case when p_1, p_2, p_3 are 6, 4, 2 is shown in Figure 15.8c.   Unlike Figure 15.7a, this is not a picture of a solid object.   Our familiarity with three-dimensional space enables us to accept the idea that the triangles in Figure 15.7a are all the same size, even though the peripheral ones are made to look smaller by perspective foreshortening.   In the case of Figure 15.8c, the smaller peripheral triangles are essentially the same shape as those in the middle (since they have the same angles), but we no longer find it easy to imagine that they are, in some sense, the same *size*.   In trying to stretch our imagination to this extent, we are taking a first step towards appreciating hyperbolic geometry, which is the subject of our next chapter.

The reader may wonder why we admit such groups as being worthy of consideration, seeing that the circle Ω contains infinitely many points of accumulation.   However, when we accept the non-Euclidean standpoint, so that the circles and inversions are regarded as lines and reflections, the consequent distortion of distance makes Ω infinitely far away, so that the points of accumulation disappear.

EXERCISES

1. If a system of concentric circles is transformed into itself by inversion in each circle, the radii are in geometric progression.

2. If three circles form a "triangle" with angles π/p_1, π/p_2, π/p_3, the inversions R_1, R_2, R_3 in its sides satisfy the relations

$$R_1{}^2 = R_2{}^2 = R_3{}^2 = (R_2R_3)^{p_1} = (R_3R_1)^{p_2} = (R_1R_2)^{p_3} = 1.$$

These relations suffice to define the abstract group generated by R_1, R_2, R_3 [Coxeter and Moser **1**, pp. 37, 55].

3. Given an angle π/p_1 at the center A of a circle Ω of unit radius, as in Figure 15.8*b*, find expressions (in terms of p_1 and p_2) for the radius of the circle q and for the distance from A to its center, in the case when $p_3 = 2$.

4. Invert Figure 15.8*c* in a circle whose center lies on Ω; that is, replace the circle Ω by a straight line, so that all the inverting circles have their centers on this line. (Such an arrangement provides an alternative proof that the group is infinite. For if its order is g, the infinite half plane is filled with g curvilinear triangles, each having a finite area!)

5. In Figure 15.8*c*, two of the small triangles (one white and one black) with a common hypotenuse form together a "curvilinear kite" having three right angles and one angle of 60°. Trace part of the figure so as to exhibit a network of such kites, alternately white and black. We now have an instance of a group generated by four inversions. Can it happen that more than four inversions are needed to generate a discrete group?

16

Hyperbolic geometry

Absolute geometry is not *categorical:* it is two geometries in one. To be precise, it leaves open the question of the existence of ultraparallel lines (see the end of § 15.2). In § 16.1 we shall compare the two possible answers, giving the unfamiliar the same status as the familiar. In § 16.2 we shall justify this action by means of a proof of *relative consistency*. Thereafter, casting aside all scruples, we shall plunge wholeheartedly into the "new universe" which Bolyai "created from nothing."

16.1 THE EUCLIDEAN AND HYPERBOLIC AXIOMS OF PARALLELISM

> *In the author there lives the perfectly purified conviction (such as he expects too from every thoughtful reader) that by the elucidation of this subject one of the most important and brilliant contributions has been made to the real victory of knowledge, to the education of the intelligence, and consequently to the uplifting of the fortunes of men.*
>
> J. Bolyai (1802-1860)
> [Carslaw **1,** p. 31]

In § 12.6, we mentioned the question whether the two rays parallel to a given line r from an outside point A are, or are not, collinear. By applying a suitable isometry, we see that the answer is independent of the position of r.

It is true, though less obvious, that, for a given r, the answer is independent of the position of A. Suppose, if possible, that the rays parallel to r from A are the two halves of a line q while the rays parallel to r from another point A' form an angle, as in Figure 16.1a. By the transitivity of parallelism, these rays from A' are parallel to q and also to the infinite sequence of parallel lines derived from q and r by applying the group D_∞ generated by reflections in q and r (Figure 3.7b). We obtain a manifest absurdity by considering any one of these lines that lies beyond A' (i.e., in such a position that A' lies between that line and r). (Strictly, this argument makes use of the so-called Axiom of Archimedes, 13.31, which is a consequence of 12.51.)

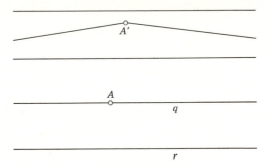

Figure 16.1a

Thus we have a clear-cut distinction between two kinds of geometry, called *Euclidean* and *hyperbolic,* which are derived from absolute geometry by adding just one of the following two alternative axioms:

THE EUCLIDEAN AXIOM.   *For some point A and some line r, not through A, there is not more than one line through A, in the plane Ar, not meeting r.*

THE HYPERBOLIC AXIOM.   *For some point A and some line r, not through A, there is more than one line through A, in the plane Ar, not meeting r.*

EXERCISE

Each of these axioms implies the stronger statement with "some point A and some line r" replaced by "any point A and any line r." The Euclidean axiom, so amended, is equivalent to the celebrated Postulate V (our 1.25). How does Postulate V break down if we assume the hyperbolic axiom?

16.2   THE QUESTION OF CONSISTENCY

> *What are we to think of the question: Is Euclidean Geometry true? It has no meaning. We might as well ask . . . if Cartesian coordinates are true and polar coordinates false. One geometry cannot be more true than another; it can only be more convenient.*
>
> H. Poincaré (1854 -1912)
> (*Science and Hypothesis,* New York, 1952)

We observe that the Euclidean and hyperbolic axioms differ by just one word: the vital word "not." It is meaningless to ask which of the two geometries is *true,* and practically impossible to decide which provides a more *convenient* basis for describing astronomical space. From the standpoint of pure mathematics, a more important question is whether either axiom is logically *consistent* with the remaining axioms of absolute geometry. Even this is difficult to answer; for according to the philosopher Gödel, there is no internal proof of consistency for a system that includes infinite sets. We have to be content with *relative* consistency: if Euclidean geometry is free from contradiction, so is hyperbolic geometry, and vice versa. Relative consistency is established by finding in each geometry a *model* of the other.

One Euclidean model of the hyperbolic plane (due to Poincaré) was mentioned in § 15.8. This uses a circle Ω, as in Figure 16.2a. Each pair of inverse points represents a hyperbolic point, and each circle orthogonal to Ω represents a hyperbolic line. The two parallels to r from A are simply the circles through A that touch r at its points of intersection with Ω. (These points are the "ends" of r.) We call this a *conformal* model because angles retain their proper values though distances are inevitably distorted.

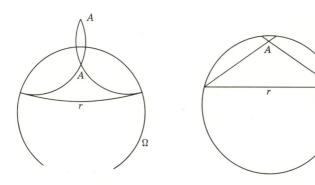

Figure 16.2a   **Figure 16.2b**

A different Euclidean model, suggested by Beltrami (1835–1900), uses another circle ω, as in Figure 16.2b. Each point inside ω represents a hyperbolic point. The two parallels to r from A are the chords joining A to the ends of the chord r. (Chords whose lines intersect outside ω represent ultraparallel lines.) We call this a *projective* model because straight lines remain straight. Nothing is lost if we replace the circle ω in the Euclidean plane by a conic in the projective plane. In fact, much is gained; for it is possible to extend the hyperbolic plane into a projective plane by means of entities defined in the hyperbolic geometry itself [Coxeter **3**, p. 196]. In this way we can prove that hyperbolic geometry is unique or *categorical* [Borsuk and Szmielew **1**, p. 345], unlike absolute geometry, which includes two contrasting possibilities.

When using models, it is desirable to have two rather than one, so as to avoid the temptation to give either of them undue prominence. Our geometric reasoning should all depend on the axioms. The models, having served their purpose of establishing relative consistency [Pedoe **1**, p. 61; Sommerville **1**, pp. 154–159], are no more essential than diagrams.

Klein [**4**, p. 296] exhibited a connection between the conformal and projective models in the manner of Figure 16.2c. A sphere, having the same radius as ω, touches the (horizontal) plane at S, the center of both ω and Ω.

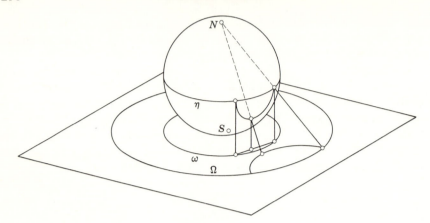

Figure 16.2c

Beginning with the projective model, we use orthogonal (vertical) projection to map ω on the "equator" η of the sphere, and each interior point on two points: one in the southern hemisphere and another (not shown) in the northern hemisphere. Every chord of ω yields a circle in a vertical plane, that is, a circle orthogonal to η. We now map the sphere back into the plane by stereographic projection, so that η projects into the larger circle Ω, concentric with ω. Because of the angle-preserving and circle-preserving nature of stereographic projection, the vertical circles yield horizontal circles orthogonal to Ω, and we have the conformal model.

Instead of stereographic projection onto the tangent plane at the "south pole" S (i.e., inversion with respect to a sphere of radius NS), we could have used stereographic projection (from the same "north pole" N) onto the equatorial plane (i.e., inversion with respect to a sphere through η) so as to make both ω and Ω coincide with η [Coxeter **3,** p. 260]. Klein's procedure is justified by its property of making the two models agree in the immediate vicinity of S. This must have seemed to him more important than making them agree "at infinity."

It must be remembered that both models are in one respect misleading: they give us the impression that the center S should play a special role, whereas, in the abstract hyperbolic plane, all points are alike.

For the sake of completeness, we should mention the problem that the inhabitants of a hyperbolic world would face in trying to visualize the Euclidean plane. One solution [Coxeter **3,** pp. 197–198] is that they could represent the Euclidean points and lines by the lines and planes parallel to a given ray in hyperbolic space!

EXERCISES

1. Reflection in a line of the hyperbolic plane appears, in the conformal model, as inversion with respect to a circle, and in the projective model as a harmonic homology. What is the corresponding transformation in the space of Klein's sphere?

2. Circles appear as circles (not meeting Ω) in the conformal model, and therefore as circles on the sphere (say, in the southern hemisphere) and as ellipses in the projective model.

16.3 THE ANGLE OF PARALLELISM

> *. . . a sea-change into something rich and strange.*
>
> W. Shakespeare (1564-1616)
>
> (*The Tempest,* Act I, Scene 2)

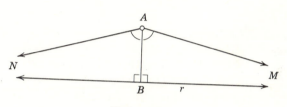

Figure 16.3a

For the rest of this chapter the geometry will be hyperbolic, that is, we shall assume the hyperbolic axiom, which implies that, for any point A and line r, not through A, the two parallels form an angle NAM, as in Figure 16.3a. From A draw AB perpendicular to r. Reflection in AB shows that $\angle BAM$ and $\angle NAB$ are equal acute angles. Following Lobachevsky, we call either of them the *angle of parallelism* corresponding to the distance AB, and write

$$\angle BAM = \Pi(AB).$$

Before we can prove that this function is monotonic, we need a few more properties of asymptotic triangles. While proving 15.26 we discovered that, if a transversal (AD in Figure 15.2f) meets two lines in such a way that the "alternate" angles are equal, then the two lines are ultraparallel. Hence [Carslaw **1**, p. 48]:

16.31 *In an asymptotic triangle EFM, the external angle at E (or F) is greater than the internal angle at F (or E).*

In other words, the sum of the angles of an asymptotic triangle is less than π. This will enable us to prove a kind of converse for Theorem 15.25, to the effect that an asymptotic triangle is determined by its two positive angles:

16.32 *If two asymptotic triangles AEM, A'E'M' have A = A' and E = E', then AE = A'E'.*

Proof [Carslaw **1**, p. 50]. If AE and $A'E'$ are not equal, one of them must

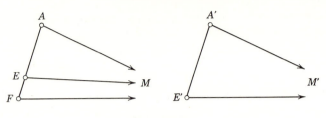

Figure 16.3b

be the greater; let it be $A'E'$, as in Figure 16.3b.   On E/A, take F so that $AF = A'E'$, and draw FM parallel to AM.   By 16.31 and 15.25, we have

$$\angle MEA > \angle MFA = \angle M'E'A' = \angle MEA,$$

which is absurd.

These results will enable us to establish the existence of *a common parallel to two given rays* forming an angle NOM, that is, a line MN which is parallel to OM at one end and to ON at the other.   From the given rays OM, ON, cut off any two equal segments OA, OA', as in Figure 16.3c.   Draw $A'M$ parallel to OM, and AN parallel to ON.   Bisect the angles NAM and $NA'M$ by lines a and a'.   We shall prove that *these lines are ultraparallel,* and that *the desired common parallel MN is perpendicular to both of them.*

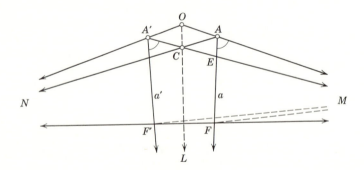

Figure 16.3c

Let $A'M$ meet AN in C, and a in E.   Since the whole figure is symmetrical by reflection in OC, the two angles at A and the two angles at A' are all equal.

If possible, let a and a' have a common point L, which is, of course, equidistant from A and A'.   Applying 15.25 to the congruent asymptotic triangles ALM and $A'LM$, we deduce that $\angle MLA = \angle MLA'$, which is absurd.

If possible, let a and a' be parallel, with a common end L.   Applying 16.32 to the congruent asymptotic triangles AEM and $A'EL$, we deduce that $AE = A'E$, whence E coincides with C, which is absurd.

We conclude that a and a' are ultraparallel.   By 15.26, they have a common perpendicular FF'. Applying 15.25 to the congruent asymptotic triangles AFM and $A'F'M$, we conclude that

$$\angle MFA = \angle MF'A'.$$

If $F'F$ were not parallel to OM, we would have an asymptotic triangle $FF'M$ whose angle sum is π, contradicting 16.31. Hence, in fact, $F'F$ is parallel to OM, and similarly FF' to ON; that is, the line FF' is a common parallel to the two rays as desired.

Moreover, this common parallel is *unique,* since two such would be parallel to each other at both ends, contradicting the "clear-cut distinction" between the Euclidean and hyperbolic properties of parallelism (Figure 16.1*a*). It follows that

16.33 *Any two ultraparallel lines have a unique common perpendicular.*

For, given a and a', we can reconstruct Figure 16.3*c* as follows: draw any common perpendicular FF', take O on its perpendicular bisector, and let the two parallels through O to the line FF' meet a in A, a' in A'.

For the sake of brevity, we have been content to assert the *existence* of a line through a given point parallel to a given ray, and of a common perpendicular to two given ultraparallel lines. Actual "ruler and compasses" constructions for these lines have been given by Bolyai and Hilbert, respectively [see Coxeter **3,** pp. 204, 191]. Hilbert apparently failed to notice that his construction for the common parallel to AM and $A'N$ remains valid if these lines meet in a point that is not equidistant from A and A', or even if they do not meet at all. In fact [Carslaw **1,** p. 76],

16.34 *Any two nonparallel rays have a unique common parallel.*

This result justifies our use of *ends* as if they were ordinary points: any two ends, M and N, determine a unique line MN.

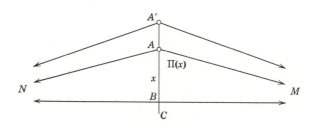

Figure 16.3d

The line through A parallel to BM (Figure 16.3*a* or *d*) determines the angle of parallelism $\Pi(AB)$. Conversely, we can now find a distance x whose angle of parallelism $\Pi(x)$ is equal to any given acute angle [Carslaw **1,** p. 77]. In other words, given an acute angle CAM, *we can find a line BM which is both perpendicular to AC and parallel to AM.* We merely have to reflect AM in AC, obtaining AN, and then draw the common parallel MN, which meets AC in the desired point B. Incidentally, since we can draw through any point a ray parallel to a given ray, it follows that

16.35 *For any two nonperpendicular lines we can find a line which is perpendicular to one and parallel to the other.*

If A' is on the ray A/B, so that $A'B > AB$ (as in Figure 16.3d), then

$$\mathrm{II}(A'B) < \mathrm{II}(AB).$$

(This is simply 16.31, applied to the asymptotic triangle $AA'M$.) It follows that the function $\mathrm{II}(x)$ decreases steadily from $\frac{1}{2}\pi$ to 0 when x increases from 0 to ∞.

We naturally call AMN a *doubly asymptotic triangle* [Coxeter **3,** p. 188]. We have seen that such a "triangle" is determined by its one positive angle; in other words,

16.36 *Two doubly asymptotic triangles are congruent if they have equal angles.*

Applying 16.34 to rays belonging to two parallel lines LM, LN, we obtain a third line parallel to both, forming a *trebly asymptotic triangle LMN*. In view of Bolyai's remark 15.24, we may regard such a triangle as a doubly asymptotic triangle whose angle is zero. Accordingly, we shall not be surprised to find that

16.37 *Any two trebly asymptotic triangles are congruent.*

Proof (due to D. W. Crowe). Given any two trebly asymptotic triangles, dissect each into two right-angled doubly asymptotic triangles by drawing an *altitude* (perpendicular to one side and parallel to another, as in 16.35). By 16.36, all the four doubly asymptotic triangles are congruent. Therefore the two trebly asymptotic triangles must be congruent.

EXERCISES

1. Draw figures for Theorems 16.33–16.35 in terms of the conformal and projective models.

2. If a quadrangle $ABED$ has right angles at D and E while $AD = BE$, then the angles at A and B are equal acute angles. (*Hint:* Draw AM and BM parallel to D/E; apply 16.31 to the asymptotic triangle ABM.)

3. The sum of the angles of any triangle is less than two right angles. (*Hint:* For a given triangle ABC, draw AD, BE, CF perpendicular to the line joining the midpoints of BC and CA.)

4. Given an asymptotic triangle ABM with acute angles at both A and B, draw AD perpendicular to BM, and BE perpendicular to AM, meeting in H. Draw HF perpendicular to AB. Then FH is parallel to AM [Bonola **1,** p. 106]. What happens if we deal similarly with rays through A and B which are not parallel but ultraparallel?

5. If two trebly asymptotic triangles have a common side, by what isometry are they related? (Of course, two trebly asymptotic triangles may have a common side without having a common altitude).

6. The inradius of a trebly asymptotic triangle is the distance whose angle of parallelism is 60°.

7. From any point on a side of a trebly asymptotic triangle, lines drawn perpendicular to the other two sides are themselves perpendicular [Bachmann **1**, p. 222].

16.4 THE FINITENESS OF TRIANGLES

> I could be bounded in a nutshell and
> count myself a king of infinite space.
>
> W. Shakespeare
>
> (Hamlet, Act II, Scene 2)

One of the most elegant passages in the literature on hyperbolic geometry since the time of Lobachevsky is the proof by Liebmann [**1**, p. 43] that the area of a triangle remains finite when all its sides are infinite. C. L. Dodgson (*alias* Lewis Carroll) could not bring himself to accept this theorem; consequently he believed non-Euclidean geometry to be nonsense.

Instead of pursuing a philosophical discussion of the meaning of *area* [Carslaw **1**, pp. 84–90], let us be content to regard it as a numerical function, defined for every simple closed polygon, invariant under isometries, and additive when two polygons are juxtaposed.

Let ABM be any asymptotic triangle. Reflect it in the bisector AF of the angle A to obtain AA_1N, as in Figure 16.4a, F being the point where the bisector meets the common parallel MN. Reflect the line BM in the bisector A_1F_1 of $\angle NA_1M$ to obtain A_2N (with A_2 on AM), and then reflect

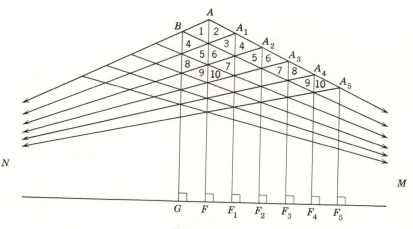

Figure 16.4a

this in AF. Continuing in this manner, we construct a network of triangles whose "vertical" sides bisect the angles at B, A, A_1, A_2, A_3, ... and are perpendicular to MN at G, F, F_1, F_2, F_3, ... These points are evenly spaced along MN, since they are all derived from F and F_1 by the group D_∞ generated by reflections in AF and A_1F_1; for instance, G is the image of F_1 in the mirror AF. The numbered triangles which fit together to fill the asymptotic triangle ABM are respectively congruent to those which fit together within the finite pentagon $ABGF_1A_1$; in fact, any two triangles that are numbered alike are related by some power of the translation from G to F_1 (or from F to F_2). Hence the area of the asymptotic triangle is less than or equal to the area of the pentagon:

16.41 *Any asymptotic triangle has a finite area.*

Since any doubly asymptotic triangle (Figure 16.3a) can be dissected into two asymptotic triangles, it follows that

16.42 *Any doubly asymptotic triangle has a finite area.*

By 16.36, the area of a doubly asymptotic triangle is a function of its angle. Comparing the triangles AMN and $A'MN$ of Figure 16.3d, we see that this is a *decreasing* function: the larger triangle has the smaller angle.

Since any trebly asymptotic triangle can be dissected into two doubly asymptotic triangles (as in the proof of 16.37), 16.42 implies

16.43 *Any trebly asymptotic triangle has a finite area.*

By 16.37, this area is a constant, depending only on our chosen unit of measurement.

16.5 AREA AND ANGULAR DEFECT

> *Gauss ... did not recognize the existence of a logically sound non-Euclidean geometry by intuition or by a flash of genius: ... on the contrary, he had spent upon this subject many laborious hours before he had overcome the inherited prejudice against it. [He] did not let any rumour of his opinions get abroad, being certain that he would be misunderstood. Only to a few trusted friends did he reveal something of his work.*
>
> R. Bonola [**1,** pp. 66-67]

János Bolyai, or Bolyai János (as it is written in Hungarian), announced his discovery of absolute geometry in an appendix to a book by his father, Bolyai Farkas, who was a friend of Gauss. When Gauss saw this book and read the appendix, he wrote a remarkable letter to his old friend, congratulating János and admitting that he himself had thought along the same lines without publishing the results. The original letter (of March 6, 1832) is lost, but the younger Bolyai's copy of it has been preserved, and it was eventually published in Gauss's collected works [Gauss **1**, vol. 8, pp. 220–225].

This letter contains a wonderful proof that the area of a triangle ABC is proportional to its *angular defect*

$$\pi - A - B - C:$$

the amount by which its angle sum falls short of two right angles. The following paraphrase fills up a few gaps in the argument, while retaining Gauss's systematic division into seven steps, numbered with Roman numerals.

I. *All trebly asymptotic triangles are congruent.* (This is our 16.37.)

II. *The area of a trebly asymptotic triangle has a finite value, say t.* (This is our 16.43.)

III. *The area of a doubly asymptotic triangle AMN is a function of its angle, NAM, say $f(\phi)$, where ϕ is the supplement of this angle.* Given the angle ϕ, we can construct the triangle in a unique fashion (Figure 16.5*a*; cf. 16.3*c*). Gauss used the supplement, rather than the angle NAM itself, to ensure that $f(\phi)$ is an *increasing* function of ϕ. (See the remark after 16.42.)

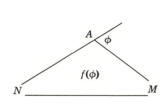

Figure 16.5a

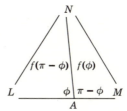

Figure 16.5b

IV. $f(\phi) + f(\pi - \phi) = t.$

This may be seen by fitting together two doubly asymptotic triangles AMN and ANL with supplementary angles, as in Figure 16.5*b*. Here it is understood that $0 < \phi < \pi$. But when ϕ approaches zero, the doubly asymptotic triangle collapses, and when ϕ approaches π it tends to become trebly asymptotic. Hence

16.51 $$f(0) = 0, \qquad f(\pi) = t,$$

and IV is valid for $0 \leqslant \phi \leqslant \pi$.

V. $f(\phi) + f(\psi) + f(\pi - \phi - \psi) = t.$

This, with $\phi > 0, \psi > 0, \phi + \psi < \pi$, may be seen by fitting together three doubly asymptotic triangles whose angles add up to 2π, as in Figure 16.5*c*. It evidently remains valid when ϕ or ψ is zero or $\phi + \psi = \pi$.

VI. $f(\phi) + f(\psi) = f(\phi + \psi)$.

This, with $\phi \geqslant 0, \psi \geqslant 0, \phi + \psi \leqslant \pi$, is obtained algebraically, by writing $\phi + \psi$ instead of ϕ in IV and then using V. It follows that $f(\phi)$ is simply a multiple of ϕ, namely,

16.52
$$f(\phi) = \mu\phi$$

where, by 16.51, $\mu = t/\pi$.

J. H. Lindsay has pointed out that this deduction can be made without assuming the function to be continuous. By VI, with $\phi = \psi$,

$$f(\phi) = \tfrac{1}{2} f(2\phi).$$

Thus 16.52 holds when $\phi = \tfrac{1}{2}\pi$, again when $\phi = \tfrac{1}{4}\pi$, and so on; that is, it holds when ϕ is π divided by any power of 2. Appealing again to VI, we deduce that $f(\phi) = \mu\phi$ whenever $\phi = n\pi$, where n is a number which terminates when expressed as a "decimal" in the scale of 2 [cf. Coxeter **3**, p. 102]. For brevity, let us call this a *binary* number.

Suppose, if possible, that, for some particular value of ϕ, $f(\phi) \neq \mu\phi$. Choose a binary number n between the two distinct real numbers ϕ/π and $f(\phi)/\mu\pi$. If $f(\phi) > \mu\phi$, so that

$$\phi < n\pi < \frac{f(\phi)}{\mu},$$

we have, since $f(\phi)$ is an increasing function,

$$f(\phi) < f(n\pi) = \mu n\pi < f(\phi),$$

which is absurd. If, on the other hand, $f(\phi) < \mu\phi$, we can argue the same way with all the inequalities reversed. Hence, in fact, $f(\phi) = \mu\phi$ for all the values of ϕ (from 0 to π).

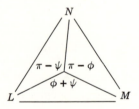

Figure 16.5c

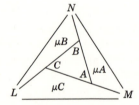

Figure 16.5d

VII. *The area Δ of any triangle ABC (with finite sides) is a constant multiple of its angular defect:*

$$\Delta = \mu(\pi - A - B - C).$$

For this final step, Gauss exhibited ABC as part of a trebly asymptotic triangle by extending its sides in cyclic order, as in Figure 16.5d. The remaining parts are doubly asymptotic triangles whose areas are μA, μB, μC. Hence

$$\Delta + \mu A + \mu B + \mu C = t = \mu\pi,$$

and the desired formula follows at once.

If we wish, we can follow Lobachevsky in using such a unit of measurement* that the area of a trebly asymptotic triangle is π. Then $\mu = 1$, and the formula is simply

16.53 $$\Delta = \pi - A - B - C.$$

This is strikingly reminiscent of the formula 6.92, which tells us that the area of a spherical triangle drawn on a sphere of radius R is

$$(A + B + C - \pi)R^2.$$

In fact, setting $R^2 = -1$, we find that Gauss's result agrees formally with the area of a triangle drawn on a sphere of radius i. Long before the time of Gauss, it was suggested by J. H. Lambert (1728–1777) that, if a non-Euclidean plane exists, it should resemble a sphere of radius i. This analogy enabled him to derive the formulas of hyperbolic trigonometry (which were later developed rigorously by Lobachevsky) from the classical formulas of spherical trigonometry. Its full significance did not appear till Minkowski (1864–1909) discovered the geometry of space-time, which provided a geometrical basis for Einstein's special theory of relativity. We know now that, in a $(2 + 1)$-dimensional space-time, the hyperbolic plane can be represented without distortion on either sheet of a *sphere of time-like radius*. In the underlying affine space, this kind of sphere is a hyperboloid of two sheets.†

EXERCISES

1. Gauss's formula 16.53 remains valid when the triangle has one or more zero angles.

2. The area of any simple p-gon is equal to its angular defect: the amount by which its angle sum falls short of that of a p-gon in the Euclidean plane. (*Hint:* Dissect the polygon into triangles. Of course, we are now assuming $\mu = 1$.) In Figure 16.4a, the area of ABM is equal to that of $ABGF_1A_1$.

3. The product of three translations along the directed sides of a triangle (through the lengths of these sides themselves) is a rotation through the angular defect of the triangle. (These translations are half as long as those in Donkin's theorem, 15.31.) [Lamb **1**, p. 7.]

4. The product of half-turns about the midpoints of the sides of a simple quadrangle (in their natural order) is a rotation through the angular defect of the quadrangle.

5. Any polygon whose angle sum is a submultiple of 2π can be repeated, by half-

---

* Coxeter, Hyperbolic triangles, *Scripta Mathematica,* **22** (1956), p. 9.

† Coxeter, A geometrical background for de Sitter's world, *American Mathematical Monthly,* **50** (1943), p. 220.

turns about the midpoints of its sides, so as to cover the whole plane without interstices [cf. Somerville **1,** p. 86, Ex. 15]. (*Hint:* See Figures 4.2*b* and *c*.)

16.6   CIRCLES, HOROCYCLES, AND EQUIDISTANT CURVES

> A circle *is the orthogonal trajectory of a pencil of lines with a real vertex* . . .
> A horocycle *is the orthogonal trajectory of a pencil of parallel lines.* . . . *The orthogonal trajectory of a pencil of lines with an ideal vertex* . . . *is called an equidistant-curve.*
>
> D. M. Y. Sommerville (1879–1934)
>
> [Sommerville **1,** pp. 51–52]

By 15.26, any two distinct lines are either intersecting, parallel, or ultra-parallel. In other words, they belong to a *pencil* of lines of one of three kinds: an ordinary pencil, consisting of all the lines through one point, a pencil of parallels, consisting of all the lines parallel to a given ray, or a *pencil of ultraparallels,* consisting of all the lines perpendicular to a given line. By 15.32, the product of reflections in the two lines is a rotation, a parallel displacement, or a translation, respectively. By fixing one of the two lines and allowing the other to vary in the pencil, we see that each of these three kinds of direct isometry can be applied as a continuous motion.

A *circle* with center O may be defined either as in § 15.1 or to be the locus of a point P which is derived from a fixed point Q (distinct from O) by continuous rotation about O, or to be the locus of the image of Q by reflection in all the lines through O. When the radius OQ becomes infinite, we have a *horocycle* with center M (at infinity): the locus of a point which is derived from a fixed point Q by a continuous parallel displacement, or the locus of the image of Q by reflection in all the lines parallel to the ray QM [Coxeter **3,** p. 213]. The rays parallel to QM are called the *diameters* of the horocycle. The first "o" in the word "horocycle" is short, as in "horror."

The locus of a point at a constant distance from a fixed line o is not a pair of parallel lines, as it would be in the Euclidean plane, but an *equidistant curve* (or "hypercycle"), having two branches, one on each side of its *axis o*. Either branch may be described as the locus of a point which is derived from a fixed point Q (not on o) by continuous translation along o, or as the locus of the image of Q by reflection in all the lines perpendicular to o.

Orthogonal to the pencil of lines through O we have a pencil of concentric circles. A rotation about O permutes the lines and slides each circle along itself. Orthogonal to the pencil of parallels with a common end M we have a pencil of *concentric horocycles*. A parallel displacement with center M permutes the parallel lines and slides each horocycle along itself. Orthogonal to the pencil of ultraparallels perpendicular to o we have a pencil of *coaxal equidistant curves*. A translation along o permutes the ultraparallel lines and slides each equidistant curve along itself.

We are now ready to fulfill the promise, made after 15.31, to show that "the product of two translations with nonintersecting axes may be a rotation." Referring to Figure 16.3d, we see that the line through C perpendicular to AB is ultraparallel to AM and AN. Therefore, it has a common perpendicular GH with AM, and a common perpendicular FE with AN, forming a pentagon $AEFGH$ with right angles at E, F, G, H as in Figure 16.6a. The remaining angle (at A) may be as small as we please; if it is zero, the pentagon is "asymptotic." The product of reflections in AE and FG is a translation along EF (through $2EF$). The product of reflections in FG and AH is a translation along GH (through $2GH$). Hence the product of these two translations is the same as the product of reflections in AE and AH, which is a rotation or, if A is an "end," a parallel displacement. Since the axes of the two translations are both perpendicular to FG, we have thus proved that the product of translations along two ultraparallel lines may be either a rotation or a parallel displacement. (Of course, it may just as easily be another translation.)

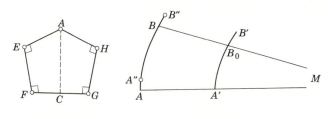

Figure 16.6a   **Figure 16.6b**

The product of translations along two parallel lines, AM and BM, leaves invariant the common end M; therefore it cannot be a rotation, but must be either a translation along another line through M or a parallel displacement with center M. We shall soon see that the latter possibility arises when the two given translations are of equal length, one towards M and the other away from M. In fact, the translation along AM from A to A' (Figure 16.6b) transforms the arc AB of a horocycle through A into an equal arc $A'B'$ of the concentric horocycle through A'. Let B_0 denote the point in which the latter arc is cut by the diameter through B. The translation along this diameter from B_0 to B transforms the arc B_0A' of the second horocycle into the equal arc BA'' of the first. Thus the product of the two translations is the parallel displacement that transforms the arc AB into $A''B''$; it slides this horocycle (and every concentric horocycle) along itself.

EXERCISES

1. The three vertices of a (finite) triangle all lie on each of three equidistant curves, whose axes join midpoints of pairs of sides, and on a fourth "cycle," which may be either a circle or a horocycle or another equidistant curve (with all three vertices on one branch). [Sommerville **1**, pp. 54, 189.]

2. The three sides of a (finite) triangle all touch a circle (the incircle) and three other "cycles," each of which may be of any one of the three kinds.

3. In Figure 15.2a, the horocycle through J with diameter p_1 passes also through L.

4. How many horocycles will pass through two given points?

5. An equidistant curve may have as many as four intersections with a circle or a horocycle or another equidistant curve.

6. Develop the analogy between conics in the affine plane and generalized circles in the hyperbolic plane. A horocycle, like a parabola, goes to infinity in one direction: if the points P and Q on it are variable and fixed, respectively, the limiting position of the line QP is the diameter through Q. An equidistant curve, like a hyperbola, has two branches.

7. Unlike the conjugate axis of a hyperbola, the axis of an equidistant curve is on the *concave* side of each branch.

16.7 POINCARÉ'S "HALF-PLANE" MODEL

> *There is a gain in simplicity when the fundamental circle is taken as a straight line, say the axis of x. . . . We may avoid dealing with pairs of points by considering only those points above the x-axis. A proper circle is represented by a circle lying entirely above the x-axis; a horocycle by a circle touching the x-axis; an equidistant-curve by the upper part of a circle cutting the x-axis together with the reflexion of the part which lies below the axis.*
>
> D. M. Y. Sommerville [**1**, pp. 188-189]

From the conformal model (Figure 16.2a) in which the lines are represented by circles (and lines) orthogonal to a fixed circle Ω, Poincaré derived another conformal model by inversion in a circle whose center lies on Ω. The inverse of Ω is a line, say a "horizontal" line, which we shall again denote by Ω. The points of the hyperbolic plane are represented by pairs of points which are images of each other by reflection in Ω, and the lines are represented by circles and lines orthogonal to Ω, that is, circles whose centers lie on Ω, and vertical lines [Burnside **1**, p. 387].

Through a pair of points which are images in Ω, we can draw an intersecting pencil of coaxal circles (like Figure 6.5a turned through a right angle) representing an ordinary pencil of lines. The orthogonal nonintersecting pencil, having Ω for its radical axis, represents a pencil of concentric circles. The limiting points of the nonintersecting pencil represent the common center of the concentric circles.

Another pencil of circles (situated as in Figure 6.5a itself) can be drawn through two points on Ω. One member of this pencil, having its center on Ω, represents a line o. The remaining circles (or strictly, pairs of them related by reflection in Ω) represent coaxal equidistant curves with axis o. For, the orthogonal nonintersecting pencil represents the pencil of ultraparallel lines perpendicular to o.

A tangent pencil of circles whose centers lie on Ω (Figure 6.5b) represents a pencil of parallels, whereas the orthogonal tangent pencil (touching Ω) represents a pencil of concentric horocycles. One particular pencil of parallels (special in the model but, of course, not special in the hyperbolic geometry itself) is represented by all the vertical lines (which pass, like Ω itself, through the point at infinity of the inversive plane). The horocycles having these lines for diameters are represented by all the horizontal lines except Ω (or strictly, pairs of such lines related by reflection in Ω). Since reflections in the vertical lines represent reflections in the parallel lines, horizontal translations represent parallel displacements. Hence the horizontal lines (other than Ω itself) represent the horocycles *isometrically:* equal segments represent equal arcs.

EXERCISES

1. What figure is represented by two lines forming an angle that is bisected by Ω?

2. When two ultraparallel lines are represented by nonintersecting circles (in either of Poincaré's conformal models), the distance between the lines, measured along their common perpendicular, appears as the *inversive distance* between the circles (see Exercise 5 of § 6.6).

3. The angle of parallelism (Figure 16.3d on page 293) is

$$\Pi(x) = 2 \arctan e^{-x}.$$

16.8 THE HOROSPHERE AND THE EUCLIDEAN PLANE

> *F. L. Wachter (1792-1817) . . . in a letter to Gauss (Dec., 1816) . . . speaks of the surface to which a sphere tends as its radius approaches infinity. . . . He affirms that even in the case of the Fifth Postulate being false, there would be a geometry on this surface identical with that of the ordinary plane.*
>
> R. Bonola [**1,** pp. 62-63]

The ideas in §§ 16.6 and 16.7 extend in an obvious manner from two to three dimensions. The locus of images of a point Q by reflection in all the planes through a point O is a *sphere* with radius OQ. As a limiting case we have a *horosphere* with center M (at infinity): the locus of images of a point Q by reflection in all the planes parallel to the ray QM [Coxeter **3,** p. 218]. The locus of images of a point Q by reflection in all the planes perpendicular to a fixed plane ω is one sheet of an *equidistant surface,* which consists of points at a constant distance from ω on either side.

There is a conformal model in inversive space in which the points of hyperbolic space are represented by pairs of points related by reflection in a fixed "horizontal" plane Ω, and the planes are represented by spheres and planes orthogonal to Ω, that is, spheres whose centers lie on Ω, and vertical planes. The representation of lines (which are intersections of planes) follows immediately. Of particular interest is the bundle of vertical lines, which represents the bundle of lines parallel to a given ray QM (special in the model, though not in the hyperbolic geometry itself). The horospheres that have these lines for diameters are represented by all the horizontal planes except Ω. Since every vertical plane provides a model (of the kind described in § 16.7) for a plane in the hyperbolic space, each horizontal plane (except Ω) represents a horosphere, and every line in the plane represents a horocycle on the horosphere. Since distances along such lines agree with distances along the corresponding horocycles, the representation of the horosphere by the Euclidean plane is *isometric:* for any figure in the Euclidean plane there is a congruent figure on the horosphere (with lines replaced by horocycles).

This astonishing theorem was discovered independently by Bolyai and Lobachevsky. For two different proofs, see Coxeter [**3,** pp. 197, 251]. It means that, along with ordinary spherical geometry, the inhabitants of a hyperbolic world would also study horospherical geometry, which is the same as Euclidean geometry!

Part IV

17

Differential geometry of curves

Differential geometry is concerned with applying the methods of analysis to geometry, especially to the study of curves and surfaces. Classically, the study is made in Euclidean space of three dimensions. But in the twentieth century other spaces, such as inversive, affine, or projective, have been used. In other words, differential geometry is still significant when there is no concept of distance. However, both distance and parallelism are usually present, in which case the notion of a vector is fundamental.

A curve, being the locus of a point P, is intimately associated with a variable vector, namely the position vector

$$\mathbf{r} = \overrightarrow{OP}$$

which goes from a fixed origin O to the point P. For simplicity we shall consider only rectifiable curves for which there is a well-defined tangent at each point [Kreyszig **1**, p. 29].

After a preliminary discussion of vectors, we shall consider the curvature of plane curves, and the curvature and torsion of twisted curves, applying the results to many important special cases such as spirals and helices.

17.1 VECTORS IN EUCLIDEAN SPACE

We have already considered, in § 13.6, the *affine* properties of vectors, such as addition and subtraction, multiplication by numbers, independence, and the unique expression

17.11
$$\mathbf{c} = x\mathbf{e} + y\mathbf{f} + z\mathbf{g}$$

for any vector $\mathbf{c}$ as a linear combination of three basic vectors $\mathbf{e}, \mathbf{f}, \mathbf{g}$. The time has now come to introduce two kinds of multiplication of vectors by one another. We shall employ the notation of J. W. Gibbs (1839–1903), although some authors, such as Birkhoff and MacLane [**1**, p. 175] and Forder [**2**], prefer that of H. Grassmann (1809–1877).

Euclidean geometry allows us to speak of the *length* (or "magnitude," or "absolute value") $|\mathbf{a}|$, of any given vector $\mathbf{a}$. If θ is the angle between $\mathbf{a}$ and another vector $\mathbf{b}$, we define the *inner* (or "scalar") product $\mathbf{a} \cdot \mathbf{b}$ and the *outer* (or "vector") product $\mathbf{a} \times \mathbf{b}$ by the formulas

$$\mathbf{a} \cdot \mathbf{b} = |\mathbf{a}|\,|\mathbf{b}|\cos\theta, \qquad \mathbf{a} \times \mathbf{b} = |\mathbf{a}|\,|\mathbf{b}|\sin\theta\,\mathbf{g},$$

where $\mathbf{g}$ is the unit vector orthogonal to the plane $\mathbf{ab}$ on the side from which θ appears as a positive angle. The introduction of the auxiliary vector $\mathbf{g}$ (orthogonal to both $\mathbf{a}$ and $\mathbf{b}$) is justified by the elegant algebra that follows.

We see at once that, if m and n are numbers,

$$m\mathbf{a} \cdot n\mathbf{b} = mn\mathbf{a} \cdot \mathbf{b}, \qquad m\mathbf{a} \times n\mathbf{b} = mn\mathbf{a} \times \mathbf{b},$$

$$\mathbf{b} \cdot \mathbf{a} = \mathbf{a} \cdot \mathbf{b}, \qquad \mathbf{b} \times \mathbf{a} = -\mathbf{a} \times \mathbf{b}.$$

Thus inner multiplication is commutative (like the multiplication of numbers), whereas outer multiplication is "anticommutative." Since $\mathbf{a} \times \mathbf{a} = \mathbf{0}$, we naturally take $\mathbf{a}^2$ to mean $\mathbf{a} \cdot \mathbf{a}$:

$$\mathbf{a}^2 = |\mathbf{a}|^2.$$

Two vectors $\mathbf{a}$ and $\mathbf{b}$ are orthogonal if $\mathbf{a} \cdot \mathbf{b} = 0$, parallel if $\mathbf{a} \times \mathbf{b} = \mathbf{0}$.

Consider two vectors $\mathbf{a} = \overrightarrow{OA}$, $\mathbf{b} = \overrightarrow{OB}$, and let BN be the perpendicular from B to OA, as in Figure 17.1a. The algebraic distance ON (negative if $\angle AOB$ is obtuse) is called the *projection* of $\mathbf{b}$ on $\mathbf{a}$. If $|\mathbf{a}| = 1$, so that $\mathbf{a}$ is a *unit* vector, this projection is clearly $\mathbf{a} \cdot \mathbf{b}$. Removing the restriction to a unit vector, we find that $\mathbf{a} \cdot \mathbf{b}$ is $|\mathbf{a}|$ times the projection. This geometrical interpretation makes it easy to establish, for inner products, the distributive law

$$\mathbf{a} \cdot (\mathbf{b} + \mathbf{b}') = \mathbf{a} \cdot \mathbf{b} + \mathbf{a} \cdot \mathbf{b}',$$

which may also be expressed as

$$(\mathbf{b} + \mathbf{b}') \cdot \mathbf{a} = \mathbf{b} \cdot \mathbf{a} + \mathbf{b}' \cdot \mathbf{a}$$

in virtue of the commutative law $\mathbf{b} \cdot \mathbf{a} = \mathbf{a} \cdot \mathbf{b}$. Replacing $\mathbf{b}'$ by $-\mathbf{b}'$, we obtain the corresponding results for differences instead of sums.

The distributive law provides a useful method for establishing certain identities. If $\mathbf{b}$ and $\mathbf{b}'$ stand for expressions which we wish to prove equal, it is sometimes helpful to introduce an arbitrary vector $\mathbf{c}$ and to compare $\mathbf{b} \cdot \mathbf{c}$ with $\mathbf{b}' \cdot \mathbf{c}$. If we find that

$$\mathbf{b} \cdot \mathbf{c} = \mathbf{b}' \cdot \mathbf{c}$$

for all choices of $\mathbf{c}$ (or even for three independent $\mathbf{c}$'s), we can safely assert that $\mathbf{b} = \mathbf{b}'$. For, since $(\mathbf{b} - \mathbf{b}') \cdot \mathbf{c} = 0$, if $\mathbf{b} - \mathbf{b}'$ is not the zero vector it

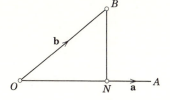

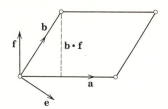

Figure 17.1a   **Figure 17.1b**

must be orthogonal to **c**; and, since **c** is arbitrary, this is impossible.

As a step towards establishing the distributive law for outer products, we compare two expressions for the area of a parallelogram, namely

$$|\mathbf{a} \times \mathbf{b}| = |\mathbf{a}| \, \mathbf{b} \cdot \mathbf{f},$$

where **f** is a unit vector orthogonal to **a** in the plane **ab** (as in Figure 17.1*b*) so that **b** · **f** is the altitude of the parallelogram from its base |**a**|. Analogously, the parallelepiped formed by three independent vectors **a**, **b**, **c** has base |**a** × **b**|, altitude **c** · **g**, and volume (with a suitable convention of sign, depending on whether the trihedron **abc** is positively or negatively oriented)

$$|\mathbf{a} \times \mathbf{b}| \, \mathbf{c} \cdot \mathbf{g} = |\mathbf{a} \times \mathbf{b}| \, \mathbf{g} \cdot \mathbf{c} = (\mathbf{a} \times \mathbf{b}) \cdot \mathbf{c}.$$

Since we could just as well regard another face of the parallelepiped as its base, the same volume is expressible as

$$(\mathbf{b} \times \mathbf{c}) \cdot \mathbf{a} = \mathbf{a} \cdot (\mathbf{b} \times \mathbf{c}).$$

Thus we can interchange the cross and the dot:

$$(\mathbf{a} \times \mathbf{b}) \cdot \mathbf{c} = \mathbf{a} \cdot (\mathbf{b} \times \mathbf{c}).$$

(This is as near as we can come to an "associative law" for products of vectors.) Since the dot and cross are interchangeable, it is convenient to use, for (**a** × **b**) · **c** or **a** · (**b** × **c**), the special symbol [**a b c**], so that the volume of the parallelepiped is

$$[\mathbf{a}\,\mathbf{b}\,\mathbf{c}] = [\mathbf{b}\,\mathbf{c}\,\mathbf{a}] = [\mathbf{c}\,\mathbf{a}\,\mathbf{b}] = -[\mathbf{c}\,\mathbf{b}\,\mathbf{a}].$$

If [**a b c**] = 0, the parallelepiped collapses, and the three vectors are coplanar, that is, dependent. Thus a necessary and sufficient condition for **a**, **b**, **c** to be independent is

$$[\mathbf{a}\,\mathbf{b}\,\mathbf{c}] \neq 0.$$

To prove the distributive law for outer products, we introduce an arbitrary vector **c** (like a catalyst) and find

$$\{(\mathbf{a} + \mathbf{a}') \times \mathbf{b}\} \cdot \mathbf{c} = (\mathbf{a} + \mathbf{a}') \cdot (\mathbf{b} \times \mathbf{c})$$
$$= \mathbf{a} \cdot (\mathbf{b} \times \mathbf{c}) + \mathbf{a}' \cdot (\mathbf{b} \times \mathbf{c})$$
$$= (\mathbf{a} \times \mathbf{b}) \cdot \mathbf{c} + (\mathbf{a}' \times \mathbf{b}) \cdot \mathbf{c}$$
$$= \{(\mathbf{a} \times \mathbf{b}) + (\mathbf{a}' \times \mathbf{b})\} \cdot \mathbf{c}.$$

Since $\mathbf{c}$ is arbitrary, we conclude that

$$(\mathbf{a} + \mathbf{a}') \times \mathbf{b} = (\mathbf{a} \times \mathbf{b}) + (\mathbf{a}' \times \mathbf{b}).$$

Since the outer product of two vectors is a vector, we might at first expect the associative law to hold. To see why $(\mathbf{a} \times \mathbf{b}) \times \mathbf{c}$ and $\mathbf{a} \times (\mathbf{b} \times \mathbf{c})$ are, in general, different, let us evaluate both expressions, using a procedure devised by Coe and Rainich.* Consider unit vectors $\mathbf{e}$ and $\mathbf{f}$ in the plane $\mathbf{ab}$, orthogonal to $\mathbf{b}$ and $\mathbf{a}$ respectively, as in Figure 17.1b. Since the vector $\mathbf{a} \times \mathbf{b}$ is perpendicular to the plane $\mathbf{ab}$ (or $\mathbf{ef}$), the two vectors

$$(\mathbf{a} \times \mathbf{b}) \times \mathbf{e}, \qquad (\mathbf{a} \times \mathbf{b}) \times \mathbf{f}$$

lie in this plane and have the same length $|\mathbf{a} \times \mathbf{b}|$, which may be expressed in either of the forms

$$\mathbf{a} \cdot \mathbf{e} \, |\mathbf{b}|, \qquad \mathbf{b} \cdot \mathbf{f} \, |\mathbf{a}|.$$

Since they have the same directions as $\mathbf{b}$, $-\mathbf{a}$, respectively, they are exactly

$$(\mathbf{a} \times \mathbf{b}) \times \mathbf{e} = \mathbf{a} \cdot \mathbf{e} \, \mathbf{b}, \qquad (\mathbf{a} \times \mathbf{b}) \times \mathbf{f} = -\mathbf{b} \cdot \mathbf{f} \, \mathbf{a}.$$

If $\mathbf{g}$ is perpendicular to the plane, we have also

$$(\mathbf{a} \times \mathbf{b}) \times \mathbf{g} = \mathbf{0}.$$

Using the three vectors $\mathbf{e}$, $\mathbf{f}$, $\mathbf{g}$ as a basis, we may express an arbitrary vector $\mathbf{c}$ in the form 17.11; thus

$$(\mathbf{a} \times \mathbf{b}) \times \mathbf{c} = (\mathbf{a} \times \mathbf{b}) \times (x\mathbf{e} + y\mathbf{f} + z\mathbf{g})$$
$$= x(\mathbf{a} \times \mathbf{b}) \times \mathbf{e} + y(\mathbf{a} \times \mathbf{b}) \times \mathbf{f} + z(\mathbf{a} \times \mathbf{b}) \times \mathbf{g}$$
$$= x(\mathbf{a} \cdot \mathbf{e})\mathbf{b} - y(\mathbf{b} \cdot \mathbf{f})\mathbf{a} = (\mathbf{a} \cdot x\mathbf{e})\mathbf{b} - (\mathbf{b} \cdot y\mathbf{f})\mathbf{a}$$
$$= \mathbf{a} \cdot (x\mathbf{e} + y\mathbf{f} + z\mathbf{g})\mathbf{b} - \mathbf{b} \cdot (x\mathbf{e} + y\mathbf{f} + z\mathbf{g})\mathbf{a},$$

since $\mathbf{a} \cdot \mathbf{f}$, $\mathbf{a} \cdot \mathbf{g}$, $\mathbf{b} \cdot \mathbf{e}$, $\mathbf{b} \cdot \mathbf{g}$ are all zero. Hence, finally,

17.12 $$(\mathbf{a} \times \mathbf{b}) \times \mathbf{c} = (\mathbf{a} \cdot \mathbf{c})\mathbf{b} - (\mathbf{b} \cdot \mathbf{c})\mathbf{a}.$$

Interchanging $\mathbf{a}$ and $\mathbf{c}$, we deduce

$$\mathbf{a} \times (\mathbf{b} \times \mathbf{c}) = (\mathbf{c} \times \mathbf{b}) \times \mathbf{a} = (\mathbf{c} \cdot \mathbf{a})\mathbf{b} - (\mathbf{b} \cdot \mathbf{a})\mathbf{c}$$
$$= (\mathbf{a} \cdot \mathbf{c})\mathbf{b} - (\mathbf{a} \cdot \mathbf{b})\mathbf{c}.$$

By considering $\{(\mathbf{a} \times \mathbf{b}) \times \mathbf{c}\} \cdot \mathbf{d}$, we find also that any four vectors $\mathbf{a}$, $\mathbf{b}$, $\mathbf{c}$, $\mathbf{d}$ satisfy *Lagrange's identity*

17.13 $$(\mathbf{a} \times \mathbf{b}) \cdot (\mathbf{c} \times \mathbf{d}) = (\mathbf{a} \cdot \mathbf{c})(\mathbf{b} \cdot \mathbf{d}) - (\mathbf{b} \cdot \mathbf{c})(\mathbf{a} \cdot \mathbf{d}).$$

* C. J. Coe and G. Y. Rainich, *American Mathematical Monthly*, **56** (1949), pp. 175–176.

It is sometimes desirable to express a vector in terms of its components in the directions of the axes of rectangular Cartesian coordinates, that is, to let the coordinate symbol (x, y, z) for the point P to be used also for the vector $\overrightarrow{OP}$, where O is the origin $(0, 0, 0)$. In other words, we use $P = (x, y, z)$ as an abbreviation for

17.14 $$\mathbf{r} = x\mathbf{i} + y\mathbf{j} + z\mathbf{k},$$

where $\mathbf{i}, \mathbf{j}, \mathbf{k}$ are unit vectors along the three axes (so that this is a special case of 17.11). Since

17.15
$$\mathbf{i}^2 = \mathbf{j}^2 = \mathbf{k}^2 = 1, \qquad \mathbf{j}\cdot\mathbf{k} = \mathbf{k}\cdot\mathbf{i} = \mathbf{i}\cdot\mathbf{j} = 0,$$
$$\mathbf{i} = \mathbf{j} \times \mathbf{k}, \qquad \mathbf{j} = \mathbf{k} \times \mathbf{i}, \qquad \mathbf{k} = \mathbf{i} \times \mathbf{j}, \qquad [\mathbf{i}\,\mathbf{j}\,\mathbf{k}] = 1,$$

we easily deduce, for any three vectors $\mathbf{r}, \mathbf{r}', \mathbf{r}''$, the products

17.151 $$\mathbf{r}\cdot\mathbf{r}' = xx' + yy' + zz',$$

$$\mathbf{r} \times \mathbf{r}' = \begin{vmatrix} y & z \\ y' & z' \end{vmatrix}\mathbf{i} + \begin{vmatrix} z & x \\ z' & x' \end{vmatrix}\mathbf{j} + \begin{vmatrix} x & y \\ x' & y' \end{vmatrix}\mathbf{k}$$

and

17.16 $$[\mathbf{r}\,\mathbf{r}'\,\mathbf{r}''] = \begin{vmatrix} x & y & z \\ x' & y' & z' \\ x'' & y'' & z'' \end{vmatrix}.$$

Since the product of two determinants (like the product of two matrices) is obtained by writing down the inner products of the rows of the first with the columns of the second, we can bring in three more vectors such as $\mathbf{q} = u\mathbf{i} + v\mathbf{j} + w\mathbf{k}$ and find

$$[\mathbf{q}\,\mathbf{q}'\,\mathbf{q}''][\mathbf{r}\,\mathbf{r}'\,\mathbf{r}''] = \begin{vmatrix} u & v & w \\ u' & v' & w' \\ u'' & v'' & w'' \end{vmatrix} \begin{Vmatrix} x & x' & x'' \\ y & y' & y'' \\ z & z' & z'' \end{Vmatrix}$$

17.17
$$= \begin{vmatrix} \mathbf{q}\cdot\mathbf{r} & \mathbf{q}\cdot\mathbf{r}' & \mathbf{q}\cdot\mathbf{r}'' \\ \mathbf{q}'\cdot\mathbf{r} & \mathbf{q}'\cdot\mathbf{r}' & \mathbf{q}'\cdot\mathbf{r}'' \\ \mathbf{q}''\cdot\mathbf{r} & \mathbf{q}''\cdot\mathbf{r}' & \mathbf{q}''\cdot\mathbf{r}'' \end{vmatrix}.$$

Returning to 17.14, we observe that

$$\mathbf{r}\cdot\mathbf{i} = x, \quad \mathbf{r}\cdot\mathbf{j} = y, \quad \mathbf{r}\cdot\mathbf{k} = z.$$

Thus we can express any vector $\mathbf{r}$ in terms of any orthogonal trihedron of unit vectors in the form

17.18 $$\mathbf{r} = (\mathbf{r}\cdot\mathbf{i})\,\mathbf{i} + (\mathbf{r}\cdot\mathbf{j})\,\mathbf{j} + (\mathbf{r}\cdot\mathbf{k})\,\mathbf{k}.$$

We shall also have occasion to use the following theorem:

17.19   *If two vectors, **a** and **b**, lie in perpendicular planes which intersect the line of a unit vector **k**, then*

$$(\mathbf{a} \cdot \mathbf{k})(\mathbf{k} \cdot \mathbf{b}) = \mathbf{a} \cdot \mathbf{b}.$$

Proof. Since the planes **ak** and **kb** are perpendicular, 17.13 yields

$$0 = (\mathbf{a} \times \mathbf{k}) \cdot (\mathbf{k} \times \mathbf{b}) = (\mathbf{a} \cdot \mathbf{k})(\mathbf{k} \cdot \mathbf{b}) - \mathbf{k}^2(\mathbf{a} \cdot \mathbf{b}) = (\mathbf{a} \cdot \mathbf{k})(\mathbf{k} \cdot \mathbf{b}) - (\mathbf{a} \cdot \mathbf{b}).$$

EXERCISES

1.   How must **a**, **b**, **c** be related in order to satisfy the associative law $(\mathbf{a} \times \mathbf{b}) \times \mathbf{c} = \mathbf{a} \times (\mathbf{b} \times \mathbf{c})$?

2.   Simplify $(\mathbf{a} \times \mathbf{b}) \times (\mathbf{c} \times \mathbf{d})$ two ways and, by equating the results, deduce an identity connecting four vectors such as $[\mathbf{a}\,\mathbf{b}\,\mathbf{c}]\,\mathbf{d}$.

3.   Simplify $(\mathbf{a} \times \mathbf{b}) \cdot (\mathbf{a} \times \mathbf{b})$, and show that the result could have been anticipated in virtue of a well known trigonometrical identity.

17.2   VECTOR FUNCTIONS AND THEIR DERIVATIVES

Vector functions can be differentiated in the same manner as numerical functions. Let the vector

$$\mathbf{a} = \mathbf{a}(s)$$

be a function of the numerical variable s, and let $\Delta\mathbf{a}$ be the increment in the vector corresponding to the increment Δs in the variable s, so that

$$\mathbf{a}(s + \Delta s) = \mathbf{a} + \Delta\mathbf{a}.$$

If the vector $\Delta\mathbf{a}/\Delta s$ tends to a limit as Δs tends to zero, the vector function $\mathbf{a}(s)$ is said to be *differentiable,* and the limit is the *derivative:*

$$\dot{\mathbf{a}} = \frac{d\mathbf{a}}{ds} = \lim_{\Delta s \to 0} \frac{\Delta\mathbf{a}}{\Delta s} = \lim_{\Delta s \to 0} \frac{\mathbf{a}(s + \Delta s) - \mathbf{a}(s)}{\Delta s}.$$

The rule for differentiating a product is the same as for ordinary functions. In fact,

$$(\mathbf{a} + \Delta\mathbf{a}) \cdot (\mathbf{b} + \Delta\mathbf{b}) - \mathbf{a} \cdot \mathbf{b} = \mathbf{a} \cdot \Delta\mathbf{b} + \Delta\mathbf{a} \cdot \mathbf{b} + \Delta\mathbf{a} \cdot \Delta\mathbf{b}$$

and therefore

$$\frac{d}{ds}(\mathbf{a} \cdot \mathbf{b}) = \mathbf{a} \cdot \dot{\mathbf{b}} + \dot{\mathbf{a}} \cdot \mathbf{b} - \lim \dot{\mathbf{a}} \cdot \dot{\mathbf{b}}\, \Delta s$$

$$= \mathbf{a} \cdot \dot{\mathbf{b}} + \dot{\mathbf{a}} \cdot \mathbf{b}.$$

Similarly,

$$\frac{d}{ds}(m\mathbf{a}) = m\dot{\mathbf{a}} + \dot{m}\mathbf{a}$$

and

$$\frac{d}{ds}(\mathbf{a} \times \mathbf{b}) = (\mathbf{a} \times \dot{\mathbf{b}}) + (\dot{\mathbf{a}} \times \mathbf{b}).$$

(Since outer products are anticommutative, we must be careful not to write the second term on the right as $\mathbf{b} \times \dot{\mathbf{a}}$.)

Since

$$\frac{d}{ds} \mathbf{a}^2 = 2\mathbf{a} \cdot \dot{\mathbf{a}},$$

a variable vector of constant length is always orthogonal to its derivative.

Since the Cartesian basic vectors $\mathbf{i}, \mathbf{j}, \mathbf{k}$ are constant, their derivatives are zero, and

$$\dot{\mathbf{r}} = \frac{d}{ds}(x\mathbf{i} + y\mathbf{j} + z\mathbf{k}) = \dot{x}\mathbf{i} + \dot{y}\mathbf{j} + \dot{z}\mathbf{k} = (\dot{x}, \dot{y}, \dot{z}).$$

Thus, when we differentiate a vector, the components of the derivative are simply the derivatives of the components.

EXERCISE

When a particle moves in a circular orbit (like a stone swung at the end of a string), its position vector from the center has constant length. In which direction is its velocity? If its speed is constant, its velocity is a vector of constant length. In which direction is its acceleration?

17.3 CURVATURE, EVOLUTES AND INVOLUTES

The simplest instance of a variable vector is the position vector $\mathbf{r} = \overrightarrow{OP}$ of a point P that moves along a curve (including a straight line as the simplest case of all). To define the length of an arc of the curve, we approximate it by a sequence of broken lines, as in §8.5. The increment $\Delta\mathbf{r}$ may be identified with the vector along one of the segments of the broken line, so that, before we pass to the limit, the corresponding increment of arc is the length $|\Delta\mathbf{r}|$.

For most purposes, the directed arc s (measured along the curve from a fixed point A to a variable point P) is the most convenient parameter to use in describing the curve. That is, we regard the vector $\mathbf{r} = \overrightarrow{OP}$ as a function of s. Since

$$\lim \frac{|\Delta\mathbf{r}|}{\Delta s} = \lim \left[\left(\frac{\Delta x}{\Delta s}\right)^2 + \left(\frac{\Delta y}{\Delta s}\right)^2 + \left(\frac{\Delta z}{\Delta s}\right)^2 \right]^{\frac{1}{2}} = (\dot{x}^2 + \dot{y}^2 + \dot{z}^2)^{\frac{1}{2}} = 1$$

[Struik **1**, p. 7], the limit of $\Delta\mathbf{r}/\Delta s$ is the unit tangent vector

$$\mathbf{t} = \dot{\mathbf{r}}.$$

If another parameter u is used instead of s, we can easily make the necessary adjustment. The derivative $d\mathbf{r}/du$ is still a tangent vector, namely

$$\frac{d\mathbf{r}}{du} = \frac{d\mathbf{r}}{ds}\frac{ds}{du} = \frac{ds}{du}\mathbf{t};$$

the connection between s and u is determined by the length, ds/du, of this vector, and $\mathbf{t}$ is the unit vector in the same direction.

For instance, in the case of the circle

$$x = \rho \cos u, \qquad y = \rho \sin u,$$

of radius ρ, we have

$$\mathbf{r} = \rho\,(\cos u,\ \sin u),$$

$$\frac{ds}{du}\,\mathbf{t} = \rho\,(-\sin u,\ \cos u),$$

$$\frac{ds}{du} = \rho, \qquad \mathbf{t} = (-\sin u,\ \cos u).$$

For any curve in the (x, y)-plane, the tangent vector is

17.31
$$\mathbf{t} = (\cos \psi,\ \sin \psi),$$

where ψ is the angle that $\mathbf{t}$ makes with the vector $\mathbf{i}$ along the x-axis. The *curvature* of the plane curve is the arc derivative of this angle:

17.32
$$\kappa = \frac{d\psi}{ds} = \dot{\psi}.$$

Since $\mathbf{t}$ is a unit vector, its derivative is in the perpendicular direction, that is, in the direction of the unit *normal* vector $\mathbf{n} = (-\sin \psi,\ \cos \psi)$, which is derived from $\mathbf{t}$ by a positive quarter-turn. Thus

$$\dot{\mathbf{t}} = \dot{\psi}(-\sin \psi,\ \cos \psi)$$

17.33
$$= \kappa\mathbf{n},$$

and we regard κ as being positive or negative according as $\mathbf{n}$ is on the concave or convex side of the curve.

The derivative of $\mathbf{n}$, being orthogonal to $\mathbf{n}$, is a certain multiple of $\mathbf{t}$. By differentiating the inner product $\mathbf{n} \cdot \mathbf{t}$, which is zero, we find the precise expression

17.34
$$\dot{\mathbf{n}} = -\kappa\mathbf{t}.$$

Applying this method to the circle

$$\mathbf{r} = \rho\,(\cos u,\ \sin u),$$

for which $\mathbf{t} = (-\sin u,\ \cos u)$, we find

$$\kappa\mathbf{n} = \dot{\mathbf{t}} = \dot{u}(-\cos u,\ -\sin u),$$

whence

$$\kappa = \dot{u} = 1/\rho \quad \text{and} \quad \mathbf{n} = -(\cos u,\ \sin u).$$

This means that the curvature of the circle is the reciprocal of its radius, Soddy's "bend" (p. 15), and its normal is towards the center along the radius.

At a point P on any plane curve, the *center of curvature* P_c is the center of the *circle of curvature*, which is the circle of "closest fit," having the same normal and the same curvature. Its "radius"

17.35 $$\rho = 1/\kappa$$

(which we allow to be positive or negative with κ) is the *radius of curvature*. The center of curvature at P, being distant ρ from P along the normal there, has the position vector

17.36 $$\mathbf{r}_c = \mathbf{r} + \rho\mathbf{n}.$$

When P moves along the given curve (which we now assume not to be a circle nor a straight line), the center of curvature P_c moves along a related curve called the *evolute,** which may be expressed parametrically in terms of its own arc length s_c. Its unit tangent $\mathbf{t}_c$ is given by

$$\dot{s}_c\,\mathbf{t}_c = \frac{d}{ds}(\mathbf{r} + \rho\mathbf{n}) = \dot{\mathbf{r}} + \rho\dot{\mathbf{n}} + \dot{\rho}\mathbf{n}$$
$$= \mathbf{t} - \rho\kappa\mathbf{t} + \dot{\rho}\mathbf{n} = \dot{\rho}\mathbf{n}.$$

Since $\mathbf{t}_c$ and $\mathbf{n}$ are unit vectors, it follows that

$$\dot{s}_c = \pm\dot{\rho} \quad \text{and} \quad \mathbf{t}_c = \pm\mathbf{n}:$$

the tangent at P_c to the evolute is the same line as the normal at P to the original curve (see Figure 17.3a). Thus the evolute, which we have defined as the locus of the center of curvature, could equally well be defined as the *envelope of normals*.

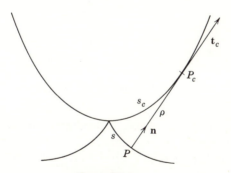

Figure 17.3a

Integrating the differential equation $ds_c = \pm d\rho$, we find that, for some constant a,

$$s_c = a \pm \rho.$$

Regarding the line PP_c as a rigid bar that rolls (without sliding) on the evolute, we now see that the end P of the bar traces out the original curve. In

* For a full discussion of this subject, see A. Ostrowski, Über die Evoluten von endlichen Ovalen, *Journal für die reine und angewandte Mathematik,* **198** (1957), pp. 14–27.

other words, the locus of P is an *involute* of the locus of P_c. We say "an involute" rather than "the involute" because different choices of the tracing point on the rolling bar yield an infinite family of "parallel" curves, each of which is an involute.

By a change of notation (from $\mathbf{r}_c$, s_c, $\mathbf{t}_c$ to $\mathbf{r}$, s, $\mathbf{t}$) we can assert that the position vector of a point which traces out an involute of a given curve is

$$\mathbf{r} + (a - s)\mathbf{t}.$$

To find $\mathbf{t}$, $\mathbf{n}$, and κ for a particular curve, the procedure that we applied to a circle (just after 17.34) is usually effective whenever the Cartesian coordinates are given in terms of a parameter. However, in the case of a central conic, the best way to obtain the evolute is as the envelope of normals. (See Ex. 3 at the end of § 8.5.)

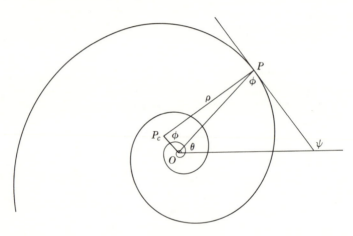

Figure 17.3b

For curves given in terms of polar coordinates, a more geometrical procedure may be desirable. For instance, to locate the center of curvature P_c at any point P on the equiangular spiral 8.71, we observe (Figure 17.3b) that $\psi = \theta + \phi$. Since also

$$\frac{dr}{ds} = \cos \phi \quad \text{and} \quad \frac{dr}{d\theta} = r \cot \phi,$$

we have

$$\kappa = \frac{d\psi}{ds} = \frac{d\theta}{ds} = \frac{d\theta}{dr}\frac{dr}{ds} = \frac{\sin \phi}{r},$$

so that

$$PP_c = \rho = r \csc \phi.$$

Thus OP_c is orthogonal to OP [Lamb **2,** p. 337] and P_c is (r_c, θ_c) where

$$r_c = r \cot \phi, \quad \theta_c = \theta + \tfrac{1}{2}\pi.$$

Since $r = r_c \tan \phi$ and $\theta = \theta_c - \frac{1}{2}\pi$, the evolute has the equation

$$r \tan \phi = a\mu^{\theta - \frac{1}{2}\pi}.$$

Since

$$\tan \phi = \mu^{\log \tan \phi / \log \mu} = \mu^{\log \tan \phi / \cot \phi} = \mu^{\tan \phi \log \tan \phi},$$

this is equivalent to

$$r = a\mu^{\theta - \frac{1}{2}\pi - \tan \phi \log \tan \phi},$$

which shows that the evolute is derived from the original spiral by a suitable rotation. (This result could have been seen from simple geometric principles, since the dilative rotation that slides the original spiral along itself must also slide the evolute along itself.)

The spiral is *its own evolute* if the "suitable rotation" consists of n whole turns, that is, if there is a positive integer n for which

$$\tfrac{1}{2}\pi + \tan \phi \log \tan \phi = 2n\pi.$$

This happens if $\tan \phi$ satisfies the transcendental equation

$$x \log x = (2n - \tfrac{1}{2})\pi.$$

From a table of natural logarithms we see that there is a unique solution for each positive integer n. The values $n = 1$ and $n = 2$ yield $\phi = 74° \ 39'$ [Cundy and Rollett **1**, p. 64] and $\phi = 80° \ 41'$. When n increases, ϕ approaches $90°$ and the spiral acquires a smaller and smaller "pitch."

EXERCISES

1. Find the evolute of the cycloid

$$x = u + \sin u, \qquad y = 1 + \cos u.$$

(*Hint:* $\mathbf{t} = (\cos \tfrac{1}{2}u, -\sin \tfrac{1}{2}u)$. A synthetic treatment is given by Lamb [**2**, pp. 351–352].)

2. Find the involute of the circle

$$x = \cos u, \qquad y = \sin u,$$

beginning at the point where $u = 0$.

3. From "simple geometric principles," the radius of curvature of an equiangular spiral is proportional to the arc s, measured from the origin. In fact,

$$\rho = s \cot \phi.$$

17.4   THE CATENARY

The *catenary* is an infinite curve, the idealized shape of a uniform chain hanging freely under the action of gravity. The curve evidently lies in a plane, which we may take to be the (x, y)-plane with the y-axis vertical, as in Figure 17.4a. Let W denote the weight of a unit length of the chain. We

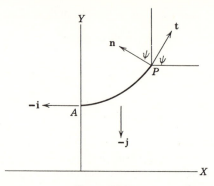

Figure 17.4a

consider the forces acting on arc AP, where A is the lowest point ($s = 0$) and P is at distance s, measured along the curve. The tangent $\mathbf{t}$ at P makes a certain angle ψ with the x-axis $\mathbf{i}$, and the normal $\mathbf{n}$ makes the same angle with the y-axis $\mathbf{j}$, so that

$$\mathbf{i} \cdot \mathbf{t} = \mathbf{j} \cdot \mathbf{n} = \cos\psi, \quad \mathbf{i} \cdot \mathbf{n} = -\sin\psi.$$

By considering various points P on the same chain, we may regard the inclination ψ as a function of the arc s, or vice versa, while conditions at A remain constant. The three forces that act on the arc AP are: the tension T at P, acting along the tangent $\mathbf{t}$, the tension Wa (equivalent to the weight of a certain length a of the chain) along the tangent $-\mathbf{i}$ at A, and the weight Ws in the direction $-\mathbf{j}$. Since these three forces are in equilibrium, we have

$$T\mathbf{t} - Wa\mathbf{i} - Ws\mathbf{j} = \mathbf{0}.$$

To eliminate the unknown (and uninteresting) tension T, we take the inner product with $\mathbf{n}$, obtaining

$$Wa \sin\psi - Ws \cos\psi = 0,$$

whence

17.41 $$s = a \tan\psi.$$

This equation, expressing the arc-length as a function of the inclination ψ, is called the *intrinsic* equation of the catenary. To deduce the Cartesian equation [cf. Lamb **2,** p. 290] we observe that

$$dx = ds \cos\psi, \qquad dy = ds \sin\psi$$

(Figure 8.5a) and make the "Gudermannian substitution"

$$\cosh u = \sec\psi, \qquad \sinh u = \tan\psi$$

(Figure 17.4b), which implies

$$\sinh u \, du = \sec\psi \tan\psi \, d\psi,$$
$$du = \sec\psi \, d\psi.$$

Differentiating 17.41, we obtain $ds = a \sec^2 \psi \, d\psi$, whence

$$dx = ds \cos \psi = a \sec \psi \, d\psi = a \, du,$$
$$dy = ds \sin \psi = a \sec \psi \tan \psi \, d\psi = a \, d(\sec \psi) = a \, d(\cosh u).$$

Taking the lowest point A (where $s = 0$, $\psi = 0$, and $u = 0$) to be $(0, a)$, we deduce

$$x = au, \qquad y = a \cosh u$$

or, in a single equation,

17.42
$$y = a \cosh \tfrac{x}{a}.$$

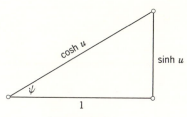

Figure 17.4b

EXERCISES

1. A uniform chain OP is held at P and hangs over a smooth peg at A, so placed that the chain just above A is horizontal and the peg gives it a right-angled bend. If the part of the chain from A to P is in the position indicated in Figure 17.4a, where is the free end O?

2. For the catenary, $s = a \sinh u$ and $\rho = a \cosh^2 u$.

3. Deduce 17.42 from $\dfrac{ds}{dx} = \sec \psi = \left[1 + \left(\dfrac{s}{a}\right)^2\right]^{\frac{1}{2}}$ and $\dfrac{dy}{dx} = \dfrac{s}{a}$.

4. Obtain intrinsic equations for (a) the cycloid $x = u + \sin u$, $y = \cos u$; (b) the parabola $y^2 = 2lx$.

5. Use the Gudermannian substitution to work out $\int \sec \psi \, d\psi$.

17.5 THE TRACTRIX

Let us now investigate the involute of the catenary, unwinding from its "lowest" point A, as in Figure 17.5a [Steinhaus **2**, pp. 212–213]. Since the position vector of the general point P on the catenary is

$$\mathbf{r} = (au, a \cosh u) = a(u, \cosh u),$$

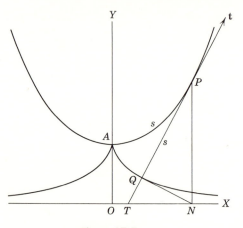

Figure 17.5a

where u is given in terms of s by the relation $s = a \sinh u$, the unit tangent vector is given by

$$a \cosh u \; \mathbf{t} = \frac{ds}{du}\frac{d\mathbf{r}}{ds} = \frac{d\mathbf{r}}{du} = a(1, \; \sinh u),$$
$$\mathbf{t} = (\operatorname{sech} u, \; \tanh u),$$

and the position vector of the general point Q on the involute is

$$\mathbf{r} - s\mathbf{t} = a(u, \; \cosh u) - a \sinh u \,(\operatorname{sech} u, \; \tanh u)$$
$$= a(u - \tanh u, \; \operatorname{sech} u).$$

Thus the involute, which is known as the *tractrix,* has the parametric equations

17.51 $x = a(u - \tanh u), \qquad y = a \operatorname{sech} u$

[Lamb **2,** p. 325], from which there is no advantage in trying to eliminate u.

Since the unit normal vector at P to the catenary is $(-\tanh u, \operatorname{sech} u)$, the unit tangent vector at Q to the tractrix is $(\tanh u, \; -\operatorname{sech} u)$, and the position vector of the point N at distance a along it is

$$a(u - \tanh u, \; \operatorname{sech} u) + a(\tanh u, \; -\operatorname{sech} u) = (au, 0).$$

Thus the length of this tangent QN, from its point of contact to its intersection with the x-axis, has the constant value a.   This is the property that gives the tractrix its name: if the (x, y)-plane is horizontal and you walk along the x-axis dragging a stone (originally at A) by means of a string of length a, the path of the stone is the tractrix.   The x-axis is clearly an asymptote.

Another way of expressing the same property is that the tractrix is an orthogonal trajectory of a system of congruent circles whose centers lie on a straight line. E. H. Lockwood* has developed this idea into an approximate construction for both the tractrix and the catenary.

EXERCISE

Compute ρ for the tractrix. What is its value at the "cusp" A, where $u = 0$?

17.6 TWISTED CURVES

We saw, in 7.52, that every displacement is a rotation or a translation or a twist (that is, the product of a rotation and a translation). As G. Mozzi remarked in 1763, this description evidently holds not only for a finite displacement but also for a continuous displacement: in the most general motion of a rigid body, there is at each instant a definite screw axis. In the case of a pure rotation, or of the motion of a screw in its nut, this axis remains invariant; but in general it is continually changing. For instance, the instantaneous axis of a wheel rolling along a road is not the line of the axle (which is moving as fast as the vehicle) but a parallel line on the road.

Any rotation may be described by its effect on a variable orthogonal trihedron of unit vectors which, for reasons that will appear a little later, we denote by **tpb,** so that

17.61
$$\mathbf{t}^2 = \mathbf{p}^2 = \mathbf{b}^2 = 1, \qquad \mathbf{p \cdot b} = \mathbf{b \cdot t} = \mathbf{t \cdot p} = 0,$$
$$\mathbf{t} = \mathbf{p \times b}, \quad \mathbf{p} = \mathbf{b \times t}, \quad \mathbf{b} = \mathbf{t \times p}, \quad [\mathbf{t\,p\,b}] = 1$$

(cf. 17.15). We regard these unit vectors as functions of a parameter s. Since the derivative of any unit vector is in a perpendicular direction, the derivative of each of **t, p, b** lies in the plane of the other two and is a linear combination of them. Differentiating the relation $\mathbf{p \cdot b} = 0$, we see that the coefficient of **p** in the expression for $\dot{\mathbf{b}}$ differs only in sign from the coefficient of **b** in the expression for $\dot{\mathbf{p}}$; similarly for the other pairs of vectors. Hence, for suitable numbers κ, λ, τ (functions of s), we have

17.62
$$\dot{\mathbf{t}} = \kappa\mathbf{p} - \lambda\mathbf{b}, \quad \dot{\mathbf{p}} = \tau\mathbf{b} - \kappa\mathbf{t}, \quad \dot{\mathbf{b}} = \lambda\mathbf{t} - \tau\mathbf{p}.$$

These derivatives are conveniently expressed in terms of *Darboux's vector*

$$\mathbf{d} = \tau\mathbf{t} + \lambda\mathbf{p} + \kappa\mathbf{b}.$$

For, we easily verify that

$$\dot{\mathbf{a}} = \mathbf{d \times a},$$

where $\mathbf{a} = \mathbf{t}$ or **p** or **b** or any other vector rigidly attached to the moving trihedron [cf. Kreyszig **1**, p. 44]. We may even omit the variable vector **a** and write "symbolically"

* *Mathematical Gazette,* **43** (1959), pp. 117–118.

17.63 $$\frac{d}{ds} = \mathbf{d} \times.$$

At any point P on a twisted curve, the unit tangent vector $\mathbf{t} = \dot{\mathbf{r}}$ can be defined in the same way as for a plane curve. But instead of a unique normal orthogonal to $\mathbf{t}$, we have a *normal plane* containing a whole flat pencil of normals. Among the unit normal vectors, we give special names to two: the *principal normal* $\mathbf{p}$, in the direction of $\dot{\mathbf{t}}$, and the *binormal*

$$\mathbf{b} = \mathbf{t} \times \mathbf{p},$$

perpendicular to the plane $\mathbf{tp}$. Since this plane contains the derivative of $\mathbf{t}$ as well as $\mathbf{t}$ itself, its order of contact with the curve is higher than that of any other plane through $\mathbf{t}$. Because of the more intimate contact, we call $\mathbf{tp}$ the *osculating plane* at the point P. (It contains the directions of velocity and acceleration of a point moving along the curve [Forder **3,** p. 131].)

The formulas 17.62 for the derivatives of $\mathbf{t}$, $\mathbf{p}$, $\mathbf{b}$ are applicable, with the simplification $\lambda = 0$ due to our choice of $\mathbf{p}$ in the direction of $\dot{\mathbf{t}}$. Thus we have the *Serret-Frenet formulas*

17.64
$$\begin{aligned} \dot{\mathbf{t}} &= \kappa \mathbf{p}, \\ \dot{\mathbf{p}} &= \tau \mathbf{b} - \kappa \mathbf{t}, \\ \dot{\mathbf{b}} &= \quad - \tau \mathbf{p}, \end{aligned}$$

which may be epitomized in the form 17.63 with

17.65 $$\mathbf{d} = \kappa \mathbf{b} + \tau \mathbf{t}.$$

The coefficients κ and τ are called the *curvature* and *torsion* of the curve (at P).

When κ is constantly zero, $\mathbf{t}$ never changes and the "curve" is a straight line. As the name "curvature" suggests, κ measures the rate at which any nonstraight curve tends to depart from its tangent. Like a plane curve, a twisted curve has a *circle of curvature* of radius $1/\kappa$, which lies in the osculating plane and has its center on the principal normal; that is, the position vector of its center is $\mathbf{r} + \rho \mathbf{p}$, where $\rho = 1/\kappa$ is the radius of curvature.

When τ is constantly zero, the osculating plane never changes, and we have a *plane* curve, with $\mathbf{n} = \mathbf{p}$. The torsion (so named by L. I. Vallée in 1825) measures the rate at which a twisted curve tends to depart from its osculating plane.

The formulas 17.64 were first given by Serret (1851) and Frenet (1852) without the vector notation, that is, as formulas for the derivatives of the direction cosines of the tangent, principal normal, and binormal. Combining them with

$$\dot{\mathbf{r}} = \mathbf{t},$$

we obtain $\qquad \ddot{\mathbf{r}} = \kappa \mathbf{p}, \qquad \dddot{\mathbf{r}} = \dot{\kappa} \mathbf{p} + \kappa(\tau \mathbf{b} - \kappa \mathbf{t}),$

whence $\qquad |\ddot{\mathbf{r}}| = \kappa, \qquad [\dot{\mathbf{r}}\, \ddot{\mathbf{r}}\, \dddot{\mathbf{r}}] = \kappa^2 \tau.$

EXERCISES

1. For a curve drawn on a sphere, the center of the circle of curvature at any point is the foot of the perpendicular from the center of the sphere upon the osculating plane at the point.

2. The tangent to the locus of the center of the circle of curvature of any curve is perpendicular to the tangent at the corresponding point on the original curve.

3. For any twisted curve, $[\dot{\mathbf{t}}\ \ddot{\mathbf{t}}\ \dddot{\mathbf{t}}] = \kappa^3(\kappa\dot{\tau} - \dot{\kappa}\tau) = \kappa^5 \dfrac{d}{ds}\left(\dfrac{\tau}{\kappa}\right),$

$$[\dot{\mathbf{b}}\ \ddot{\mathbf{b}}\ \dddot{\mathbf{b}}] = \tau^3(\dot{\kappa}\tau - \kappa\dot{\tau}) = \tau^5 \dfrac{d}{ds}\left(\dfrac{\kappa}{\tau}\right).$$

17.7   THE CIRCULAR HELIX

As we saw in § 8.7, the locus of a point moving in a plane under the action of a continuous dilative rotation is an equiangular spiral. Analogously, the locus of a point moving in space under the action of a continuous twist is a *circular helix* (§ 11.5). In terms of *cylindrical* coordinates (r, θ, z), defined by

$$x = r \cos \theta, \quad y = r \sin \theta, \quad z \text{ as usual,}$$

a twist along and around the z-axis is

$$(r, \theta, z) \rightarrow (r, \quad \theta + u, \quad z + uc):$$

the product of the rotation $\theta \rightarrow \theta + u$ and the translation $z \rightarrow z + uc$. Applying this twist to the point $(a, 0, 0)$, we obtain (a, u, uc). Thus the circular helix has the parametric equations

$$r = a, \qquad \theta = u, \qquad z = uc,$$

or

$$x = a \cos u, \quad y = a \sin u, \quad z = cu$$

[Weatherburn **2,** p. 16]. In other words, the helix, which is the shape of the rail of a "spiral" staircase, has the equations

$$r = a, \qquad z = c\theta$$

or

$$x^2 + y^2 = a^2, \qquad \frac{y}{x} = \tan \frac{z}{c},$$

which express it as the curve of intersection of two surfaces: the circular cylinder

$$r = a, \quad \text{or} \quad x^2 + y^2 = a^2,$$

and the *helicoid*

$$z = c\theta, \quad \text{or} \quad \frac{y}{x} = \tan \frac{z}{c},$$

which is the shape of the ceiling of the staircase (or of a propellor blade) [Steinhaus **2**, p. 196].

Differentiating

$$\mathbf{r} = (a \cos u, a \sin u, cu)$$

with respect to s, we obtain

$$\mathbf{t} = \dot{u} (-a \sin u, a \cos u, c).$$

Since this must be a *unit* vector, we have

$$\dot{u} = 1/\sqrt{a^2 + c^2},$$

and we shall find it convenient to retain the symbol $\dot{u}$ as a temporary abbreviation for this constant. The Serret-Frenet formulas yield

$$\kappa\mathbf{p} = \dot{\mathbf{t}} = \dot{u}^2 (-a \cos u, -a \sin u, 0) = -\dot{u}^2 a (\cos u, \sin u, 0),$$

$$\kappa = \dot{u}^2 a = \frac{a}{a^2 + c^2},$$

$$\mathbf{p} = -(\cos u, \sin u, 0) \qquad \text{(perpendicular to the z-axis)},$$

$$\mathbf{b} = \mathbf{t} \times \mathbf{p} = \dot{u} (c \sin u, -c \cos u, a),$$

$$-\tau\mathbf{p} = \dot{\mathbf{b}} = \dot{u}^2 c (\cos u, \sin u, 0),$$

$$\tau = \dot{u}^2 c = \frac{c}{a^2 + c^2}.$$

Thus both κ and τ are constant, a result which could have been seen from first principles without any appeal to calculus, since the twist that slides the circular helix along itself transforms the curvature and torsion at one point into the same properties at another point. Conversely, since every displacement is a twist, every curve whose curvature and torsion are constant is a circular helix if we include, as limiting cases, the straight line ($\kappa = 0$, $a = 0$) and the circle ($\tau = 0$, $c = 0$).

When κ and τ are constant, Darboux's vector 17.65, being rigidly attached to the moving trihedron, is one of the vectors to which 17.63 is applicable. Thus

$$\dot{\mathbf{d}} = \mathbf{d} \times \mathbf{d} = \mathbf{0},$$

and $\mathbf{d}$ is constant. In fact, just as the motion of the tangent at a point describing a plane curve is, at each instant, a rotation about the center of curvature, so the motion of the **tpb** trihedron at a point describing a twisted curve is, at each instant, a twist about a certain line (*screw axis*) in the direction of Darboux's vector. In the case of the plane curve, the center of curvature appeared as the center of the circle of "closest fit": having the same tangent and curvature as the given curve. Analogously for the twisted curve, the screw axis can be obtained as the axis of the circular helix of closest fit: having the same **tpb** and the same curvature and torsion. Thus the screw axis is the line in the direction $\kappa\mathbf{b} + \tau\mathbf{t}$ through the point whose position vector is

$$\mathbf{r} + a\mathbf{p},$$

where a, being the radius of the circular cylinder containing the helix, is obtained by eliminating c from the equations

$$\kappa = \frac{a}{a^2 + c^2}, \qquad \tau = \frac{c}{a^2 + c^2};$$

in fact,

$$a = \frac{\kappa}{\kappa^2 + \tau^2}.$$

In the case of a plane curve we have $\tau = 0$, $a = \rho$, the position vector $\mathbf{r} + a\mathbf{p}$ becomes $\mathbf{r} + \rho\mathbf{n}$, and Darboux's vector becomes $\kappa\mathbf{b}$: perpendicular to the plane of the curve.

EXERCISES

1. The orthogonal projection of the circular helix on a plane through its axis, such as the plane $x = 0$, is the *sine curve*

$$y = a \sin \frac{z}{c}.$$

2. Describe the surface formed by the midpoints of all the chords of a circular helix.

3. The locus of centers of circles of curvature of a circular helix H is another circular helix H', and the locus of centers of circles of curvature of H' is H itself. For what value of c/a (or τ/κ) will H and H' be congruent? (It is, of course, sufficient to consider a single point on H, such as the point where $u = 0$.)

17.8 THE GENERAL HELIX

We have seen that the circular helix is characterized by its property of having constant curvature and constant torsion. It is a special case of the general *helix*, which may be defined either as a curve whose curvature and torsion are in a constant ratio or as a curve whose tangent makes a constant angle with a fixed vector. We proceed to prove the equivalence of these two definitions.

Suppose first that the curvature and torsion are in a constant ratio (i.e., a ratio independent of s), say

$$\tau = c\kappa.$$

Then

$$\dot{\mathbf{t}} = \kappa\mathbf{p}, \qquad \dot{\mathbf{b}} = -\tau\mathbf{p} = -c\kappa\mathbf{p},$$
$$c\dot{\mathbf{t}} + \dot{\mathbf{b}} = \mathbf{0}.$$

Since this is the derivative of $c\mathbf{t} + \mathbf{b}$, the latter is a fixed vector, say $\mathbf{a}$, which makes a constant angle with $\mathbf{t}$ since

$$\mathbf{a} \cdot \mathbf{t} = (c\mathbf{t} + \mathbf{b}) \cdot \mathbf{t} = c.$$

Conversely, suppose $\mathbf{t}$ makes a constant angle β with a fixed unit vector $\mathbf{k}$. Differentiating the equation

$$\mathbf{t} \cdot \mathbf{k} = \cos \beta,$$

we obtain
$$\kappa \mathbf{p} \cdot \mathbf{k} = 0.$$

Assuming that $\kappa \neq 0$, we have $\mathbf{p} \cdot \mathbf{k} = 0$, so that the constant vector $\mathbf{k}$ lies in the **bt** plane and makes complementary angles with $\mathbf{b}$ and $\mathbf{t}$. Since $\mathbf{t} \cdot \mathbf{k} = \cos \beta$, we have also

$$\mathbf{b} \cdot \mathbf{k} = \sin \beta.$$

Differentiating the equation $\mathbf{p} \cdot \mathbf{k} = 0$, we obtain

$$(\tau \mathbf{b} - \kappa \mathbf{t}) \cdot \mathbf{k} = 0,$$

$$\tau \sin \beta - \kappa \cos \beta = 0,$$

$$\frac{\kappa}{\tau} = \tan \beta.$$

Lines in the constant direction $\mathbf{k}$ through all the points of the curve generate a (general) cylinder. Thus the helix can alternatively be described as a curve drawn on a cylinder in such a way as to cut the generators at a constant angle. In other words, it can be obtained by drawing a straight line obliquely on a sheet of paper and then wrapping the paper on the cylinder.

EXERCISES

1. Using Darboux's operator 17.63 to differentiate the constant vector $\mathbf{k}$, obtain

$$\mathbf{d} \times \mathbf{k} = \mathbf{0}.$$

Deduce that Darboux's vector $\mathbf{d} = \kappa \mathbf{b} + \tau \mathbf{t}$ is parallel to $\mathbf{k}$: its direction is constant (though its length, $\sqrt{\kappa^2 + \tau^2}$, may vary).

2. Find κ and τ for the curve

$$x = 3u - u^3, \quad y = 3u^2, \quad z = 3u + u^3,$$

and deduce that this curve is a helix.

17.9 THE CONCHO-SPIRAL

The spirals described on shells, and called concho-spirals, are such as would result from winding plane logarithmic spirals on cones.

Henry Moseley (1801 -1872)

[Moseley **1**, p. 301]

The two most interesting helices are as follows: (1) the circular helix, which is the locus of a point under the action of a continuous twist, so that its curvature and torsion are constant; and (2) the *concho-spiral*, which is the locus of a point under the action of a continuous dilative rotation, so that its curvature and torsion are both inversely proportional to its arc s, measured from the apex O of the cone on which it evidently lies (cutting the generators at a constant angle). A considerable arc of this curve can be seen on the shell *Turritella duplicata* [Weyl **1**, p. 68]. Architectural applications appear on spires in Copenhagen, notably that of the Stock Exchange Building, where the tails of four dragons are twisted together.

In terms of cylindrical coordinates, a dilative rotation round the z-axis, say

$$(r, \theta, z) \to (\mu^u r, \quad \theta + u, \quad \mu^u z),$$

applied to the point $(a, 0, c)$, yields the concho-spiral

$$r = \mu^u a, \quad \theta = u, \quad z = \mu^u c.$$

To see how the circular helix can arise as a limiting form of the concho-spiral, we change the origin by writing $z + c$ for z, and then make c tend to infinity and μ to 1 in such a manner that $(\mu - 1)c$ approaches a finite number b. Instead of $r = \mu^u a$ and $z = \mu^u c$, we have $r = a$ and

$$z = (\mu^u - 1) c = \frac{\mu^u - 1}{\mu - 1} (\mu - 1)c \to ub.$$

Thus the limiting form is the circular helix

$$r = a, \quad \theta = u, \quad z = ub.$$

EXERCISES

1. Express the parametric equations for the concho-spiral in terms of Cartesian coordinates.

2. Verify from these equations that the tangent **t** to the concho-spiral makes a constant angle with the z-axis.

3. Obtain a formula for the angle at which the concho-spiral cuts the generators of the cone

$$\frac{r}{a} = \frac{z}{c}.$$

4. A familiar model for a cone of revolution is obtained by cutting out a circular sector from a sheet of paper and rolling it up so that the center of the circle becomes the vertex of the cone. The angle α of the sector and the semivertical angle β of the cone are connected by the formula

$$\alpha = 2\pi \sin \beta;$$

for example, the sector is a semicircle if $\beta = \pi/6$. If $\sin \beta = 1/n$, where n is an integer greater than 1, the unfolded form of any concho-spiral on the cone consists of a sequence of arcs belonging to n equiangular spirals.

5. Like any other helix, the concho-spiral lies on a cylinder and cuts the generators at a constant angle. What kind of cylinder is this in the present case?

18

The tensor notation

In this interlude between the differential geometry of curves (Chapter 17) and the differential geometry of surfaces (Chapter 19) we introduce Ricci's famous notation, which is both suggestive and economical. (Without its aid the general theory of relativity could hardly have been formulated.) One of its simplest applications has no direct connection with differential geometry: "reciprocal lattices" are used in both x-ray crystallography (§ 18.3) and the geometry of numbers (§ 18.4).

18.1 DUAL BASES

The tensor method . . . has the great advantage that it is not a new notation, but a concise way of writing the ordinary notation.

Harold Jeffreys (1891 -)

[Jeffreys **1,** Preface]

As a basis for our vector space (or a frame for affine coordinates), instead of **e**, **f**, **g** (as in 17.11) or **i**, **j**, **k** (as in 17.14), it is more systematic to write $\mathbf{r}_1$, $\mathbf{r}_2$, $\mathbf{r}_3$. Along with this set of three independent vectors we use also the *dual basis* $\mathbf{r}^1$, $\mathbf{r}^2$, $\mathbf{r}^3$, defined in terms of $\mathbf{r}_1$, $\mathbf{r}_2$, $\mathbf{r}_3$ by the equation

18.11 $$\mathbf{r}^\alpha \cdot \mathbf{r}_\beta = \delta^\alpha_\beta,$$

where the *Kronecker delta* δ^α_β is a useful symbol which means 1 or 0 according as α and β are equal or unequal. (The "1, 2, 3" of the dual basis are not exponents but "upper indices" or "superscripts," analogous to the subscripts used in the original basis.) Thus $\mathbf{r}^1$ is perpendicular to the plane $\mathbf{r}_2\mathbf{r}_3$ and its length is adjusted so that $\mathbf{r}^1 \cdot \mathbf{r}_1 = 1$; similarly for $\mathbf{r}^2$ and $\mathbf{r}^3$. Each $\mathbf{r}^\alpha$, being perpendicular to two $\mathbf{r}_\beta$'s, may be expressed as an outer product:

18.12 $$\mathbf{r}^1 = \frac{\mathbf{r}_2 \times \mathbf{r}_3}{J}, \quad \mathbf{r}^2 = \frac{\mathbf{r}_3 \times \mathbf{r}_1}{J}, \quad \mathbf{r}^3 = \frac{\mathbf{r}_1 \times \mathbf{r}_2}{J},$$

where, since $\mathbf{r}^\alpha \cdot \mathbf{r}_\alpha = 1$,

18.13 $$J = [\mathbf{r}_1 \ \mathbf{r}_2 \ \mathbf{r}_3].$$

Since the basic vectors $\mathbf{r}_\alpha$ are independent, $J \neq 0$. By 17.17, we have

$$[\mathbf{r}^1\,\mathbf{r}^2\,\mathbf{r}^3]\,[\mathbf{r}_1\,\mathbf{r}_2\,\mathbf{r}_3] = \begin{vmatrix} \mathbf{r}^1\cdot\mathbf{r}_1 & \mathbf{r}^1\cdot\mathbf{r}_2 & \mathbf{r}^1\cdot\mathbf{r}_3 \\ \mathbf{r}^2\cdot\mathbf{r}_1 & \mathbf{r}^2\cdot\mathbf{r}_2 & \mathbf{r}^2\cdot\mathbf{r}_3 \\ \mathbf{r}^3\cdot\mathbf{r}_1 & \mathbf{r}^3\cdot\mathbf{r}_2 & \mathbf{r}^3\cdot\mathbf{r}_3 \end{vmatrix} = \begin{vmatrix} 1 & 0 & 0 \\ 0 & 1 & 0 \\ 0 & 0 & 1 \end{vmatrix} = 1,$$

so that

18.14 $[\mathbf{r}^1\,\mathbf{r}^2\,\mathbf{r}^3] = J^{-1}$

and the dual basic vectors are likewise independent. Interchanging "uppers and lowers" in 18.12, we obtain

18.15 $\mathbf{r}_1 = J\mathbf{r}^2 \times \mathbf{r}^3, \quad \mathbf{r}_2 = J\mathbf{r}^3 \times \mathbf{r}^1, \quad \mathbf{r}_3 = J\mathbf{r}^1 \times \mathbf{r}^2.$

EXERCISE

Deduce 18.14 from 18.12 without using 17.17. (*Hint:* Apply 17.13 to $\mathbf{r}_2, \mathbf{r}_3, \mathbf{r}^2, \mathbf{r}^3$.)

18.2   THE FUNDAMENTAL TENSOR

Any vector $\mathbf{u}$ may be expressed as $\sum u_\alpha\,\mathbf{r}^\alpha$ (meaning $u_1\,\mathbf{r}^1 + u_2\,\mathbf{r}^2 + u_3\,\mathbf{r}^3$) or as $\sum u^\alpha\,\mathbf{r}_\alpha$. The *covariant** components u_α and *contravariant* components u^α are simply the inner products $\mathbf{u}\cdot\mathbf{r}_\alpha$ and $\mathbf{u}\cdot\mathbf{r}^\alpha$; for

$$\mathbf{u}\cdot\mathbf{r}_\beta = \sum u_\alpha\,\mathbf{r}^\alpha\cdot\mathbf{r}_\beta = \sum u_\alpha\,\delta^\alpha_\beta = u_\beta$$

and

$$\mathbf{u}\cdot\mathbf{r}^\beta = \sum u^\alpha\,\mathbf{r}_\alpha\cdot\mathbf{r}^\beta = \sum u^\alpha\,\delta^\beta_\alpha = u^\beta$$

(In such expressions as $\sum u_\alpha\,\mathbf{r}^\alpha\cdot\mathbf{r}_\beta$, it is understood that the summation is taken over the index α that appears twice, once "up" and once "down," and not over the index β which only appears once. The sum $\sum\delta^\alpha_\beta\,u_\alpha$ involves three values of α, one of which must be equal to β, so the "incomplete symbol" $\sum\delta^\alpha_\beta$ serves as a substitution operator transforming u_α into u_β.)

Thus

18.21 $\mathbf{u} = \sum\mathbf{u}\cdot\mathbf{r}_\beta\,\mathbf{r}^\beta = \sum\mathbf{u}\cdot\mathbf{r}^\beta\,\mathbf{r}_\beta.$

In particular, this holds when $\mathbf{u} = \mathbf{r}_\alpha$ or $\mathbf{r}^\alpha$, in which cases the components $u_\beta = \mathbf{u}\cdot\mathbf{r}_\beta$ and $u^\beta = \mathbf{u}\cdot\mathbf{r}^\beta$ are denoted by

18.22 $g_{\alpha\beta} = \mathbf{r}_\alpha\cdot\mathbf{r}_\beta, \qquad g^{\alpha\beta} = \mathbf{r}^\alpha\cdot\mathbf{r}^\beta,$

so that

18.221 $\mathbf{r}_\alpha = \sum g_{\alpha\beta}\,\mathbf{r}^\beta, \qquad \mathbf{r}^\alpha = \sum g^{\alpha\beta}\,\mathbf{r}_\beta.$

For the whole story of the subtleties that underlie the terms *covariant* and *contravariant*, see Kreyszig [1**, p. 88]. The present treatment was suggested by G. Hessenberg, Vektorielle Begründung der Differentialgeometrie, *Mathematische Annalen,* **78** (1917), 187–217.*

(In the expression $\Sigma g_{\alpha\beta} \, \mathbf{r}^\beta$, it is β that appears twice, so we sum over β, obtaining $g_{\alpha 1} \, \mathbf{r}^1 + g_{\alpha 2} \, \mathbf{r}^2 + g_{\alpha 3} \, \mathbf{r}^3$.) The commutativity of inner products shows that

$$g_{\alpha\beta} = g_{\beta\alpha}, \qquad g^{\alpha\beta} = g^{\beta\alpha}.$$

The connection between the *covariant tensor* $g_{\alpha\beta}$ and the *contravariant tensor* $g^{\alpha\beta}$ is found as follows:

$$\Sigma g_{\alpha\gamma} \, g^{\alpha\beta} = \Sigma g_{\alpha\gamma} \, \mathbf{r}^\alpha \cdot \mathbf{r}^\beta = \mathbf{r}_\gamma \cdot \mathbf{r}^\beta$$

18.23
$$= \delta_\gamma^\beta.$$

Thus the two symmetric matrices $\| g_{\alpha\beta} \|$ and $\| g^{\alpha\beta} \|$ have as their product the unit matrix. The two corresponding determinants cannot vanish, since their product is 1. When the coefficients $g_{\alpha\gamma}$ are given, we have, in 18.23 for each value of β, a set of three linear equations

$$g_{1\gamma} g^{1\beta} + g_{2\gamma} g^{2\beta} + g_{3\gamma} g^{3\beta} = \delta_\gamma^\beta \qquad (\gamma = 1, 2, 3)$$

to be solved for the three unknowns $g^{1\beta}, g^{2\beta}, g^{3\beta}$. By Cramer's rule [Birkhoff and MacLane **1**, p. 286], the solution is

$$g^{\alpha\beta} = (\text{cofactor of } g_{\alpha\beta} \text{ in } G)/G, \qquad G = \det(g_{\alpha\beta}).$$

In particular, if $g_{23} = g_{31} = g_{12} = 0$, we have

$$g^{\alpha\alpha} = 1/g_{\alpha\alpha}$$

and $g^{23} = g^{31} = g^{12} = 0$.

To express either kind of components of a vector $\mathbf{u}$ in terms of the other kind, we have

$$u_\alpha = \mathbf{u} \cdot \mathbf{r}_\alpha = \mathbf{u} \cdot \Sigma g_{\alpha\beta} \, \mathbf{r}^\beta$$

18.24
$$= \Sigma g_{\alpha\beta} \, u^\beta,$$

and similarly, $u^\alpha = \Sigma g^{\alpha\beta} u_\beta$. The inner product of two vectors

$$\mathbf{u} = \Sigma u^\alpha \, \mathbf{r}_\alpha = \Sigma u_\alpha \, \mathbf{r}^\alpha \quad \text{and} \quad \mathbf{v} = \Sigma v^\beta \, \mathbf{r}_\beta = \Sigma v_\beta \, \mathbf{r}^\beta$$

may be expressed in various ways as a bilinear form:

$$\mathbf{u} \cdot \mathbf{v} = \Sigma u^\alpha \, \mathbf{r}_\alpha \cdot \mathbf{v} = \Sigma u^\alpha \, v_\alpha,$$
$$\mathbf{u} \cdot \mathbf{v} = \Sigma u_\alpha \, \mathbf{r}^\alpha \cdot \mathbf{v} = \Sigma u_\alpha \, v^\alpha,$$
$$\mathbf{u} \cdot \mathbf{v} = \Sigma u^\alpha \, \mathbf{r}_\alpha \cdot \Sigma v^\beta \, \mathbf{r}_\beta = \Sigma\Sigma g_{\alpha\beta} \, u^\alpha \, v^\beta,$$
$$\mathbf{u} \cdot \mathbf{v} = \Sigma u_\alpha \, \mathbf{r}^\alpha \cdot \Sigma v_\beta \, \mathbf{r}^\beta = \Sigma\Sigma g^{\alpha\beta} \, u_\alpha \, v_\beta.$$

In particular, the length $|\mathbf{u}|$ is given by

$$|\mathbf{u}|^2 = \mathbf{u} \cdot \mathbf{u} = \Sigma u_\alpha u^\alpha$$

18.25
$$= \Sigma\Sigma g_{\alpha\beta} \, u^\alpha \, u^\beta = \Sigma\Sigma g^{\alpha\beta} \, u_\alpha u_\beta.$$

Let us regard **u** and **v** as the position vectors of points having contravariant coordinates (u^1, u^2, u^3) and covariant coordinates (v_1, v_2, v_3), respectively. If **u** is fixed while **v** varies, the two equations

18.26 $$\mathbf{u} \cdot \mathbf{v} = 0, \qquad \mathbf{u} \cdot \mathbf{v} = 1$$

define respectively the plane through the origin perpendicular to **u**, and the parallel plane whose distance from the origin, being the length of the projection of **v** on the unit vector $\mathbf{u}/|\mathbf{u}|$, is

$$\frac{\mathbf{u}}{|\mathbf{u}|} \cdot \mathbf{v} = \frac{1}{|\mathbf{u}|}.$$

The latter plane, passing through the inverse of (u^1, u^2, u^3) in the unit sphere

$$\mathbf{v} \cdot \mathbf{v} = 1,$$

is the *polar plane of* (u^1, u^2, u^3) with respect to the sphere. (See § 8.8.)

It is sometimes convenient to express the basic vectors in terms of Cartesian coordinates:

$$\mathbf{r}_\alpha = x_\alpha \mathbf{i} + y_\alpha \mathbf{j} + z_\alpha \mathbf{k}.$$

Then, by 18.22 and 17.151,

18.27 $$g_{\alpha\beta} = \mathbf{r}_\alpha \cdot \mathbf{r}_\beta = x_\alpha x_\beta + y_\alpha y_\beta + z_\alpha z_\beta,$$

18.28 $$J = [\mathbf{r}_1\, \mathbf{r}_2\, \mathbf{r}_3] = \begin{vmatrix} x_1 & y_1 & z_1 \\ x_2 & y_2 & z_2 \\ x_3 & y_3 & z_3 \end{vmatrix},$$

and the determinant of the fundamental tensor is

18.29 $$G = \begin{vmatrix} g_{11} & g_{12} & g_{13} \\ g_{21} & g_{22} & g_{23} \\ g_{31} & g_{32} & g_{33} \end{vmatrix} = \begin{vmatrix} x_1 & y_1 & z_1 \\ x_2 & y_2 & z_2 \\ x_3 & y_3 & z_3 \end{vmatrix} \begin{vmatrix} x_1 & x_2 & x_3 \\ y_1 & y_2 & y_3 \\ z_1 & z_2 & z_3 \end{vmatrix} = J^2.$$

EXERCISES

1. $\mathbf{u} \cdot \mathbf{v} = u^1 v_1 + u^2 v_2 + u^3 v_3.$

2. $|\mathbf{u}|^2 = g_{11}(u^1)^2 + g_{22}(u^2)^2 + g_{33}(u^3)^2 + 2g_{23}\, u^2 u^3 + 2g_{31}\, u^3 u^1 + 2g_{12}\, u^1 u^2.$

Give the corresponding expression for $\mathbf{u} \cdot \mathbf{v}$.

3. Use 18.12, 18.221 and 18.15 to prove 17.12 in the form

$$(\mathbf{r}_1 \times \mathbf{r}_2) \times \mathbf{r}_3 = g_{13}\mathbf{r}_2 - g_{23}\mathbf{r}_1.$$

4. $\Sigma\, \mathbf{r}^\alpha \times \mathbf{r}_\alpha = \mathbf{0}.$

5. Express $\det (g^{\alpha\beta})$ in terms of $G = \det (g_{\alpha\beta})$.

18.3   RECIPROCAL LATTICES

It can fairly be said that the reciprocal lattice provides one of the most important tools available in the study of the diffraction of x-rays by crystals.

M. J. Buerger [**1**, p. 107]

The study of x-ray diffraction has confirmed the notion that the symmetrical appearance of a crystal is a result of the symmetrical pattern formed by its atoms or molecules.   In other words, there is an infinite group of symmetry operations transforming the pattern (regarded as extending throughout the whole space) into itself.   These operations may or may not include rotations, reflections, glide reflections, rotatory reflections or screw displacements, but in any case the translations contained in the symmetry group form a nonempty normal subgroup. This translation subgroup determines a lattice whose unit cell contains one or more atoms.   The arrangement of atoms in the first cell determines their arrangement in all the other cells derived by the translations.   If the cell contains only one atom, we naturally choose the origin at the center of such an atom; then instead of an atom inside each cell, we shall have one at every vertex, that is, at every lattice point.   But if the cell contains several atoms there will be several superposed lattices of atoms.   Thus a given crystal has a perfectly definite translation group, and the lattice becomes definite as soon as we have chosen an origin (at the center of an atom or elsewhere).   There is still a theoretically unlimited choice of unit cells, though in practice we tend to use basic vectors of roughly equal lengths.   However, the volume of the unit cell is definite, since it depends on the number of lattice points in a crystal of a given size.   In fact, any three independent vectors generate a parallelepiped, and this is a unit cell whenever it has a lattice point at each of its eight vertices but none on its edges, nor on its faces, nor inside.

The affine theory shows that a sequence of "rational planes" 13.93 can be chosen in infinitely many ways.   In Euclidean geometry these are no longer indistinguishable: each sequence has its *interplanar spacing*, which can be measured as the distance from the origin to the "first" plane

18.31 $$Xx + Yy + Zz = 1.$$

Each sequence of planes contains all the lattice points.   Hence, when we compare one such sequence with another, the interplanar spacing is proportional to the density of distribution of lattice points in one plane of the sequence.   This idea is physically important because the face planes and cleavage planes of a crystal naturally tend to occur among the rational planes of high density.   Accordingly, we are chiefly interested in the sequences that have a relatively large interplanar spacing.   On the other hand, the most interesting visible points are those at a relatively small distance from the origin.   The two lattices which we are going to consider are related in such a way that the visible points of either are in directions perpendicular to the rational planes of the other and at distances reciprocal to the interplanar spacings [Buerger **1**, p. 117]. Thus the more important planes of either lattice will correspond to the more important points of the other.

The definition in terms of a basis is extremely simple [Coxeter **1**, p. 181] and the result is easily seen to be independent of the chosen basis. If a given lattice consists of the points whose contravariant coordinates are integers, the *reciprocal* lattice consists of the points whose covariant coordinates are integers. In other words, the position vectors are

$$\mathbf{u} = \Sigma\, u^\alpha\, \mathbf{r}_\alpha \quad \text{and} \quad \mathbf{v} = \Sigma\, v_\alpha\, \mathbf{r}^\alpha,$$

respectively, where u^α and v_α are integers and $\mathbf{r}^\alpha \cdot \mathbf{r}_\beta = \delta^\alpha_\beta$. The equation $\mathbf{u} \cdot \mathbf{v} = 1$ or

$$u^1 v_1 + u^2 v_2 + u^3 v_3 = 1$$

(see 18.26) implies that the three u's are coprime, and likewise the three v's. For each visible point (u^1, u^2, u^3) of the given lattice, this equation may be identified with 18.31: it represents a first rational plane of the reciprocal lattice; perpendicular to $\mathbf{u}$ and at the reciprocal distance $|\mathbf{u}|^{-1}$. Since the distinction between "covariant" and "contravariant" is made by an arbitrary choice, the relation between the two lattices is symmetric: the first rational planes of either are the polar planes (with respect to the unit sphere) of the visible points of the other.

The shapes of the unit cells of the two lattices are determined by the inner products 18.22. For the edge lengths of these parallelepipeds, it is convenient to use the abbreviations

18.32 $$g_\alpha = \sqrt{g_{\alpha\alpha}} = |\mathbf{r}_\alpha|, \quad g^\alpha = \sqrt{g^{\alpha\alpha}} = |\mathbf{r}^\alpha|,$$

so that the angles between pairs of adjacent edges are the angles whose cosines are

$$\frac{g_{23}}{g_2 g_3}, \frac{g_{31}}{g_3 g_1}, \frac{g_{12}}{g_1 g_2}; \frac{g^{23}}{g^2 g^3}, \frac{g^{31}}{g^3 g^1}, \frac{g^{12}}{g^1 g^2}.$$

By 18.13 and 18.14, their volumes are J and J^{-1} where, by 18.29,

$$J = \sqrt{G}, \quad G = \det\,(g_{\alpha\beta}).$$

The simplest special case is the *cubic* lattice consisting of the points whose rectangular Cartesian coordinates are integers. In this case

$$\mathbf{r}_1 = \mathbf{r}^1 = \mathbf{i}, \quad \mathbf{r}_2 = \mathbf{r}^2 = \mathbf{j}, \quad \mathbf{r}_3 = \mathbf{r}^3 = \mathbf{k},$$

the distinction between covariant and contravariant disappears, and the lattice is its own reciprocal. Other important lattices are obtained as sublattices of the simple cubic lattice, that is, by putting suitable restrictions on these integral Cartesian coordinates. By making the three coordinates of each point have an even sum, we obtain the *face-centered* cubic lattice; and by making them either all even or all odd, the *body-centered* cubic lattice. (These names refer to a larger simple lattice whose points have only even coordinates. In this larger lattice the center of a face has two odd coordinates and the center of a cell or "body" has three odd coordinates.) Each of these two lattices is similar to the reciprocal of the other; for, the bases

18.33 $\mathbf{r}_1 = (0, 1, 1)$, $\mathbf{r}_2 = (1, 0, 1)$, $\mathbf{r}_3 = (1, 1, 0)$,

18.34 $\mathbf{r}^1 = (-1, 1, 1)$, $\mathbf{r}^2 = (1, -1, 1)$, $\mathbf{r}^3 = (1, 1, -1)$

evidently satisfy a trivially modified form of 18.11, namely,

$$\mathbf{r}^\alpha \cdot \mathbf{r}_\beta = 2\,\delta^\alpha_\beta.$$

This means that they are reciprocal with respect to a sphere of radius $\sqrt{2}$. To make them reciprocal with respect to a unit sphere, we merely have to divide all the coordinates of one lattice by 2 (or all the coordinates of both by $\sqrt{2}$).

For comparison with Buerger [**1**, pp. 117–127], it is perhaps worth while to point out that his

$$\mathbf{a}, \mathbf{b}, \mathbf{c}, \mathbf{a}^*, \mathbf{b}^*, \mathbf{c}^*, a, \ b, \ c, \ a^*, b^*, c^*, \ d_{(hkl)}, \ \sigma_{hkl}, \ V, \ V^*$$

are our

$$\mathbf{r}_1, \mathbf{r}_2, \mathbf{r}_3, \mathbf{r}^1, \ \mathbf{r}^2, \ \mathbf{r}^3, \ g_1, g_2, \ g_3, \ g^1, \ g^2, \ g^3, \ |\mathbf{v}|^{-1}, \ \mathbf{v}, \quad J, J^{-1}.$$

EXERCISES

1. Consider two plane lattices derived from one another by a quarter-turn about the origin. Exhibit them as "reciprocal lattices"

$$u^1\mathbf{r}_1 + u^2\mathbf{r}_2 \quad \text{and} \quad v_1\mathbf{r}^1 + v_2\mathbf{r}^2.$$

(*Hint*: $\mathbf{r}^1$ is perpendicular to $\mathbf{r}_2$, and $\mathbf{r}^2$ to $\mathbf{r}_1$; also $\mathbf{r}^1 \cdot \mathbf{r}_1 = \mathbf{r}^2 \cdot \mathbf{r}_2$.)

2. Write out the fundamental tensors for the face-centered and body-centered cubic lattices with bases 18.33 and 18.34. Sketch the unit cells, which are rhombohedra. (The former may be regarded as a solid octahedron with a regular tetrahedron stuck on to each of two opposite faces.)

3. A lattice, in three dimensions as in two (§ 4.1), has a Dirichlet region (or Voronoi polyhedron) consisting of all the points that are as near to the origin as to any other lattice point. For the simple cubic lattice, this is a cube; for the face-centered lattice it is a rhombic dodecahedron, whose faces are twelve equal rhombi; and for the body-centered lattice it is a truncated octahedron, whose faces consist of six squares and eight regular hexagons [Steinhaus **2**, pp. 152, 157].

18.4   THE CRITICAL LATTICE OF A SPHERE

> Given a bounded region K . . . which contains the origin O as an inner
> point, we consider all those lattices which have no point except O in
> the interior of K.   The lower bound of their determinants is called the
> critical determinant of K . . . and the lattices for which this lower bound
> is attained are called the critical lattices of K.
>
> Harold Davenport (1907-1969)*

Every lattice has a certain minimum distance c between pairs of lattice points, and a certain volume J for its unit cell (§ 13.9).   This J is called the *determinant* of the lattice, because it is the determinant of the Cartesian components of the three basic vectors $\mathbf{r}_\alpha$.   We proceed to prove that, for a given value of c, the minimum value of J occurs when the lattice is face-centered cubic.†

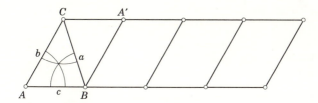

Figure 18.4a

Consider any point A of a given lattice whose unit cell has volume J. Choose a lattice point B at the minimum distance c from A, and a lattice point C outside the line AB, at the shortest distance $b\ (\geqslant c)$ from A.   These points can always be chosen so that the angle A and sides a, b, c of the triangle ABC satisfy

$$A \leqslant \tfrac{1}{2}\pi, \qquad a \geqslant b \geqslant c,$$

and therefore

$$b^2 + c^2 \geqslant a^2$$

(Figure 18.4a).   Let Δ and R denote the area and circumradius of this triangle, so that, by 1.53 and 1.55,

$$16\Delta^2 = -a^4 - b^4 - c^4 + 2b^2c^2 + 2c^2a^2 + 2a^2b^2, \quad 16\Delta^2 R^2 = b^2c^2a^2.$$

The plane ABC is, of course, a rational plane of the lattice.   In one of the two nearest parallel planes of the same system, there is a lattice point D

  * Recent progress in the geometry of numbers, *Proceedings of the International Congress of Mathematicians*, 1950, vol. I, p. 166.

  † A. P. Dempster, The minimum of a definite ternary quadratic form, *Canadian Journal of Mathematics*, **9** (1957), pp. 232–234.

whose orthogonal projection D_1 on the plane ABC is not outside the parallelogram $ABA'C$. Replacing the triangle ABC, if necessary, by the triangle $A'BC$, we may assume that D_1 is not outside the triangle ABC. (The possible change of notation involves the central inversion that interchanges B and C.) Denoting by d the distance DD_1 from D to the plane ABC, we have

$$J = 2\Delta d.$$

Since none of AD, BD, CD is parallel to AB, all of them are greater than or equal to the next shortest distance b. Since the triangle ABC has no obtuse angle, circles of radius R with centers at the vertices overlap in such a way that every interior point of the triangle except the circumcenter is inside at least one of these circles. Therefore the distance of D_1 from at least one vertex is less than R, except that it is equal to R when D_1 is the circumcenter. Thus at least one of AD, BD, CD has its square less than or equal to $R^2 + d^2$, and consequently

$$R^2 + d^2 \geqslant b^2,$$

with equality only when D_1 is the circumcenter (and possibly not even then). Hence

$$
\begin{aligned}
J^2 &= (2\Delta d)^2 \\
&\geqslant 4\Delta^2(b^2 - R^2) = \tfrac{1}{4}b^2(-a^4 - b^4 - c^4 + 2b^2c^2 + c^2a^2 + 2a^2b^2) \\
&= \tfrac{1}{2}c^6 + \tfrac{1}{4}c^2(b^2 - c^2)(3b^2 + 2c^2) + \tfrac{1}{4}b^2(a^2 - b^2)(b^2 + c^2 - a^2) \\
&\geqslant \tfrac{1}{2}c^6,
\end{aligned}
$$

with equality only when

$$R^2 + d^2 = b^2, \qquad b = c,$$

and either

$$\text{(i) } a = b \quad \text{or} \quad \text{(ii) } b^2 + c^2 = a^2.$$

Thus there are apparently two "critical" lattices for which the ratio J/c^3 attains its minimum value $\sqrt{\tfrac{1}{2}}$. However, we shall soon see that these two are merely different aspects of one: the face-centered cubic lattice.

In case (i) (Figure 18.4b) the tetrahedron $ABCD$ is regular, and we may choose Cartesian coordinates

$$(0, 0, 0), \quad (0, 1, 1), \quad (1, 0, 1), \quad (1, 1, 0)$$

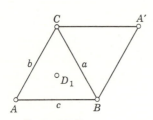

Figure 18.4b

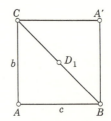

Figure 18.4c

for its vertices, in agreement with the basis 18.33 for the face-centered cubic lattice. In case (ii) (Figure 18.4c) $ABA'C$ is a square, the base of a pyramid whose sloping faces (such as ABD) are equilateral triangles. Choosing A, B and D as before, we now have C at $(0, 1, -1)$, yielding the alternative basis

18.41 $\mathbf{r}_1 = (0, 1, 1), \quad \mathbf{r}_3 - \mathbf{r}_2 = (0, 1, -1), \quad \mathbf{r}_3 = (1, 1, 0)$

for the same lattice. Thus we have proved that the face-centered cubic lattice (whose points have integral Cartesian coordinates with an even sum) is really the only "critical" lattice.

By 18.25, the square of the length of the lattice vector

$$\mathbf{u} = \Sigma\, u^\alpha\, \mathbf{r}_\alpha$$

is

$$\Sigma\Sigma\, g_{\alpha\beta}\, u^\alpha\, u^\beta,$$

and c^2 is the minimum value of this positive definite ternary quadratic form when the coordinates u^1, u^2, u^3 are restricted to integral values other than $0, 0, 0$. Hence, among all such forms with a given minimum value c^2, the minimum determinant $G = J^2 = \frac{1}{2}c^6$ occurs when the basic vectors are given by 18.33, so that the form is

$$\begin{aligned}
(\Sigma\, u^\alpha\, \mathbf{r}_\alpha)^2 &= (u^2 + u^3, \, u^3 + u^1, \, u^1 + u^2)^2 \\
&= (u^2 + u^3)^2 + (u^3 + u^1)^2 + (u^1 + u^2)^2 \\
&= 2\{(u^1)^2 + (u^2)^2 + (u^3)^2 + u^2 u^3 + u^3 u^1 + u^1 u^2\}.
\end{aligned}$$

In other words, every "extreme" form in three variables is equivalent to

$$(u^1)^2 + (u^2)^2 + (u^3)^2 + u^2 u^3 + u^3 u^1 + u^1 u^2.$$

This is a famous result, first proved by Gauss [**1**, vol. 2, pp. 192–196].*

EXERCISE

Using the basis 18.41 instead of 18.33, obtain the equivalent form

$$(u^1)^2 + (u^2)^2 + (u^3)^2 + u^2 u^3 + u^3 u^1.$$

18.5   GENERAL COORDINATES

The position of a point in Euclidean space may be specified by three numbers in many ways. Rectangular Cartesian coordinates (x, y, z) are the most familiar; but we have seen (e.g., in § 17.7) that other systems, such as cylindrical coordinates, are sometimes more convenient. Let us use the notation (u^1, u^2, u^3) for general coordinates. The essential requirements are that, within a suitable range of variation, x, y, z are single-valued differentiable functions of u^1, u^2, u^3, while u^1, u^2, u^3 are equally well-behaved functions of x, y, z. For instance, if (u^1, u^2, u^3) are cylindrical coordinates, we have

---

* For an account of the history of extreme forms up to 1951, see Coxeter, *Canadian Journal of Mathematics*, **3** (1951), p. 393.

$$x = u^1 \cos u^2, \qquad\qquad y = u^1 \sin u^2, \qquad\qquad z = u^3;$$

$$u^1 = \sqrt{x^2 + y^2}, \qquad\qquad u^2 = \arctan \frac{y}{x}, \qquad\qquad u^3 = z.$$

For arbitrary constants a, b, c, the surfaces

18.51 $u^1 = a, \qquad u^2 = b, \qquad u^3 = c$

are called *level surfaces,* and their three curves of intersection such as

$$u^2 = b, \quad u^3 = c$$

are called *level curves.* Through a given point there is usually one level surface of each kind, and one level curve of each kind, but exceptions are allowed. For instance, in the case of cylindrical coordinates the level surfaces are cylinders $u^1 = a$ or $x^2 + y^2 = a^2$, planes $u^2 = b$ or $y = x \tan b$ through the z-axis, and planes $z = c$ orthogonal to the z-axis. The z-axis itself is exceptional because each of its points lies on infinitely many planes $u^2 = b$ (in fact, on all of them).

According to the ordinary meaning of partial differentiation, the partial derivatives of the position vector

$$\mathbf{r} = x\mathbf{i} + y\mathbf{j} + z\mathbf{k}$$

are the unit vectors along the Cartesian axes:

$$\frac{\partial \mathbf{r}}{\partial x} = \mathbf{i}, \quad \frac{\partial \mathbf{r}}{\partial y} = \mathbf{j}, \quad \frac{\partial \mathbf{r}}{\partial z} = \mathbf{k}.$$

The differential of $\mathbf{r}$, representing displacement in any given direction, is

$$d\mathbf{r} = \mathbf{i}\,dx + \mathbf{j}\,dy + \mathbf{k}\,dz = (dx, dy, dz),$$

and the element of arc of any curve in this direction is ds, where

18.52 $(ds)^2 = |d\mathbf{r}|^2 = d\mathbf{r} \cdot d\mathbf{r} = (dx)^2 + (dy)^2 + (dz)^2.$

Instead of regarding $\mathbf{r}$ as a function of x, y, z, we may regard it as a function of u^1, u^2, u^3. Using a subscript α to indicate a partial derivative with respect to u^α (so that $x_\alpha = \partial x / \partial u^\alpha$, etc.), we have

18.53 $$\mathbf{r}_\alpha = \frac{\partial \mathbf{r}}{\partial u^\alpha} = x_\alpha \mathbf{i} + y_\alpha \mathbf{j} + z_\alpha \mathbf{k}$$

and $$d\mathbf{r} = \mathbf{r}_1\,du^1 + \mathbf{r}_2\,du^2 + \mathbf{r}_3\,du^3 = \Sigma\,\mathbf{r}_\alpha\,du^\alpha.$$

For a displacement along the level curve $u^2 = b$, $u^3 = c$ we have $du^2 = 0$, $du^3 = 0$, and $d\mathbf{r} = \mathbf{r}_1\,du^1$. Thus $\mathbf{r}_1$ is a *tangent* vector to this level curve. Similarly $\mathbf{r}_2$ is a tangent vector to the curve $u^3 = c$, $u^1 = a$, and $\mathbf{r}_3$ is a tangent vector to $u^1 = a$, $u^2 = b$. At a general point in space we thus have a definite trihedron $\mathbf{r}_1 \mathbf{r}_2 \mathbf{r}_3$ depending on the coordinate system. These basic vectors are not necessarily of unit length, and not necessarily orthogonal (though they do happen to be orthogonal in the case of cylindrical coordinates). The derivatives of the Cartesian coordinates are expressible in terms of them:

18.54 $$x_\alpha = \mathbf{r}_\alpha \cdot \mathbf{i}, \quad y_\alpha = \mathbf{r}_\alpha \cdot \mathbf{j}, \quad z_\alpha = \mathbf{r}_\alpha \cdot \mathbf{k}.$$

At a general point in space, any two of the three basic vectors determine a tangent plane to one of the three level surfaces through the point. For instance, the plane $\mathbf{r}_2\mathbf{r}_3$ touches the surface $u^1 = a$ since it contains tangents to two curves lying in that surface. Hence the dual basic vectors 18.12, orthogonal to the tangent planes $\mathbf{r}_2\mathbf{r}_3$, $\mathbf{r}_3\mathbf{r}_1$, $\mathbf{r}_1\mathbf{r}_2$, are *normals* to the level surfaces 18.51: not, in general, of unit length, but adjusted so that

$$\mathbf{r}^\alpha \cdot \mathbf{r}_\alpha = 1 \qquad\qquad (\alpha = 1, 2, 3).$$

The notation 18.22 provides the following formula for the element of arc ds in the direction of a given displacement $d\mathbf{r}$:

18.55
$$\begin{aligned}(ds)^2 = d\mathbf{r} \cdot d\mathbf{r} &= \Sigma\, \mathbf{r}_\alpha\, du^\alpha \cdot \Sigma\, \mathbf{r}_\beta\, du^\beta \\ &= \Sigma\Sigma\, g_{\alpha\beta}\, du^\alpha\, du^\beta \\ &= g_{11}\,(du^1)^2 + g_{22}\,(du^2)^2 + g_{33}\,(du^3)^2 \\ &\quad + 2g_{23}\, du^2\, du^3 + 2g_{31}\, du^3\, du^1 + 2g_{12}\, du^1\, du^2.\end{aligned}$$

In the special case when u^1, u^2, u^3 are x, y, z, this reduces to 18.52. In general, the coefficients $g_{\alpha\beta}$ are not constants but functions of the coordinates and their derivatives (see 18.27).

To deal with any given system of coordinates, we work out $g_{\alpha\beta} = \mathbf{r}_\alpha \cdot \mathbf{r}_\beta$ from the derivatives 18.53, then obtain $g^{\alpha\beta}$ by taking the cofactor of $g_{\alpha\beta}$ in the determinant G and dividing by G itself.

Our use of the letter J in 18.13 and 18.28 commemorates the German mathematician C. G. J. Jacobi (1804–1851). In fact, for transforming the triple integral of a function

$$f(x, y, z) = F(u^1, u^2, u^3)$$

from Cartesian to other coordinates, we use the formula

$$\iiint f(x, y, z)\, dx\, dy\, dz = \iiint F(u^1, u^2, u^3)\, \frac{\partial(x, y, z)}{\partial(u^1, u^2, u^3)}\, du^1\, du^2\, du^3,$$

which involves the *Jacobian*

$$\frac{\partial(x, y, z)}{\partial(u^1, u^2, u^3)} = \begin{vmatrix} x_1 & y_1 & z_1 \\ x_2 & y_2 & z_2 \\ x_3 & y_3 & z_3 \end{vmatrix} = [\mathbf{r}_1\, \mathbf{r}_2\, \mathbf{r}_3] = J.$$

EXERCISES

1. If u^1, u^2, u^3 are *affine* coordinates, they are the components of $\mathbf{r}$ with reference to three fixed independent vectors $\mathbf{r}_1$, $\mathbf{r}_2$, $\mathbf{r}_3$, so that

18.56 $\mathbf{r} = \Sigma u^\alpha \, \mathbf{r}_\alpha.$

(This notation is appropriate since it makes $\mathbf{r}_\alpha = \partial \mathbf{r}/\partial u^\alpha$.) The components of $\mathbf{r}$ with reference to $\mathbf{i}, \mathbf{j}, \mathbf{k}$ are

$$ x = \Sigma \, x_\alpha \, u^\alpha, \qquad y = \Sigma \, y_\alpha \, u^\alpha, \qquad z = \Sigma \, z_\alpha \, u^\alpha; $$

$\mathbf{r}_\alpha = x_\alpha \mathbf{i} + y_\alpha \mathbf{j} + z_\alpha \mathbf{k}$ is a constant vector for each α, and all the $g_{\alpha\beta}$ are constants.

2. *Oblique* Cartesian coordinates are affine coordinates with the same unit of measurement along all three axes, so that $|\mathbf{r}_\alpha| = 1$. In this case $g_{\alpha\alpha} = 1$, and $g_{\alpha\beta}$ is the cosine of the angle between $\mathbf{r}_\alpha$ and $\mathbf{r}_\beta$ (which are the axes of the coordinates u^α and u^β).

3. *Rectangular* Cartesian coordinates (with axes rotated to new positions without changing the origin) arise when $\mathbf{r}_1\mathbf{r}_2\mathbf{r}_3$ is an orthogonal trihedron of unit vectors, like $\mathbf{i}\,\mathbf{j}\,\mathbf{k}$, so that

18.57 $g_{\alpha\beta} = \delta_{\alpha\beta}$

(meaning 1 or 0 according as α and β are equal or unequal). By 18.54, x_α is the cosine of the angle between the new axis $\mathbf{r}_\alpha$ and the old axis $\mathbf{i}$; similarly for y_α and z_α. Interchanging the roles of the new and old axes in the relation

$$ \mathbf{r}_\alpha = x_\alpha \mathbf{i} + y_\alpha \mathbf{j} + z_\alpha \mathbf{k}, $$

we obtain

18.58 $\mathbf{i} = \Sigma \, x_\alpha \, \mathbf{r}_\alpha, \qquad \mathbf{j} = \Sigma \, y_\alpha \, \mathbf{r}_\alpha, \qquad \mathbf{k} = \Sigma \, z_\alpha \, \mathbf{r}_\alpha.$

(Since now $\mathbf{r}^\alpha = \mathbf{r}_\alpha$, the distinction between covariant and contravariant disappears.) From 18.56 deduce

$$ u^\alpha = \mathbf{r}_\alpha \cdot \mathbf{r} = x_\alpha x + y_\alpha y + z_\alpha z $$

so that (paradoxically)

$$ \frac{\partial u^\alpha}{\partial x} = x_\alpha = \frac{\partial x}{\partial u^\alpha}. $$

From 18.27 and 18.57 deduce

$$ x_\alpha x_\beta + y_\alpha y_\beta + z_\alpha z_\beta = \delta_{\alpha\beta}. $$

With the help of 18.58, evaluate $\Sigma \, x_\alpha{}^2$ and $\Sigma \, y_\alpha z_\alpha$. (In technical language, these properties make

$$ \left\| \begin{matrix} x_1 & y_1 & z_1 \\ x_2 & y_2 & z_2 \\ x_3 & y_3 & z_3 \end{matrix} \right\| $$

an "orthogonal matrix.")

4. Find $g_{\alpha\beta}$ in the case of *cylindrical* coordinates. Verify that $G = J^2$.

5. Find $g_{\alpha\beta}$ in the case of *spherical polar* coordinates, defined by

$$ x = u^3 \sin u^1 \cos u^2, \quad y = u^3 \sin u^1 \sin u^2, \quad z = u^3 \cos u^1. $$

Describe the level surfaces.

6. Find $g_{\alpha\beta}$ in the case of *confocal* coordinates, defined by

$$ x^2 = \frac{(A - u^1)(A - u^2)(A - u^3)}{(A - B)(A - C)}, \quad y^2 = \frac{(B - u^1)(B - u^2)(B - u^3)}{(B - C)(B - A)}, $$

$$ z^2 = \frac{(C - u^1)(C - u^2)(C - u^3)}{(C - A)(C - B)}, $$

where $u^1 < C < u^2 < B < u^3 < A$ (and x^2 means "x squared"). In this case the level surfaces are the central quadrics

18.59
$$\frac{x^2}{A - \lambda} + \frac{y^2}{B - \lambda} + \frac{z^2}{C - \lambda} = 1,$$

which are ellipsoids $u^1 = \lambda$ if $\lambda < C$, one-sheet hyperboloids $u^2 = \lambda$ if $C < \lambda < B$, and two-sheet hyperboloids $u^3 = \lambda$ if $B < \lambda < A$. In fact, u^1, u^2, u^3 are the roots of 18.59, regarded as a cubic equation for λ [Weatherburn **2**, p. 211].

18.6 THE ALTERNATING SYMBOL

As a kind of counterpart of the Kronecker delta, we shall find it convenient to use the "alternating epsilon"

$$\varepsilon^{\alpha\beta\gamma} = \varepsilon_{\alpha\beta\gamma} = \tfrac{1}{2}(\beta - \gamma)(\gamma - \alpha)(\alpha - \beta),$$

which is 1 if $\alpha\beta\gamma$ is an even permutation of 123, -1 if it is an odd permutation, and 0 otherwise. This trick provides one of the best ways to introduce the theory of determinants [Jeffreys **1**, p. 13]:

18.61
$$\begin{vmatrix} x_1 & y_1 & z_1 \\ x_2 & y_2 & z_2 \\ x_3 & y_3 & z_3 \end{vmatrix} = \Sigma\Sigma\Sigma \, \varepsilon^{\alpha\beta\gamma} \, x_\alpha \, y_\beta \, z_\gamma,$$

$$\begin{vmatrix} g_{11} & g_{12} & g_{13} \\ g_{21} & g_{22} & g_{23} \\ g_{31} & g_{32} & g_{33} \end{vmatrix} = \Sigma\Sigma\Sigma \, \varepsilon^{\alpha\beta\gamma} \, g_{1\alpha} \, g_{2\beta} \, g_{3\gamma}.$$

From 18.12–18.15 we deduce

$$[\mathbf{r}_\alpha \, \mathbf{r}_\beta \, \mathbf{r}_\gamma] = \varepsilon_{\alpha\beta\gamma} \, J, \qquad [\mathbf{r}^\alpha \, \mathbf{r}^\beta \, \mathbf{r}^\gamma] = \varepsilon^{\alpha\beta\gamma} \, J^{-1},$$
$$J^{-1} \, \mathbf{r}_\alpha \times \mathbf{r}_\beta = \Sigma \, \varepsilon_{\alpha\beta\gamma} \, \mathbf{r}^\gamma, \qquad J \, \mathbf{r}^\alpha \times \mathbf{r}^\beta = \Sigma \, \varepsilon^{\alpha\beta\gamma} \, \mathbf{r}_\gamma.$$

Since $\Sigma\Sigma \, \varepsilon^{\alpha\beta\gamma} \, g_{\alpha\beta} = 0$, it follows that

$$\Sigma \, \mathbf{r}^\alpha \times \mathbf{r}_\alpha = \Sigma \, \mathbf{r}^\alpha \times \Sigma \, g_{\alpha\beta}\mathbf{r}^\beta = \Sigma\Sigma\Sigma \, \varepsilon^{\alpha\beta\gamma} \, g_{\alpha\beta} \, \mathbf{r}_\gamma / J$$
$$= \mathbf{0}.$$

The analogous "two-dimensional" symbol is

$$\varepsilon^{ij} = \varepsilon_{ij} = j - i \qquad\qquad (i = 1 \text{ or } 2, \; j = 1 \text{ or } 2)$$

which enables us to write

$$\begin{vmatrix} g_{11} & g_{12} \\ g_{21} & g_{22} \end{vmatrix} = \Sigma\Sigma \, \varepsilon^{ij} \, g_{1i} \, g_{2j}.$$

(We use Latin or Greek indices according as the range of values is 12 or 123.)

EXERCISES

1. Use 18.61 to obtain a formula for the cofactor of x_α. Work this out for the case when $\alpha = 3$.

2. If $c^{ij} = c^{ji}$, $\Sigma\Sigma \, \varepsilon_{ij} \, c^{ij} = 0$. Use the same idea to justify the step $\Sigma\Sigma \, \varepsilon^{\alpha\beta\gamma} \, g_{\alpha\beta} = 0$ in the above evaluation of $\Sigma \, \mathbf{r}^\alpha \times \mathbf{r}_\alpha$.

19

Differential geometry of surfaces

The present chapter extends the notion of *curvature* from curves to surfaces. This extension is achieved by considering plane sections of a given surface, especially normal sections. Through the normal at a given point we can draw infinitely many planes; in fact, we can imagine such a plane to rotate continuously about the normal. In general, the curvature of the section varies continuously. For one of the planes the curvature attains its maximum value, for another, its minimum. We shall see that these two planes are at right angles, and that the product of the two "principal curvatures" determines the essential nature of the surface. For instance, this "Gaussian curvature" is positive for an oval surface such as an ellipsoid, zero for a developable surface such as a cylinder or cone, and negative for a saddle-shaped surface such as a hyperbolic paraboloid.

19.1 THE USE OF TWO PARAMETERS ON A SURFACE

To fix the position of a point on the earth's surface, we may give its latitude and longitude. . . . Through points on the equator draw meridians; through points on the Greenwich meridian, draw parallels of latitude. The position of a point . . . is then given by the two curves, one of each family, which go through it. . . . Each point, except the poles, acquires two definite coordinates. We can generalize this method to any surface, or rather to a piece of any surface; we take two families of curves on the surface, such that through each point goes just one curve of each family . . . as if a fine fishing-net were thrown over the surface.

H. G. Forder [**3,** p. 133]

A surface $f(x, y, z) = 0$ is often conveniently represented by a set of three parametric equations

$$x = x(u^1, u^2), \quad y = y(u^1, u^2), \quad z = z(u^1, u^2),$$

from which the single equation $f = 0$ could be derived by eliminating the *parameters* u^1, u^2. We shall assume, as before, that the functions involved are continuous and possess all the continuous derivatives that we need.

A simple instance arises when we regard x and y as the parameters, so that the three equations become

$$x = u^1, \quad y = u^2, \quad z = F(u^1, u^2),$$

where the expression for z is the result of solving the equation $f(x, y, z) = 0$ for z in terms of x and y. Such an equation

$$z = F(x, y)$$

is called *Monge's form* of the equation for a surface. For instance, the sphere $x^2 + y^2 + z^2 = 1$ becomes

$$z = \pm \sqrt{1 - x^2 - y^2}.$$

The square root makes this a clumsy way to investigate the sphere. It is far better to take u^1 and u^2 to be colatitude and longitude, so that

19.11 $\qquad x = \sin u^1 \cos u^2, \qquad y = \sin u^1 \sin u^2, \qquad z = \cos u^1.$

(Colatitude u^1 means latitude $\frac{1}{2}\pi - u^1$.)

The vector equation for a surface is

19.12 $\qquad\qquad\qquad \mathbf{r} = \mathbf{r}(u^1, u^2),$

just as the vector equation for a curve is $\mathbf{r} = \mathbf{r}(u)$. The essential difference is that the curve has only one parameter, whereas the surface has two independent parameters.

One way to explore a surface is to investigate families of curves lying on it. Among these are the *parametric* curves

$$u^1 = a \quad \text{and} \quad u^2 = b,$$

where a and b are arbitrary constants. Through a given point there is usually one parametric curve of each kind, but exceptions are allowed. For instance, when colatitude and longitude are used on the unit sphere, the curves $u^1 = a$ are circles called *parallels of latitude* and the curves $u^2 = b$ are great circles called *meridians*. Almost every point on the sphere lies on one "parallel" and one meridian, but the north and south poles lie on all the meridians.

The position vector $\mathbf{r}$, of a point on the surface, is a vector function of u^1 and u^2. Using a subscript i to indicate a partial derivative with respect to u^i, we have

$$d\mathbf{r} = \mathbf{r}_1 \, du^1 + \mathbf{r}_2 \, du^2 = \Sigma \, \mathbf{r}_i \, du^i,$$

where $\qquad\qquad\qquad \mathbf{r}_i = \dfrac{\partial \mathbf{r}}{\partial u^i}.$

The differential $d\mathbf{r}$ may be regarded as a displacement along a given curve on the surface or, more precisely, a displacement along a tangent. In the case of the parametric curve $u^2 = b$, we have $du^2 = 0$, so that $d\mathbf{r} = \mathbf{r}_1\,du^1$. Thus $\mathbf{r}_1$ is a tangent vector to this curve, and similarly $\mathbf{r}_2$ is a tangent vector to the other parametric curve $u^1 = a$. It follows that the plane $\mathbf{r}_1\mathbf{r}_2$, spanned by these two covariant basic vectors, is the *tangent plane* to the surface at the point considered, which is the point $(u^1, u^2) = (a, b)$. In the same plane we define the two contravariant basic vectors $\mathbf{r}^i$ to be normal to the parametric curves, adjusted so that

$$\mathbf{r}^i \cdot \mathbf{r}_j = \delta^i_j.$$

In the tangent plane, there are tangent vectors going out from the point of contact in all directions. Any such vector

19.13
$$\mathbf{t} = \Sigma\, a_i\, \mathbf{r}^i = \Sigma\, a^i\, \mathbf{r}_i$$

is said to have covariant components a_i and contravariant components a^i. It is easily verified (cf. § 18.2) that

19.14
$$a_j = \mathbf{t} \cdot \mathbf{r}_j, \qquad a^j = \mathbf{t} \cdot \mathbf{r}^j.$$

In particular, the basic vectors themselves have components g_{ij}, g^{ij}, such that

19.15
$$\mathbf{r}_i = \Sigma\, g_{ij}\, \mathbf{r}^j, \qquad \mathbf{r}^i = \Sigma\, g^{ij}\, \mathbf{r}_j,$$

$$g_{ij} = \mathbf{r}_i \cdot \mathbf{r}_j, \qquad g^{ij} = \mathbf{r}^i \cdot \mathbf{r}^j.$$

In terms of the *fundamental magnitudes of the first order*

$$g_{11}, \quad g_{12} = g_{21}, \quad g_{22},$$

defined by $g_{ij} = \mathbf{r}_i \cdot \mathbf{r}_j$, we have the following formula for the element of arc ds (of any curve on the surface):

19.16
$$
\begin{aligned}
ds^2 = d\mathbf{r} \cdot d\mathbf{r} &= \Sigma\, \mathbf{r}_i\, du^i \cdot \Sigma\, \mathbf{r}_j\, du^j \\
&= \Sigma\Sigma g_{ij}\, du^i\, du^j \\
&= g_{11}(du^1)^2 + 2g_{12}\, du^1\, du^2 + g_{22}\,(du^2)^2.
\end{aligned}
$$

The fundamental magnitudes (which are functions of the parameters u^i) are spoken of collectively as a covariant tensor. The corresponding contravariant tensor g^{ij} is given by the last part of 19.15, which implies

$$\Sigma\, g_{ik}\, g^{ij} = \delta^j_k.$$

For each value of j, this is a pair of equations to be solved for the two unknowns g^{ij} $(i = 1, 2)$. The solution is

$$g^{ij} = (\text{cofactor of } g_{ij} \text{ in } g)/g,$$

where
$$g = \begin{vmatrix} g_{11} & g_{12} \\ g_{21} & g_{22} \end{vmatrix} = g_{11}g_{22} - (g_{12})^2.$$

Since the number of rows (or columns) in this determinant is only 2, the cofactors are single elements, and we have the explicit expressions

19.17 $$g^{11} = \frac{g_{22}}{g}, \quad g^{12} = g^{21} = -\frac{g_{12}}{g}, \quad g^{22} = \frac{g_{11}}{g}.$$

It is now easy to derive the covariant components of the tangent vector **t** from its contravariant components, or vice versa:

$$a_j = \mathbf{t} \cdot \mathbf{r}_j = \Sigma\, a^i\, \mathbf{r}_i \cdot \mathbf{r}_j = \Sigma\, g_{ij}\, a^i,$$
$$a^j = \mathbf{t} \cdot \mathbf{r}^j = \Sigma\, a_i\, \mathbf{r}^i \cdot \mathbf{r}^j = \Sigma\, g^{ij}\, a_i.$$

Of course, we are free to interchange i and j, obtaining

19.18 $$a_i = \Sigma\, g_{ij}\, a^j, \qquad a^i = \Sigma\, g^{ij}\, a_j.$$

Since $g_{12} = \mathbf{r}_1 \cdot \mathbf{r}_2$, where $\mathbf{r}_1$ and $\mathbf{r}_2$ are tangents to the parametric curves, the condition for the two families of parametric curves to intersect at right angles is

$$g_{12} = 0.$$

Thus in the case of orthogonal parametric curves we have simply

$$g = g_{11}\, g_{22}, \quad g^{12} = 0, \quad g^{ii} = 1/g_{ii},$$

whence by 19.15,

$$\mathbf{r}^i = g^{ii}\, \mathbf{r}_i = \mathbf{r}_i / g_{ii}.$$

EXERCISES

1. The basic vector $\mathbf{r}_i$ has covariant components g_{ij}, contravariant components δ^j_i.
2. $\Sigma\, \mathbf{r}^j \times \mathbf{r}_j = \mathbf{0}$. Interpret this geometrically in terms of areas of triangles.
3. Find $\mathbf{r}^1$ and $\mathbf{r}^2$ for the general surface of revolution

$$\mathbf{r} = (u^1 \cos u^2, \quad u^1 \sin u^2, \quad z),$$

where z is a function of u^1 alone.

4. Find g_{ij} and g^{ij} for the unit sphere expressed in terms of colatitude and longitude.

19.2 DIRECTIONS ON A SURFACE

Just as a curve in the (x, y)-plane is given by an equation connecting x and y, a curve on the surface 19.12 is given by an equation connecting u^1 and u^2. A differential equation determines a family of curves. In general, a first-order, first-degree differential equation

$$\Sigma\, c_i\, du^i = 0$$

determines a one-parameter family of curves: one curve through each point of general position on the surface, going out from that point in a direction determined by $du^2/du^1 = -c_1/c_2$; for instance, the equation

$$du^2 = 0$$

determines the "first" family of parametric curves. On the other hand, a first-order, second-degree equation

19.21 $$\Sigma\Sigma\, c_{ij}\, du^i\, du^j = 0,$$

where $c_{12} = c_{21}$ and $c_{11}\, c_{22} < c_{12}^2$, determines a *net* of curves: two curves through a general point on the surface; for example, the quadratic equation

$$du^1\, du^2 = 0$$

determines a net consisting of the two families of parametric curves taken together.

We have seen that the vectors $\mathbf{r}_i$ are in the directions of the tangents to the parametric curves. Since

$$\mathbf{r}_i{}^2 = \mathbf{r}_i \cdot \mathbf{r}_i = g_{ii},$$

their lengths are $\sqrt{g_{ii}}$. Because of its frequent occurrence, we shall use the abbreviation g_i for this square root; thus

$$g_i = \sqrt{g_{ii}} = |\mathbf{r}_i|,$$

and analogously

$$g^i = \sqrt{g^{ii}} = |\mathbf{r}^i|.$$

In this notation, the *unit* tangent vectors to the parametric curves (touching $du^2 = 0$ and $du^1 = 0$, respectively) are

$$\mathbf{r}_1/g_1 \quad \text{and} \quad \mathbf{r}_2/g_2.$$

The angle ϕ at which the two parametric curves intersect is given by

19.22 $$\cos\phi = \frac{\mathbf{r}_1 \cdot \mathbf{r}_2}{g_1\, g_2} = \frac{g_{12}}{g_1\, g_2}.$$

We see from 19.17 that $g^1 = g_2/\sqrt{g}$, $g^2 = g_1/\sqrt{g}$; therefore

$$\sin\phi = \frac{\sqrt{g}}{g_1\, g_2} = \frac{1}{g_1\, g^1} = \frac{1}{g_2\, g^2},$$

that is, $$g_1 g^1 = g_2 g^2 = \csc\phi.$$

It follows from the definition of an outer product that the length of the vector $\mathbf{r}_1 \times \mathbf{r}_2$ is

19.23 $$|\mathbf{r}_1 \times \mathbf{r}_2| = g_1\, g_2 \sin\phi = \sqrt{g},$$

and that the element of *area* on the surface (naturally defined as the element of area in the tangent plane) is

19.24 $$dS = |\mathbf{r}_1\, du^1 \times \mathbf{r}_2\, du^2| = \sqrt{g}\, du^1\, du^2$$

[Kreyszig **1**, pp. 111–117]. The equation 19.23, in the form

$$g = (\mathbf{r}_1 \times \mathbf{r}_2)^2,$$

is sometimes useful as a means of computing g without first finding g_{ij}.

A displacement along any curve on the surface is given by

$$d\mathbf{r} = \Sigma\, \mathbf{r}_i\, du^i.$$

If s is the arc of this curve, the unit tangent vector is

19.25 $$\mathbf{t} = \frac{d\mathbf{r}}{ds} = \Sigma\, a^i\, \mathbf{r}_i = a^1\, \mathbf{r}_1 + a^2\, \mathbf{r}_2,$$

where

19.251 $$a^i = \frac{du^i}{ds} \qquad\qquad (i = 1, 2).$$

Thus the arc derivatives of the parameters are the contravariant components of $\mathbf{t}$. We shall not attempt to find a corresponding interpretation for the covariant components a_i, given by 19.14 or 19.18. On the other hand, it is easy to give a *geometrical* interpretation for both kinds of component. Let $\overrightarrow{PQ_1}$ and $\overrightarrow{PQ_2}$ be tangent vectors to the parametric curves, of such lengths that PT, representing $\mathbf{t}$, is a diagonal of the parallelogram PQ_1TQ_2, as in Figure 19.2a. Let $\mathbf{t}$ divide the angle $\phi = \angle\, Q_1PQ_2$ into the two parts θ and $\phi - \theta$. Let PR_1TR_2 be a parallelogram whose sides are perpendicular to the tangents. Since

$$\overrightarrow{PT} = \mathbf{t} = a^1\, \mathbf{r}_1 + a^2\, \mathbf{r}_2 = \overrightarrow{PQ_1} + \overrightarrow{PQ_2}$$

$$= a_1\, \mathbf{r}^1 + a_2\, \mathbf{r}^2 = \overrightarrow{PR_1} + \overrightarrow{PR_2},$$

the lengths of the various lines are:

$$PQ_i = g_i\, a^i, \quad PR_i = g^i\, a_i, \quad PS_i = \mathbf{t}\cdot\mathbf{r}_i/g_i = a_i/g_i.$$

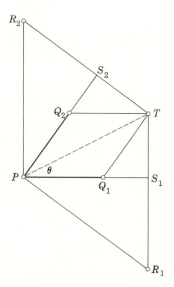

Figure 19.2a

The angles are given by

$$\cos\theta = PS_1, \qquad \cos(\phi - \theta) = PS_2.$$

By taking the inner product of 19.13 with itself in various ways, we can express the obvious relation $\mathbf{t}^2 = 1$ in the equivalent forms

19.26 $$\Sigma\Sigma\, g^{ij}\, a_i\, a_j = \Sigma\, a_i\, a^i = \Sigma\Sigma\, g_{ij}\, a^i\, a^j = 1.$$

In virtue of 19.251, the last of these relations is a restatement of 19.16.

Similarly, by working out the inner product of two such unit tangent vectors

$$\Sigma\, a_i\, \mathbf{r}^i = \Sigma\, a^i\, \mathbf{r}_i \quad \text{and} \quad \Sigma\, b_j\, \mathbf{r}^j = \Sigma\, b^j\, \mathbf{r}_j,$$

we obtain various expressions for the cosine of the angle between them:

19.27 $$\Sigma\Sigma\, g^{ij}\, a_i\, b_j = \Sigma\, a_i\, b^i = \Sigma\, a_j\, b^j = \Sigma\Sigma\, g_{ij}\, a^i\, b^j = \Sigma\, a^i\, b_i.$$

Eliminating ds from the two equations 19.251, we obtain the differential equation

$$a^2\, du^1 - a^1\, du^2 = 0$$

for a family of curves whose typical tangent is given by 19.25. Another family, cutting all the members of the first family at right angles, has the differential equation

19.28 $$b^2\, du^1 - b^1\, du^2 = 0,$$

where, by 19.27,

$$\Sigma\Sigma\, g_{ij}\, a^i\, b^j = 0.$$

Writing this relation in the form $\Sigma\, a_j\, b^j = 0$ or

$$a_1\, b^1 + a_2\, b^2 = 0,$$

where $a_j = \Sigma\, g_{ij}\, a^i$, we find that the equation 19.28 is equivalent to

$$a_1\, du^1 + a_2\, du^2 = 0.$$

In other words,

The orthogonal trajectories of the curves $a^2\, du^1 - a^1\, du^2 = 0$ are the curves

$$\Sigma\, a_i\, du^i = 0.$$

It follows also that the two families of curves

$$\Sigma\, a_i\, du^i = 0, \qquad \Sigma\, b_j\, du^j = 0$$

are orthogonal if and only if

19.29 $$\Sigma\Sigma\, g^{ij}\, a_i\, b_j = 0.$$

The net of curves given by the quadratic differential equation 19.21 is an *orthogonal* net if and only if

19.291 $$\Sigma\Sigma\, g^{ij}\, c_{ij} = 0.$$

For, this condition is the same as 19.29 if the quadratic expression factorizes in the form

$$\Sigma\Sigma\, c_{ij}\, du^i\, du^j = \Sigma\, a_i\, du^i \cdot \Sigma\, b_j\, du^j,$$

so that

$$c_{ij} + c_{ji} = a_i\, b_j + a_j\, b_i.$$

EXERCISES

1. Use 17.13 to prove that $(\mathbf{r}_1 \times \mathbf{r}_2)^2 = g$ (cf. 19.23).
2. Express $\tan \phi$ in terms of the fundamental magnitudes g_{ij} and their determinant g.
3. In Figure 19.2a, $TS_1 = a^2/g^2$ and $TS_2 = a^1/g^1$.
4. Reconcile the formulas

$$\cos \theta = a_1/g_1, \quad \cos (\phi - \theta) = a_2/g_2,$$
$$\sin \theta = a^2/g^2, \quad \sin (\phi - \theta) = a^1/g^1$$

with 19.22.

5. The net of curves bisecting the angles between the parametric curves $(du^1\, du^2 = 0)$ is given by the differential equation

$$g_{11}\, (du^1)^2 - g_{22}\, (du^2)^2 = 0.$$

(*Hint*: Find the condition for the parallelogram PQ_1TQ_2 to be a rhombus.)

6. Interpret the equation 19.21 in the case when $c_{11} = c_{22} = 0$. What does the condition 19.291 tell us in this case? What does it tell us about the curves described in the preceding exercise?

7. Use 19.24 to prove that the area of the unit sphere 19.11 is 4π.

19.3 NORMAL CURVATURE

The unit *normal* vector $\mathbf{n}$ at a point P on the surface is naturally defined as the unit vector perpendicular to the tangent plane in such a direction that the three vectors $\mathbf{r}_1\, \mathbf{r}_2\, \mathbf{n}$ form a right-handed trihedron. In virtue of 19.23, we have $\mathbf{r}_1 \times \mathbf{r}_2 = \sqrt{g}\, \mathbf{n}$, so that

19.31
$$\mathbf{n} = \frac{\mathbf{r}_1 \times \mathbf{r}_2}{\sqrt{g}},$$

$$[\mathbf{r}_1\, \mathbf{r}_2\, \mathbf{n}] = \sqrt{g},$$

and

19.32
$$\mathbf{r}_i \times \mathbf{r}_j = \varepsilon_{ij} \sqrt{g}\, \mathbf{n} \qquad (\varepsilon_{ij} = j - i).$$

Identifying this trihedron $\mathbf{r}_1\, \mathbf{r}_2\, \mathbf{n}$ with the $\mathbf{r}_1\, \mathbf{r}_2\, \mathbf{r}_3$ of § 18.1, we see from 18.12 that

$$\mathbf{r}^1 = \frac{\mathbf{r}_2 \times \mathbf{n}}{\sqrt{g}}, \quad \mathbf{r}^2 = \frac{\mathbf{n} \times \mathbf{r}_1}{\sqrt{g}}, \quad \mathbf{r}^3 = \frac{\mathbf{r}_1 \times \mathbf{r}_2}{\sqrt{g}} = \mathbf{n} = \mathbf{r}_3.$$

Thus

19.33
$$\mathbf{n} \times \mathbf{r}_i = \Sigma\, \varepsilon_{ij} \sqrt{g}\, \mathbf{r}^j.$$

The tangent plane at P contains a flat pencil of tangents

$$\mathbf{t} = \Sigma\, a^i \mathbf{r}_i$$

each of which determines a normal plane $\mathbf{tn}$. The section of the surface by such a plane is called a normal section; it is a plane curve whose curvature κ at P is called the *normal curvature* at P in the direction $\mathbf{t}$. Differentiating with respect to the arc s of the normal section, we obtain

$$\dot{\mathbf{t}} = \frac{d}{ds} \Sigma\, a^i \mathbf{r}_i = \Sigma \frac{da^i}{ds} \mathbf{r}_i + \Sigma\, a^i \frac{d}{ds} \mathbf{r}_i\,.$$

By 17.33, this is $\kappa\mathbf{n}$. Since $\mathbf{n}$ is perpendicular to $\mathbf{r}_i$, inner multiplication by $\mathbf{n}$ will eliminate the first sum on the right. (To differentiate $\mathbf{r}_i$ we use the operator

$$\frac{d}{ds} = \Sigma\, a^j \frac{\partial}{\partial u^j},$$

where $a^j = du^j/ds$.) Thus we are left with

$$\kappa = \kappa\mathbf{n} \cdot \mathbf{n} = \dot{\mathbf{t}} \cdot \mathbf{n} = \Sigma \left(a^i \frac{d}{ds} \mathbf{r}_i \right) \cdot \mathbf{n}$$

$$= \Sigma \left(a^i \Sigma\, a^j \frac{\partial}{\partial u^j} \mathbf{r}_i \right) \cdot \mathbf{n} = \Sigma\Sigma\, a^i a^j \frac{\partial^2 \mathbf{r}}{\partial u^i\, \partial u^j} \cdot \mathbf{n} = \Sigma\Sigma\, a^i a^j \mathbf{r}_{ij} \cdot \mathbf{n}.$$

Introducing the important notation

19.34 $$b_{ij} = \mathbf{r}_{ij} \cdot \mathbf{n} \qquad\qquad (i, j = 1, 2)$$

we now have the simple formula

19.35 $$\kappa = \Sigma\Sigma\, b_{ij} a^i a^j$$

for the normal curvature in the direction $\Sigma\, a^i \mathbf{r}_i$. Since $\mathbf{r}_{ij}$ is a second derivative,

$$b_{ij} = b_{ji}.$$

The three functions b_{11}, b_{12}, b_{22} are known as *fundamental magnitudes of the second order*. Like those of the first order, they occur as coefficients in a quadratic differential form:

19.36 $$\kappa\, ds^2 = \Sigma\Sigma\, b_{ij}\, du^i\, du^j$$
$$= b_{11}(du^1)^2 + 2b_{12}\, du^1\, du^2 + b_{22}(du^2)^2.$$

(It must be remembered that the normal curvature κ depends on the direction of the tangent, and therefore on $du^1 : du^2$.)

Differentiating the identity $\mathbf{r}_i \cdot \mathbf{n} = 0$, we obtain

$$\mathbf{r}_{ij} \cdot \mathbf{n} + \mathbf{r}_i \cdot \mathbf{n}_j = 0,$$

whence $$b_{ij} = -\mathbf{r}_i \cdot \mathbf{n}_j = -\mathbf{n}_i \cdot \mathbf{r}_j.$$

Along with the "covariant tensor" b_{ij}, we shall sometimes find it convenient to consider the "mixed tensor"

19.37 $$b_j^k = \Sum g^{ik} b_{ij} = -\Sum g^{ik} \mathbf{r}_i \cdot \mathbf{n}_j = -\mathbf{r}^k \cdot \mathbf{n}_j$$

and the "contravariant tensor"

19.371
$$b^{ik} = \Sum g^{ij} b_j^k = \Sum\Sum g^{ij} g^{kl} b_{jl}$$
$$= \Sum\Sum g^{kl} g^{ji} b_{jl} = \Sum g^{kl} b_l^i = b^{ki}.$$

The derivative $\mathbf{n}_i$, being perpendicular to the normal $\mathbf{n}$, is a tangent vector, capable of being expressed as a linear combination of the basic vectors $\mathbf{r}^j$ or $\mathbf{r}_j$. Since its covariant and contravariant components are

$$\mathbf{n}_i \cdot \mathbf{r}_j = -b_{ij}, \qquad \mathbf{n}_i \cdot \mathbf{r}^j = -b_i^j,$$

the expressions are

19.38 $$\mathbf{n}_i = -\Sum b_{ij} \mathbf{r}^j = -\Sum b_i^j \mathbf{r}_j.$$

We have thus established the "Weingarten equations"

$$\mathbf{n}_1 = -b_1^1 \mathbf{r}_1 - b_1^2 \mathbf{r}_2,$$
$$\mathbf{n}_2 = -b_2^1 \mathbf{r}_1 - b_2^2 \mathbf{r}_2,$$

which express the derivatives of the normal $\mathbf{n}$ in terms of the derivatives of the position vector $\mathbf{r}$.

EXERCISES

1. Evaluate $\mathbf{r}^1 \times \mathbf{r}^2$.

2. Work out b_{ij} for the unit sphere in terms of colatitude and longitude. Verify that the normal curvature is the same in all directions and the same at all points on the sphere. (*Hint*: Since $\mathbf{n} = \mathbf{r}$, $b_{ij} = -g_{ij}$.)

19.4 PRINCIPAL CURVATURES

> Take a unit sphere and draw the radius parallel to the normal at a point P of [a given] surface. The radius meets the sphere in the spherical representation of P. Clearly we must distinguish between the two sides of the surface, and draw the normal on the selected side. By this representation, to a curve on the surface corresponds, in general, a curve on the sphere, and to a piece, a piece. But as normals to the surface at different points may be parallel, pieces on the sphere might overlap even when they correspond to non-overlapping pieces on the surface. But if we take pieces on the surface, not too large, this will not occur. . . . To a small piece round P on the surface will correspond a small piece on the sphere, and the ratio of the area of the latter to the area of the former, as these areas shrink to zero, tends to the total curvature at P.
>
> H. G. Forder [**3**, pp. 139-140]

Consider a variable plane through the normal at a point P on a given surface. For each position of the plane, the normal section has curvature

κ given by 19.35. In exceptional cases (e.g., at the north and south poles of a spheroid) it may happen that κ remains constant; such a point P is called an *umbilic*. If P is not an umbilic, a continuous rotation of the normal plane **nt** makes κ vary in such a way as to return to its original value as soon as a half-turn has been completed. With the help of 19.26 we may express 19.35 in the homogeneous form

$$\kappa \Sigma\Sigma g_{ij} a^i a^j = \Sigma\Sigma b_{ij} a^i a^j$$

or

19.41
$$\Sigma\Sigma(b_{ij} - \kappa g_{ij})a^i a^j = 0.$$

This exhibits κ as a continuous function of the ratio

$$\frac{a^2}{a^1} = \frac{du^2}{du^1},$$

which determines the direction of the tangent

$$\mathbf{t} = \Sigma a^i \mathbf{r}_i.$$

In the course of its continuous variation, the normal curvature κ must attain at least one maximum and at least one minimum. We proceed to prove that there is just one of each, and that they occur in perpendicular directions. The maximum and minimum values of κ are called the *principal curvatures*, the positions of **t** in which they occur are called the *principal directions*, and the curves whose direction is always principal are called (perhaps not too happily) the *lines of curvature*.

As a temporary abbreviation, we write

$$c_{ij} = b_{ij} - \kappa g_{ij},$$

so that $c_{ij} = c_{ji}$. To find the principal curvatures and principal directions, we may differentiate 19.41 and then set $d\kappa = 0$ or, more conveniently, differentiate 19.41 regarding κ as a constant. Since b_{ij} and g_{ij} depend only on the fixed point P, this means that we differentiate

$$\Sigma\Sigma c_{ij} a^i a^j = 0$$

treating the coefficients c_{ij} as constants. Differentiating partially with respect to a^k, we find

$$\frac{\partial}{\partial a^k} \Sigma\Sigma c_{ij} a^i a^j = \Sigma\Sigma c_{ij} \left(\frac{\partial a^i}{\partial a^k} a^j + a^i \frac{\partial a^j}{\partial a^k} \right)$$

$$= \Sigma\Sigma c_{ij}(\delta_k^i a^j + a^i \delta_k^j) = \Sigma c_{kj} a^j + \Sigma c_{ik} a^i$$

$$= \Sigma(c_{ki} + c_{ik})a^i = 2\Sigma c_{ik} a^i.$$

Restoring the proper expression for c_{ik}, we deduce

19.42 $$\Sigma (b_{ik} - \kappa g_{ik}) a^i = 0 \qquad (k = 1, 2).$$

Multiplying by g^{jk} (and summing over k) to eliminate the coefficient g_{ik}, we obtain

19.43
$$\Sigma \, (b_i^j - \kappa \delta_i^j) \, a^i = 0,$$

that is,

19.44
$$\Sigma_i \, b_i^j \, a^i - \kappa a^j = 0 \qquad\qquad (j = 1, 2).$$

From these two equations we can find the principal curvatures by eliminating a^2/a^1, and the principal directions by eliminating κ.

When written out in detail, the equations are:

$$(b_1^1 - \kappa) \, a^1 + b_2^1 \, a^2 = 0,$$
$$b_1^2 \, a^1 + (b_2^2 - \kappa) \, a^2 = 0.$$

Eliminating a^2/a^1, we obtain

$$\begin{vmatrix} b_1^1 - \kappa & b_2^1 \\ b_1^2 & b_2^2 - \kappa \end{vmatrix} = 0,$$

that is,

19.45
$$\kappa^2 - \Sigma \, b_i^i \kappa + \det (b_i^j) = 0.$$

This quadratic equation has for its two roots the principal curvatures $\kappa_{(1)}$, $\kappa_{(2)}$, whose product and arithmetic mean are known as the *Gaussian curvature K* and the *mean curvature H*. Thus $\kappa_{(1)}$ and $\kappa_{(2)}$ are the roots of the equation

$$\kappa^2 - 2H\kappa + K = 0,$$

where

19.46
$$2H = \kappa_{(1)} + \kappa_{(2)} = \Sigma_i \, b_i^i = b_1^1 + b_2^2$$

and

19.47
$$K = \kappa_{(1)} \, \kappa_{(2)} = \det (b_i^j) = b_1^1 \, b_2^2 - b_2^1 \, b_1^2.$$

Since

$$gK = \begin{vmatrix} g_{11} & g_{12} \\ g_{21} & g_{22} \end{vmatrix} \begin{vmatrix} b_1^1 & b_2^1 \\ b_1^2 & b_2^2 \end{vmatrix} = \begin{vmatrix} b_{11} & b_{12} \\ b_{21} & b_{22} \end{vmatrix} = b,$$

say, another expression for K is the ratio of the two fundamental determinants:

19.471
$$K = \frac{b}{g}.$$

When K is positive, the normal curvature (never going outside the range from $\kappa_{(1)}$ to $\kappa_{(2)}$) has the same sign in all directions; the tangent plane at P meets the surface "instantaneously" at P and not anywhere else in the neighborhood of P. The surface is then said to be *synclastic* (or "oval"). Ellipsoids, elliptic paraboloids, and hyperboloids of two sheets are everywhere synclastic.

When K is negative, the normal curvature changes sign twice (during the rotation of the normal plane through a half-turn about the normal at P); therefore it is zero in the directions of two special tangents, called the *inflectional* tangents at P. The surface crosses its tangent plane, and its section by this plane is a pair of curves that cross each other at P, the two tangents at this "node" being the inflectional tangents.

A practical instance is the general shape of the ground at the top of a mountain pass. The tangent plane is the horizontal plane, which touches the curve of the footpath and cuts into the ground on both sides. The fact that the tangent section has a node is seen in a map on which contour lines are marked. The mountain pass occurs where one of the contour lines crosses itself [Hardy **1**, p. 65].

Such a surface is said to be *anticlastic* (or "saddle-shaped"). Nondegenerate ruled quadrics (namely, hyperbolic paraboloids and hyperboloids of one sheet) are everywhere anticlastic.

Surfaces more complicated than quadrics may be synclastic in some regions and anticlastic in others. Regions of the two kinds are then separated by a locus of *parabolic* points, at which $K = 0$. Hilbert and Cohn-Vossen [**1**, p. 197, Fig. 204] show a bust of Apollo on which the curves of parabolic points have been drawn. They are quite complicated, especially round the nose and mouth.

Surfaces on which $K = 0$ everywhere are said to be *developable*. Such surfaces include cones and cylinders, and also the surface traced out by the tangents of any twisted curve.

The Weingarten equations 19.38 provide a useful expression for the Gaussian curvature as a triple product:

19.48
$$K = [\mathbf{n}\, \mathbf{n}_1\, \mathbf{n}_2]/\sqrt{g}.$$

In fact,

$$[\mathbf{n}\, \mathbf{n}_1\, \mathbf{n}_2] = [\mathbf{n}\ \Sigma\, b_1^j \mathbf{r}_j\ \Sigma\, b_2^k \mathbf{r}_k] \quad = \Sigma\Sigma\, b_1^j b_2^k [\mathbf{n}\, \mathbf{r}_j\, \mathbf{r}_k]$$

$$= \Sigma\Sigma\, b_1^j b_2^k \varepsilon_{jk} \sqrt{g} \quad = \det{(b_i^j)} \sqrt{g}$$

$$= K \sqrt{g}.$$

Another expression, involving an arbitrary unit tangent vector $\mathbf{t}$, was discovered by A. J. Coleman:

19.49
$$\sqrt{g}\, K = \Sigma\Sigma\, \varepsilon^{ij} [\mathbf{n}\, \mathbf{t}\, \mathbf{t}_i]_j,$$

where the final subscript indicates differentiation with respect to u^j. This is deduced from Lagrange's identity 17.13 by introducing another unit tangent vector $\mathbf{m} = \mathbf{n} \times \mathbf{t}$, so that $\mathbf{n} = \mathbf{t} \times \mathbf{m}$ and

$$[\mathbf{n}\ \mathbf{n}_1\ \mathbf{n}_2] = (\mathbf{t} \times \mathbf{m}) \cdot (\mathbf{n}_1 \times \mathbf{n}_2)$$
$$= \mathbf{t} \cdot \mathbf{n}_1\ \mathbf{m} \cdot \mathbf{n}_2 - \mathbf{t} \cdot \mathbf{n}_2\ \mathbf{m} \cdot \mathbf{n}_1$$
$$= \Sigma\Sigma\ \varepsilon^{ij}\ \mathbf{t} \cdot \mathbf{n}_i\ \mathbf{m} \cdot \mathbf{n}_j.$$

Differentiating $\mathbf{t} \cdot \mathbf{n} = 0$, $\mathbf{m} \cdot \mathbf{n} = 0$, and using 17.19, we see that

$$\mathbf{t} \cdot \mathbf{n}_i\ \mathbf{m} \cdot \mathbf{n}_j = \mathbf{t}_i \cdot \mathbf{n}\ \mathbf{m}_j \cdot \mathbf{n} = \mathbf{t}_i \cdot \mathbf{n}\ \mathbf{n} \cdot \mathbf{m}_j$$
$$= \mathbf{t}_i \cdot \mathbf{m}_j = (\mathbf{t}_i \cdot \mathbf{m})_j - \mathbf{t}_{ij} \cdot \mathbf{m}.$$

Since $\Sigma\Sigma\ \varepsilon^{ij}\ \mathbf{t}_{ij} = 0$, it follows that

$$\sqrt{g}\,K = \Sigma\Sigma\ \varepsilon^{ij}\ (\mathbf{t}_i \cdot \mathbf{m})_j = \Sigma\Sigma\ \varepsilon^{ij}\ (\mathbf{m} \cdot \mathbf{t}_i)_j$$
$$= \Sigma\Sigma\ \varepsilon^{ij}\ [\mathbf{n}\ \mathbf{t}\ \mathbf{t}_i]_j.$$

(We have interchanged the $\mathbf{t}$ and $\mathbf{m}$ of Kreyszig [**1**, p. 146].)

Since $\mathbf{n}_1 \times \mathbf{n}_2$ is parallel to $\mathbf{n}$, 19.48 may be expressed in the form

$$|\mathbf{n}_1 \times \mathbf{n}_2| = |K|\sqrt{g},$$

which can be used to establish Gauss's geometrical interpretation for K. To obtain his *spherical representation* of a surface, Gauss considered the locus of the end Q of a vector

$$\overrightarrow{OQ} = \mathbf{n},$$

where O is a fixed point and $\mathbf{n}$ is the unit normal at a point P which varies on the given surface [Hilbert and Cohn-Vossen **1**, pp. 193–196]. When P travels over a sufficiently small region F, bounded by a simple closed curve on the surface, Q travels over a corresponding region G of the unit sphere with center O. Gauss defined the *total curvature* of the surface at P to be the limit of the ratio of the areas of G and F when these regions are shrunk to single points. By 19.24, the area of F is

$$\iint |\mathbf{r}_1\ du^1 \times \mathbf{r}_2\ du^2| = \iint \sqrt{g}\ du^1\ du^2.$$

Analogously, the area of G is

$$\iint |\mathbf{n}_1\ du^1 \times \mathbf{n}_2\ du^2| = \iint |K|\sqrt{g}\ du^1\ du^2.$$

Hence the limit of the ratio is $|K|$.

The characteristic property of a developable surface is that, instead of a two-parameter family of tangent planes, it only has a one-parameter family of tangent planes, and so also a one-parameter family of normals. In this case G is not a proper region but merely an arc, and therefore $K = 0$.

If the parametric curves are orthogonal, so that $g_{12} = 0$, we have $g^{11} = 1/g_{11}$, $g^{12} = 0$, $g^{22} = 1/g_{22}$, whence, by 19.37,

$$b_i^j = \Sigma\ g^{jk}\ b_{ik} = g^{jj}\ b_{ij} = b_{ij}/g_{jj}.$$

In this case the mean curvature H is given by

$$2H = \Sigma \, b^j_j = \frac{b_{11}}{g_{11}} + \frac{b_{22}}{g_{22}}.$$

EXERCISES

1.　Find the mean curvature H at a given point on the helicoid　$y = x \tan (z/c)$, parametrized in the form

$$x = u^1 \cos u^2, \quad y = u^1 \sin u^2, \quad z = cu^2.$$

2.　Find the mean curvature H at a given point on the general surface of revolution

$$\mathbf{r} = (u^1 \cos u^2, \quad u^1 \sin u^2, \quad z),$$

where z is a function of u^1 alone. This mean curvature is zero when z is given by

$$u^1 = a \cosh (z - c)/a,$$

i.e., when the surface is a catenoid.

3.　Locate the curves of parabolic points on the torus

$$x = (a + b \cos u^1) \cos u^2, \quad y = (a + b \cos u^1) \sin u^2, \quad z = b \sin u^1.$$

4.　The tangents $\mathbf{t} = \dot{\mathbf{r}}$ of a twisted curve $\mathbf{r} = \mathbf{r}\,(s)$ generate a surface

$$\mathbf{r}\,(s, u) = \mathbf{r}\,(s) + u\,\mathbf{t}\,(s).$$

Using s and u as parameters, obtain the fundamental magnitudes

$$b_{11} = \kappa \tau u, \qquad b_{12} = b_{22} = 0.$$

Deduce that $K = 0$ everywhere.

5.　The mean curvature and Gauss curvature are connected by the inequality

$$H^2 \geqslant K.$$

At what kind of point do we find $H^2 = K$?　(*Hint:* $H^2 - K = \frac{1}{4} (\kappa_{(1)} - \kappa_{(2)})^2$.)

6.　Derive 19.48 another way, by applying Lagrange's identity to $(\mathbf{r}_1 \times \mathbf{r}_2) \cdot (\mathbf{n}_1 \times \mathbf{n}_2)$.

7.　Derive 19.49 another way, by applying 17.62 in the form

$$\mathbf{t}_i = \nu_i \mathbf{m} - \mu_i \mathbf{n}, \quad \mathbf{m}_i = \lambda_i \mathbf{n} - \nu_i \mathbf{t}, \quad \mathbf{n}_i = \mu_i \mathbf{t} - \lambda_i \mathbf{m}$$

(where λ_i, μ_i, ν_i are functions of u^1 and u^2), so that

$$[\mathbf{n} \ \mathbf{n}_1 \ \mathbf{n}_2] = \lambda_1 \mu_2 - \lambda_2 \mu_1 = \mathbf{m}_2 \cdot \mathbf{t}_1 - \mathbf{m}_1 \cdot \mathbf{t}_2.$$

19.5　PRINCIPAL DIRECTIONS AND LINES OF CURVATURE

Returning to 19.44, which may be expressed as

$$\Sigma \, b^k_j \, a^j = \kappa a^k \qquad\qquad (k = 1, 2),$$

we find that the easiest way to eliminate κ is to multiply by $\Sigma \, \varepsilon_{ik} \, a^i$ and sum over k, obtaining

$$\Sigma\Sigma\Sigma\, \varepsilon_{ik}\, b_j^k\, a^i\, a^j = \kappa\, \Sigma\Sigma\, \varepsilon_{ik}\, a^i\, a^k = 0.$$

(This sum is zero because the only nonvanishing terms involve $\varepsilon_{12}\, a^1\, a^2$ and $\varepsilon_{21}\, a^2\, a^1$, which cancel.) Thus the principal directions $\Sigma\, a^i\, \mathbf{r}_i$ are given by the roots of the quadratic equation

$$\Sigma\Sigma\Sigma\, \varepsilon_{ik}\, b_j^k\, a^i\, a^j = 0$$

for $a^1 : a^2$. In other words, the lines of curvature are determined by the differential equation

19.51 $$\Sigma\Sigma\Sigma\, \varepsilon_{ik}\, b_j^k\, du^i\, du^j = 0$$

or

19.52 $$-b_1^2\, (du^1)^2 + (b_1^1 - b_2^2)\, du^1\, du^2 + b_2^1\, (du^2)^2 = 0.$$

To prove that the lines of curvature form an *orthogonal* net, we may apply the criterion 19.291 either to 19.51 or to 19.52, using

$$c_{ij} = \Sigma\, \varepsilon_{ik}\, b_j^k.$$

We obtain, in the notation of 19.371,

$$\Sigma\Sigma\, g^{ij}\, c_{ij} = \Sigma\Sigma\Sigma\, \varepsilon_{ik}\, g^{ij}\, b_j^k = \Sigma\Sigma\, \varepsilon_{ik}\, b^{ik} = 0.$$

It follows that, at any point P on a surface, *the two principal directions are perpendicular.*

Another consequence of 19.44 is Rodrigues's formula

19.53 $$d\mathbf{n} + \kappa\, d\mathbf{r} = \mathbf{0},$$

which shows what happens to the normal $\mathbf{n}$ when $\mathbf{r}$ is displaced in a principal direction. (The coefficient κ is the corresponding principal curvature.) In fact, by combining the Weingarten equations

$$\mathbf{n}_i = -\Sigma\, b_i^j\, \mathbf{r}_j$$

(19.38) with 19.44 in the form

19.54 $$\Sigma\, b_i^j\, du^i = \kappa\, du^j,$$

we obtain

$$d\mathbf{n} = \Sigma\, \mathbf{n}_i\, du^i = -\Sigma\Sigma\, b_i^j\, \mathbf{r}_j\, du^i$$

$$= -\kappa\Sigma\, \mathbf{r}_j\, du^j = -\kappa\, d\mathbf{r}.$$

(Olinde Rodrigues, 1794–1851.)

It follows that, if $d\mathbf{r}$ is in a principal direction, $d\mathbf{n}$ is in the same (or the opposite) direction. Moreover, the principal directions are the only directions in which this happens. For, if $d\mathbf{n}$ is parallel to $d\mathbf{r}$, the analysis given above shows that, for some number λ,

$$\Sigma\Sigma\, b_i^j\, \mathbf{r}_j\, du^i = \lambda\, \Sigma\, \mathbf{r}_j\, du^j.$$

Writing this in the form

$$\Sigma(\Sigma\, b_i^j\, du^i - \lambda\, du^j)\, \mathbf{r}_j = \mathbf{0},$$

we see that it implies

$$\Sigma_i\, b_i^j\, du^i - \lambda\, du^j = 0 \qquad\qquad (j = 1, 2).$$

Eliminating λ, the way we eliminated κ before, we again obtain 19.51, which is the differential equation for the lines of curvature.

If the parametric curves are orthogonal, so that $b_i^j = b_{ij}/g_{jj}$, the equation 19.52 becomes

19.55 $$-\frac{b_{12}}{g_{22}}(du^1)^2 + \left(\frac{b_{11}}{g_{11}} - \frac{b_{22}}{g_{22}}\right) du^1\, du^2 + \frac{b_{12}}{g_{11}}(du^2)^2 = 0.$$

For the investigation of a given surface, any net of curves on the surface may be used as parametric curves. The lines of curvature provide a standard net which is always available. Comparing 19.52 with $du^1\, du^2 = 0$, we see that

19.56 $$b_1^2 = b_2^1 = 0.$$

Hence

The parametric curves are the lines of curvature if and only if b_1^2 and b_2^1 are identically zero.

In this case the equation 19.45 reduces to

$$(\kappa - b_1^1)\,(\kappa - b_2^2) = 0,$$

so the two principal curvatures are b_1^1 and b_2^2. To see which is which, we apply Rodrigues's formula 19.53 to displacements along the parametric curves. The "first" principal direction is naturally the one along which u^1 varies while u^2 remains constant, that is, the direction of $\mathbf{r}_1$; and the "second" is the direction of $\mathbf{r}_2$. Thus

19.57 $$\mathbf{n}_i + \kappa_{(i)}\, \mathbf{r}_i = \mathbf{0}.$$

Taking inner products with $\mathbf{r}^j$ and $\mathbf{r}_j$, we deduce

$$-b_i^j + \kappa_{(i)}\, \delta_i^j = 0, \quad -b_{ij} + \kappa_{(i)}\, g_{ij} = 0.$$

Hence the two principal curvatures are precisely

19.58 $$\kappa_{(1)} = b_1^1 = \frac{b_{11}}{g_{11}}, \quad \kappa_{(2)} = b_2^2 = \frac{b_{22}}{g_{22}},$$

and we see also that

$$b_{12} = \kappa_{(1)}\, g_{12} = 0$$

(since the lines of curvature are orthogonal).

Conversely, if the parameters on any surface are so chosen that

19.59 $g_{12} = 0, \qquad b_{12} = 0,$

then the parametric curves are the lines of curvature; for, since

$$g^{21} = -\frac{g_{12}}{g} = 0,$$

we have

$$b_1^2 = \Sigma \, g^{2j} b_{1j} = g^{22} b_{12} = 0$$

and likewise $b_2^1 = 0$. In fact, the conditions 19.59 are equivalent to 19.56.

EXERCISES

1. Apply 19.291 to 19.52, so as to prove the orthogonality of the principal directions without using the symbol ε_{ik}.

2. The differential equation for the lines of curvature may be expressed in the form

$$\begin{vmatrix} g_{11} & g_{12} & g_{22} \\ b_{11} & b_{12} & b_{22} \\ (du^2)^2 & -du^1 \, du^2 & (du^1)^2 \end{vmatrix} = 0.$$

3. Find the lines of curvatures on the hyperbolic paraboloid $x^2 - y^2 = 2z$, parametrized in the form

$$x = \sinh u^1 + \sinh u^2, \quad y = \sinh u^1 - \sinh u^2, \quad z = 2 \sinh u^1 \sinh u^2.$$

4. Find the lines of curvature on the helicoid $y = x \tan (z/c)$, parametrized in the form

$$x = u^1 \cos u^2, \quad y = u^1 \sin u^2, \quad z = cu^2.$$

5. The equations 19.56 or 19.59, holding at a particular point (but not necessarily identically), are conditions for the parametric directions to coincide with the principal directions at the point considered. The formulas 19.58 still hold at this point.

19.6   UMBILICS

An umbilic is a point at which the normal curvature κ is the same in all directions. At such a point, the equations 19.42, 19.43 are satisfied for all values of $a^1 : a^2$, and therefore

$$b_{ik} = \kappa g_{ik}, \qquad b_i^j = \kappa \delta_i^j.$$

In fact, we have two alternative sets of conditions for an umbilic: one set is

$$b_{11} : b_{12} : b_{22} = g_{11} : g_{12} : g_{22},$$

and the other,

19.61 $b_1^2 = b_2^1 = 0, \qquad b_1^1 = b_2^2.$

If a surface is symmetrical by reflection in a plane, its section by the plane is a line of curvature. To see why this happens, consider a point P on the section. If the principal directions at P were oblique to the plane, either of them would reflect into another principal direction associated with the same principal curvature. Such an abundance of principal directions would make P an umbilic. But a curve consisting entirely of umbilics is, trivially, a line of curvature.

In particular, *on any surface of revolution, the meridians and parallels are lines of curvature.* An interesting special case arises when we rotate a plane curve about one of the normals of its evolute; that is, if C is the center of curvature for a point P on the plane curve, we rotate about the line through C parallel to the tangent at P. In this case the locus of P is an "equator" whose radius is equal to the radius of curvature of the meridian. The equator, like the meridian, is both a line of curvature and a normal section. Since the two principal curvatures are equal, every point on the equator is an umbilic.

At an umbilic, Rodrigues's formula 19.53 holds for all displacements on the surface. In particular, we can apply it to displacements along the parametric curves, obtaining

19.62 $$\mathbf{n}_j + \kappa \mathbf{r}_j = \mathbf{0} \qquad\qquad (j = 1, 2).$$

Hence

19.63   *If every point is an umbilic, the surface is either a plane or a sphere.*

For, in this case 19.62 holds everywhere. Differentiating, we deduce

$$\mathbf{n}_{ij} + \kappa \mathbf{r}_{ij} + \kappa_i\, \mathbf{r}_j = \mathbf{0}, \qquad \kappa_1\, \mathbf{r}_2 = \kappa_2\, \mathbf{r}_1,$$

and $\kappa_1 = \kappa_2 = 0$. Thus κ is constant, and 19.62 yields $(\mathbf{n} + \kappa \mathbf{r})_j = \mathbf{0}$, so that $\mathbf{n} + \kappa \mathbf{r}$ is constant. If $\kappa = 0$, $\mathbf{n}$ is constant, and we have a plane. If $\kappa \neq 0$, a suitable origin makes $\mathbf{r} = -\kappa^{-1}\mathbf{n}$, $|\mathbf{r}| = |\kappa|^{-1}$, and we have a sphere.

EXERCISES

1. How can the equation 19.52 be used to derive the conditions

$$b_1^2 = b_2^1 = 0, \qquad b_1^1 = b_2^2$$

for an umbilic?

2. What happens to the equation 19.45 when these conditions are satisfied?

3. Anticlastic surfaces have no umbilics.

4. The surface

$$x = \sqrt{2}\cos u^1, \quad y = \sqrt{2}\cos u^2, \quad z = \sin u^1 \sin u^2$$

has a curve of umbilics lying on the sphere $x^2 + y^2 + z^2 = 4$.

5. Does every umbilic lie on infinitely many lines of curvature?

19.7   DUPIN'S THEOREM AND LIOUVILLE'S THEOREM

> *Dupin investigated triply orthogonal families of surfaces, not as a bar-*
> *ren exercise in the differential calculus but because certain instances of*
> *such families are of the first importance in . . . mathematical physics.*
> *[They] were the occasion for one of Darboux' more famous works,*
> *extending to 567 pages.*
>
> E. T. Bell [**2**, p.331]

In the exercises at the end of § 18.5, we found several instances of three-dimensional coordinates having the special property

$$g_{23} = g_{31} = g_{12} = 0,$$

so that the level surfaces all cut one another at right angles. In such a case the three systems of surfaces are said to be *mutually orthogonal.*

Differentiating the equation $\mathbf{r}_1 \cdot \mathbf{r}_2 = 0$ with respect to u^3, we obtain

$$\mathbf{r}_{13} \cdot \mathbf{r}_2 + \mathbf{r}_1 \cdot \mathbf{r}_{23} = 0.$$

From this and two other such equations (derived by cyclic permutation of 123) we deduce

$$\mathbf{r}_1 \cdot \mathbf{r}_{23} = \mathbf{r}_2 \cdot \mathbf{r}_{31} = \mathbf{r}_3 \cdot \mathbf{r}_{12} = 0.$$

Since $\mathbf{r}_3$ is normal to the surface $u^3 = c$ of the third system, this surface satisfies not only $g_{12} = 0$ but also, since $\mathbf{r}_3 \cdot \mathbf{r}_{12} = 0$,

$$b_{12} = \mathbf{n} \cdot \mathbf{r}_{12} = 0,$$

as in 19.59. Hence the parametric curves $u^1 = a$ and $u^2 = b$ on this surface are lines of curvature. Since a similar result holds for a surface of either of the other systems, we have now proved

DUPIN'S THEOREM.   *In three mutually orthogonal systems of surfaces, the lines of curvature on any surface in one of the systems are its intersections with the surfaces of the other two systems.*

Moreover, any surface may be exhibited as a member of one of three mutually orthogonal systems. This can actually be done in many ways. One way is to use a system of *parallel* surfaces, defined as the loci of points at constant distances along the normals of the given surface. Using the lines of curvature as parametric curves, we may express the position vector of a typical parallel surface in the form

$$\bar{\mathbf{r}} = \mathbf{r} + u^3\, \mathbf{n}.$$

We see from 19.57 that the directions of the parametric curves on the new surface are given by

$$\bar{\mathbf{r}}_i = (\mathbf{r} + u^3\mathbf{n})_i = \mathbf{r}_i + u^3\mathbf{n}_i$$
$$= \mathbf{r}_i - u^3 \kappa_{(i)}\, \mathbf{r}_i = (1 - \kappa_{(i)}\, u^3)\, \mathbf{r}_i,$$

that is, they are parallel to those on the original surface. Therefore, both surfaces have the same normal: $\bar{\mathbf{n}} = \mathbf{n}$. Since

$$\bar{g}_{12} = \bar{\mathbf{r}}_1 \cdot \bar{\mathbf{r}}_2 = 0$$

and

$$\bar{b}_{12} = -\bar{\mathbf{n}}_1 \cdot \bar{\mathbf{r}}_2 = -\mathbf{n}_1 \cdot \bar{\mathbf{r}}_2 = \kappa_{(1)} \mathbf{r}_1 \cdot \bar{\mathbf{r}}_2 = 0,$$

the parametric curves on the new surface are again lines of curvature [Weatherburn **2**, p. 159]. Allowing u^3 to take various values, we obtain a whole system of parallel surfaces. The other two orthogonal systems are traced out by the normals at points that run along the lines of curvature [La Vallée Poussin **2**, p. 447].

In § 6.7 and § 7.7 we analysed the general circle-preserving and sphere-preserving transformations. In § 9.7 we indicated how the theory of functions of a complex variable could be employed to prove that the circle-preserving transformations are the only angle-preserving transformations of the whole inversive plane into itself. It is very remarkable that the analogous theorem in three (or more) dimensions is elementary! We merely have to observe that, if a transformation of space preserves the angles at which surfaces cut, it transforms mutually orthogonal systems of surfaces into mutually orthogonal systems of surfaces. Hence, if it transforms a surface σ into another surface σ', it transforms the lines of curvature on σ into the lines of curvature on σ'. Since a sphere (including a plane as a special case) is characterized by the property that *all* directions on it are principal directions, we can immediately deduce

LIOUVILLE'S THEOREM. *Every angle-preserving transformation is a sphere-preserving transformation.*

Taking this along with 7.71, we see that every angle-preserving transformation is either a similarity or the product of an inversion and an isometry [Forder **3,** pp. 137–138].

EXERCISES

1. When the surfaces $u^1 = a$ and $u^2 = b$ are traced out by normals along lines of curvature, while the surfaces $u^3 = c$ are "parallel," the fundamental magnitudes for $u^1 = a$ are naturally denoted by $g_{22}, g_{23}, g_{33}, b_{22}, b_{23}, b_{33}$. Dupin's theorem shows that $g_{23} = b_{23} = 0$. Compute b_{33}, and deduce that for this surface $K = 0$.

2. The central quadrics

$$\frac{x^2}{A - \lambda} + \frac{y^2}{B - \lambda} + \frac{z^2}{C - \lambda} = 1 \qquad (A > B > C)$$

(which are ellipsoids when $\lambda < C$, hyperboloids of one sheet when $C < \lambda < B$, hyperboloids of two sheets when $B < \lambda < A$) are said to be *confocal*. At a point (x, y, z) on such a quadric, the direction of the normal is

$$\left(\frac{x}{A - \lambda}, \frac{y}{B - \lambda}, \frac{z}{C - \lambda} \right).$$

When x, y, z are given, 18.59 is a cubic equation for λ, whose roots u^1, u^2, u^3 (numbered in increasing order) satisfy

$$u^1 < C < u^2 < B < u^3 < A.$$

Deduce that any point for which $xyz \neq 0$ lies on three quadrics of the system (one of each kind), cutting one another orthogonally [La Vallée Poussin **2**, p. 448].

3. Where are the lines of curvature on an ellipsoid? [Hilbert and Cohn-Vossen **1**, p. 189.]

19.8   DUPIN'S INDICATRIX

Although the indicatrix was not invented by Dupin, he made more effective use than had his predecessors of this suggestive conic in which a plane parallel to, and "infinitesimally near to," the tangent plane at any point of a surface intersects the surface.

E. T. Bell [**2**, p. 331]

When a surface is given in Monge's form $z = F(x, y)$, it is often convenient to use the coordinates x, y themselves as parameters, so that z is a function of x and y with derivatives

$$z_1 = \frac{\partial z}{\partial x}, \quad z_2 = \frac{\partial z}{\partial y}, \quad z_{11} = \frac{\partial^2 z}{\partial x^2}, \quad z_{12} = \frac{\partial^2 z}{\partial x\, \partial y}, \quad z_{22} = \frac{\partial^2 z}{\partial y^2}.$$

Differentiating $\mathbf{r} = (x, y, z)$, we obtain

$$\mathbf{r}_1 = (1, 0, z_1), \quad \mathbf{r}_2 = (0, 1, z_2), \quad \mathbf{r}_{ij} = (0, 0, z_{ij}),$$

whence

$$g_{11} = 1 + z_1{}^2, \quad g_{12} = z_1 z_2, \quad g_{22} = 1 + z_2{}^2,$$

$$g = g_{11}\, g_{22} - g_{12}{}^2 = 1 + z_1{}^2 + z_2{}^2,$$

$$\sqrt{g}\, \mathbf{n} = \mathbf{r}_1 \times \mathbf{r}_2 = (-z_1, -z_2, 1),$$

$$\sqrt{g}\, b_{ij} = \sqrt{g}\, \mathbf{n} \cdot \mathbf{r}_{ij} = z_{ij}.$$

If the coordinate axes are so chosen that the point under consideration is the origin and the normal there is the z-axis, we have

$$\mathbf{n} = (0, 0, 1),$$

whence $z_1 = z_2 = 0$, $g_{11} = 1$, $g_{12} = 0$, $g_{22} = 1$, $g = 1$, $b_{ij} = z_{ij}$.

Since z is a function of x and y, we can expand it in a Maclaurin series

$$z = z(0, 0) + z_1\, x + z_2\, y + \tfrac{1}{2}(z_{11}\, x^2 + 2z_{12}\, xy + z_{22}\, y^2) + \tfrac{1}{6}(z_{111}\, x^3 + \dots) + \dots$$

$$= \tfrac{1}{2}(b_{11}\, x^2 + 2b_{12}\, xy + b_{22}\, y^2) + \text{(terms of higher degree in } x \text{ and } y).$$

The "terms of higher degree" become important if $b_{11} = b_{12} = b_{22} = 0$, in which case the origin is a parabolic umbilic. In all other cases the section by a plane $z = \varepsilon$, parallel to the tangent plane $z = 0$ at a small distance $|\varepsilon|$, resembles the conic

$$b_{11} x^2 + 2b_{12} xy + b_{22} y^2 = 2\varepsilon,$$

which is similar to *Dupin's indicatrix*

19.81 $$b_{11} x^2 + 2b_{12} xy + b_{22} y^2 = \pm 1.$$

We shall find that this conic (or pair of conics) indicates, in a remarkably simple manner, the normal curvature in every direction. We first observe that this part of the surface is synclastic, anticlastic, or parabolic, according as

$$b > 0, \quad b < 0, \quad \text{or} \quad b = 0,$$

that is, according as the indicatrix is an ellipse, two conjugate hyperbolas, or two parallel lines. (In the synclastic case the ambiguous sign in the equation 19.81 must agree with the sign of b_{11}; otherwise the plane $z = \varepsilon$ would fail to meet the surface. In the anticlastic case we need both signs: one for each of the two conjugate hyperbolas.)

In the plane $z = 0$, which is the tangent plane at the origin, the vector $(\cos \theta, \sin \theta, 0)$, making an angle θ with the x-axis $(1, 0, 0)$, may be identified with the tangent

$$\overrightarrow{PT} = \mathbf{t} = \Sigma \, a^i \, \mathbf{r}_i = (a^1, a^2, 0)$$

of Figure 19.2*a*. Thus the contravariant components of $\mathbf{t}$ are $a^1 = \cos \theta$, $a^2 = \sin \theta$, and, by 19.35, the normal curvature in this direction is

$$\kappa = \Sigma\Sigma \, b_{ij} \, a^i \, a^j$$

19.82 $$= b_{11} \cos^2 \theta + 2b_{12} \cos \theta \sin \theta + b_{22} \sin^2 \theta.$$

Expressing the indicatrix 19.81 in polar coordinates, we obtain

$$b_{11} r^2 \cos^2 \theta + 2b_{12} r^2 \cos \theta \sin \theta + b_{22} r^2 \sin^2 \theta = \pm 1,$$

that is, $\kappa r^2 = \pm 1$, or

$$r = |\kappa|^{-\frac{1}{2}}.$$

In other words [La Vallée Poussin **2,** p. 427],

The radius of the indicatrix in any direction is equal to the square root of the radius of normal curvature in this direction.

Another way of expressing the same idea is to remark that the surface is approximated by the paraboloid or parabolic cylinder

$$2z = b_{11} x^2 + 2b_{12} xy + b_{22} y^2.$$

In any direction (at the origin) the given surface and the quadric have the same normal curvature.

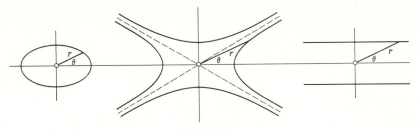

Figure 19.8a

If we choose the x- and y-axes along the principal directions at the origin, as in Figure 19.8a, so that $b_{12} = 0$ and, by 19.58, $\kappa_{(i)} = b_{ii}$, the indicatrix 19.81 is simply

$$\kappa_{(1)} x^2 + \kappa_{(2)} y^2 = \pm 1,$$

and 19.82 yields *Euler's formula*

$$\kappa = \kappa_{(1)} \cos^2 \theta + \kappa_{(2)} \sin^2 \theta$$

for the normal curvature in a direction making an angle θ with the first principal direction.

EXERCISES

1. For which directions on a surface is the normal curvature equal to the arithmetic mean of the two principal curvatures?

2. For the surface $z = F(x, y)$,

$$2H = \frac{g_{22} b_{11} - 2g_{12} b_{12} + g_{11} b_{22}}{g} = \frac{(1 + z_2{}^2)z_{11} - 2z_1 z_2 z_{12} + (1 + z_1{}^2)z_{22}}{(1 + z_1{}^2 + z_2{}^2)^{3/2}}.$$

3. The surface $xyz = 1$ has umbilics at the four vertices of a regular tetrahedron [Salmon **2**, p. 300].

4. The surface $z = x(x^2 - 3y^2)$ has a parabolic umbilic at the origin. Sketch the section by a plane $z = \varepsilon$, where ε is a small number, positive or negative.

This surface is called the *monkey saddle* because it would be the right kind of saddle for a monkey riding a bicycle: one way down for each hind leg and a third for the tail. Hilbert and Cohn-Vossen [**1**, p. 202, Fig. 213] made a nice drawing of the surface but a very misleading one of its generalized indicatrix [*ibid.*, p. 192, Fig. 200]. For the true shape, see the second edition of Struik [**1**, p. 85].

20

Geodesics

Imagine a two-dimensional creature, sufficiently intelligent to make precise measurements on the surface he inhabited, but unable to conceive of a third dimension. His world might be a plane or a parabolic cylinder; he could not tell the difference. If it were a circular cylinder, local measurements would still give the same results, though an expedition all the way round would reveal a topological peculiarity. In all these cases his conclusion would be that his surface had zero curvature: $K = 0$. If, on the other hand, the surface were a sphere, he could detect its positive curvature by measurements within easy reach of his home, even if the radius of the sphere were relatively large. This information is a consequence of Gauss's formula 20.16, which expresses K in terms of the fundamental magnitudes of the first order. In § 20.3 we shall see how Gauss's complicated expression can be replaced by his simple one involving the three angles of a triangle (like our formula 6.92 for the area of a spherical triangle). In § 20.4 we shall extend these local measurements to global measurements, which would enable our intelligent ant to determine the topological nature of his world: an idea which we shall investigate more systematically in Chapter 21.

The remaining sections of the present chapter deal with the differential-geometric approach to the non-Euclidean planes, which may be regarded as surfaces of constant curvature.

20.1 THEOREMA EGREGIUM

The Christoffel symbols are called after Erwin Bruno Christoffel (1829 - 1901), who introduced [them] in 1869, . . . denoting our Γ^i_{jk} by $\{^{jk}_i\}$.

The change to our present notation has been made under the influence of tensor theory.

D. J. Struik (1894 -   )

[Struik **1,** p. 108]

For an adequate discussion of "geodesics," which are the most important curves on a surface, we need one more notational device: the *Christoffel symbols*

20.11 $$\Gamma_{ij,k} = \mathbf{r}_{ij} \cdot \mathbf{r}_k, \quad \Gamma_{ij}^k = \mathbf{r}_{ij} \cdot \mathbf{r}^k.$$

In virtue of 19.15, these symbols are related as follows:

20.12 $$\Gamma_{ij,k} = \Sigma g_{kl} \Gamma_{ij}^l, \quad \Gamma_{ij}^k = \Sigma g^{kl} \Gamma_{ij,l}$$

Clearly, $$\Gamma_{21,k} = \Gamma_{12,k} \quad \text{and} \quad \Gamma_{21}^k = \Gamma_{12}^k.$$

Since the derivatives of the fundamental magnitudes g_{ij} are

$$(g_{ij})_k = \frac{\partial}{\partial u^k} (\mathbf{r}_i \cdot \mathbf{r}_j) = \mathbf{r}_{ik} \cdot \mathbf{r}_j + \mathbf{r}_i \cdot \mathbf{r}_{jk}$$

20.13 $$= \Gamma_{ik,j} + \Gamma_{jk,i},$$

we can compute the Christoffel symbols of the first kind by means of the formula

20.14 $$\Gamma_{ij,k} = \tfrac{1}{2} \{(g_{jk})_i + (g_{ik})_j - (g_{ij})_k\}$$

and then deduce the Christoffel symbols of the second kind by means of the latter half of 20.12.

Applying 18.21 with $\mathbf{u} = \mathbf{r}_{ij}$ and $\mathbf{r}_3 = \mathbf{r}^3 = \mathbf{n}$, we obtain

$$\mathbf{r}_{ij} = \mathbf{r}_{ij} \cdot \mathbf{r}^1 \mathbf{r}_1 + \mathbf{r}_{ij} \cdot \mathbf{r}^2 \mathbf{r}_2 + \mathbf{r}_{ij} \cdot \mathbf{r}^3 \mathbf{r}_3$$

$$= \Gamma_{ij}^1 \mathbf{r}_1 + \Gamma_{ij}^2 \mathbf{r}_2 + \mathbf{r}_{ij} \cdot \mathbf{n} \, \mathbf{n}$$

20.15 $$= \Sigma \Gamma_{ij}^k \mathbf{r}_k + b_{ij} \mathbf{n}.$$

These expressions for the second derivatives of $\mathbf{r}$ are known as "the equations of Gauss."

Another of Gauss's discoveries, which pleased him so much that he named it *Theorema egregium,* is an expression for K in terms of the g's and their derivatives. This means that the Gaussian curvature can be computed by measurements made on the surface itself, without reference to the three-dimensional space in which it may lie. In other words, K is a "bending invariant," unchanged by the kind of distortion that takes place when a flat sheet of paper is rolled up to make a cylinder or a cone. The expression appears in the literature in various forms, not obviously identical. One of the neatest, discovered by Liouville,* is

20.16 $$\sqrt{g} \, K = \frac{\partial}{\partial u^2} \left(\frac{\sqrt{g}}{g_{11}} \Gamma_{11}^2 \right) - \frac{\partial}{\partial u^1} \left(\frac{\sqrt{g}}{g_{11}} \Gamma_{12}^2 \right).$$

This can be derived by applying 19.49 to the unit tangent vector $\mathbf{t} = \mathbf{r}_1/g_1$. Since $[\mathbf{n} \, \mathbf{r}_1 \, \mathbf{r}_1] = 0$ and (by 19.33) $\mathbf{n} \times \mathbf{r}_1 = \sqrt{g} \, \mathbf{r}^2$, we have

---

* J. Liouville, Sur la théorie générale des surfaces, *Journal de Mathématiques* **16** (1851), pp. 130–132.

$$[\mathbf{n}\,\mathbf{t}\,\mathbf{t}_i] = \begin{bmatrix} \mathbf{n} & \dfrac{\mathbf{r}_1}{g_1} & \dfrac{\mathbf{r}_{1i}}{g_1} \end{bmatrix} = (\mathbf{n} \times \mathbf{r}_1) \cdot \dfrac{\mathbf{r}_{1i}}{g_{11}}$$

$$= \sqrt{g}\,\mathbf{r}^2 \cdot \dfrac{\mathbf{r}_{1i}}{g_{11}} = \dfrac{\sqrt{g}}{g_{11}}\,\Gamma_{1i}^2\,.$$

Hence
$$\sqrt{g}\,K = \Sigma\Sigma\varepsilon^{ij}\left(\dfrac{\sqrt{g}}{g_{11}}\,\Gamma_{1i}^2\right)_{\!j}$$

$$= \left(\dfrac{\sqrt{g}}{g_{11}}\,\Gamma_{11}^2\right)_{\!2} - \left(\dfrac{\sqrt{g}}{g_{11}}\,\Gamma_{12}^2\right)_{\!1}.$$

Theorema egregium takes a more symmetrical form when the parametric curves are orthogonal, so that

$$g_{12} = 0, \quad g^{ii} = 1/g_{ii} = 1/(g_i)^2,$$

$$\dfrac{\sqrt{g}}{g_{11}}\,\Gamma_{12}^2 = \dfrac{g_1 g_2}{g_{11}}\dfrac{(g_{22})_1}{2g_{22}} = \dfrac{g_2}{g_1}\dfrac{2g_2(g_2)_1}{2g_2{}^2} = \dfrac{(g_2)_1}{g_1}$$

and

$$\dfrac{\sqrt{g}}{g_{11}}\,\Gamma_{11}^2 = \dfrac{g_1 g_2}{g_{11}}\left(-\dfrac{(g_{11})_2}{2g_{22}}\right) = -\dfrac{g_2}{g_1}\dfrac{2g_1(g_1)_2}{2g_2{}^2} = -\dfrac{(g_1)_2}{g_2}\,.$$

In fact,

$$-g_1 g_2 K = \left(\dfrac{\sqrt{g}}{g_{11}}\,\Gamma_{12}^2\right)_{\!1} - \left(\dfrac{\sqrt{g}}{g_{11}}\,\Gamma_{11}^2\right)_{\!2}$$

20.17
$$= \left(\dfrac{(g_2)_1}{g_1}\right)_{\!1} + \left(\dfrac{(g_1)_2}{g_2}\right)_{\!2}.$$

[Weatherburn **2**, p. 98; Struik **1**, p. 113].

EXERCISES

1. Obtain a variant of 20.16 by taking $\mathbf{t} = \mathbf{r}_2/g_2$.

2. For the case when $g_{12} = 0$, express all the Christoffel symbols in terms of g_{11}, g_{22}, and their derivatives.

3. Compute the Christoffel symbols for polar coordinates in the plane.

4. For a surface in Monge's form $z = F(x, y)$,
$$\Gamma_{ij}^k = z_{ij}\,z_k/(1 + z_1{}^2 + z_2{}^2).$$

5. $\Sigma\,\Gamma_{ij}^i = (\log \sqrt{g})_j$.

6. *Compute K for a surface on which
$$g_{11} = \dfrac{(u^2)^2 + a^2}{\{(u^1)^2 + (u^2)^2 + a^2\}^2}, \quad g_{12} = \dfrac{-u^1 u^2}{\{(u^1)^2 + (u^2)^2 + a^2\}^2},$$

$$g_{22} = \dfrac{(u^1)^2 + a^2}{\{(u^1)^2 + (u^2)^2 + a^2\}^2}\,.$$

* E. Beltrami, *Annali di Matematica* (1) **7** (1866), pp. 197–198.

20.2 THE DIFFERENTIAL EQUATIONS FOR GEODESICS

Every sufficiently small portion of a geodesic is the shortest path on the surface connecting the end-points of the portion. . . . All the intrinsic properties of a surface (such as its Gaussian curvature) can be determined by drawing geodesics and measuring their arc lengths. . . . We can obtain an approximation to the geodesics by moving a very small buggy along the surface on two wheels, the wheels being rigidly fastened to their common axis so that their speeds of rotation are equal. . . . The entire course of a geodesic is determined if one of its points and its direction at this point are given. . . . The straightest lines may also be characterized by the geometric requirement that the osculating plane of the curve is to contain the normal to the surface at every point of the curve.

Hilbert and Cohn-Vossen

[**1,** pp. 220-221]

Consider the possibility of a curve on a surface having all its principal normals normal to the surface. As we saw in § 19.3, any curve on a surface satisfies

$$\kappa \mathbf{p} = \dot{\mathbf{t}} = \frac{d}{ds} \Sigma \dot{u}^i \mathbf{r}_i = \Sigma \ddot{u}^i \mathbf{r}_i + \Sigma\Sigma \dot{u}^i \dot{u}^j \mathbf{r}_{ij}.$$

In the present case, since $\mathbf{p} = \mathbf{n}$ is perpendicular to $\mathbf{r}^k$, we have

$$(\Sigma \ddot{u}^i \mathbf{r}_i + \Sigma\Sigma \dot{u}^i \dot{u}^j \mathbf{r}_{ij}) \cdot \mathbf{r}^k = 0.$$

Since $\mathbf{r}_i \cdot \mathbf{r}^k = \delta_i^k$ and $\mathbf{r}_{ij} \cdot \mathbf{r}^k = \Gamma_{ij}^k,$ these equations reduce to

20.21 $$\ddot{u}^k + \Sigma\Sigma \Gamma_{ij}^k \, \dot{u}^i \dot{u}^j = 0 \qquad (k = 1, 2),$$

meaning

$$\frac{d^2 u^k}{ds^2} + \Sigma\Sigma \Gamma_{ij}^k \frac{du^i}{ds} \frac{du^j}{ds} = 0$$

[Struik **1,** p. 132]. Theoretically, we could eliminate s from these two equations so as to obtain a single differential equation; but it is usually more convenient either to use both equations or to use one of them along with 19.16.

The curves so determined are called *geodesics* [Weatherburn **2,** p. 100]. Since the equations express the second derivatives of u^k in terms of the first derivatives, there is a geodesic through any given point A (on the surface) in any given direction. There is also, in general, a unique geodesic joining two given points A and B. In these respects the geodesics on a surface resemble the straight lines in a plane; in fact, as we shall see, they are straight when the surface is a plane.

When $g_{12} = 0$, so that the parametric curves are orthogonal, the differential equations take the form

20.22 $g_{11} \ddot{u}^1 + \frac{1}{2}(g_{11})_1 (\dot{u}^1)^2 + (g_{11})_2 \dot{u}^1 \dot{u}^2 - \frac{1}{2}(g_{22})_1 (\dot{u}^2)^2 = 0,$

20.23 $g_{22} \ddot{u}^2 - \frac{1}{2}(g_{11})_2 (\dot{u}^1)^2 + (g_{22})_1 \dot{u}^1 \dot{u}^2 + \frac{1}{2}(g_{22})_2 (\dot{u}^2)^2 = 0.$

The latter shows that the parametric curves $\dot{u}^2 = 0$ occur among the geodesics if and only if g_{11} is a function of u^1 alone (not involving u^2), so that $(g_{11})_2 = 0$. In this case the curves $\dot{u}^2 = 0$ are a one-parameter system of geodesics and the curves $\dot{u}^1 = 0$ are their orthogonal trajectories. Since g_{11} is a function of u^1 alone, the differential form

$$ds^2 = g_{11} (du^1)^2 + g_{22} (du^2)^2$$

can be simplified by changing the notation so that $\int g_1 \, du^1$ is called u^1. Then $g_{11} = 1, \quad g = g_{22}, \quad$ and

20.24 $$ds^2 = (du^1)^2 + g(du^2)^2.$$

The effect of this change of notation is to make u^1 measure the arc of each geodesic $u^2 = $ constant. The differential equations are now

20.25 $$\ddot{u}^1 - \frac{1}{2}(g)_1 (\dot{u}^2)^2 = 0,$$

20.26 $$\frac{d}{ds} (g\dot{u}^2) - \frac{1}{2}(g)_2 (\dot{u}^2)^2 = 0.$$

In particular, we obtain *geodesic polar* coordinates (analogous to ordinary polar coordinates in the plane) by measuring u^1 from A along all the geodesics through A, and defining u^2 to be the angle that such a geodesic $u^2 = $ constant makes with an "initial" geodesic $u^2 = 0$.

The length of any curve from A to a point B (of general position) is obtained by integrating ds along the curve. The equation 20.24 shows that $\int ds \geqslant \int du^1$, with equality only when $du^2 = 0$. Hence the geodesic AB is the *shortest path* from A to B. In fact, it is the curve along which a tightly stretched string would lie on the smooth convex side of a material surface. Since the only forces acting on an "element" of the string are the tensions at its two ends and the reaction of the surface along the normal $\mathbf{n}$, these three forces must be in equilibrium. Hence $\mathbf{n}$ must lie in the plane of the two tensions, which is the osculating plane of the curve. These considerations provide a statical explanation for the equation $\mathbf{p} = \mathbf{n}$ which started this investigation.

The curves $u^1 = $ constant (which, of course, are not geodesics) are called "geodesic circles." The circumference of such a "circle" is obtained by integrating ds (given by 20.24 with $du^1 = 0$); thus it is

$$\int_0^{2\pi} \sqrt{g} \, du^2.$$

When the radius u^1 is small, this circumference is approximated by both $2\pi\sqrt{g}$ and $2\pi u^1$. Hence the first term in the Maclaurin series for $\sqrt{g}$ is simply u^1, that is,

20.27 If $u^1 = 0$, then $(\sqrt{g})_1 = 1$.

Since geodesics are curves of shortest length, the geodesics on a sphere are the great circles, and the geodesics in a plane are the straight lines. We can adapt Figure 8.5*b* to geodesic polar coordinates by writing u^1 and u^2 for r and θ, so that 20.24 expresses Pythagoras's theorem for the infinitesimal triangle $PP'N$, and the angle ϕ between the geodesics OP' and PP' is given by

$$\cos \phi = \lim \frac{NP'}{PP'} = \dot{u}^1 \quad \text{or} \quad \sin \phi = \lim \frac{NP}{PP'} = \sqrt{g}\,\dot{u}^2.$$

Differentiating $\cos \phi$ and using 20.25, we obtain

$$-\sin \phi\, \dot{\phi} = \ddot{u}^1 = \tfrac{1}{2}(g)_1\,(\dot{u}^2)^2 = \frac{(g)_1}{2\sqrt{g}}\,(\sqrt{g}\,\dot{u}^2)\dot{u}^2$$

$$= (\sqrt{g})_1(\sin \phi)\dot{u}^2.$$

Thus

20.28 $$d\phi = -(\sqrt{g})_1\,du^2.$$

EXERCISES

1. In the case of the general surface of revolution

$$\mathbf{r} = (u^1 \cos u^2, \quad u^1 \sin u^2, \quad z),$$

where z is a function of u^1 alone, the differential equation 20.23 for geodesics reduces to

$$\frac{d}{ds}\left[(u^1)^2\, \dot{u}^2 \right] = 0.$$

One solution is $du^2 = 0$, showing that *the meridians are geodesics*. In all other cases the constant value of $(u^1)^2\, \dot{u}^2$ may be denoted by $1/h$, so that

$$ds = h\,(u^1)^2\, du^2.$$

Comparing this with 19.16, obtain the complete integral

$$u^2 = C \pm \int \left[\frac{1 + z_1{}^2}{h^2(u^1)^2 - 1} \right]^{\frac{1}{2}} \frac{du^1}{u^1}$$

[Weatherburn **2,** p. 102].

2. The geodesics on a cylinder are helices.
3. The geodesics on the cone

$$\mathbf{r} = (u^1 \cos u^2, \quad u^1 \sin u^2, \quad u^1 \cos \alpha)$$

are given by

$$au^1 = \sec(\beta + u^2 \sin \alpha),$$

where α and β are constants. Are these curves concho-spirals?

20.3　THE INTEGRAL CURVATURE OF A GEODESIC TRIANGLE

> *The integral curvature of a region on the surface is equal to the area of its spherical image. . . . This property of the integral curvature was already known to the French school of Monge before Gauss pointed out its significance for the intrinsic geometry of a surface.*
>
> D. J. Struik [**1**, p. 156]

Formulas 6.92 and 16.53, for the areas of spherical and hyperbolic triangles, are special cases of a beautiful formula which Gauss discovered for the *integral curvature* of a triangle formed by arcs of three geodesics on any smooth surface. We proceed to establish this formula

20.31
$$\iint_{ABC} K \, dS = A + B + C - \pi.$$

Setting $g_1 = 1$ and $g_2 = \sqrt{g}$ in 20.17, we see that, when geodesic polar coordinates are used, the formula for K is simply

20.32
$$\sqrt{g} \, K = -(\sqrt{g})_{11}.$$

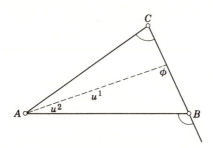

Figure 20.3a

Consider a geodesic triangle ABC with its side AB along the initial geodesic $u^2 = 0$, as in Figure 20.3a. Integrating K over the area of this triangle with the help of 19.24, we obtain

$$\iint K \, dS = \iint K \sqrt{g} \, du^1 \, du^2 = -\iint (\sqrt{g})_{11} \, du^1 \, du^2$$

$$= -\int (\sqrt{g})_1 \Big|_0^{u^1} \, du^2.$$

By 20.27, $(\sqrt{g})_1 = 1$ when $u^1 = 0$; and by 20.28, $-(\sqrt{g})_1 \, du^2 = d\phi$ for any point on the geodesic BC. Hence

$$\iint_{ABC} K \, dS = \int_0^A \{1 - (\sqrt{g})_1\} \, du^2 = \int_0^A du^2 + \int_{\pi-B}^C d\phi$$

$$= A + C - (\pi - B) = A + B + C - \pi$$

[Weatherburn **2**, p. 117].

EXERCISES

1. Obtain 20.32 directly from 19.49 with $\mathbf{t} = \mathbf{r}_1$.

2. On the unit sphere, colatitude and longitude serve as geodesic polar coordinates with $g = \sin^2 u^1$. What happens to 20.32 in this case? For convenience, write r, θ for u^1, u^2. The differential equation 20.26 (with $(g)_2 = 0$) has, as a first integral,

$$(\sin^2 r)\dot{\theta} = 1/h$$

(an arbitrary constant). Combining this formula for $\dot{\theta} = d\theta/ds$ with

$$ds^2 = dr^2 + \sin^2 r \, d\theta^2,$$

deduce $dr = \sin r \sqrt{h^2 \sin^2 r - 1} \, d\theta$,

$$\theta = \int \frac{\csc^2 r \, dr}{\sqrt{h^2 - \csc^2 r}} = \theta_0 + \arccos (k \cot r)$$

where $k = 1/\sqrt{h^2 - 1}$, and

$$k \cot r = \cos(\theta - \theta_0).$$

Expressing this solution in terms of the Cartesian coordinates

$$x = \sin r \cos \theta, \quad y = \sin r \sin \theta, \quad z = \cos r,$$

verify that the geodesics on a sphere are the great circles (lying in planes through the origin).

20.4 THE EULER-POINCARÉ CHARACTERISTIC

As we saw on page 281, any polyhedron inscribed in a sphere can be projected from the center onto the surface of the sphere so as to form a *map*. In fact, the V vertices are joined in pairs by E geodesic arcs (which we still call edges), decomposing the spherical surface into F polygonal regions (which we still call faces). More generally, a map may be obtained by drawing a sufficient number of geodesic arcs on any closed surface. We can insist that the points ("vertices") be so placed and so joined that every face is simply connected, that is, that the boundary of the face can be continuously shrunk to a single point without leaving the surface.

In § 10.3 we used a Schlegel diagram to prove Euler's formula. We could just as well have used the corresponding map on a sphere. The same argument, applied to a map on the general surface, shows that the Euler-Poincaré characteristic

$$\chi = V - E + F$$

is essentially a property of the surface, that is, that it has the same value for all maps drawn on the given surface. It is a remarkable fact that this property of the surface can be expressed very simply in terms of the integral curvature (which is obtained by integrating K over the whole surface).

Consider first a sphere, on which a map (with $V = E = 3$ and $F = 2$) is obtained by taking, as vertices, three points on a great circle. Each hemisphere is bounded by three arcs of the great circle, forming a spherical "triangle" whose three angles are π, π, π. By 20.31, each hemisphere has integral curvature

$$\pi + \pi + \pi - \pi = 2\pi.$$

Hence the integral curvature of the whole sphere is 4π (as it obviously must be, since the "spherical image" of any sphere is a unit sphere, whose surface area is 4π). The general formula, of which this is a very special case, is

20.41 $\iint K \, dS = 2\chi\pi,$

where the integration is taken over the whole of any given surface of characteristic χ.

To establish 20.41, we consider a map formed by E geodesic arcs on the given surface, choosing the V vertices in such positions that no face has a re-entrant angle (i.e., an angle greater than π). The map can then be "triangulated" by selecting a new vertex inside each face and joining it by new geodesic arcs to all the vertices of that face. This procedure yields a new map having $V + F$ vertices and $2E$ triangular faces. Since the sum of all the angles of all these $2E$ triangles amounts to 2π for each of the $V + F$ vertices, the integral curvature of the whole surface is

$$\begin{aligned} \iint K \, dS &= \Sigma\,(A + B + C - \pi) \\ &= 2\pi(V + F) - 2E\pi = 2\pi(V + F - E) \\ &= 2\pi\chi. \end{aligned}$$

It follows that the integral curvature of a closed surface is not altered by topological transformation. For instance, the value 4π is maintained when a sphere is deformed into an ellipsoid or any other oval surface. The deformation may even be continued so as to bring in anticlastic regions.

EXERCISES

1. The torus 8.88 (where $a > b$) is constructed by revolving a circle of radius b about a line (in its plane) distant a from the center. On this surface we can draw two circles, of radii b and $a + b$, which are geodesics having just one common point. These form a map in which $V = F = 1$, $E = 2$. Hence the integral curvature is zero. (The positive integral curvature of the "outer" synclastic part of the torus is exactly balanced by the negative integral curvature of the "inner" anticlastic part.)

2. Describe two further geodesics on the torus so that the four geodesics make a map in which $V = F = 4$, $E = 8$.

20.5   SURFACES OF CONSTANT CURVATURE

> *When Gauss was nineteen his mother asked his mathematical friend Wolfgang Bolyai whether Gauss would ever amount to anything. When Bolyai exclaimed, "The greatest mathematician in Europe!" she burst into tears.*
>
> E. T. Bell [**1**, p. 252]

When we study a surface by means of the fundamental magnitudes of the first order and the consequent Christoffel symbols, we are treating it "intrinsically," exploring it like the hypothetical two-dimensional creature who could not imagine any direction outside the surface itself. Such a creature could measure distances by means of the formula 19.16, distinguish geodesics as the shortest paths from place to place, and measure the Gaussian curvature K at any point.

One of the most elegant approaches to non-Euclidean geometry is to regard the elliptic or hyperbolic plane as a nondevelopable surface which is homogeneous (all positions alike) and isotropic (all directions alike). Since the surface is homogeneous, its Gaussian curvature is constant. By using a suitable unit of distance, we may take the constant value of K to be 1 or -1 according as it is positive or negative. We shall find it convenient to use geodesic polar coordinates. Since the surface is homogeneous and isotropic, the expression 20.24 will be the same wherever we place the pole $u^1 = 0$ and the initial geodesic $u^2 = 0$, and g will be a function of u^1 alone, independent of u^2. The "straight lines" of the geometry are the geodesics on the surface, and *it is not necessary to regard the surface as being embedded in a 3-space.*

Setting $K = \pm 1$ in 20.32, we obtain the differential equation

$$(\sqrt{g})_{11} = \mp\sqrt{g},$$

which yields

$$\sqrt{g} = A \sin u^1 + B \cos u^1 \quad \text{or} \quad A \sinh u^1 + B \cosh u^1.$$

At the pole, $ds \,(= du^1)$ is independent of u^2; therefore $g = 0$ when $u^1 = 0$; that is, $B = 0$. Also, by 20.27, $A = 1$. Hence

$$\sqrt{g} = \sin u^1 \quad \text{or} \quad \sinh u^1$$

and

$$ds^2 = (du^1)^2 + \sin^2 u^1 \, (du^2)^2 \quad \text{or} \quad (du^1)^2 + \sinh^2 u^1 \, (du^2)^2.$$

EXERCISE

Compute the circumference of the geodesic circle $u^1 = r$ (i) in the elliptic plane, (ii) in the hyperbolic plane.

20.6 THE ANGLE OF PARALLELISM

In the elliptic case, we can identify u^1 and u^2 with colatitude and longitude on the unit sphere (as in Ex. 2 at the end of § 20.3). Accordingly, we now restrict consideration to the hyperbolic case, in which

$$g = \sinh^2 u^1.$$

For convenience, let us write r, θ for u^1, u^2, so that*

$$ds^2 = dr^2 + \sinh^2 r \, d\theta^2.$$

To determine the straight lines of the hyperbolic plane, we use the differential equation 20.26 with $g = \sinh^2 r$, namely,

$$\frac{d}{ds} (\sinh^2 r \, \dot\theta) = 0.$$

For a first integral we obtain $\sinh^2 r \, \dot\theta = h^{-1}$ (a constant), whence

$$dr^2 + \sinh^2 r \, d\theta^2 = ds^2 = (h \sinh^2 r \, d\theta)^2,$$

so that $dr^2 = \sinh^2 r \, (h^2 \sinh^2 r - 1)d\theta^2$ and

$$\theta = \int \frac{dr}{\sinh r \, \sqrt{h^2 \sinh^2 r - 1}} = \int \frac{\operatorname{csch}^2 r \, dr}{\sqrt{h^2 - \operatorname{csch}^2 r}}$$

$$= - \int \frac{d \coth r}{\sqrt{h^2 + 1 - \coth^2 r}} = \theta_0 + \arccos (k \coth r),$$

where $k = 1/\sqrt{h^2 + 1}$. Hence, finally, the lines are given by

$$k \coth r = \cos(\theta - \theta_0).$$

When h tends to infinity, so that k tends to zero, we have the radial lines, for which θ is constant. The line through $(a, 0)$ perpendicular to the initial line $\theta = 0$ (being a geodesic which is unchanged when θ is replaced by $-\theta$), is

$$\tanh a \coth r = \cos \theta.$$

We can use these results to find relations between the sides and angles of a right-angled triangle ABC with its right angle at C and its side BC along the initial line, as in Figure 20.6a. Since the equations for the lines BC, AB, and CA are

$$\theta = 0, \quad \theta = B, \quad \text{and} \quad \tanh a = \tanh r \cos \theta,$$

the lengths of the sides $BC = a$ and $AB = c$ are related by the equation

$$\tanh a = \tanh c \cos B$$

[Carslaw **1**, p. 109; Coxeter **3**, pp. 238, 282]. (Another formula of the same kind can be obtained by changing a and B into b and A.)

* E. Beltrami, *Giornale di Matematiche*, **6** (1868), p. 298 (12).

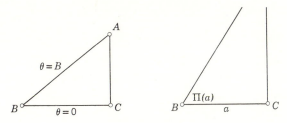

Figure 20.6a

The angle of parallelism $\mathrm{II}(a)$ is the value of B that makes c infinite; that is,

$$\mathrm{II}(a) = B,$$

where
$$\cos B = \tanh a, \qquad \sin B = \operatorname{sech} a,$$
$$\csc B = \cosh a, \qquad \cot B = \sinh a,$$
$$\csc B - \cot B \;=\; \cosh a - \sinh a,$$
$$\tan \tfrac{1}{2} B \;=\; e^{-a}.$$

We have thus established Lobachevsky's formula

$$\mathrm{II}(a) = 2 \arctan e^{-a}$$

[Coxeter **3**, p. 208]. This is a precise expression for the function that we studied tentatively in § 16.3.

EXERCISE

Compute $\mathrm{II}(a)$ for a few suitably chosen values of a, and sketch the curve $y = \mathrm{II}(x)$. Where does $\mathrm{II}(u)$ occur in Figure 17.4*b*?

20.7 THE PSEUDOSPHERE

Having obtained the hyperbolic plane as a surface of constant negative curvature, it is natural for us to ask whether such a surface can be embedded in Euclidean space. In other words, can the hyperbolic plane, or a finite part of it, be represented isometrically by a surface in ordinary space, in some such manner as the elliptic plane is represented (twice over) by a sphere? The answer is No and Yes: there is no such representation of the whole hyperbolic plane* but there are certain surfaces that will serve for a portion of finite area [Klein **4**, p. 286]. The simplest instance, which Liouville named the *pseudosphere,* is one half of the "tractroid" formed by revolving the tractrix 17.51 about its asymptote.

Writing z for x, x for y, and setting $a = 1$, we obtain the tractrix

$$x = \operatorname{sech} u^1, \qquad z = u^1 - \tanh u^1$$

---

* G. Lütkemeyer, *Ueber den analytischen Charakter der Integrale von partiellen Differentialgleichungen* (Göttingen, 1902).

in the plane $y = 0$. Revolution about the z-axis yields the tractroid

$$x = \operatorname{sech} u^1 \cos u^2, \qquad y = \operatorname{sech} u^1 \sin u^2, \qquad z = u^1 - \tanh u^1,$$

which has a cuspidal edge where $u^1 = 0$. The pseudosphere is the horn-shaped surface given by $u^1 \geqslant 0$. Differentiating the position vector $\mathbf{r} = (x, y, z)$, we obtain

$$g_{11} = \tanh^2 u^1, \qquad g_{12} = 0, \qquad g_{22} = \operatorname{sech}^2 u^1.$$

Setting $g_1 = \tanh u^1$, $g_2 = \operatorname{sech} u^1$ in 20.17, we deduce

$$-\tanh u^1 \operatorname{sech} u^1 K = (-\operatorname{sech} u^1)_1 = \operatorname{sech} u^1 \tanh u^1,$$

whence $$K = -1.$$

Since the pseudosphere has the same Gaussian curvature as the hyperbolic plane, the geodesics on the former represent lines in the latter isometrically. Among these geodesics are the meridians (for which u^2 is constant), representing a pencil of parallels. Orthogonal to them, we find the circles for which $u^1 =$ constant, representing arcs of concentric horocycles (§ 16.6). The longest of these horocyclic arcs is of length 2π, since it is represented by the circle

$$x = \cos u^2, \qquad y = \sin u^2$$

in the plane $u^1 = 0$. Thus the whole pseudosphere represents the *horocyclic sector* bounded by this arc of length 2π and the diameters at its two ends. The horocyclic sector is wrapped round the pseudosphere so that the two diameters are brought together to form a single meridian.

This limitation to a horocyclic sector renders the pseudosphere utterly useless as a means for drawing significant hyperbolic figures. Every geodesic that is not merely a meridian winds itself round the "horn" as it proceeds in one direction, whereas in the opposite direction it is abruptly cut off by the cuspidal edge. Thus we cannot even draw such a simple arrangement of lines as Figure 16.3a! These remarks are needed to counteract the widespread but mistaken idea that hyperbolic geometry can be identified with the intrinsic geometry of the pseudosphere.

EXERCISE

Use 20.23 to obtain an equation for the geodesics on the pseudosphere.

21

Topology of surfaces

In Chapter 4 we considered various tessellations of the Euclidean plane (including, in § 4.6, *regular* tessellations). These may be regarded as infinite "maps." In § 15.7 we considered the analogous tessellations of a sphere, which are finite maps. In § 10.3 we proved Euler's formula

$$V - E + F = 2,$$

which connects the numbers of vertices, edges (or arcs), and faces (or regions) of any map drawn on a sphere. In § 20.4 we extended this to

$$V - E + F = \chi \leqslant 2$$

for a map on any closed surface, the Euler-Poincaré characteristic χ being the same for all maps on the given surface. In § 6.9 we identified antipodal points of a sphere so as to obtain the real projective plane; for a centrally symmetrical map on the sphere, this identification naturally halves V, E, and F, thus reducing χ from 2 to 1. In Figure 10.5a we considered reciprocal polyhedra which, when regarded as spherical tessellations, are a special case of *dual* maps. In § 10.1 we defined the Schläfli symbol $\{p, q\}$, which is appropriate for a map of p-gons, q at each vertex; and in 10.31 we obtained the equations $qV = 2E = pF$. In § 15.4 we discussed groups of permutations of the faces of a map. In § 15.3 we found that the theory of translations and glide reflections belongs to *absolute* geometry; that is, that it belongs not only to Euclidean geometry but also to hyperbolic geometry.

The present chapter applies all these ideas to a discussion of the topological properties of surfaces, including the conjecture of P. J. Heawood that, for the coloring of any map on a surface of characteristic χ,

$$[\tfrac{7}{2} + \tfrac{1}{2}\sqrt{49 - 24\chi}]$$

colors suffice. What makes this conjecture remarkable is that, although in 1890 he established its truth for every $\chi < 2$, it still remains an open question for maps on the ordinary sphere or plane. Another conjecture is that Heawood's formula is "best possible" in the sense that, for each χ, a map requiring the full number of colors can be drawn.

21.1   ORIENTABLE SURFACES

> *The group of transformations of greatest importance in present-day mathematics, namely, the group of topological transformations, is far wider than the projective group.   Here we are dealing with a topological space; that is, with a set of elements, call them points, for which the concept of a neighborhood is defined. . . .   Any transformation preserving neighborhoods is called topological.*
>
> *Topology may be visualized as rubber-sheet geometry, since a topological transformation permits any amount of stretching or compressing (without tearing).*
>
> S. H. Gould [**1,** p. 304]

In Chapter 5 we mentioned Klein's famous classification of geometries according to the groups of transformations under which their theorems remain true.   In this sense, projective geometry is characterized by the group of collineations and correlations, and hyperbolic geometry by the subgroup of collineations leaving invariant a conic (the locus of points at infinity). Topology, sometimes described as "the most general of all geometries," is characterized by the group of *continuous* transformations.   For instance, since a polyhedron can be continuously transformed into the corresponding spherical tessellation, topology does not recognize any distinction between the polyhedron and the tessellation.   Again, in § 20.4 we defined the characteristic

21.11 $$\chi = V - E + F$$

in terms of a map formed by geodesic arcs on the given surface, but the value of χ will not change if we replace the geodesic arcs by any continuous arcs which join the same pairs of points without crossing one another. In other words, the edges of the map are not necessarily geodesics like the boundary between Colorado and Utah; they can just as well be "wild" like the boundary between Indiana and Kentucky.

In the same spirit, the torus 8.88 is topologically equivalent to a sphere with a *handle* (like the handle of a teacup), and we can derive more compli-

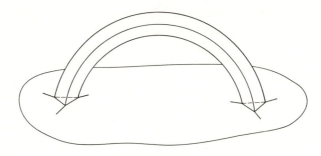

Figure 21.1a

cated surfaces by adding any number of further handles. The operation of adding a handle to a given surface reduces the value of χ by 2. For, since the handle may be chosen in the form of a bent triangular prism joining two triangular faces of a suitable map, as in Figure 21.1a, its insertion leaves V unchanged, increases E by 3, and increases F by $3 - 2$. Knowing that a a sphere has $\chi = 2$, we deduce that a sphere with p handles has

21.12 $$\chi = 2 - 2p.$$

This is called a surface of *genus* p. In particular, a sphere is a surface of genus 0 and a torus is a surface of genus 1.

From a rectangular rubber sheet, we can make a model of a torus by identifying, or bringing together, each pair of opposite sides. The first identification produces a tube, and the second an "inner tube." Conversely, by cutting a torus along two circles that have only one common point, we can unfold it (after some distortion) to make a rectangle whose pairs of opposite sides arise from the two cuts. More generally, given a surface of genus p, a suitable set of 2p cuts, all beginning and ending at a single point, enables us to unfold the surface into a 4p-gon whose pairs of opposite sides arise from the 2p cuts [Coxeter and Moser **1**, p. 25], as in Figure 21.1b.

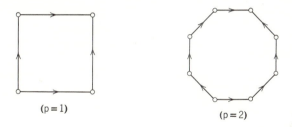

(p = 1)   (p = 2)

Figure 21.1b

If we regard the 2p cuts on the surface as the 2p edges of a map, we find that this map has one face and one vertex, in agreement with the formula

$$\chi = V - E + F = 1 - 2p + 1 = 2 - 2p.$$

When $p = 1$, the rectangle is conveniently taken to be a square, and this square may be regarded as one of the infinitely many faces of the regular tessellation $\{4, 4\}$ (Figure 4.6a) or as the unit cell of a lattice generated by two translations in perpendicular directions. The identification of opposite sides may be achieved by identifying points in corresponding positions in all the squares, that is, by pretending that the translations have no effect. In technical language, the Euclidean plane is the *universal covering surface* of the torus. Similarly, when $p > 1$, the 4p-gon is conveniently taken to be a regular 4p-gon of angle $\pi/2p$ in the hyperbolic plane, and this $\{4p\}$ may be regarded as a face of a regular hyperbolic tessellation $\{4p, 4p\}$. Opposite sides of the $\{4p\}$ are related by 2p translations, which generate a group having the {4p} for a fundamental region. The infinite hyperbolic plane (which is the universal covering surface) is reduced to the given finite surface by identifying each point in one {4p} with the corresponding points in all the other {4p}'s. The group of translations is called the *fundamental group* of the surface [Coxeter and Moser **1**, pp. 24–27, 58–60].

EXERCISE

A *graph* is a set of points (called *vertices*), certain pairs of which are joined by arcs (called *edges*). In particular, the vertices and edges of a map form a graph. Conversely, any connected graph can be drawn on a surface so as to form a map covering the surface.* A graph is said to be *planar* if it can be drawn on a sphere without any edges crossing one another, in which case it can just as easily be drawn in the (inversive) plane, provided we allow one face of the consequent map to be infinite. A vertex is said to have *valency* (or "degree") q if it belongs to q edges. A graph is said to be *trivalent* if every vertex belongs to three edges. In this case $3V = 2E$; therefore V is even. The *Thomsen graph*† has six vertices $P_1, \ldots, P_6$ and nine edges $P_i P_j$, where $i + j$ is odd. This is the simpest nonplanar trivalent graph. Can it be drawn on a torus?

21.2 NONORIENTABLE SURFACES

> A surface is non-orientable if and only if there exists on the surface some closed curve . . . such that a small oriented circle whose center traverses the curve continuously will arrive at its starting point with its orientation reversed.
>
> Hilbert and Cohn-Vossen
> [**1**, p. 306]

Each of the surfaces discussed in § 21.1 is *orientable*, that is, a positive sense of rotation can be defined consistently everywhere. More precisely, the faces of any map on the surface can be regarded as directed polygons in such a way that the two directions thus assigned to each edge disagree, or cancel out. A surface is said to be *nonorientable* if it admits one map which cannot be oriented in this manner. The most famous instance is the *Möbius strip*, which can be illustrated by taking a strip of paper *ABAB*, several times longer than it is wide, and sticking the two ends together after twisting one of them by a half-turn. Its nonorientability can be checked by means of a map consisting of a single row of squares. It is *one-sided* in the sense that an ant could crawl along the whole length of the strip, without crossing the bounding edge, and find himself at the starting point on the "other side." If two wheels in a machine are connected by a belt of such a shape (e.g., for the purpose of conveying hot or abrasive materials), the substance of the belt will wear out equally on both sides. A patent for

---

* J. H. Lindsay, Jr., Elementary treatment of the imbedding of a graph in a surface, *American Mathematical Monthly*, **66** (1959), pp. 117–118.

† W. Blaschke and G. Bol, *Geometrie der Gewebe* (Berlin, 1938), p. 35.

this practical application of the Möbius strip has been acquired by the Good-rich Company.*

Unlike the closed surfaces considered in § 21.1, the Möbius strip is *bounded.* The boundary is a simple closed curve, but physically it cannot be shrunk away because, if we make it coincide with the circumference of a circle, the interior of the circle must intersect the surface of the strip. This practical difficulty arises because the model is embedded in Euclidean space. Theoretically, no such embedding is needed. When the boundary has been shrunk away, the resulting closed surface is topologically a real projective plane! In other words, *the Möbius strip is the real projective plane with a hole cut out of it.* For we may regard the projective plane (§ 6.9) as a sphere with antipodal points identified. When cutting out a circular hole round the north pole we must, of course, also cut out an equal hole round the south pole. What remains of the sphere is a zone bounded by two parallels of latitude such as the Tropics of Cancer and Capricorn. But the identification of antipodes has the effect that only half the zone is needed, say the "visible" half (Figure 21.2a). This half-zone, with its ends AB identified, is evidently a Möbius strip.

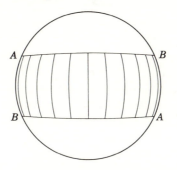

Figure 21.2a

Instead of a whole sphere with every pair of antipodal points identified, we may regard the projective plane as a hemisphere (say the "southern" hemisphere) with identification of diametrically opposite points on the peripheral equator. In the spirit of topology, the hemispherical surface can be stretched until it covers almost the whole sphere, and the periphery (with opposite points identified) is reduced to a very small circle round the north pole. In other words, the projective plane is topologically equivalent to a sphere with a *cross-cap*, which may be described as a small circular hole having the magic property that, as soon as the crawling ant reaches it, he finds himself leaving the same hole from its diametrically opposite point (inside, instead of outside, the sphere).

* U.S. Patents 1,442,632 (1923), 2,479,929 (1949), 2,784,834 (1957).

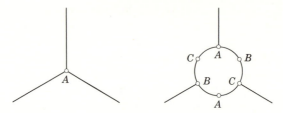

Figure 21.2b

We can derive more complicated surfaces by adding any number of cross-caps, each of which reduces the value of χ by 1. For, if a map on a given surface has a vertex A belonging to three faces, we can replace A by a cross-cap $ABCABC$ as in Figure 21.2b. Since this requires the formation of two new vertices B, C, and three new edges BC, CA, AB, the insertion of the cross-cap increases V by 2, E by 3, and leaves F unchanged. (The faces on the left and right now meet twice: along the original edge through A and again along the new edge BC.) Knowing that a sphere has $\chi = 2$, we deduce that a sphere with q cross-caps has

21.21 $$\chi = 2 - q.$$

When q = 1 this is, as we have seen, the real projective plane. When q = 2 it is the *Klein bottle* (or "nonorientable torus") [Hilbert and Cohn-Vossen **1**, p. 308].

A suitable set of q cuts, all beginning and ending at a single point and each passing through a different cross-cap, enables us to unfold the surface into a 2q-gon such that q pairs of adjacent sides arise from the q cuts [Coxeter and Moser **1**, pp. 25–28, 56–58] as in Figure 21.2c. These cuts are the q edges of a map having one face and one vertex, in agreement with the formula

$$\chi = V - E + F = 1 - q + 1 = 2 - q.$$

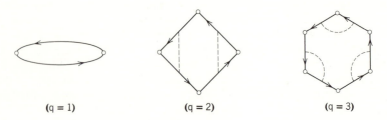

(q = 1)   (q = 2)   (q = 3)

Figure 21.2c

When q = 1, the 2q-gon is a digon which may be regarded as one of the two faces of the spherical tessellation {2, 2} (see Ex. 1 at the end of § 15.7). In fact, the universal covering surface of the projective plane is the sphere, and its fundamental group is of order 2, generated by the central inversion.

When q = 2, the 2q-gon may be regarded as a face of the Euclidean tessellation $\{4, 4\}$, so that the universal covering surface is the Euclidean plane. Unlike the torus, whose fundamental group **p1** is generated by two translations, the Klein bottle has the fundamental group **pg**, generated by two glide reflections (see § 4.3 and Plate I). Similarly, when $q > 2$, so that the universal covering surface is the hyperbolic plane, the 2q-gon is a face of the hyperbolic tessellation $\{2q, 2q\}$, and the fundamental group is generated by q glide reflections [Coxeter and Moser **1,** pp. 56–58].

EXERCISES

1. The projective plane is topologically equivalent to a disk* with diametrically opposite points identified.

2. How can the Thomsen graph (see the end of § 21.1) be drawn on a sphere with a cross-cap (or a disk with opposite points identified)?

3. What happens to the vertices and edges of a regular hexagonal prism when we project it centrally onto its circumsphere and then identify antipodes?

4. Is a sphere with p handles and q cross-caps topologically equivalent to a sphere with 2p + q cross-caps?

21.3 REGULAR MAPS

> We first give a method of reducing any two-dimensional manifold to one of the known polygonal normal forms. The method used is one by which a polygon on which the manifold is represented is subjected to a series of transformations by cutting it apart in a simple manner and then joining it together again so as to obtain a new polygon representing the same manifold.
>
> H. R. Brahana (1895 -)
> (Annals of Mathematics, **23** (1921), p. 144)

It can be proved† that every closed surface is topologically equivalent either to a sphere with p ($\geqslant 0$) handles (if the surface is orientable) or to a sphere with q (> 0) cross-caps. In virtue of 21.12 and 21.21, this means that, from the standpoint of topology, there is just one orientable closed surface for each of the values

$$\chi = 2, \quad 0, -2, -4, \ldots,$$

namely a sphere with $1 - \frac{1}{2}\chi$ handles, and there is just one nonorientable closed surface for each of the values

$$\chi = 1, \quad 0, -1, -2, \ldots,$$

---

* A disk is a circle plus its interior. For other topological properties of the disk and sphere, see A. W. Tucker, *Proceedings of the First Canadian Mathematical Congress* (University of Toronto Press, 1946).

† Brahana's original proof has been simplified by Lefschetz [**1,** pp. 72–85] and others. One of the best expositions is by R. C. James, Combinatorial topology of surfaces, *Mathematics Magazine,* **29** (1955), pp. 1–39.

namely, a sphere with $2 - \chi$ cross-caps. We have already described, on such a surface, a very simple map having one vertex, $2 - \chi$ edges, and one face. In the orientable case, this map is "regular" in the following sense.

The vertices, edges, and faces of a map (on any closed surface, orientable or nonorientable) may conveniently be called the *elements* of the map. Those permutations of the elements which preserve all the relations of incidence are called *automorphisms* of the map. The automorphisms form a group (of order 1 or more) called *the group of the map*. This is a natural generalization of the symmetry group of a polyhedron or tessellation (§ 15.7), but metrical ideas are no longer used. A map is said to be *regular* if its automorphisms include the cyclic permutation of the edges (and vertices) belonging to any one face and also the cyclic permutation of the edges (and faces) that meet at any one vertex of this face. Such a map is "of type $\{p, q\}$" if p edges belong to a face, and q to a vertex. The *dual* map, whose edges cross those of the original map, is of type $\{q, p\}$. (The letters p and q used here have no connection with our previous use of p and q for the numbers of handles and cross-caps.)

The equations 10.31 remain valid. Combining them with 21.11, we obtain a generalization of 10.32:

21.31 $$V = 2pr, \quad E = pqr, \quad F = 2qr,$$

where, if $\chi \neq 0$,

21.32 $$r = \frac{\chi}{2p + 2q - pq}.$$

If $\chi = 0$, so that $2p + 2q = pq$ as in § 4.6, there are infinitely many possible values for r, as we shall soon see.

If $\chi = 1$ or 2, the possible values for p and q are given by 10.33 without the restrictions $p > 2, q > 2$. Thus the regular maps on a sphere ($\chi = 2$) are just the spherical tessellations

21.33 $$\{p, 2\}, \quad \{2, p\}, \quad \{3, 3\},$$
$$\{4, 3\}, \quad \{3, 4\}, \quad \{5, 3\}, \quad \{3, 5\},$$

namely: the *dihedron* whose p vertices are evenly spaced along the equator, the *hosohedron** whose edges and faces are p meridians and p lunes, and "blown-up" variants of the five Platonic solids. All these are centrally symmetrical, except the dihedron and its dual with p odd, and the tetrahedron $\{3, 3\}$. In the centrally symmetrical cases we can identify antipodes to obtain the regular tessellations of the elliptic plane ($\chi = 1$):

21.34 $$\{p, 2\}/2 \quad \text{and} \quad \{2, p\}/2 \quad (p \text{ even}),$$
$$\{4, 3\}/2, \quad \{3, 4\}/2, \quad \{5, 3\}/2, \quad \{3, 5\}/2$$

[Coxeter and Moser **1**, p. 111]. For instance, identifying opposite elements

---

* This term (literally "any number of faces") was coined by Vito Caravelli (1724–1800), whose *Traité des hosoèdres* was published in Paris (1959) by the Librairie Scientifique et Technique.

of the cube (Figure 10.5b), we see that $\{4, 3\}/2$ is a partition of the elliptic plane (or of the real projective plane) into three "squares," say $ABCD$, $ACDB$, $ADBC$. (These squares are just the three handkerchiefs which Lady Muriel began to sew together in her attempt to make the Purse of Fortunatus [Dodgson **4**, pp. 100–104]. The first two, joined along their common side CD, form a Möbius strip whose boundary is $ADBC$.) Likewise, $\{5, 3\}/2$ is a partition of the elliptic plane into six pentagons, each of which is surrounded by the remaining five.

The regular maps on the torus are derived from infinite regular maps on its universal covering surface, which is the Euclidean plane. As we saw in § 4.6, these infinite maps are the regular tessellations

21.35 $$\{6, 3\}, \quad \{4, 4\}, \quad \{3, 6\}.$$

The necessary identifications are determined by subgroups of the translation groups of these tessellations.

The vertices of $\{4, 4\}$ may be taken to be the lattice of points whose Cartesian coordinates (x, y) are integers. The torus is derived by identifying opposite sides of a square, one of whose sides goes from $(0, 0)$ to (b, c), where b and c are positive integers or zero (but not *both* zero). Since the area of this square is $b^2 + c^2$, the part of the original $\{4, 4\}$ that lies inside it consists of $b^2 + c^2$ unit squares. We thus find, on the torus, a map

$$\{4, 4\}_{b,c}$$

in which

$$V = b^2 + c^2, \qquad E = 2V, \qquad F = V$$

[Coxeter and Moser **1**, p. 103]. In particular, $\{4, 4\}_{1,0}$ is the map (having one vertex and one face) which we used in § 21.1 when we unfolded the torus after cutting it along the two edges of the map. (If it seems paradoxical for a map of type $\{4, 4\}$ to have only one vertex, we must recognize that the face is still quadrangular even though its four vertices all coincide with the single vertex of the map.) The map $\{4, 4\}_{2,1}$, whose five faces are each surrounded by the remaining four, is shown in Figure 21.3a.

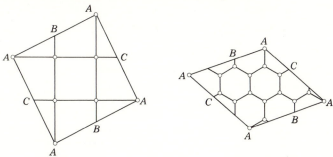

Figure 21.3a   **Figure 21.3b**

Similarly, the vertices of {3, 6} may be taken to be the lattice of points whose oblique coordinates, with axes inclined at 60°, are integers. The torus is derived by identifying opposite sides of a rhombus of angle 60°, one of whose sides goes from $(0, 0)$ to (b, c). Since the area of this rhombus is $b^2 + bc + c^2$ times that of the unit cell of the lattice (consisting of two adjacent faces of {3, 6}), the part of {3, 6} that lies inside it consists of $2(b^2 + bc + c^2)$ equilateral triangles. We thus find, on the torus, a map

$$\{3, 6\}_{b,c}$$

in which

$$V = b^2 + bc + c^2, \qquad E = 3V, \qquad F = 2V$$

[Coxeter and Moser **1**, p. 107]. (For an affine variant of one-half of $\{3, 6\}_{2,1}$, see Figure 13.5c.) The dual map

$$\{6, 3\}_{b,c}$$

has $b^2 + bc + c^2$ hexagonal faces. In particular, $\{6, 3\}_{2,1}$ (Figure 21.3b)* is Heawood's partition of the torus into seven hexagons, each of which is surrounded by the remaining six.

Thus we see that the torus admits infinitely many regular maps of each of the three types 21.35. On the other hand, there are no regular maps on the Klein bottle [Coxeter and Moser **1**, p. 116].

If a regular map has more than one vertex and more than one face, every edge joins two vertices and separates two faces. If there is an automorphism which interchanges these two vertices without interchanging the two faces (in which case there is another automorphism which does vice versa), the map is said to be *reflexible* [Ball **1**, p. 129; Coxeter and Moser **1**, p. 101]. Clearly, all the regular maps on the sphere and all those on any nonorientable surface are reflexible, but those on the torus are reflexible only if $bc(b - c) = 0$. It was suggested by Coxeter and Moser [**1**, p. 102] that possibly all regular maps on more complicated surfaces (i.e., on surfaces of negative characteristic) are reflexible. However, this conjecture is refuted by J. R. Edmonds' discovery of a nonreflexible regular map† of type {7, 7} and genus 7, having 8 vertices, 28 edges, and 8 heptagonal faces

FDCGBEH,   GEDACFH,   AFEBDGH,   BGFCEAH,   CAGDFBH,
DBAEGCH,   ECBFADH,   ABCDEFG.

EXERCISES

1. Describe the maps {2, 1} and {1, 2} on the sphere. (The former has one face, a digon {2}; the latter has two faces which are monogons {1}.)

2. The elliptic tessellation $\{2q, 2\}/2$ has q vertices and q edges, all on one line,

---

* For other ways of drawing Heawood's map, see Coxeter, Map-coloring problems, *Scripta Mathematica,* **23** (1957), pp. 19–21, and The four-color map problem, 1840–1890, *Mathematics Teacher,* **52** (1959), pp. 288–289.
† See also Robert Frucht, *Canadian Journal of Mathematics,* **4** (1952), p. 247.

and one face. Its dual, $\{2, 2q\}/2$, has one vertex, q complete lines for its edges, and q angular regions for its faces. Describe the remaining regular tessellations of the elliptic plane.

3. As we have seen, the standard decomposition of the torus provides a one-faced map $\{4, 4\}_{1,0}$. The standard decomposition of the Klein bottle (as in the second part of Figure 21.2c) provides another one-faced map of type $\{4, 4\}$, but this is not regular. In both cases, the one vertex and the two edges form a very simple graph, which may be described roughly as a figure of eight. The same graph can be drawn on the projective plane to form $\{2, 4\}/2$, or on the sphere to form an irregular map whose faces consist of a digon and two monogons.

4. The standard decomposition of a sphere with three cross-caps (as in the third part of Figure 21.2c) provides an irregular one-faced map of type $\{6, 6\}$. Its one vertex and three edges form a "clover-leaf" having three loops. The same graph can be drawn on the Klein bottle as an irregular map of type $\{3, 6\}$, on the torus as $\{3, 6\}_{1,0}$, on the projective plane as $\{2, 6\}/2$, and on the sphere as an irregular map whose faces consist of a triangle and three monogons.

5. Describe the reflexible maps

$$\{4, 4\}_{1,1}, \quad \{4, 4\}_{2,0}, \quad \{3, 6\}_{1,1}, \quad \{6, 3\}_{1,1}, \quad \{6, 3\}_{2,0}.$$

6. The vertices and edges of $\{6, 3\}_{1,1}$ form the Thomsen graph (see the end of § 21.1). Those of $\{4, 4\}_{2,2}$ form an analogous graph having eight vertices instead of six.

7. A graph is called a *complete V-point* if every two of its V vertices are joined by an edge. The vertices and edges of the following maps form complete V-points (for which values of V?):

$$\{3, 2\}, \quad \{3, 3\}, \quad \{4, 3\}/2, \quad \{4, 4\}_{2,1}, \quad \{3, 5\}/2, \quad \{3, 6\}_{2,1}.$$

8. The vertices and edges of Edmonds' map form a complete 8-point.

9. There is no map of type $\{1, 1\}$. (*Hint:* Set $p = q = 1$ in 21.31 and 21.32.)

21.4   THE FOUR-COLOR PROBLEM

> "I doubt it," said the Carpenter,
> And shed a bitter tear.
>
> Lewis Carroll
> [Dodgson **2**, Chap. 4]

The theory of maps on surfaces may be said to have begun in 1840, when Möbius puzzled his students with the problem of dividing a country into five districts in such a way that every two would have a common boundary line (not merely a common point). The impossibility of such a partition led naturally to the question whether four colors always suffice for coloring a map when we stipulate that different colors are needed wherever two districts share a boundary line or, in mathematical terms, wherever two faces share an edge. It must be emphasized that each face is simply connected (i.e., topologically a disk). Thus the geographical problem applies to a single island or continent: the ocean and all the other islands and continents are to be taken together as forming one more face, and if this is a blue face the same color must be allowed for some of the other faces. For instance, in a map of Europe, we need three different colors (say green, red, and yellow) for Belgium, France, and Germany; Holland may have the same color as France, but Luxembourg must be blue, like the sea.

Figures 15.4a, b, c illustrate the use of four colors for the tetrahedron and octahedron, and five for the icosahedron. For the tetrahedron, four is the only possible number, since each face meets all the others. Apart from this simplest case, *any map whose faces are triangles can be colored in three colors.** Moreover, any map having an even number of faces at each vertex (such as the octahedron) can be colored in two colors, like a chessboard.

The problem of deciding whether four colors suffice for coloring any map on a plane or a sphere is sometimes called *Guthrie's problem,* after Francis Guthrie, who took his B.A. in London in 1850 and his LL.D. in 1852. Between these dates the problem occurred to him while he was coloring a map of England. He tried in vain to prove that four colors are always sufficient. On October 23, 1852, his younger brother Frederick communicated the conjecture to Augustus De Morgan (author of *A Budget of Paradoxes*). In 1878, Cayley revived interest in the problem at a meeting of the London Mathematical Society by asking whether anyone had proved the conjecture. In 1880, Cayley's challenge was answered by A. B. Kempe and P. G. Tait, who published plausible arguments, which were accepted for ten years (even by Klein himself), as proving that four colors will always suffice. In 1890, Heawood drew attention to the fallacy in Kempe's argument, using for a counterexample a particular map having 18 faces. The number of faces can actually be reduced to 9, so as to reveal the fallacy more quickly.†

In §§ 21.5–21.7 we shall describe the valid part of Kempe's work and also Heawood's extension to maps on the torus and other multiply connected surfaces.

EXERCISES

1. In how many essentially different ways can a cube be colored with three given colors, a dodecahedron with four?

2. In Figure 15.4c, the icosahedron is colored with five colors so that each face and its three neighbors have four different colors. Replace each "*e*" by the one remaining color, thus reducing the number of colors to four. Starting afresh, color the icosahedron with three colors [Ball **1**, pp. 238–241].

3. Try to draw a map that is difficult to color with four colors.

---

* R. L. Brooks, On coloring the nodes of a network, *Proceedings of the Cambridge Philosophical Society,* **37** (1941), pp. 194–197.

† Coxeter, *Mathematics Teacher,* **52** (1959), pp. 283–286. For the fallacy in Tait's argument, see Ball [**1**, pp. 224–226].

21.5 THE SIX-COLOR THEOREM

> *Even the moonlit track ahead of him faded from his consciousness, for into his head had come a theorem which might be true or might be false, and his mind darted hither and thither seeking proofs to establish its truth and counter-examples to show that it could not possibly be true.*
>
> J. L. Synge [**2**, p. 165]

The equations 10.31, which apply to a regular map of type $\{p, q\}$, remain valid for a general map having various kinds of face and various numbers of faces at a vertex, provided we interpret p as the *average* number of vertices (or edges) of a face, and q as the *average* number of faces (or edges) at a vertex. Since a vertex belonging to only two edges can be omitted by combining the two edges, there is no real loss of generality in assuming that every vertex belongs to at least three edges. Thus $q \geqslant 3$ and

$$2E = qV \geqslant 3V,$$

whence, by 21.11, $E \leqslant 3(E - V) = 3(F - \chi)$ and

21.51
$$p = \frac{2E}{F} \leqslant 6 \left(1 - \frac{\chi}{F}\right).$$

This proves that $p < 6$ whenever $\chi > 0$, that is, for the sphere ($\chi = 2$) or the projective plane ($\chi = 1$). Hence

21.52 *Every map on the sphere or the projective plane has at least one face whose number of edges is less than 6.*

We can now prove, by induction over the number of faces,

THE SIX-COLOR THEOREM. *To color any map on the sphere or the projective plane requires at most six colors.*

We take "any map" to mean "any map having F faces," for each particular value of F. When $F \leqslant 6$ there is no problem: we can assign distinct

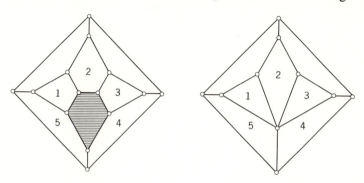

Figure 21.5a **Figure 21.5b**

colors to all the faces. The theorem will be proved if we can deduce the case of 7 faces from the case of 6, and then 8 from 7, and so on. Accordingly, we make the inductive assumption that the theorem holds for every map of $F - 1$ faces, and then proceed to investigate a given F-faced map, paying particular attention to the face (or one of the faces) having 5 or fewer edges (see 21.52). For definiteness, we assume this face to be a pentagon, like the shaded face in Figure 21.5a. (The same arguments can be carried through with trivial changes if it is a quadrangle, triangle, or digon.) Figure 21.5b shows a modified map in which this pentagonal face has shrunk to a point, that is, in which its territory has been ceded to its five neighbors. By the inductive assumption, the modified map, having only $F - 1$ faces, can be colored with six colors. Let this be done, and let the same coloring be applied to the original map. Then, even if the five neighbors need five distinct colors, there is still a sixth color left for the pentagonal face itself.

Since this argument can be applied with $F = 7$, then with $F = 8$, and so on, the six-color theorem holds for all values of F.

Can the number 6 be replaced by 5? For the projective plane it cannot, as we shall soon see. For the sphere it can, by a subtler argument depending on the topological theorem that a circle on the sphere decomposes it into two separate regions [Ball **1**, p. 229]. But the gap between the five colors that are always sufficient and the four that are usually necessary has never been bridged. Heawood himself continued to investigate the problem for the rest of his life, reducing it to pure algebra. Other authors have gradually increased the lower bound of the number of faces for a map that might possibly require five colors.

It is almost certainly a mere coincidence that the numbers 4 and 5 play an analogous role in arithmetic. According to Mordell [**1**, p. 19], it is "very easy to prove that every integer is the sum of at most five integer cubes, positive or negative, and there is an unproved conjecture that four cubes suffice."

For the projective plane, on the other hand, there is no gap to be bridged: six colors are both necessary and sufficient. The simplest map that needs

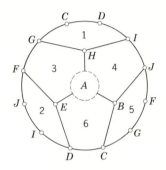

Figure 21.5c

all six is $\{5, 3\}/2$ (see p. 387), which is drawn in Figure 21.5c as a disk with diametrically opposite points identified. By cutting out a hole round the vertex A, we obtain H. Tietze's six-color map on the Möbius strip (Figure 21.5d). Here, as on the whole projective plane, six colors are both necessary and sufficient.

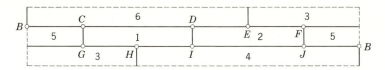

Figure 21.5d

EXERCISE

Make a model of the Möbius strip and color it as indicated in Figure 21.5d. (Since this is a "one-sided" surface, the paper must everywhere have the same color on both sides.)

21.6 A SUFFICIENT NUMBER OF COLORS FOR ANY SURFACE

> [Huck Finn to Tom Sawyer in their flying boat:] *"We're right over Illinois yet. And you can see for yourself that Indiana ain't in sight. . . . Illinois is green, Indiana is pink. You show me any pink down here, if you can. No, sir; it's green."*
>
> *"Indiana pink? Why, what a lie!"*
>
> *"It ain't no lie; I've seen it on the map, and it's pink."*
>
> Mark Twain ($=$S. L. Clemens, 1835-1910)
>
> (*Tom Sawyer Abroad*, Harper, New York, 1896, Chap. 3)

The problem of coloring maps on a more complicated surface is not difficult, as on the sphere, but easy, as on the projective plane. In fact, we can now prove

HEAWOOD'S THEOREM. *To color any map on a surface of characteristic* $\chi < 2$ *requires at most* $[N]$ *colors, where*

$$N = \frac{7 + \sqrt{49 - 24\chi}}{2}.$$

Since the case $\chi = 1$ has already been proved in § 21.5, we shall suppose that

$$\chi \leqslant 0.$$

Since the theorem is obviously true when $F \leqslant N$ (which implies $F \leqslant [N]$), we shall suppose also that

$$F > N$$

and use induction over the number of faces, assuming that $[N]$ colors suffice for any map having $F - 1$ faces. Since N satisfies the quadratic equation

$$N^2 - 7N + 6\chi = 0$$

or

$$6 \left(1 - \frac{\chi}{N} \right) = N - 1,$$

the inequality 21.51 yields

$$p \leqslant 6 \left(1 - \frac{\chi}{F} \right) \leqslant 6 \left(1 - \frac{\chi}{N} \right) = N - 1.$$

Hence there is at least one face having $[N] - 1$ or fewer edges (cf. 21.52). We continue as in § 21.5, using $[N]$ instead of 6, and conclude that $[N]$ colors suffice for the given map.

Although this proof would break down if χ were positive, we note that Heawood's expression for N not only yields the correct value 6 when $\chi = 1$ but also yields the conjectured value 4 when $\chi = 2$.

EXERCISE

Tabulate $[N]$ for values of χ from 2 down to -9.

21.7 SURFACES THAT NEED THE FULL NUMBER OF COLORS

"Suppose there's a brown calf and a big brown dog, and an artist is making a picture of them. . . . He has got to paint them so you can tell them apart the minute you look at them, hain't he? Of course. Well, then, do you want him to go and paint both of them brown? Certainly you don't. He paints one of them blue, and then you can't make no mistake. It's just the same with maps. That's why they make every state a different color. . . ."

Mark Twain (*ibid.*)

Heawood's theorem (with $\chi = 0$) tells us that every map on the torus can be colored with seven colors. His regular map $\{6, 3\}_{2,1}$, whose seven faces all meet one another, shows that at least one map on the torus really needs seven colors. Since Heawood's expression for N depends only on χ, it yields the same number 7 for the Klein bottle. However, Philip Franklin has proved a *six-color theorem for the Klein bottle*.[*]

Ringel [**1**, p. 124] proved that the Klein bottle is the *only* nonorientable surface not needing as many as $[N]$ colors. For instance, inserting a cross-cap at one vertex of Heawood's seven-faced map on the torus, as in Figure 21.2*b*, we obtain a seven-faced map on the surface of characteristic -1. This map still needs seven colors, since all its faces meet one another. (In fact, some pairs of faces meet twice.)

* Coxeter, *Scripta Mathematica,* **23** (1957), pp. 21–23.

In the case of an orientable surface of genus $p = 1 - \frac{1}{2}\chi$, Heawood's number $[N]$ is given by

$$N = \frac{7 + \sqrt{48p + 1}}{2},$$

and it is appropriate to give the name "The Heawood conjecture" to the statement that, for every p, the orientable surface of genus p carries a map of $[N]$ faces, all meeting one another. This conjecture became a theorem in 1968, when Ringel and Youngs* found such a map for every p. This breakthrough was the climax of a long story. The investigation was begun by L. Heffter in 1891 and was then neglected until 1952, when Ringel resumed it. He collaborated with Youngs from 1966 on, and some helpful ideas were contributed by W. Gustin, C. M. Terry, and L. R. Welch. The values of p up to 32 were disposed of independently by Jean Mayer (a professor of French literature) in 1967. The four most difficult cases (p = 59, 83, 158, 257) were achieved by Youngs and Richard Guy in 1968.

EXERCISES

1. Replacing the "hole" in Figure 21.5c by a cross-cap, obtain a six-color map (of 3 pentagons and 3 heptagons) on the Klein bottle.

2. Draw an eight-color map on the surface of genus two [Ball **1**, p. 237].

---

* Gerhard Ringel and J. W. T. Youngs, Solution of the Heawood map-colouring problem, *Proceedings of the National Academy of Sciences* (U.S.A.), 1968.

22

Four-dimensional geometry

The idea of four-dimensional space has long been surrounded by an attractive aura of mystery. The axiomatic approach (12.44) dispels the mystery without reducing the fascination. Having become accustomed to non-Euclidean geometries, we are no longer disconcerted by the possibility that two planes may have a common point without having a common line. More simply, we may regard the points of Euclidean 4-space as having four Cartesian coordinates instead of the usual two or three. Any two distinct points determine a line, the three vertices of a triangle determine a plane, and the four vertices of a tetrahedron determine a *hyperplane,* which is given by a single linear equation connecting the four coordinates.

In §§ 22.1–22.3 we describe the four-dimensional analogues of the Platonic solids. We shall see that there are six of these regular *polytopes.* Each consists of a finite number of solid cells in distinct hyperplanes, so arranged that every face of each cell belongs also to another cell. All these regular polytopes were discovered by Schläfli before 1855.

Just as we can make flat pictures of solids by projecting them orthogonally onto a plane, so we can make flat or solid "pictures" of hypersolids by projecting them either onto a plane or onto a hyperplane. Instances of the former procedure are shown in Figures 22.1a, b and 22.3b; for an example of the latter, see Plate III on page 404.

In § 22.4 we consider certain honeycombs (or "solid tessellations," or "degenerate polytopes") consisting of infinitely many solid cells in the same 3-space. In § 22.5 we see how these ideas help to explain some experimental results on the packing of equal spheres.

The geometry of this chapter is Euclidean. But all the other kinds of geometry can similarly be extended to spaces of any number of dimensions. As L. Fejes Tóth remarks in one of his books, we are able "to create an infinite set of new universes, the laws of which are within our reach, though we can never set foot in them."

22.1   THE SIMPLEST FOUR-DIMENSIONAL FIGURES

> That ye . . . may be able to comprehend with all saints what is the
> breadth, and length, and depth, and height.
>
> Ephesians III, 17 -18

> Spirits have four dimensions.
>
> Henry More (1614 -1687)

> If an inhabitant of flatland was able to move in three dimensions, he
> would be credited with supernatural powers by those who were unable
> so to move; for he could appear or disappear at will, could (so far as
> they could tell) create matter or destroy it. . . . We may go one step
> lower, and conceive of a world of one dimension—like a long tube—
> in which the inhabitants could move only forwards and backwards. . . .
> Life in line-land would seem somewhat dull. . . . An inhabitant could
> know only two other individuals; namely, his neighbours, one on each
> side.
>
> W. W. Rouse Ball
> (Mathematical Recreations and Essays, 9th edition, 1920, p. 426)

When trying to appreciate the idea of Euclidean 4-space, we are helped by imagining the efforts of a hypothetical two-dimensional being to visualize a three-dimensional world.*   In solid geometry we can find a line ("the third dimension") which is perpendicular to both of two intersecting lines and consequently perpendicular to every line in their plane.   Analogously, in 4-space we can find a line ("the fourth dimension") which is perpendicular to all three edges of a trihedral solid angle, such as a corner of a cube, and consequently perpendicular to every line in the 3-space that contains the solid angle.   It follows that two 3-spaces that have a common point have a common plane, and the product of reflections in them is a rotation about this plane, analogous to the familiar rotation about a line in three dimensions or about a point in two dimensions.

After accepting the idea of a fourth dimension, we can soon imagine a pyramid or a prism whose "base" is a solid.   For instance, a regular tetrahedron $ABCD$ may serve as the base of a pyramid $ABCDE$ (Figure 22.1a) whose apex E is along the fourth dimension through the center of $ABCD$. If E is so chosen that its distances from A, B, C, D are all equal to the edge AB, we have a *regular simplex,* which may be regarded in five ways as a pyramid, each vertex in turn serving as the apex while the remaining four form the base.

Figure 22.1b is merely an octagon with a square drawn inwards on each

* Flatland: A Romance of Many Dimensions, by A Square (E. A. Abbott), Boston, 1885 and 1928.

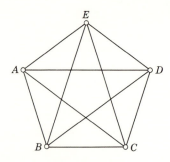

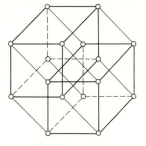

Figure 22.1a   **Figure 22.1b**

side, or an octagon and an octagram with corresponding sides joined by squares. It may be regarded as a picture of a *hypercube* (or "8-cell," or "tesseract," or "measure polytope") which is a prism whose base is a cube, the "height" of the prism being equal to the edge of the base. Just as a cube can be traced out by moving a square along the third dimension, so a hypercube can be traced out by moving a cube along the fourth dimension. In Figure 22.1*b* the initial and final positions of the moving cube have been drawn in heavy lines. There are altogether eight cubes: these two, and six others traced out by the six faces. Each of the 24 squares (which appear in the figure as either squares or rhombi) belongs to two of the cubes, not lying in the same 3-space but rotated about the plane of the square until the two 3-spaces are at right angles.

The regular simplex and the hypercube are the two simplest instances

$$\{3, 3, 3\}, \qquad \{4, 3, 3\}$$

of a *regular polytope* $\{p, q, r\}$, which is a configuration of equal Platonic solids $\{p, q\}$, called *cells,* fitting together in such a way that each face $\{p\}$ belongs to two cells, and each edge to r cells. It follows that the arrangement of the cells at a vertex corresponds to the arrangement of the faces of a $\{q, r\}$, in the sense that each face of the $\{q, r\}$ is a vertex figure of the corresponding cell. This $\{q, r\}$, whose vertices are the midpoints of the edges at one vertex of $\{p, q, r\}$, is naturally called the *vertex figure* of the polytope. In fact, the three-digit Schläfli symbol $\{p, q, r\}$ is derived by "telescoping" the two-digit symbols $\{p, q\}$ and $\{q, r\}$ which denote the cell and the vertex figure.

We can now complete the first two rows of Table IV (on p. 414), in which the numbers of vertices, edges, faces, and cells are denoted by N_0, N_1, N_2, N_3. Although there is no easy formula for any of these numerical properties as a function of p, q, r, we can readily find their mutual ratios by arguments analogous to those that led to 10.31. In fact, if V, E, F refer to the cell $\{p, q\}$, and V', E', F' to the vertex figure $\{q, r\}$, we have

$$FN_3 = 2N_2, \quad VN_3 = F'N_0, \quad V'N_0 = 2N_1, \quad EN_3 = rN_1 = pN_2 = E'N_0.$$

For instance, the first equation comes from the observation that the F faces of the N_3 cells are just the N_2 faces, counted twice because each belongs to two cells.

EXERCISES

1. The numbers $N_0 : N_1 : N_2 : N_3$ are proportional to

$$\frac{1}{q} + \frac{1}{r} - \frac{1}{2} \; : \; \frac{1}{r} \; : \; \frac{1}{p} \; : \; \frac{1}{p} + \frac{1}{q} - \frac{1}{2}.$$

Thus $\{p, q, r\}$ satisfies Schläfli's four-dimensional analogue of Euler's theorem:

$$N_0 - N_1 + N_2 - N_3 = 0.$$

2. A hypercube of edge 1, with one vertex at the origin and 4 edges along the Cartesian axes, has the 16 vertices (x_1, x_2, x_3, x_4), where each of the four x's is either 0 or 1, independently.

3. A hypercube of edge 2, with its center at the origin and its edges parallel to the Cartesian axes, has the 16 vertices

$$(\pm 1, \pm 1, \pm 1, \pm 1).$$

4. Where is the center of the dilatation that relates the hypercubes described in the two preceding exercises?

22.2 A NECESSARY CONDITION FOR THE EXISTENCE OF $\{p, q, r\}$

> "... Space ... is spoken of as having three dimensions, which one may call Length, Breadth, and Thickness. ... But some philosophical people have been asking why three dimensions particularly—why not another direction at right angles to the other three? ... I do not mind telling you I have been at work upon this geometry of Four Dimensions for some time. ... "
>
> H. G. Wells
> (The Time Machine, 1895, p. 5)

It was apparently Kepler who first thought of the regular tessellations (§ 4.6) as infinite polyhedra. Analogously, the three-dimensional honeycomb of cubes (whose vertices may be taken to be all the points (x, y, z) for which x, y, z are integers) is the infinite polytope $\{4, 3, 4\}$: its cell is the cube $\{4, 3\}$, and its vertex figure is the octahedron $\{3, 4\}$ whose eight faces are the vertex figures of the eight cubes that surround a vertex, one in each "octant." The final 4 in the symbol $\{4, 3, 4\}$ means that there are four cells surrounding an edge. These four cubes fit together without any interstices because the dihedral angle of the cube is exactly a right angle. On the other hand, the hypercube $\{4, 3, 3\}$ is a finite polytope because the total angle at an edge is only three right angles, allowing the cells to be rotated out of the 3-space the way one derives a polyhedron by folding up its net (only now the angular deficiency is not related in any simple way to the number of vertices).

Similarly, since the dihedral angle of the tetrahedron $\{3, 3\}$ is slightly less than $71°$ (see Table II), we may place three, four, or five (but no more) tetrahedra together at a common edge, so as to begin the construction of $\{3, 3, 3\}$, $\{3, 3, 4\}$, or $\{3, 3, 5\}$. Again, since the dihedral angles of the octahedron and dodecahedron are between $90°$ and $120°$, we may place just three of either together at an edge to obtain $\{3, 4, 3\}$ and $\{5, 3, 3\}$. But the icosahedron cannot be used in this manner, as its dihedral angle is greater than $120°$. We have thus proved that the only possible finite regular polytopes in four dimensions are

$$\{3, 3, 3\}, \quad \{3, 3, 4\}, \quad \{3, 3, 5\}, \quad \{4, 3, 3\}, \quad \{3, 4, 3\}, \quad \{5, 3, 3\}.$$

The condition for $\{p, q, r\}$ to be a finite polytope may be expressed in general terms by recalling (from 10.43) that the dihedral angle of the Platonic solid $\{p, q\}$ is

$$2 \arcsin \left(\cos \frac{\pi}{q} \middle/ \sin \frac{\pi}{p} \right).$$

If r such angles together make less than 2π, each must be less than $2\pi/r$. Hence

$$\arcsin \left(\cos \frac{\pi}{q} \middle/ \sin \frac{\pi}{p} \right) < \frac{\pi}{r},$$

that is,

22.21 $$\cos \frac{\pi}{q} < \sin \frac{\pi}{p} \sin \frac{\pi}{r}.$$

Similarly, the condition for $\{p, q, r\}$ to be an infinite honeycomb filling three-dimensional space is

22.22 $$\cos \frac{\pi}{q} = \sin \frac{\pi}{p} \sin \frac{\pi}{r} :$$

an equation for which the only solution in integers greater than 2 is $\{4, 3, 4\}$.

EXERCISES

1. The condition 22.21 implies both 10.33 and the analogous inequality with p replaced by r. *Hint:*

$$\sin \frac{\pi}{p} \sin \frac{\pi}{r} < \sin \frac{\pi}{p}.$$

2. Obtain the Schläfli symbol for the regular polytope whose eight vertices are

$$(\pm 1, 0, 0, 0), \quad (0, \pm 1, 0, 0), \quad (0, 0, \pm 1, 0), \quad (0, 0, 0, \pm 1),$$

that is, for the polytope

$$|x_1| + |x_2| + |x_3| + |x_4| \leqslant 1.$$

22.3   CONSTRUCTIONS FOR REGULAR POLYTOPES

Though analogy is often misleading, it is the least misleading thing we have.

Samuel Butler (1835 -1902)

(Music, Pictures, and Books)

We have seen that the inequality 22.21 is a *necessary* condition for the existence of a finite polytope $\{p, q, r\}$. The sufficiency of the condition requires an actual construction for each of the six figures. We know that r cells can fit together at an edge, but it is not obvious that the addition of further cells will ultimately yield a closed configuration in which every face of every cell belongs also to another cell.

As the complete story of such constructions is very long [Coxeter **1,** pp. 145–153], we must be content with a brief sketch, aided by analogy with what happens in three dimensions.

We recall that alternate vertices of a cube $\{4, 3\}$ belong to an inscribed tetrahedron $\{3, 3\}$ whose four faces correspond in an obvious manner to the four omitted vertices of the cube, whereas its six edges are diagonals of the six faces of the cube (*one* diagonal of each face). Moreover, the midpoints of these six edges, being the centers of the faces of the cube, are the vertices of an octahedron $\{3, 4\}$.

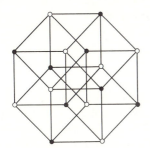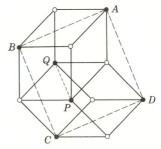

Figure 22.3a

Analogously, by selecting alternate vertices of the hypercube $\{4, 3, 3\}$ we obtain a polytope which has 8 vertices (the black points in Figure 22.3a) and 16 cells: one tetrahedron (such as $BCPQ$) corresponding to each of the 8 omitted vertices, and another (such as $ABPQ$) inscribed in each of the 8 cubic cells. This "16-cell" has 24 edges, which are diagonals of the 24 square faces of the hypercube (one diagonal of each face). Each of these 24 edges belongs to 4 tetrahedra (2 of each type, occurring alternately); for example, the edge PQ belongs to the 4 tetrahedra

$$ABPQ, \ BCPQ, \ CDPQ, \ DAPQ,$$

the first and third of which are inscribed in two adjacent cubes whose common face has PQ for one of its diagonals. We have thus proved that *the 16-cell is* $\{3, 3, 4\}$.

Completing the third line of Table IV, we observe that the numerical properties of $\{3, 3, 4\}$ are just those of $\{4, 3, 3\}$ in the reverse order. In fact, instead of obtaining the vertices of the 16-cell as alternate vertices of the hypercube, we could have obtained the vertices of another (similar) 16-cell as the centers of the cells of the hypercube. In other words, the hypercube and the 16-cell are *reciprocal* polytopes [Coxeter **1**, p. 127], like the cube and the octahedron. More generally, *the reciprocal of* $\{p, q, r\}$ *is* $\{r, q, p\}$.

The midpoints of the 24 edges of $\{3, 3, 4\}$ are the 24 vertices of a polytope whose cells are 24 octahedra: the vertex figures at the 8 vertices of $\{3, 3, 4\}$, and 16 inscribed in the 16 tetrahedra. Since all its cells are octahedra $\{3, 4\}$, this "24-cell" is $\{3, 4, 3\}$ [Hilbert and Cohn-Vossen **1**, p. 152, Fig. 172].

By suitably dividing the 12 edges of an octahedron in the ratio $\tau : 1$, we obtain the 12 vertices of an icosahedron (see § 11.2). By dividing the 96 edges of the 24-cell $\{3, 4, 3\}$ in this same ratio, we obtain the 96 vertices of a semiregular polytope s$\{3, 4, 3\}$ (the "snub 24-cell"), whose cells consist of 24 icosahedra and 120 tetrahedra: namely, at each vertex of the 24-cell, a set of 5 tetrahedra consisting of 1 surrounded by 4 others (like a partially folded "net" for the regular simplex $\{3, 3, 3\}$). When each icosahedral cell of s$\{3, 4, 3\}$ is capped by an icosahedral pyramid (the way an icosahedron is derived from a pentagonal antiprism by adding two pentagonal pyramids), we obtain a new polytope having a cluster of 20 tetrahedra to replace each of the 24 icosahedra, making a total of

$$24 \cdot 20 + 120 = 600$$

tetrahedra. The 120 vertices of this polytope consist of the 96 vertices of s$\{3, 4, 3\}$ and the 24 apices of the 24 icosahedral pyramids. (These 24 points, corresponding to the cells of the original $\{3, 4, 3\}$, are the vertices of a reciprocal $\{3, 4, 3\}$.) By careful examination [Coxeter **1**, pp. 152–153] we find that every edge belongs to 5 of the 600 tetrahedra. Hence the 600-cell is $\{3, 3, 5\}$ [Coxeter **1**, frontispiece].

Finally, the 120-cell $\{5, 3, 3\}$ (Figure 22.3*b* and Plate III) can be constructed as the reciprocal of $\{3, 3, 5\}$: its 600 vertices are the centers of the 600 tetrahedra. This information enables us to complete Table IV.

It is interesting to record that the snub 24-cell s$\{3, 4, 3\}$, which plays such a useful role in the above construction for $\{3, 3, 5\}$, was discovered by Thorold Gosset in 1897.* Figure 22.3*b* was drawn by B. L. Chilton. Plate III is a photograph of a wire model made by P. S. Donchian.

* Gosset was born in 1869 and died in 1962 [see Coxeter **1**, pp. 162–164].

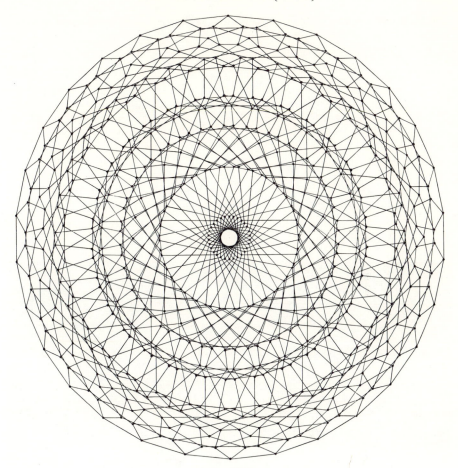

Figure 22.3b

Hartley [**1,** Nos. 56, 60] has given instructions for making models of a tetrahedron with an icosahedron placed on each face and of a dodecahedron with a dodecahedron placed on each face. When completed, these models show how we might begin to make solid nets for s{3, 4, 3} and {5, 3, 3}, respectively.

EXERCISES

1. Locate the centers of the 8 cubic cells of the hypercube $(\pm1, \pm1, \pm1, \pm1)$.

2. Locate the midpoints of the 24 edges of the 16-cell

$$(\pm2, 0, 0, 0), \quad (0, \pm2, 0, 0), \quad (0, 0, \pm2, 0), \quad (0, 0, 0, \pm2).$$

3. Verify that the 96 vertices of s{3, 4, 3}, which are

$$(\pm\tau, \pm1, \pm\tau^{-1}, 0),$$

evenly permuted, divide the 96 edges of the 24-cell $(\pm\tau, \pm\tau, 0, 0)$ (permuted) in the ratio $\tau : 1$.

4. The 120 vertices of the 600-cell $\{3, 3, 5\}$ are the 96 vertices of the above polytope $s\{3, 4, 3\}$, along with the 24 extra points

$$(\pm2, 0, 0, 0) \quad \text{(permuted)} \quad \text{and} \quad (\pm1, \pm1, \pm1, \pm1).$$

5. The 600 vertices of the 120-cell $\{5, 3, 3\}$ are the permutations of

$$(\pm2, \pm2, 0, 0), \quad (\pm\sqrt{5}, \pm1, \pm1, \pm1),$$
$$(\pm\tau, \pm\tau, \pm\tau, \pm\tau^{-2}), \quad (\pm\tau^2, \pm\tau^{-1}, \pm\tau^{-1}, \pm\tau^{-1})$$

along with the even permutations of

$$(\pm\tau^2, \pm\tau^{-2}, \pm1, 0), \quad (\pm\sqrt{5}, \pm\tau^{-1}, \pm\tau, 0), \quad (\pm2, \pm1, \pm\tau, \pm\tau^{-1}).$$

(This corrects an error in the first edition of Coxeter **1,** p. 157.)

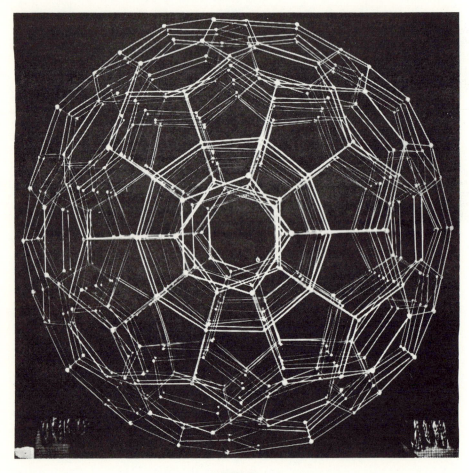

PLATE III

22.4   CLOSE PACKING OF EQUAL SPHERES

As the foot presses upon the sand when the falling tide leaves it firm, that portion of it immediately surrounding the foot becomes momentarily dry. . . . The pressure of the foot causes dilatation of the sand, and so more water is [drawn] through the interstices of the surrounding sand . . . , leaving it dry until a sufficient supply has been obtained from below, when it again becomes wet. On raising the foot we generally see that the sand under and around it becomes wet for a little time. This is because the sand contracts when the distorting forces are removed, and the excess of water escapes at the surface.

Osborne Reynolds (1842-1913)

(British Association Report, Aberdeen, 1885, p. 897).

Of all the two hundred thousand million men, women, and children who, from the beginning of the world, have ever walked on wet sand, how many, prior to the British Association Meeting at Aberdeen in 1885, if asked, "Is the sand compressed under your foot?" would have answered otherwise than "Yes!"? (Contrast with this the case of walking over a bed of wet sea-weed!)

Lord Kelvin (1824-1907)

(Baltimore Lectures, 1904, p. 625)

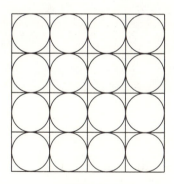

Figure 22.4a

Figure 22.4b

Figures 22.4*a* and *b* show two possible ways of packing equal circles in a plane: the incircles of the faces of the regular tessellations $\{4, 4\}$ and $\{6, 3\}$ (§ 4.6). It is intuitively obvious that the latter is the more "economical" packing. To make this idea precise, we consider the incircles of the faces of the general regular tessellation $\{p, q\}$, and define the *density* of the packing to be the ratio of the area of a circle to the area of the $\{p\}$ in which it is inscribed. The density so defined is evidently less than 1, and the closest packing will have the greatest density, that is, the density nearest to 1. If

PLATE IV

the face is a p-gon of side $2l$, its inradius is $r = l \cot \pi/p$ and its area is plr (see 2.91, 2.92); therefore, the density is

$$\frac{\pi r^2}{plr} = \frac{\pi}{p}\frac{r}{l} = \frac{\pi}{p}\cot\frac{\pi}{p} = \frac{\pi}{p}\Big/\tan\frac{\pi}{p}.$$

This is an increasing function of p, and tends to 1 when p tends to infinity. But since the p-gon is a face of a regular tessellation, the only relevant values of p are 3, 4, 6. Therefore the "best" value of p is 6, and the closest regular packing consists of the incircles of the faces of $\{6, 3\}$, the density being

$$\frac{\pi}{6}\cot\frac{\pi}{6} = \frac{\pi}{6}\sqrt{3} = \frac{\pi}{2\sqrt{3}} = 0.9069\ldots$$

[Hilbert and Cohn-Vossen **1**, p. 47].

It can easily be proved that this is still the closest packing when we abandon the requirement of regularity but insist instead that the centers of the circles form a lattice [Hilbert and Cohn-Vossen **1**, pp. 33–35]. Actually, even this restriction can be abandoned [Darwin **1**, p. 345; Fejes Tóth **1**, p. 58], as the bees discovered millions of years ago (Plate IV).

An analogous packing of spheres in three-dimensional space may be obtained by taking the inspheres of the cells of a honeycomb of equal polyhedra. The density is naturally defined as the ratio of the volume of a sphere to the volume of the cell in which it is inscribed. In the case of $\{4, 3, 4\}$, the honeycomb of cubes of edge $2l$, this is

$$\frac{\frac{4}{3}\pi l^3}{(2l)^3} = \frac{\pi}{6} = 0.5236. \ldots$$

A greater density can be obtained by using the *midspheres* (§ 10.4) of *alternate* cells, as we shall soon see.

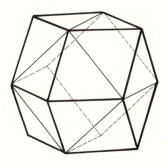

Figure 22.4c

If we imagine the cells of the cubic honeycomb to be colored alternately black and white, like a three-dimensional chessboard, we may dissect each white cube into six square pyramids (by planes joining pairs of opposite edges) and attach each pyramid to the neighboring black cube. Each black cube is now covered with six white pyramids, one on each face, to form a *rhombic dodecahedron* (Figure 22.4c), whose twelve rhombic faces have the twelve edges of the black cube for their shorter diagonals [Steinhaus **2**, p. 152]. Thus the insphere of the rhombic dodecahedron is the midsphere of the cube, of radius $\sqrt{2}\, l$, and the volume of the rhombic dodecahedron is twice that of the cube, namely, $2(2l)^3 = 16l^3$. In the honeycomb of such larger cells, each insphere is the midsphere of a black cube, and such spheres touch one another at the centers of the rhombic faces, that is, at the midpoints of the edges of the original honeycomb of cubes. Thus each sphere touches *twelve* others, the points of contact being the midpoints of the twelve edges of a cube. The density of this ·*cubic close packing* is evidently

$$\frac{\frac{4}{3}\pi(\sqrt{2}l)^3}{16l^3} = \frac{\pi}{3\sqrt{2}} = 0.74048\ldots$$

[Hilbert and Cohn-Vossen **1**, p. 47].

The rhombic dodecahedron occurs in nature as a crystal of garnet, and the three-dimensional chessboard occurs as the arrangement of atoms in a crystal of common salt, with a sodium atom in each black cube and a chlorine atom in each white cube (or vice versa). The centers of the black cubes, which are the centers of the spheres in cubic close packing, are easily seen to form the face-centered cubic lattice. It follows from § 18.4 that this is the densest possible packing of spheres whose centers form a lattice.

In old war memorials we often see a pyramidal pile of cannon balls: one at the apex resting on four others which, in turn, rest on nine, and so on. Each interior ball touches 12 others: 4 in its own layer, 4 above, and 4 below. In fact, these cannon balls are arranged in cubic close packing [Kepler **1**, pp. 268–269]. The base of the square pyramid consists of (say) n^2 balls arranged like the circles in Figure 22.4a. When n is large, the shape of the whole pyramid is essentially the "top" half of a regular octahedron (regarded as a square dipyramid); each sloping face is an equilateral triangle formed by $1 + 2 \ldots + n$ balls.

By turning the pyramid over so that such a sloping face becomes horizontal, we obtain a different aspect of the same packing. In this aspect we begin with a horizontal layer of spheres whose "equators" are the incircles of the hexagons of $\{6, 3\}$, as in Figure 22.4b. The next higher layer is just like this but shifted slightly to the right (say), so that each sphere rests on three, its center being vertically above a vertex of $\{6, 3\}$ from which an edge goes off to the right. Since all the centers form a three-dimensional lattice, the spheres in the third layer (resting on the second) are shifted again to the right, so that each center is vertically above a vertex of $\{6, 3\}$ from which an edge goes off to the left. The fourth layer is vertically above the first, and thereafter the sequence recurs.

In 1883, the crystallographer Barlow described an equally dense packing in which the centers do not form a lattice. This can be derived by taking the same horizontal layers in a different order. More precisely, we discard the "third layer" just described and substitute a new third layer vertically above the first. Then we add a fourth layer vertically over the second, and so on; the shifting from one layer to the next is alternately to the right and left, like a zigzag. This nonlattice packing is called *hexagonal close packing* [Ball **1**, p. 150; Hilbert and Cohn-Vossen **1**, p. 46; Steinhaus **2**, p. 170; Fejes Tóth **1**, pp. 172–173].

Helpful models to illustrate these ideas are provided by fourteen golf balls and two shallow trays of dimensions 5 in. × 5 in. and 4.6 in. × 5.8 in., respectively. Either tray will hold nine balls in three rows of three. In the square tray, a pyramidal "cannon ball" arrangement can be completed by adding four more above and the remaining one at the top. In the oblong tray, the four balls in the second layer should have their centers at the vertices of a rhombus, not a rectangle. The third layer is again represented by just one ball, but now there are two possible positions for it: one belongs to cubic close packing and the other to hexagonal close packing.

Since hexagonal close packing has the same density as cubic close packing, namely 0.74048 . . . , it is natural to ask whether some still less systematic packing (without any straight rows of spheres) may have a greater density. This remains an open question. The best theoretical approach to an answer is the proof by Rogers* that, if such a packing exists, its density must be less than 0.7797. . . .

Experiments in this direction began as long ago as 1727, when Stephen Hales stated, in his *Vegetable Staticks,*

I compressed several fresh parcels of Pease in the same Pot, with a force equal to 1600, 800, and 400 pounds; in which Experiments, tho' the Pease dilated, yet they did not raise the lever, because what they increased in bulk was, by the great incumbent weight, pressed into the interstices of the Pease, which they adequately filled up, being thereby formed into pretty regular Dodecahedrons.

Hales presumably reached his conclusion by observing some pentagonal faces on his dilated peas. They could not all have been regular dodecahedra. For, since the dihedral angle of the regular dodecahedron is less than 120° (see Table II on p. 413), three such solids with a common edge will leave an angular gap of about 10° 19′. In fact, dodecahedra {5, 3} are the cells of the configuration {5, 3, 3}, which is not an infinite three-dimensional honeycomb but a finite four-dimensional polytope.

In 1939, the botanists J. W. Marvin and E. B. Matzke repeated Hales's experiment, replacing his peas by lead shot, "carefully selected under a microscope for uniformity of size and shape," in a steel cylinder, compressed with a steel plunger at a sufficient pressure (40,000 pounds) to eliminate all interstices.† When the shot were stacked in cannon-ball fashion and compressed, they became nearly perfect rhombic dodecahedra. But "if the shot were just poured into the cylinder the way Hales presumably put his peas into the iron pot, irregular 14-faced bodies were formed." Almost all the faces were either quadrangles, pentagons, or hexagons, with pentagons predominating. Another botanist examined cells in undifferentiated vegetable tissues, and concluded that the internal cells have an average of approximately 14 faces, though the most prevalent shape (occurring 32 times among the 650 cells examined) had 13 faces: 3 quadrangles, 6 pentagons, and 4 hexagons. The few cells that had only 12 faces were neither rhombic dodecahedra nor regular dodecahedra.

Matzke also made a microscopic examination of a froth of 1900 measured bubbles. "For 600 central bubbles examined, the average number of contacts was 13.7." The commonest shape had again 13 faces: 1 quadrangle, 10 pentagons, and 2 hexagons.

* C. A. Rogers, The packing of equal spheres, *Proceedings of the London Mathematical Society* (3), **8** (1958), pp. 609–620.

† E. B. Matzke, In the twinkling of an eye, *Bulletin of the Torrey Botanical Club,* **77** (1950), pp. 222–227.

In 1959, Professor Bernal* confirmed the prevalence of pentagonal faces by a remarkably simple experiment in which equal balls of "Plasticene" (oily modeling clay) were rolled in powdered chalk, packed together irregularly, and pressed into one solid lump. The resulting polyhedra were found to have an average of 13.3 faces.

To test the possibility that a random packing of equal spheres might attain a density between 0.7405 and 0.7797, G. D. Scott poured thousands of ball bearings into spherical flasks of various sizes, gently shaking each flask as it was being filled. Assuming that the exceptional situation at the surface of the container will make the density

$$\rho - \varepsilon N^{-1/3}$$

for N balls, where ρ and ε are constants, he found from these experiments a closest random packing with

$$\rho = 0.6366, \qquad \varepsilon = 0.33.$$

By careful filling of the flasks without shaking, a loosest incompressible random packing was found with

$$\rho = 0.60, \qquad \varepsilon = 0.37.$$

Since, for the closest random packing, ρ falls far short of 0.7405, it seems unlikely that any greater density can be maintained throughout a region that extends indefinitely in all directions.

If we could fill a spherical flask with N ball bearings in cubic close packing, we would expect the density to be expressible as a series beginning with the two terms

$$0.7405 - \varepsilon N^{-1/3}.$$

But this experiment does not seem to be feasible. A scholarly book has been written on the theory of lattice points in spheres† without throwing any light on the value of ε in this 3-dimensional case, although considerable progress has been made on the analogous problem in spaces of other numbers of dimensions, such as 2 or 4.

Whatever the closest random packing may be, it is clear from Osborne Reynolds's experiment on the seashore that any small disturbance increases the size of the interstices. The same principle may explain a Hindu fakir's magic trick, which was mentioned by Martin Gardner. A cylindrical jar with a rather narrow opening at the top is filled with uncooked rice, gently shaken down so as to be well packed. A table knife is plunged repeatedly into the jar, to a greater depth each time. After about a dozen plunges, the knife will suddenly bind so that, when raised by the handle, it will support the whole jar of rice.

* J. D. Bernal, A geometrical approach to the structure of liquids, *Nature*, 183 (1959), pp. 141–147.

† Arnold Walfisz, *Gitterpunkte in mehrdimensionalen Kugeln*, Warsaw, 1957.

EXERCISES

1. Is the arrangement of incircles of all the faces of the tessellation {4, 4} (Figure 22.4*a*) any less dense than that of the circumcircles of alternate faces, i.e., the circumcircles of the black squares of a chessboard?

2. Is it possible to arrange seven equal non-overlapping spheres in such a way that two of them touch each other and both touch all the remaining five, while these five (a) form a ring in which each touches two others? (b) do not touch one another at all?

3. Is it possible to arrange thirteen equal non-overlapping spheres in such a way that one of them touches all the remaining twelve while these twelve do not touch one another at all?

4. A pyramidal pile with *n* layers contains $n(n + 1)(2n + 1)/6$ cannon balls [Ball **1**, p. 59]; a tetrahedral pile contains $n(n + 1)(n + 2)/6$. In both cases the arrangement is cubic close packing.

22.5 A STATISTICAL HONEYCOMB

> *The fluidity of a liquid is a consequence of its molecular irregularity.*
>
> J. D. Bernal (1901 -   )

Three equal circles in a plane are packed as closely as possible when they all touch one another. The two-dimensional problem of close packing is easy because any number of further circles can be added in such a way as to continue the pattern systematically over the whole plane. This is, as we have seen, the pattern formed by the incircles of the faces of the regular tessellation of hexagons, {6, 3} (Figure 22.4*b*).

Analogously in space, four equal spheres are packed as closely as possible when they all touch one another, and some further spheres can be added so as to form the beginning of a pattern apparently consisting of the inspheres of the cells of a regular honeycomb {*p*, 3, 3}. Although the equation 22.22 has no integral solution when $q = r = 3$, we naturally conclude that a compressed random packing of equal lead shot, a nearly homogeneous aggregate of vegetable cells, and a froth of equal bubbles, are all somehow trying to approximate to a honeycomb {*p*, 3, 3} in which *p* lies between 5 and 6. The fractional value of *p* means that this "honeycomb" can exist only in a statistical sense, but the agreement with experiment is striking.

When $q = r = 3$, the equation 22.22 actually becomes

22.51
$$\sin \frac{\pi}{p} = \cot \frac{\pi}{3} = \sqrt{\frac{1}{3}}.$$

This shows that the angle $180°/p$ is 35° 15′ 52″, which is half the dihedral angle of the regular tetrahedron {3, 3} (see Table II). (In fact, we may regard *p* as the number of regular tetrahedra {3, 3} that can be placed together around a common edge, as if we were beginning to construct the dual honeycomb {3, 3, *p*} whose vertices are the centers of the spheres.) Thus

$$p = \frac{180}{35.264\ldots} = 5.1044\ldots,$$

in agreement with Matzke's observation that pentagons are prevalent (especially in a froth) whereas hexagons are more frequent than quadrangles. The cell $\{p, 3\}$ has an average of F faces and V vertices where, by 10.32 with $q = 3$,

$$F = \frac{12}{6 - p} = 13.398\ldots, \qquad V = \frac{4}{(6/p) - 1} = 22.796\ldots,$$

in reasonably close agreement with Matzke's 13.7, with Bernal's 13.3, and with one of the two theoretical models proposed by Meijering,* who used intricate statistical methods to obtain $V = 22.56.\ldots$ A fourth theoretical model [Coxeter **4,** p. 756] yields

$$F = \tfrac{1}{3}(23 + \sqrt{313}) = 13.564\ldots, \qquad V = \tfrac{2}{3}(17 + \sqrt{313}) = 23.128.\ldots$$

EXERCISE

In the "twisted prism" formed by 28 regular tetrahedra

$$A_0A_1A_2A_3, \quad A_1A_2A_3A_4, \quad \ldots, \quad A_{27}A_{28}A_{29}A_{30},$$

the broken line $A_0A_3A_6A_9\ldots A_{30}$ consists of 10 equal chords of a circular helix. Taking the axis of this helix to be vertical, do we find the vertex A_{30} exactly above A_0? (A model can be conveniently made by fastening together 87 equal sticks from the "D-stix Pre-engineering Kit 701," manufactured in Yardley, Wash.) For the whole story, we regard the tetrahedra as being 28 cells of the "honeycomb" $\{3, 3, p\}$, where p is given by 22.51. ($A_0A_1A_2\ldots$ is a "Petrie polygon" of this honeycomb.) Setting

$$\cos^2 \frac{\pi}{p} = \frac{2}{3}$$

and $q = r = 3$ in the equation 12.35 of Coxeter [**1,** p. 221], we obtain $\xi_1 = 0$ and $\cos \xi_2 = -\tfrac{2}{3}$. The angle between the planes joining the axis to A_0 and A_{30} is

$$30(\xi_2 - 120°) = 354° \, 20'.$$

(This ξ_2, being nearly $131° \, 49'$, is remarkably close to the corresponding property of the four-dimensional polytope $\{3, 3, 5\}$, which is exactly $132°$ [Coxeter **1,** p. 247].)

---

*J. L. Meijering, *Philips Research Reports,* **8** (1953), p. 282. The value $V = 22.79 \ldots$ was first obtained by C. S. Smith, *Acta Metallurgica,* **1** (1953), p. 299. See also E. N. Gilbert, *Annals of Mathematical Statistics,* **33** (1962), pp. 958–972, and R. E. Williams, *Science,* **161** (1968), pp. ⌐76–277.

Table I
The 17 Space Groups of Two-Dimensional Crystallography (§ 4.3)

Symbol	Generators
p1	Two translations
p2	Three half-turns
pm	Two reflections and a translation
pg	Two parallel glide reflections
cm	A reflection and a parallel glide reflection
pmm	Reflections in the four sides of a rectangle
pmg	A reflection and two half-turns
pgg	Two perpendicular glide reflections
cmm	Two perpendicular reflections and a half-turn
p4	A half-turn and a quarter-turn
p4m	Reflections in the three sides of a (45°, 45°, 90°) triangle
p4g	A reflection and a quarter-turn
p3	Two rotations through 120°
p3m1	A reflection and a rotation through 120°
p31m	Reflections in the three sides of an equilateral triangle
p6	A half-turn and a rotation through 120°
p6m	Reflections in the three sides of a (30°, 60°, 90°) triangle

Table II
The Five Platonic Solids (§ 10.3)

Name	Schläfli Symbol	V	E	F	Dihedral Angle
Tetrahedron	$\{3, 3\}$	4	6	4	70° 32′ −
Cube	$\{4, 3\}$	8	12	6	90°
Octahedron	$\{3, 4\}$	6	12	8	109° 28′ +
Dodecahedron	$\{5, 3\}$	20	30	12	116° 34′ −
Icosahedron	$\{3, 5\}$	12	30	20	138° 11′ +

Table III
The Finite Groups of Isometries (§ 15.5)

Rotation Groups			Direct Products		Mixed Groups	
Name	Symbol	Order	Symbol	Order	Symbol	Order
Cyclic	C_n	n	$C_n \times \{I\}$	$2n$	$C_{2n}C_n$	$2n$
Dihedral	D_n	$2n$	$D_n \times \{I\}$	$4n$	D_nC_n	$2n$
Tetrahedral	A_4	12	$A_4 \times \{I\}$	24	$D_{2n}D_n$	$4n$
Octahedral	S_4	24	$S_4 \times \{I\}$	48	S_4A_4	24
Icosahedral	A_5	60	$A_5 \times \{I\}$	120		

Table IV
The Regular Polytopes $\{p, q, r\}$ (§ 22.2)

Name	Schläfli Symbol	N_0	N_1	N_2	N_3
Regular simplex	$\{3, 3, 3\}$	5	10	10	5
Hypercube	$\{4, 3, 3\}$	16	32	24	8
16-cell	$\{3, 3, 4\}$	8	24	32	16
24-cell	$\{3, 4, 3\}$	24	96	96	24
120-cell	$\{5, 3, 3\}$	600	1200	720	120
600-cell	$\{3, 3, 5\}$	120	720	1200	600
Cubic honeycomb	$\{4, 3, 4\}$	∞	∞	∞	∞

References

E. Artin 1. *Geometric Algebra.* Interscience, New York, 1957.

F. Bachmann 1. *Aufbau der Geometrie aus dem Spiegelungsbegriff* (Grundlehren der mathematischen Wissenschaften, 96), Springer, Berlin, 1959.

H. F. Baker 1. *An Introduction to Plane Geometry.* Cambridge University Press, London, 1943.

W. W. R. Ball 1. *Mathematical Recreations and Essays* (11th ed.). Macmillan, London, 1959.

—— 2. *A Short Account of the History of Mathematics.* Macmillan, London, 1927.

E. T. Bell 1. *Men of Mathematics.* Simon and Schuster, New York, 1937.

—— 2. *The Development of Mathematics.* McGraw-Hill, New York, 1940.

R. J. T. Bell 1. *An Elementary Treatise on Coordinate Geometry of Three Dimensions.* Macmillan, London, 1926.

G. Birkhoff and S. MacLane 1. *A Survey of Modern Algebra* (3rd ed.). Macmillan, New York, 1965.

W. Blaschke 1. *Analytische Geometrie.* Birkhäuser, Basel, 1954.

—— 2. *Projektive Geometrie.* Birkhäuser, Basel, 1954.

J. Bolyai 1. La science absolue de l'espace. *Mémoires de la Société des Sciences de Bordeaux,* 5 (1867), pp. 207–248.

R. Bonola 1. La geometria non-euclidea, Bologna, 1906.

K. Borsuk and W. Szmielew 1. *Foundations of Geometry.* North-Holland, Amsterdam, 1960.

O. Bottema 1. *Hoofdstukken uit de Elementaire Meetkunde.* N. V. Servire, The Hague, 1944.

D. Brewster 1. *A Treatise on the Kaleidoscope.* Constable, Edinburgh, 1819.

M. Brückner 1. *Vielecke und Vielflache.* Teubner, Leipzig, 1900.

M. J. Buerger 1. *X-Ray Crystallography.* Wiley, New York, 1942.

J. J. Burckhardt 1. *Die Bewegungsgruppen der Kristallographie.* Birkhäuser, Basel, 1957.

W. Burnside 1. *Theory of Groups of Finite Order* (2nd ed.). Cambridge University Press, London, 1911.

H. Busemann 1. *Geometry of Geodesics.* Academic Press, New York, 1955.

H. S. Carslaw 1. *The Elements of Non-Euclidean Plane Geometry and Trigonometry.* Longmans, London, 1916.

J. Casey 1. *A Sequel to the First Six Books of the Elements of Euclid* (6th ed.). Hodges Figgis, Dublin, 1892.

A. H. Church 1. *The Relation of Phyllotaxis to Mechanical Laws.* Williams and Norgate, London, 1904.

J. L. Coolidge 1. *A History of the Conic Sections and Quadric Surfaces.* Clarendon, Oxford, 1945.

R. Courant 1. *Differential and Integral Calculus,* Vol. 1. Blackie, London, 1934.

—— and H. E. Robbins 1. *What is Mathematics?* Oxford University Press, New York, 1953.

N. A. Court 1. *College Geometry* (1st ed.). Johnson, Richmond, Va., 1925.

—— 2. *College Geometry* (2nd ed.). Barnes and Noble, New York, 1952.

H. S. M. Coxeter 1. *Regular Polytopes.* (2nd ed.). Macmillan, New York, 1963.

—— 2. *The Real Projective Plane* (2nd ed.). Cambridge University Press, London, 1961.

————3. *Non-Euclidean Geometry* (5th ed.). University of Toronto Press, Toronto, 1968.

————4. Close-packing and froth. *Illinois Journal of Mathematics,* 2 (1958), pp. 746–758.

————and S. L. Greitzer 1. *Geometry Revisited.* Random House, New York, 1967.

————and W. O. J. Moser 1. *Generators and Relations for Discrete Groups* (Ergebnisse der Mathematik und ihrer Grenzgebiete, 14). Springer, Berlin, 1965.

H. M. Cundy and A. P. Rollett 1. *Mathematical Models.* Oxford University Press, London, 1961.

C. Darwin 1. *Origin of Species.* Appleton, New York, 1927.

F. Denk and J. E. Hofmann 1. *Ebene Geometrie.* Blutenburg, Munich, 1957.

C. L. Dodgson (*alias* Lewis Carroll) 1. *Alice's Adventures in Wonderland.* Macmillan, London, 1865.

———— 2. *Through the Looking-glass, and What Alice Found There.* Macmillan, London, 1872.

————2a. *The Hunting of the Snark,* Macmillan, London, 1876.

————3. *Euclid and his Modern Rivals.* Macmillan, London, 1879.

———— 4. *Sylvie and Bruno Concluded.* Macmillan, London, 1893.

H. Dörrie 1. *Mathematische Miniaturen.* Breslau, 1943.

L. Fejes Tóth 1. *Lagerungen in der Ebene, auf der Kugel und im Raum.* (Grundlehren der mathematischen Wissenschaften, 65). Springer, Berlin, 1953.

L. R. Ford 1. *Automorphic Functions* (2nd ed.). McGraw-Hill, New York, 1929.

H. G. Forder 1. *The Foundations of Euclidean Geometry.* Cambridge University Press, London, 1927; Dover, New York, 1958.

———— 2. *The Calculus of Extension.* Cambridge University Press, London, 1941; Chelsea, New York, 1960.

———— 3. *Geometry* (2nd ed.). Hutchinson's University Library, London; 1960.

R. Fricke and F. Klein 1. *Vorlesungen über die Theorie der automorphen Funktionen,* Vol. 1. Teubner, Leipzig, 1897.

C. F. Gauss 1. *Werke,* Königliche Gesellschaft der Wissenschaften, Göttingen, 1900.

S. H. Gould 1. Origins and Development of Concepts of Geometry. *Twenty-third Year Book of the National Council of Teachers of Mathematics,* 1957, Chapter IX.

W. C. Graustein 1. *Introduction to Higher Geometry.* Macmillan, New York, 1930.

G. H. Hardy 1. *Pure Mathematics* (10th ed.). Cambridge University Press, London, 1955.

———— 2. *A Mathematician's Apology.* Cambridge University Press, London, 1940.

———— and E. M. Wright 1. *An Introduction to the Theory of Numbers* (2nd ed.). Clarendon Press, Oxford, 1945.

M. C. Hartley 1. *Patterns of Polyhedrons.* Edwards Bros., Ann Arbor, Mich., 1951.

T. L. Heath 1, 2, 3. *The Thirteen Books of Euclid's Elements* (3 vols.) Cambridge University Press, London, 1908; Dover, New York, 1956.

D. Hilbert 1. *The Foundations of Geometry.* Open Court, Chicago, 1902.

———— and S. Cohn-Vossen 1. *Geometry and the Imagination.* Chelsea, New York, 1952.

H. Hilton 1. *Mathematical Crystallography and the Theory of Groups of Movements.* Clarendon Press, Oxford, 1903.

E. W. Hobson 1. *A Treatise on Plane Trigonometry,* Cambridge University Press, London, 1925.

L. Infeld 1. *Whom The Gods Love.* Whittlesey House, New York, 1948.

H. Jeffreys 1. *Cartesian Tensors.* Cambridge University Press, London, 1931.

R. Johnson 1. *Modern Geometry.* Houghton Mifflin, Boston, 1929.

O. Jones 1. *Grammar of Ornament.* Quaritch, London, 1868.

E. Kasner and J. Newman 1. *Mathematics and The Imagination.* Simon and Schuster, New York, 1940.

J. Kepler 1. *Gesammelte Werke.* Vol. 4. Beck, Munich, 1941.

B. Kerékjártó 1. *Les Fondements de la Géométrie.* Akadémiai Kiadó, Budapest, 1955.

F. Klein 1, 2. *Elementary Mathematics from an Advanced Standpoint* (2 vols.). Macmillan, New York, 1939.

———— 3. *Lectures on the Icosahedron* (2nd ed.). Kegan Paul, London, 1913.

———— 4. *Vorlesungen über Nicht-Euklidische Geometrie.* Springer, Berlin, 1928.

E. Kreyszig 1. *Differential Geometry.* University of Toronto Press, Toronto, 1959.

R. Lachlan 1. *An Elementary Treatise on Modern Pure Geometry.* Macmillan, London, 1927.

H. Lamb 1. *Higher Mechanics.* Cambridge University Press, London, 1920.

———— 2. *Infinitesimal Calculus* (2nd ed.). Cambridge University Press, London, 1924.

C. J. de La Vallée Poussin 1, 2. *Cours d'Analyse Infinitésimale* (2 vols., 5th ed.). Libraire Universitaire, Louvain, 1925.

S. Lefschetz 1. *Introduction to Topology.* Princeton University Press, Princeton, 1949.

H. Liebmann 1. *Nichteuklidische Geometrie* (3rd ed.). Berlin, 1923.

J. E. Littlewood 1. *A Mathematical Miscellany,* Methuen, London, 1953.

L. J. Mordell 1. *Reflections of a Mathematician.* Canadian Mathematical Congress, Montreal, 1958.

M. Moseley 1. On conchyliometry, *Philosophical Magazine* (3), 21 (1842), pp. 300–305.

E. H. Neville 1. *Jacobian Elliptic Functions.* Clarendon Press, Oxford, 1944.

L. Pacioli (*alias* Paccioli) 1. *De Divina Proportione.* Venice, 1509; Milan, 1956.

M. Pasch and M. Dehn 1. *Vorlesungen über neuere Geometrie* (2nd ed., Grundlehren der mathematischen Wissenschaften, 23). Springer, Berlin, 1926.

D. Pedoe 1. *Circles.* Pergamon, New York, 1957.

———— 2. *An Introduction to Projective Geometry.* Pergamon, New York, 1963.

H. Rademacher and O. Toeplitz 1. *The Enjoyment of Mathematics.* Princeton University Press, Princeton, 1957.

G. Ringel 1. *Färbungsprobleme auf Flächen und Graphen.* VeB Deutscher Verlag der Wissenschaften, Berlin, 1959.

G. de B. Robinson 1. *The Foundations of Geometry* (4th ed.). University of Toronto Press, Toronto, 1959.

A. Robson 1, 2. *An Introduction to Analytical Geometry* (2 vols.). Cambridge University Press, London, 1940, 1947.

E. J. Routh 1. *A Treatise on Analytical Statics, with Numerous Examples,* Vol. 1 (2nd ed.). Cambridge University Press, 1896.

B. Russell 1. *The Principles of Mathematics.* Cambridge University Press, London, 1903.

———— 2. *A Critical Exposition of the Philosophy of Leibniz* (2nd ed.). Allen and Unwin, London, 1937.

G. Salmon 1. *A Treatise on Higher Plane Curves* (3rd ed.). Hodges, Dublin, 1879.

———— 2. *A Treatise on the Analytic Geometry of Three Dimensions,* vol. 1 (6th ed.). Longmans Green, London, 1914.

Dorothy L. Sayers and R. Eustace 1. *The Documents in the Case.* Penguin, London, 1937.

L. Schläfli 1. *Gesammelte mathematische Abhandlungen* (vol. 1). Birkhäuser, Basel, 1950.

D. E. Smith 1, 2. *History of Mathematics* (2 vols.). Ginn, Boston, 1910; Dover, New York, 1958.

D. M. Y. Sommerville 1. *The Elements of Non-Euclidean Geometry.* Bell, London, 1914.

———— 2. *An Introduction to the Geometry of n Dimensions.* Methuen, London, 1929.

A. Speiser 1. *Theorie der Gruppen von endlicher Ordnung* (4th ed.). Birkhäuser, Basel, 1956.

J. Steiner 1, 2. *Gesammelte Werke* (2 vols.). Reimer, Berlin, 1882.

H. Steinhaus 1. *Mathematical Snapshots* (1st ed.). Stechert, New York, 1938.

———— 2. *Mathematical Snapshots* (2nd ed.). Oxford University Press, London, 1950.

D. J. Struik 1. *Classical Differential Geometry.* Addison-Wesley, Cambridge, Mass., 1961.

J. L. Synge 1. *Science: Sense and Nonsense.* Norton, New York, 1951.

———— 2. *Kandelman's Krim.* Jonathan Cape, London, 1957.

D'Arcy W. Thompson 1, 2. *On Growth and Form* (2 vols.). Cambridge University Press, London, 1952.

W. Thomson (*alias* Lord Kelvin) and P. G. Tait 1. *Treatise on Natural Philosophy* (vol. 1.1). Cambridge University Press, London, 1888.

E. C. Titchmarsh 1. *Mathematics for the General Reader.* Hutchinson's University Library, London, 1943.

J. V. Uspensky and M. A. Heaslet 1. *Elementary Number Theory.* Macmillan, New York, 1939.

O. Veblen 1. *The Foundations of Geometry.* Chapter I of J. W. Young's *Monographs on Topics of Modern Mathematics,* Longmans Green, New York, 1911.

———— and J. W. Young 1, 2. *Projective Geometry* (2 vols.). Ginn, Boston, 1910, 1918.

C. E. Weatherburn 1. *Elementary Vector Analysis.* Bell, London, 1921.

———— 2. *Differential Geometry of Three Dimensions.* Cambridge University Press, London, 1931.

H. Weyl 1. *Symmetry.* Princeton University Press, Princeton, 1952.

I. M. Yaglom 1. *Geometric Transformations* (Vol. 1 of a translation by A. Shields). Random House, New York, 1962.

———— 2. *Geometricheskie Preobrazovaniya,* Vol. 2. Gosudarstvennoe Izdatel'stvo Tekhniko-Teoreticheskoĭ Literaturȳ, Moscow, 1956.

Answers to Exercises

§1.3

1. The reflection in the line $x = y$ interchanges x and y.

2. If the line BA meets the circle in P (beyond A) and P' (between A and B), we have $BC^2 = BP \times BP' = (BA + AC)(BA - AC)$.

3. The triangle CDF is equilateral; so is ABC.

4. This result was conjectured by Paul Erdös and first proved by L. J. Mordell (see the *American Mathematical Monthly*, **44** (1937), p. 252, problem 3740, or Fejes Tóth **1**, pp. 12–14). In 1960, Mordell discovered the following simpler proof. For convenience, let OA, OB, OC, OP, OQ, OR be denoted by x, y, z, p, q, r, so that the theorem to be proved is

$$x + y + z \geqslant 2(p + q + r).$$

Also let p', q', r' denote the lengths (within the triangle) of the bisectors of the angles

$$2\alpha = \angle BOC, \qquad 2\beta = \angle COA, \qquad 2\gamma = \angle AOB.$$

By comparing the area of the triangle OBC with the two parts into which it is dissected by the bisector p', and using the well-known inequality $y + z \geqslant 2\sqrt{yz}$ which comes from $(\sqrt{y} - \sqrt{z})^2 \geqslant 0$, we find

$$yz \sin 2\alpha = p'(y + z) \sin \alpha \geqslant 2p'\sqrt{yz} \sin \alpha,$$

with equality only when $y = z$. Hence $\sqrt{yz} \cos \alpha \geqslant p'$, and similarly $\sqrt{zx} \cos \beta \geqslant q'$, $\sqrt{xy} \cos \gamma \geqslant r'$. Since

$$x + y + z - 2\sqrt{yz} \cos \alpha - 2\sqrt{zx} \cos \beta - 2\sqrt{xy} \cos \gamma$$
$$= (\sqrt{x} - \sqrt{y} \cos \gamma - \sqrt{z} \cos \beta)^2 + (\sqrt{y} \sin \gamma - \sqrt{z} \sin \beta)^2 \geqslant 0,$$

it follows that

$$x + y + z \geqslant 2(p' + q' + r') \geqslant 2(p + q + r).$$

The inequality $x + y + z \geqslant 2(p' + q' + r')$, which may be regarded as an extended form of the Erdös-Mordell theorem, was first noticed by D. F. Barrow. See also O. Bottema, R. Ž. Djordjević, R. R. Janić, D. S. Mitrinović and P. M. Visać, *Geometric Inequalities* (Wolters-Noordhoff, Groningen, The Netherlands, 1969), p. 139.

5. Equality occurs only if $x = y = z$ and $\sqrt{y} \sin \gamma - \sqrt{z} \sin \beta = 0$, etc., so that $\alpha = \beta = \gamma$. The triangle is equilateral, and O is its center.

6. Let h_a denote the "altitude" from A to BC, and Δ the area of the triangle ABC.

419

Since $x + p \geqslant h_a$, we have

$$a(x + p) \geqslant ah_a = 2\Delta = ap + bq + cr,$$

whence

$$ax \geqslant bq + cr.$$

Consideration of similar triangles shows that this remains true when the line AO is extended (even if O is on the far side of the base BC). Applying the same inequality to the image of O by reflection in the internal bisector of $\angle BAC$, we obtain

$$ax \geqslant br + cq.$$

Adding these two inequalities for ax, we obtain

$$2ax \geqslant (b + c)(q + r).$$

Multiplying together this and two other inequalities of the same kind,

$$8\,abcxyz \geqslant (b + c)(c + a)(a + b)(q + r)(r + p)(p + q).$$

Since $b + c \geqslant 2\sqrt{bc}$, etc., we have $(b + c)(c + a)(a + b) \geqslant 8abc$. Hence

$$xyz \geqslant (q + r)(r + p)(p + q).$$

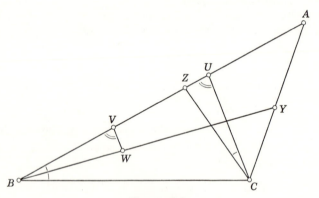

Figure I-3e

7. Let B be the smaller of the two different angles B and C of the triangle ABC. Let BY and CZ be the internal bisectors of these angles, as in Figure 1.3e. Take U on AZ so that $\angle ZCU = \frac{1}{2}B$. Since the triangle UBC has a smaller angle at B than at C, $BU > CU$. Take V on BU so that $BV = CU$. Take W on BY so that $\angle BVW = \angle CUZ$. By the angle-side-angle criterion, BVW and CUZ are congruent triangles, and $BW = CZ$. But W and Y are on opposite sides of the line CU. Hence $BY > BW$, that is, $BY > CZ$.

This theorem was proposed in 1840 by C. L. Lehmus, and proved by Jacob Steiner. For its history, see J. A. McBride, *Edinburgh Mathematical Notes*, **33** (1943), pp. 1–13. McBride asserts that more than sixty proofs have been given. The simple one given above came in a letter from H. G. Forder; it excels most by being "absolute" in the sense of §12.1. For the original proofs by Steiner and Lehmus, respectively, see *Journal für die reine und angewandte Mathematik*, **28** (1844), p. 376, and *Archiv der Mathematik und Physik*, **15** (1850), p. 225.

§1.4

1. Apply Euclid I.5 to the isosceles triangle GBC and then I.4 to the two triangles $B'BC$, $C'CB$.

2. Add together three inequalities such as

$$\tfrac{2}{3}BB' + \tfrac{2}{3}CC' > BC.$$

Complete the parallelogram $CABK$ and observe that twice the median from A is $AK < AC + CK = b + c$.

§1.5

1. The circle through P with center O.

2. Use 1.52.

3. Let the tangents to the incircle from A, B, C be t_a, t_b, t_c. Then

$$t_b + t_c = a, \qquad t_c + t_a = b, \qquad t_a + t_b = c;$$

therefore

$$t_a = \tfrac{1}{2}(b + c - a) = s - a.$$

4. By Euclid III.20, if the angle at the circumference is greater than $90°$, the angle at the center is greater than $180°$.

5. At the midpoint of the hypotenuse.

6. Make repeated use of Pythagoras's theorem.

7. Since $bc + ca + ab = \{(s - a)(s - b)(s - c) + abc + s^3\}/s = r^2 + 4Rr + s^2$ and $abc = 4R\Delta = 4Rrs$, we have

$$(2R - a)(2R - b)(2R - c) = 8R^3 - 8R^2s + 2R(r^2 + 4Rr + s^2) - 4Rrs$$
$$= 2R(2R + r - s)^2.$$

Alternatively, in the notation of ex. 3,

$$(t_a - r)(t_b - r)(t_c - r) = (s - r - a)(s - r - b)(s - r - c)$$
$$= (s - r)^3 - 2s(s - r)^2 + (r^2 + 4Rr + s^2)(s - r) - 4Rrs$$
$$= 2r^2(s - r - 2R).$$

This yields the desired criterion, since the angle A is acute or right or obtuse according as $t_a - r$ is positive or zero or negative. In fact, we can conclude further that *the triangle has an obtuse angle if and only if $r + 2R > s$.* Corrado Ciamberlini [*Bolletino della Unione Matematica Italiana* (2), **5** (1943), pp. 37–41] observed that

$$4R^2 \cos A \cos B \cos C = s^2 - (r + 2R)^2.$$

8. Since $2\eta_1 = -\epsilon_1 + \epsilon_2 + \epsilon_3 + \epsilon_4$, $\epsilon_1 + \eta_1 = \tfrac{1}{2}\Sigma\epsilon_i$. Also $\Sigma\epsilon_i\eta_i = \Sigma\epsilon_i(\epsilon_i + \eta_i) - \Sigma\epsilon_i^2 = \tfrac{1}{2}(\Sigma\epsilon_i)^2 - \Sigma\epsilon_i^2 = 0$.

9. If $\epsilon_1 = \epsilon_2 = 0$, Beecroft's equations imply $\eta_3 = \eta_4 = 0$ and $\epsilon_3 = \epsilon_4 = \eta_1 = \eta_2$, so that the configuration consists of four lines E_1, H_3, E_2, H_4 forming a square, and four circles H_2, E_4, H_1, E_3 having the sides of the square for diameters. This is what happens when $k + l = m + n = 0$. In any other case, we can assign arbitrary values to the three bends ϵ_1, ϵ_2, η_3, subject only to the condition $\epsilon_1 + \epsilon_2 > 0$ (which ensures that if E_1 and E_2 have internal contact, the larger circle is the one whose bend is negative). Then η_4 is determined by the simple equation

$$\epsilon_1 + \epsilon_2 = \eta_3 + \eta_4$$

(which follows from $\epsilon_1 + \epsilon_2 - \epsilon_3 + \epsilon_4 = 2\eta_3$, $\epsilon_1 + \epsilon_2 + \epsilon_3 - \epsilon_4 = 2\eta_4$), and the remaining bends are

$$\epsilon_3 = \frac{\eta_4^2 - \epsilon_1\epsilon_2}{\epsilon_1 + \epsilon_2}, \qquad \epsilon_4 = \frac{\eta_3^2 - \epsilon_1\epsilon_2}{\epsilon_1 + \epsilon_2}, \qquad \eta_1 = \frac{\epsilon_2^2 - \eta_3\eta_4}{\epsilon_1 + \epsilon_2}, \qquad \eta_2 = \frac{\epsilon_1^2 - \eta_3\eta_4}{\epsilon_1 + \epsilon_2}.$$

The proposed parametrization is obtained by choosing

$$k = \epsilon_1/\sqrt{\epsilon_1 + \epsilon_2}, \qquad\qquad l = \epsilon_2/\sqrt{\epsilon_1 + \epsilon_2},$$
$$m = -\eta_3/\sqrt{\epsilon_1 + \epsilon_2}, \qquad n = -\eta_4/\sqrt{\epsilon_1 + \epsilon_2}.$$

10. Using 1.59, 1.58, 1.52, and 1.56 in turn, we obtain

$$\epsilon_4 = \epsilon_1 + \epsilon_2 + \epsilon_3 - 2\eta_4 = \frac{1}{s-a} + \frac{1}{s-b} + \frac{1}{s-c} \pm \frac{2}{r}$$
$$= \frac{r_a + r_b + r_c \pm 2s}{\Delta} = \frac{r + 4R \pm 2s}{\Delta}.$$

These two circles (which touch three mutually tangent circles) became known as *Soddy's circles* before anyone noticed the earlier work of Descartes and Steiner [1, pp. 60–63, 524]. As Descartes used the letters d, e, f, x for the radii of four mutually tangent circles, he undoubtedly saw that he had obtained a quadratic equation for x in terms of d, e, f.

11. Let $A_1B_1C_1$ be the feet of the perpendiculars from P to BC, CA, AB. Applying Euclid III.21 or 22 to each of the cyclic quadrangles PA_1B_1C, $PABC$, PA_1BC_1, show that $\angle PA_1B_1$ and $\angle PA_1C_1$ are either equal or supplementary.

12. $\angle C_3B_3A_3 = \angle C_3P_3P + \angle PB_3A_3 = \angle CBP + \angle PBA = \angle CBA$.

§1.6

1. The circumcenter of the new triangle is the orthocenter of ABC.

3. At the vertex where the right angle occurs.

4. Because $\sin B = \sin C$.

5. On $B'E$, take C so that $GC = GB$, and A so that $AB' = B'C$.

6. It is $b \sin C$, and $b = 2R \sin B$.

7. One-third of the altitude.

8. If the Euler line passes through A, and if A is not a right angle, the altitude line AH is a median.

9. $R \cos A = \frac{2}{3}R \sin B \sin C$.

§1.7

1. (a) One pair. (b) Three pairs.

2. If $A > B > C$, the order is $EA''FC'B''DA'C''B'$.

3. The angular measures of the relevant arcs of the nine-point circle (with $A > B > C$, as in Figure 1.7a) are

$$A'E = A'F = 2A, \qquad B'F = B'D = 2B, \qquad C'D = C'E = 2C,$$

whence

$$DA' = 2(\pi - 2C - A) \qquad A'B' = 2(\pi - A - B),$$
$$B'E = 2(2A + B - \pi), \qquad EF = 2(\pi - 2A), \qquad FC' = 2(2A + C - \pi).$$

4. The internal and external bisectors of an angle are perpendicular.

5. The nine-point center of $I_aI_bI_c$ is the circumcenter of its orthic triangle ABC.

6. Each circumradius is twice the radius of the nine-point circle.

§1.8

1. $U'V$, which passes through W, is the image of UV by reflection in AC.

2. When there is a right angle at A, V and W coincide with A. When there is an obtuse angle at A, this degenerate triangle UAA is still "better" than any proper triangle.

3. The lines joining pairs of centers, being perpendicular to the common chords AP, BP, CP, make angles of $60°$ with one another, and thus form an equilateral triangle.

4. At the Fermat point. (H. G. Forder uses analogous considerations to prove that, if $ABCD$ is a tetrahedron and $PA + PB + PC + PD$ is a minimum, the angles APB and CPD are equal and their bisectors lie on one line.)

5. They join pairs of villages to the ends of a short road in the middle.

6. The "best" point for the "very obtuse" triangle is A itself. For the convex quadrangle it is the point of intersection of the diagonals.

7. P is the incenter of $P'BC$.

8. Let Z, X, U be the centers of the squares on three consecutive sides AB, BC, CD of the parallelogram $ABCD$. The triangle XBZ is derived from XCU by a quarter-turn (i.e., rotation through a right angle) about X.

9. Let M be the midpoint of CA. By ex. 8, the segments MZ and MX are congruent and perpendicular. The same can obviously be said of MY and MA. Therefore the triangle MAX is derived from MYZ by a quarter-turn about M.

10. As in ex. 9, the segments MZ and MX are congruent and perpendicular. Similarly (by considering the triangle CDA instead of ABC), the segments MU and MV are congruent and perpendicular. Therefore the triangle MXV is derived from MZU by a quarter-turn about M.

§1.9

1. These lines are the medians of the equilateral triangle PQR.

2. (i) $\alpha = \beta = \gamma = 40°$; (ii) $\alpha = 30°$, $\beta = \gamma = 45°$.

3. Since $\angle CP_1Q = \angle CPQ = \gamma + \alpha = \angle QRA$, the circumcircle of AQR passes through P_1, and likewise through P_2. Since

$$P_1Q = PQ = QR = RP = RP_2,$$

the points P_1, Q, R, P_2 are evenly spaced along this circle. In the special case, each of the arcs P_1Q, QR, RP_2 subtends $20°$ at A, and the triangle AQR is isosceles.

§2.1

1. Since

$$\frac{ON_1}{N_1P_0} = \frac{OD}{DP_0} = \frac{1}{\sqrt{5}},$$

$$\frac{ON_1}{OP_1} = \frac{ON_1}{ON_1 + N_1P_0} = \frac{1}{1 + \sqrt{5}} = \frac{\sqrt{5} - 1}{4} = \cos 72°.$$

2. $\frac{6}{17} - \frac{1}{3} = \frac{1}{51}$.

§2.2

If s is odd, $x^s + 1$ is divisible by $x + 1$, and therefore $2^{rs} + 1$ by $2^r + 1$.

§2.4

1. Suppose a given isometry of period 2 interchanges A and A', and interchanges B and B', where B does not lie on AA'. The midpoints of AA' and BB' are invariant points. If they are distinct, the isometry is a reflection (by 2.31). If they coincide, it is a half-turn.

2. (a) (i) $(x, y) \to (-x, -y)$;   (ii) $(r, \theta) \to (r, \theta + 180°)$.

   (b) (i) $(x, y) \to (-y, x)$;   (ii) $(r, \theta) \to (r, \theta + 90°)$.

§2.5

(i) $(r, \theta) \to (r, \theta + \alpha)$;

(ii) $(x, y) \to (x \cos \alpha - y \sin \alpha + y \cos \alpha)$.

The transformed curve is $f(r, \theta - \alpha) = 0$.

§2.6

1. If O is the center of a suitable one of the two squares that can be drawn on BC, the first quarter-turn is the product of reflections in CO and CB whereas the second is the product of reflections in BC and BO.

2. P is transformed into A by a quarter-turn about C, and thence into S by a quarter-turn about B.

§2.7

1. (a) C_1,   (b) D_1,   (c) D_1,   (d) C_2,

   (e) D_2,   (f) D_2,   (g) D_2.

2. $RTT^{-1} = STT^{-1}$.

3. $R_1R_2 = R_1R_2R_1{}^2 = R_2R_1R_2R_1 = R_2{}^2R_1R_2$; therefore $R_2{}^2 = 1$

and

$$(R_1R_2)^3 = R_1R_2R_1 \cdot R_2R_1R_2 = (R_1R_2R_1)^2 = 1.$$

4. The periods of the elements of C_n are divisors of n.

§2.8

2. If the angles are all equal, sides of two different lengths can only occur alternately, and this is impossible if their number is odd.

3. $108°$, $36°$, $140°$, $100°$, $20°$.

4. Circumradii $l\sqrt{4 \pm 2\sqrt{2}}$, $l\sqrt{2(\sqrt{3} \pm 1)}$; inradii $l(\sqrt{2} \pm 1)$, $l(2 \pm \sqrt{3})$; vertex figures $l\sqrt{2 \pm \sqrt{2}}$, $l(\sqrt{3} \pm 1)/\sqrt{2}$.

5. $\left(1, \dfrac{2k\pi}{n}\right)$.

6. Yes. Make every cut from the center. If the perimeter is divided into equal parts, the area is automatically divided into equal parts.

§3.1

1. Rotation, translation.

2. Reflection. Yes.

3. If the perpendicular bisectors are distinct, find where they intersect. (If they are parallel, the segments are not related by a rotation!) If they coincide, find where the lines AB and $A'B'$ intersect. If these lines also coincide, the center is the midpoint of AA'.

4. Rotate the first two mirrors until the second (in its new position) coincides with the third.

§3.2

1. T^{-1}.

2. It can be any line perpendicular to the direction of the translation.

3. A translation.

4. Translate the first two mirrors till the second (in its new position) coincides with the third.

5. Any translation is expressible as the product of two half-turns, one of which may be arbitrarily assigned. Therefore, if H_1, H_2, H_3 are half-turns,

$$H_2H_3 = H_1H_4$$

for a suitable H_4; that is, $H_1H_2H_3 = H_4$.

7. $(x, y) \rightarrow (x + a, y)$. The transformed curve is $f(x - a, y) = 0$; for instance, the unit circle with center $(a, 0)$ is

$$(x - a)^2 + y^2 - 1 = 0.$$

§3.3

1. (i) Half-turn about B, or reflection in the perpendicular line through B.

   (ii) Translation from A to B, or a glide reflection.

2. Two is the only even number less than or equal to 3. The product of a reflection and a half-turn is a reflection or a glide reflection according as the center of the half-turn does or does not lie on the mirror.

3. Reflection in the perpendicular line through O.

4. Half-turn.

5. An opposite transformation.

6. The relation $R_1R_2R_3 = R$ is equivalent to $R_1R_2 = RR_3$, which means that R_1R_2 and RR_3 are either equal rotations or equal translations.

7. $G^2 = TR_1R_1T = T^2$.

8. A glide reflection. An equation requires an expression for the *old* coordinates in terms of the *new*.

§3.4

(a) $O_1O_2O_3O_4$ is a parallelogram (possibly collapsing like $OO'Q'Q$ in Figure 3.2a).

(b) If m_1 and m_2 intersect, m bisects one of the angles between them. If instead they are parallel, m is parallel to them and midway between them.

§3.5

1. S is a glide reflection.

2. If a translation commutes with a reflection, its direction must be along the mirror.

§3.7

1. (i), (ii), (iii), (iv), (v).

2. (iii), (v).

§4.1

1. Each side of a Dirichlet region joins the circumcenters of two congruent triangles having a common side.

§4.2

1. Because two opposite vertices of each quadrangle are related by a translation.

§4.3

2. Place the two parallelograms in such a position that a side of one is part of a side of the other (with a common vertex at one end).

§4.4

No. A "Procrustean stretch" [Coxeter and Greitzer **1**, p. 102], which doubles vertical distances while halving horizontal distances, can be applied to three of the patterns in Figure 4.4a, namely those illustrating the groups **p2, cmm, pmg**.

§4.5

1. Rotation through the same angle about P'.
2. If T is a translation and S is a rotation whose period is greater than 2, then $S^{-1}TS$ is a translation in a new direction.

§4.6

1. Since the vertex figure is regular, two adjacent faces are alike; therefore any two faces are alike. Since the face is regular, two adjacent vertices are surrounded alike; therefore any two vertices are surrounded alike.
3. No. To complete the lattice we would also need the centers of the hexagons.

§4.7

1. If Q were not between P_2 and P_3, we could obtain another pair having a smaller distance than P_1Q.
2. Use induction over n deriving a set of $n - 1$ points by omitting one of two whose join contains no others.
3. A complete quadrangle with its diagonal points.

§5.1

1. $O(\lambda^{-1})$.
2. It divides O_1O_2 in the ratio $(\lambda_2 - 1):(\lambda_1 - 1)\lambda_2$.
3. (a) $(r, \theta) \rightarrow (\lambda r, \theta)$,
    (b) $(x, y) \rightarrow (\lambda x, \lambda y)$.
4. By similar triangles, $OP'/OP = OA'/OA$.
5. By taking O between A and A'.

§5.2

2. If a common tangent TT' meets the line of centers in O, the dilatation $O(OT'/OT)$ transforms the first circle into the second.
3. It divides $O\,O_1$ in the ratio $(\lambda_1 - 1):(1 - \lambda)$.

§5.3

1. $0, -2, -3, -6$.
2. The nine-point center is the same for all.

§5.4

Since any invariant point of a transformation is also an invariant point of the inverse transformation, we lose no generality by considering a similarity $ABC \rightarrow A'B'C'$ in which ABC is the larger of the two given similar triangles. (If it were the smaller, we would alter the notation and consider the inverse similarity instead.) If A and A' coincide, we have already found an invariant point. If not, suppose the similarity transforms A' into A'', A'' into A''', and so on. Let μ denote the ratio of magnification, so that $A'B' = \mu AB$, $0 < \mu < 1$. Then $A'A'' = \mu AA'$, $A''A''' = \mu A'A''$, and so on. The circle with center A and radius $(1 - \mu)^{-1} AA'$ is transformed into the circle with center A' and radius $\mu(1 - \mu)^{-1} AA'$. Since

$$1 + \frac{\mu}{1 - \mu} = \frac{1}{1 - \mu},$$

the former circle encloses the latter. Continuing, we obtain an infinite sequence of circles whose radii $\mu^n(1 - \mu)^{-1} AA'$ tend to zero as n tends to infinity. Since these circles are "nested," their centers A, A', A'', ... converge to a *point of accumulation O*. Since the similarity transforms $AA'A'' \cdots$ into $A'A''A''' , \ldots$, which is essentially the same sequence, the point O is invariant.

§5.5

1. A dilative rotation, possibly reducing to a dilatation or a rotation.

2. Let P be the point of intersection of the corresponding lines AB, $A'B'$. Let the circles $AA'P$, $BB'P$, which have the common point P, meet again in O. The triangles ABO, $A'B'O$ (possibly collapsing into triads of collinear points) are easily seen to be directly similar. Hence, this point O is the invariant point of the direct similarity $AB \to A'B'$.

If AA' and BB' are parallel, O coincides with P (and the two circles have a common tangent at that point). In any other case, an alternative construction makes use of the point T where AA' meets BB'. The circles ABT, $A'B'T$, which have the common point T, meet again in O.

It follows that the four circles $AA'P$, $BB'P$, ABT, $A'B'T$ all pass through one point [Baker **1**, p. 110].

§5.6

1. This is the invariant point of a similarity.

2. The segments AB and $A'B'$ are related by a direct similarity and an opposite similarity. The latter is a dilative reflection whose two axes divide each segment PP' in the ratio $AB:A'B'$ (one internally and the other externally; see Figure 5.6a). If A_1 coincides with B_1, or A_2 with B_2, the direct similarity is a dilatation, and the analogous points P_1 or P_2 (respectively) all coincide.

3. If S is reflection, S^2 is the identity. If S is a glide reflection, S^2 is a translation, which is one kind of dilatation. If S is a dilative reflection, S^2 is a central diltation.

4. (a) A dilative rotation, possibly reducing to a dilatation or a rotation. (b) A dilative reflection or a glide reflection, possibly reducing to a reflection.

5. As we saw in §5.4, the invariant lines of the dilative reflection meet in the invariant point O and are the internal and external bisectors of $\angle AOA'$. By Euclid VI.3 and its "external" analogue, they are the lines OA_1 and OA_2. The same reasoning can be repeated using the B's instead of the A's.

6. (a) A dilative rotation. (b) A dilative reflection.

§6.1

2. Draw two circles with centers O, A and radius OA, meeting in C, C'. The circle with center C and radius CC' determines B on the circle OCC'.

3. Find the inverse of a point distant $\frac{2}{3}k$ from O and double its distance from O. For a point whose distance from O lies between $k/2n$ and $k/(2n-1)$, apply the dilatation $O(n)$, invert and then apply $O(n)$ again.

4. To bisect OA, construct B as in ex. 2, and use three more circles to construct its inverse in the circle with center O.

5. To divide OA into n equal parts, transform A by the dilatation $O(n)$, and invert with respect to the circle with center O and radius OA.

§6.3

1. Compare Figure 6.3b with Figure 5.2a. In the case of equal intersecting circles, one of the inversions is replaced by a reflection.

2. Let Q denote the center of the rhombus $APBP'$. Then

$$OP \times OP' = (OQ - PQ)(OQ + PQ) = OQ^2 - PQ^2$$
$$= OQ^2 + AQ^2 - (AQ^2 + PQ^2)$$
$$= OA^2 - PA^2.$$

3. Let N be the midpoint of BD, and H the foot of the perpendicular from A on the line BD. Suppose $AO = \mu AB$, so that $OP = \mu BD$ and $OP' = (1 - \mu)AC$. Then

$$BD \times AC = (HD - HB)(HD + HB) = HD^2 - HB^2$$
$$= AD^2 - AB^2$$

and

$$OP \times OP' = \mu BD \times (1 - \mu)AC = \mu(1 - \mu)(AD^2 - AB^2).$$

4. Let d be the distance from O to the center of γ. Comparing the diameter through O of γ with the corresponding diameter of the inverse circle, we see that the latter is of length

$$\frac{k^2}{d-r} - \frac{k^2}{d+r} = \frac{2k^2r}{d^2 - r^2} = \frac{2k^2r}{p}.$$

§6.5

1. Orthogonal circles invert into orthogonal circles, and any circle orthogonal to the circle of inversion inverts into itself.

2. The two limiting points are the common points of any two members of the orthogonal pencil.

3. Let α_1, α_2 be the two given circles, and β_1, β_2 any two circles that cut them both. Let l_{ij} be the radical axis of α_i and β_j. Let P_j be the point where l_{1j} meets l_{2j}. Then P_1P_2 is the radical axis of α_1 and α_2.

4. One is the center and the other is the point at infinity.

5. Invert the whole figure in a circle whose center is one of the points of contact. Two circles of the ring become parallel lines, say a and b. The rest have their centers and points of contact on a line l, perpendicular to a and b. The original circles become equal circles both touching a and b. The line l serves as a mirror reflecting these two circles into each other. The inversion transforms this reflection into an inversion.

The centers of the circles lie on an ellipse.

6. Inverting in a circle whose center is the point of contact of the tangent circles with centers A and B, we obtain two parallel lines E_1, E_2, and a circle E_3 sandwiched between them (as in the first part of the answer to ex. 9 of §1.5). In this very simple case, Soddy's circles (in the same sandwich) are congruent to E_3 and tangent to it on opposite sides. Their radical axis H_4, joining the points of contact of E_3 with E_1 with E_2, is the inverse of the incircle.

§6.6

1. Each circle of Apollonius is orthogonal to all the circles through A and A'.

2. With its center where the perpendicular bisector of AA' meets l, draw a circle through A (and A'), meeting l in P_1 (near A) and P_2 (near A'). Consider the value of the ratio

$$\mu = \frac{A'P}{AP}$$

for various positions of P. Since P_1P_2 is a diameter of the circle drawn through A and A', two of the circles of Apollonius touch l at P_1 and P_2 respectively. Let μ_1 and μ_2 be the values of μ on these two circles.

On all circles of Apollonius inside the one through P_1, we have $\mu > \mu_1$, and on all circles outside, $\mu < \mu_1$. Therefore, among the various positions of P on l, P_1 has the maximal μ, namely $\mu = \mu_1$. Similarly, the circle of Apollonius through P_2 has $\mu < \mu_2$ inside and $\mu > \mu_2$ outside; therefore, among all positions of P on l, P_2 has the minimal μ, namely $\mu = \mu_2$.

3. $\mu/(1 - \mu^2)$.

4. Since O and $\bar{O}$ are the common points of two circles of Apollonius, we have

$$\frac{OA'}{OA} = \frac{OB'}{OB} = \frac{A'B'}{AB}$$

and the same with O replaced by $\bar{O}$. If $A' = B$, use ex. 3 on p. 76.

5. Inverting in a circle whose center is either of the limiting points of the coaxal pencil, we obtain three *concentric* circles whose radii satisfy either $a_1 < a_2 < a_3$ or $a_1 > a_2 > a_3$. Choosing the limiting point that yields the former order, we find

$$(\alpha_1, \alpha_2) + (\alpha_2, \alpha_3) = \log\frac{a_2}{a_1} + \log\frac{a_3}{a_2} = \log\frac{a_3}{a_1}$$

$$= (\alpha_1, \alpha_3).$$

For further details see Coxeter and Greitzer **1**, pp. 123–131.

6. The circle of similitude is a circle of Apollonius, namely, the locus of a point whose distances from the centers of the two given circles are proportional to their radii, say $OA:OA' = r:r'$. The direct or opposite similarity that transforms OA into OA' also transforms the given circle with center A into the given circle with center A'. It follows that the locus of points from which the two given circles *subtend equal angles* is their circle of similitude or, if the two circles intersect, it is the part of their circle of similitude that lies outside them.

For two *equal* circles, the locus reduces to their radical axis.

7. Let the given circles have centers A, B and radii a, b. Let P, Q be the inverses of a point W on the circle of similitude. Then

$$\frac{AP \times AW}{BQ \times BW} = \frac{a^2}{b^2} = \left(\frac{a}{b}\right)^2 = \left(\frac{AW}{BW}\right)^2,$$

$$\frac{AP}{BQ} = \frac{AW}{BW},$$

and PQ is parallel to AB. Since both the given circles are orthogonal to the circle WPQ, their radical axis is a diameter of the latter, namely the diameter perpendicular to AB. Since this diameter is also perpendicular to PQ, P and Q are images of each other by reflection in it.

§6.7

$J_kS = S \cdot S^{-1}J_kS = SJ'_k$, where J'_k is the inversion in the circle with center O^S and radius k. The two inversions are the same if and only if the isometry S leaves O invariant, that is, if and only if T interchanges O and O'.

§6.8

1. $OA \times OA' = OB \times OB'$ and $\angle AOB = \angle A'OB'$.

2. The ratio of magnification is

$$\frac{OA'}{OB} = \frac{OA \times OA'}{OA \times OB} = \frac{k^2}{ab}.$$

3. Let $a = OA$, $b = OB$, $c = OC$, $d = OD$. Then

$$\frac{A'B' \times C'D'}{A'C' \times B'D'} = \frac{(k^2/ab)AB \times (k^2/cd)CD}{(k^2/ac)AC \times (k^2/bd)BD} = \frac{AB \times CD}{AC \times BD}.$$

4. Spheres through O invert into planes. Two spheres that touch each other at O have no other common point. Two planes that have no common point are parallel.

5. After inversion we have a sphere γ "sandwiched" between two parallel planes α and β. All the spheres $\sigma_1, \sigma_2, \ldots$ are congruent to γ. The section of the figure by the plane midway between α and β is a circle touching a ring of six congruent circles.

§6.9

1. Consider, for instance, two circles of radius $\frac{1}{3}\pi$ whose centers are distant $\frac{1}{2}\pi$. When each circle is represented by a pair of parallel small circles on a sphere, the points of intersecton are the vertices of 2 squares. (The common radius of the circles may be taken to have any value between $\frac{1}{4}\pi$ and $\frac{1}{2}\pi$. The result is most evident when this value is nearly $\frac{1}{2}\pi$.)

2. Prove this first for a triangle and then dissect the p-gon into triangles.

§7.1

Let R denote the reflection in the plane of the two lines, R_1 and R_2 the reflections in the planes through the respective lines perpendicular to that plane. Then the half-turns may be expressed as R_1R and RR_2, so that their product is R_1R_2.

§7.2

The identity.

§7.3

The reflection in another parallel plane.

§7.4

1. The reflection in another plane through the same line.

2. The two tetrahedra $OABC$ and $OA'B'C'$, being congruent, are related either by a rotation or by a rotatory-inversion. In the former case any point on the axis of rotation would be, like O, equidistant from A and A', from B and B', and from C and C'.

§7.5

1. (a) Reflection, (b) Quarter-turn, (c) Translation, (d) Twist, (e) Glide reflection, (f) Rotatory inversion.

2. In the notation of Figure 7.5a, the half-turns are $R'_1 R'_2$, $R'_3 R'_4$, and their product is the twist $R'_1 R'_3 \cdot R'_2 R'_4$.

§7.6

1. It is transformed into

$$(\mu x \cos \alpha - \mu y \sin \alpha, \ \mu x \sin \alpha + \mu y \cos \alpha, \ \mu z).$$

2. Axis $x = y = z$. Angle $2\pi/3$.

3. This is a dilative rotation with angle π and ratio $-\lambda$.

4. Yes. Simply use spheres instead of circles. The same proof is applicable in a space of any number of dimensions.

§7.7

An isometry is the product of four or fewer reflections. Any other similarity is the product of a rotation and a dilatation. If the ratio of magnification happens to be negative, we can use instead a rotatory inversion and a direct dilatation. Since a direct dilatation is the product of inversions in two concentric spheres, this makes, altogether, two or three reflections and two inversions.

Finally, the product of an inversion and an isometry is the product of an inversion and r reflections, $r \leqslant 4$.

§8.1

3. If P_i is (x_i, y_i), M_{ij} is

$$\left(\frac{x_i + x_j}{2}, \ \frac{y_i + y_j}{2} \right)$$

and the midpoint of $M_{12} M_{34}$ is

$$\left(\frac{x_1 + x_2 + x_3 + x_4}{4}, \ \frac{y_1 + y_2 + y_3 + y_4}{4} \right).$$

§8.2

1. $r_1^2 + r_2^2 - 2r_1 r_2 \cos(\theta_2 - \theta_1)$.

2. (r, θ), where

$$r^2 = \tfrac{1}{4}[r_1^2 + r_2^2 + 2r_1 r_2 \cos(\theta_1 + \theta_2)]$$

and

$$\tan \theta = \frac{r_1 \sin \theta_1 + r_2 \sin \theta_2}{r_1 \cos \theta_1 + r_2 \cos \theta_2}.$$

3. $\theta = \alpha$.

4. The respectively parallel lines through the origin are $ax + by = 0$ and $a'x + b'y = 0$, or

$$\frac{y}{x} = -\frac{a}{b} \quad \text{and} \quad \frac{y}{x} = -\frac{a'}{b'}.$$

The condition is derived from 8.22 by writing it in the form

$$\frac{y}{x}\frac{y'}{x'} = -1, \qquad \text{that is,} \qquad \frac{a}{b}\frac{a'}{b'} = -1.$$

5. Replacing x and y by $x \cos \alpha - y \sin \alpha$ and $x \sin \alpha + y \cos \alpha$, where $\alpha = \frac{1}{2}\arctan\left(-\frac{24}{7}\right) = \arctan\left(-\frac{3}{4}\right)$, we obtain

$$4(4x + 3y)^2 + 24(4x + 3y)(-3x + 4y) + 11(-3x + 4y)^2 = 125,$$

which reduces to

$$-x^2 + 4y^2 = 1.$$

§8.3

1. $c(x^2 + y^2) + k^2(ax + by) = 0,$
$\quad c(x^2 + y^2) + 2k^2(gx + fy) + k^4 = 0.$
2. $x^2 + y^2 = k^2.$
3. $(x^2 + y^2)^2 = 2a^2(x^2 - y^2),$
$\quad\quad r^2 = 2a^2 \cos 2\theta.$
6. From

$$r \cos \theta + b = 2b \cos t - b \cos 2t = 2b \cos t (1 - \cos t) + b$$

and

$$r \sin \theta = 2b \sin t - b \sin 2t = 2b \sin t (1 - \cos t)$$

we deduce $r^2 = (r \cos \theta)^2 + (r \sin \theta)^2 = 4b^2(1 - \cos t)^2$ and $\tan \theta = \tan t$. When θ is replaced by $-\theta$, r changes from $2b(1 - \cos \theta)$ to $2b(1 + \cos \theta)$. The sum of these distances is $4b$, for all values of θ.

§8.4

1. A parabola.
2. A hyperbola.
3. The half-turn about the center yields a second focus and a second directrix for any central conic. For the ellipse, the two foci lie between the two directrices; for the hyperbola, the foci lie beyond the directrices. Let O and O' denote the foci and K' the foot of the perpendicular from P to the second directrix. Since

$$OP = \epsilon PK \qquad \text{and} \qquad O'P = \epsilon PK'$$

we have, for the ellipse,

$$OP + O'P = \epsilon(PK + PK') = \epsilon KK',$$

and for the hyperbola,

$$O'P - OP = \epsilon(PK' - PK) = \epsilon KK'.$$

4. $\epsilon = \sqrt{1 \mp b^2/a^2},\ \sqrt{2}.$
5. Since the circumcenter must be equidistant from A and C, we have

$$x^2 + (\tfrac{2}{3}y)^2 = 1 + (\tfrac{1}{3}y)^2,$$
$$x^2 + \tfrac{1}{3}y^2 = 1.$$

6. By the theory of quadratic equations, F is the product of two linear forms if it is indefinite, and a perfect square if it is semidefinite.

7. $2xy = a^2$.

8. For each point P on the ellipse there is a corresponding point P' on the *auxiliary* circle whose diameter is the major axis, PP' being perpendicular to that axis. The radius through P' makes an angle t with the major axis.

9. Both branches are included.

10. Replace r by l^2/r.

§8.5

1. The parabola $x = 2lt^2$, $y = 2lt$ meets the line $Xx + Yy + Z = 0$ in points given by the roots of the quadratic equation

$$X \cdot 2lt^2 + Y \cdot 2lt + Z = 0.$$

The sum and product of the roots, say t and t', are

$$-\frac{Y}{X} = t + t', \qquad \frac{Z}{2lX} = tt'.$$

2. The secant of the ellipse $x = a \cos t$, $y = b \sin t$ is

$$\begin{vmatrix} x & y & 1 \\ a \cos (\alpha + \beta) & b \sin (\alpha + \beta) & 1 \\ a \cos (\alpha - \beta) & b \sin (\alpha - \beta) & 1 \end{vmatrix} = 0.$$

For the hyperbola, the tangent is

$$\frac{x}{a} \cosh t - \frac{y}{b} \sinh t = 1.$$

3. The envelope of the line $Xx + Yy + Z = 0$, whose coefficients X, Y, Z are functions of a parameter t, is the locus of its point of intersection with

$$(X + dX)x + (Y + dY)y + (Z + dZ) = 0,$$

or with $X'x + Y'y + Z' = 0$, where $X' = \partial X/\partial t$, etc. Differentiating

$$ax \sec t - by \csc t = a^2 - b^2$$

and dividing by $\sin t \cos t$, we obtain

$$\frac{ax}{\cos^3 t} = \frac{-by}{\sin^3 t} = \frac{ax \sec t - by \csc t}{\cos^2 t + \sin^2 t} = a^2 - b^2.$$

Thus the envelope of normals is the locus of (x, y) where

$$\frac{ax}{a^2 - b^2} = \cos^3 t, \qquad \frac{by}{a^2 - b^2} = -\sin^3 t.$$

§8.6

1. πab.

2. $\frac{1}{2}tab$.

§8.7

1. $\mu^{2\pi} r = a\mu^{\theta + 2\pi}$. This inversion has the same effect as reflection in the initial line.

2.   (i)   (ii)

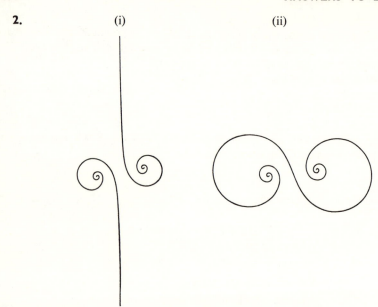

These drawings were made by Ryszard Krasnodębski. See also Coxeter, *Mathematical Gazette*, **52** (1968), p. 5; *Aequationes Mathematicae*, **1** (1968), pp. 112–114.

§8.8

1. Eliminate $X:Y:Z:T$ from the four equations

$$Xx_i + Yy_i + Zz_i = T \quad \text{and} \quad Xx + Yy + Zz = T.$$

The effect of replacing the ith point by a direction is the same as the effect of replacing it by (X_it, Y_it, Z_it), so that the ith row becomes

$$X_it \quad Y_it \quad Z_it \quad 1$$

or (equally well)

$$X_i \quad Y_i \quad Z_i \quad 1/t,$$

and then making $1/t$ tend to zero.

5. The condition for the radii to any common point (x, y, z) of the two spheres to be perpendicular is

$$(x + u)(x + u') + (y + v)(y + v') + (z + w)(z + w') = 0.$$

The desired condition is obtained by doubling this and subtracting the equations of the spheres.

6. The polar plane of (X, Y, Z) passes through (X', Y', Z') if

$$XX' + YY' + ZZ' = k^2.$$

The symmetrical nature of this condition shows that then the polar plane of (X', Y', Z') passes through (X, Y, Z). (Two such points are said to be *conjugate* with respect to the sphere.) The special case asked for arises when (X', Y', Z') lies on the sphere and (X, Y, Z) lies in the tangent plane at that point.

7. Factorizing both sides of the equation

$$\left(\frac{x}{a}\cos\alpha + \frac{y}{b}\sin\alpha\right)^2 - \left(\frac{z}{c}\right)^2 = -\left(\frac{y}{b}\cos\alpha - \frac{x}{a}\sin\alpha\right)^2 + 1,$$

we see that, for each value of α, every point on the line

$$\frac{x}{a}\cos\alpha + \frac{y}{b}\sin\alpha = \frac{z}{c}, \qquad \frac{y}{b}\cos\alpha - \frac{x}{a}\sin\alpha = 1$$

lies on the hyperboloid. Reversing the sign of Z, we obtain the other system of generators. The general generator of the first system meets the special generator

$$\frac{x}{a} = -\frac{z}{c}, \qquad \frac{y}{b} = 1$$

of the second where

$$\sin\alpha = \frac{z}{c}(1 + \cos\alpha), \qquad \frac{z}{c}\sin\alpha = 1 - \cos\alpha.$$

The consistency of these equations shows that generators of opposite systems intersect (or else, if $\alpha = \pi$, are parallel). On the other hand, any common point of the general generator of the first system and the special generator

$$\frac{x}{a} = \frac{z}{c}, \qquad \frac{y}{b} = 1$$

of the *same* system would have to satisfy both the equations

$$\sin\alpha = \frac{z}{c}(1 - \cos\alpha), \qquad -\frac{z}{c}\sin\alpha = 1 - \cos\alpha,$$

which can happen only when the two generators coincide.

§9.3

1. $z = 2 \pm i$.

2. $u + vi = 0$ means that the point (u, v) coincides with the origin $(0, 0)$.

3. $(a + bi)^{-1} = x + yi$, where

$$x = \frac{a}{a^2 + b^2}, \qquad y = -\frac{b}{a^2 + b^2}.$$

4. (i) The dilative rotation reduces to a dilatation, and the two shaded triangles are homothetic.

(ii) The dilative rotation reduces to a rotation, and the two shaded triangles are congruent.

§9.4

(a) $3 + 4i = 5(\cos\alpha + i\sin\alpha)$, where $\cos\alpha = \frac{3}{5}$, so that $\cos\frac{1}{2}\alpha = \sqrt{\frac{4}{5}}$ and $\sin\frac{1}{2}\alpha = \sqrt{\frac{1}{5}}$. Hence one square root is

$$(3 + 4i)^{\frac{1}{2}} = 5^{\frac{1}{2}}(\cos\tfrac{1}{2}\alpha + i\sin\tfrac{1}{2}\alpha) = 2 + i$$

and the other is $-(2 + i) = -2 - i$.

(b) $1 = \cos 0 + i \sin 0 = \cos 2\pi + i \sin 2\pi = \cos 4\pi + i \sin 4\pi$. Hence the three cube roots are

$$\cos 0 + i \sin 0 = 1,$$
$$\cos \tfrac{2}{3}\pi + i \sin \tfrac{2}{3}\pi = \tfrac{1}{2}(-1 + i\sqrt{3}),$$
$$\cos \tfrac{4}{3}\pi + i \sin \tfrac{4}{3}\pi = \tfrac{1}{2}(-1 - i\sqrt{3}).$$

(c) ± 1, $\pm\omega$, $\pm\omega^2$.

(d) The same and also $\pm i$, $\pm i\omega$, $\pm i\omega^2$.

§9.5

1. $e^{\frac{1}{2}\pi i} = \cos \tfrac{1}{2}\pi + i \sin \tfrac{1}{2}\pi = i$. Yes.

§9.6

By Pythagoras,

$$\left(\frac{24}{7}\right)^2 = (x + 1)^2 + \left(1 + \frac{1}{x}\right)^2 = \left(x + 1 + \frac{1}{x}\right)^2 - 1,$$

$$x + 1 + \frac{1}{x} = \frac{25}{7}, \qquad x = \frac{9 + \sqrt{32}}{7}.$$

§9.7

1. The angle is am a

2. The angle is am a.

§10.1

1. The octahedron is a square dipyramid with equilateral side faces.

2. A triangular dipyramid.

3. (i) A square, (ii) a hexagon, (iii) a decagon.

4. The cutting plane, for either acute corner, passes through the midpoints of the three edges that meet there. The rhombohedron, like any parallelepiped, can be repeated by translations to fill the whole space without interstices.

§10.2

1. The bases appear as two pentagons: a large one with a small one oppositely placed inside.

2. Seven.

3. Eight.

§10.3

1. Use 10.32.

2. If a polyhedron has $p \geqslant 4$ for every face and $q \geqslant 4$ for every vertex,

$$4(V - E + F) \leqslant \Sigma q - 4E + \Sigma p = 2E - 4E + 2E$$
$$= 0.$$

3. The only possibility is a tessellation of rhombi whose vertices form a lattice.

§10.5

1. The edges at a vertex are mutually orthogonal and of equal length.

2. One vertex in each "octant."

3. $(1, 1, 1)$.

4. From the cube in ex. 1 we derive the tetrahedron

$$(0, 0, 0)(0, 1, 1)(1, 0, 1)(1, 1, 0),$$

with face planes $x + y + z = 2$, $x = y + z$, $y = z + x$, $z = x + y$. To normalize these equations we divide by $\sqrt{3}$. The cosine of the internal angle between two of the planes is $\frac{1}{3}$.

5. $(\pm 1, 0, 0)$, $(0, \pm 1, 0)$, $(0, 0, \pm 1)$. The face planes are $\pm x \pm y \pm z = 1$. The edge joining $(1, 0, 0)$ and $(0, 1, 0)$ belongs to the planes $x + y \pm z = 1$, which make an angle whose cosine is $-\frac{1}{3}$.

6. Points between the parallel planes $x + y + z = \pm 1$ satisfy $x + y + z < 1$ and $-x - y - z < 1$; similarly for the other pairs.

7. $120°$. The regular tetrahedron

$$(-1, -1, -1)(-1, 1, 1)(1, -1, 1)(1, 1, -1)$$

has its center at the origin. The planes joining the origin to pairs of vertices are

$$y \pm z = 0, \qquad z \pm x = 0, \qquad x \pm y = 0.$$

§11.1

1. $\tau \sin \dfrac{\pi}{5} = \sin \dfrac{2\pi}{5} = 2 \cos \dfrac{\pi}{5} \sin \dfrac{\pi}{5}$.

2. Center S, radius QU.

§11.2

1. The four points $(\pm \tau, \pm 1, 0)$ evidently form a golden rectangle in the plane $z = 0$.

2. The segment $(0, 0, \tau^2)(0, \tau^2, 0)$ is divided in the ratio $\tau:1$ by the point $(0, \tau, 1)$.

3. $(0, \pm \tau^{-1}, \pm \tau)$, $(\pm \tau, 0, \pm \tau^{-1})$, $(\pm \tau^{-1}, \pm \tau, 0)$, $(\pm 1, \pm 1, \pm 1)$. Thus the 20 vertices belong to 3 "doubly golden" rectangles (whose sides are in the ratio $\tau^2:1$) and a cube.

§11.3

1. Corresponding sides of the two rectangles meet in the points B, D, F, H. The lines BF and DH meet in O.

2. The points I, G, C, A, having polar coordinates $(\tau^{-2}, -\pi)$, $(\tau^{-1}, -\frac{1}{2}\pi)$, $(\tau, \frac{1}{2}\pi)$, (τ^2, π), have Cartesian coordinates $(-\tau^{-2}, 0)$, $(0, -\tau^{-1})$, $(0, \tau)$, $(-\tau^2, 0)$. Hence the lines IC, GA are

$$\tau^3 x - y + \tau = 0, \qquad x + \tau^3 y + \tau^2 = 0;$$

and H, their point of intersection, is $(-\frac{1}{2}, -\frac{1}{2})$ or, in polar coordinates, $(\sqrt{\frac{1}{2}}, -\frac{3}{4}\pi)$. Thus the points J, H, F, D, B, given by

$$r = 2^{-\frac{1}{2}}\tau^n, \qquad \theta = \frac{1}{4}(2n - 3)\pi \qquad (n = -1, 0, 1, 2, 3),$$

lie on the spiral 8.71, where $a = 2^{-\frac{1}{2}}\tau^{\frac{3}{2}}$. This is derived from the spiral $r = \mu^\theta$ by the dilatation $O(a)$ or by rotation through the angle

$$\frac{\log a}{\log \mu}.$$

§11.4

1. $f_{n+2} - 1$.

2. Using induction, assume $f_{n-2}f_n - f_{n-1}^2 = (-1)^{n-1}$. Then

$$f_{n-1}f_{n+1} - f_n^2 = f_{n-1}(f_{n-1} + f_n) - f_n(f_{n-2} + f_{n-1})$$
$$= f_{n-1}^2 - f_{n-2}f_n = (-1)^n.$$

3. Working modulo 10, we have $5 + 8 \equiv 3$, $8 + 3 \equiv 1$, and so on.

4. Setting $k + j = n$, we see that the coefficient of t^n is

$$\Sigma \binom{k}{j}$$

where $k = n - j$ and, since $2j \leqslant j + k = n$, $0 \leqslant j \leqslant \frac{1}{2}n$.

5. $1.010203050813213455\ldots$.

6. By 11.48,

$$\frac{g_{n+1}}{g_n} = \frac{\tau^{n+1} + (-\tau)^{-n-1}}{\tau^n + (-\tau)^{-n}} = \frac{\tau + (-1)^{n+1}\tau^{-2n-1}}{1 + (-1)^n\tau^{-2n}}.$$

Hence the limit is τ.

§11.5

When $k = 1$, 11.51 yields $h = \tau^{-3/2} = 0.48587\ldots$.

§12.1

1. Both.

2. The sum of the three angles of a triangle is equal to two right angles.

3. Affine geometry.

4. (a) Affine, (b) absolute, (c) absolute.

§12.2

1. By Axiom 12.22, there is a point C with $[ABC]$, also a point D with $[BCD]$, and so on forever.

2. Theorem 12.271.

3. If $[FDE]$, we could apply 12.27 to the triangle BFD with $[DCB]$, obtaining Z on EC with $[BZF]$. Since $Z = A$, this contradicts $[AFB]$.

4. Any line not belonging to the set contains an infinite number of points, among which only a finite number can lie on lines of the set (at most one on each).

5. Use 12.278 (and Fig. 12.2d) with D, A, B, F, C, L replaced by A, B, C, L, M, N.

6. Take A' on A/B, B' on B/C, and apply Axiom 12.27 to the triangle $A'B'B$ with $[B'BC]$ and $[BAA']$.

7. For *any* such A' and B', the line $A'B'$ meets A/C; therefore it does not meet C/A.

§12.3

The $n - 1$ points P_i $(i > 1)$ are joined in pairs by at most $\binom{n-1}{2}$ lines, some or all of which may meet P_1Q. In Figure 12.3b, the six joins P_2P_4, P_4P_1, P_1P_2, P_3P_6, P_6P_1, P_1P_3 all make the same contribution as P_1P_5.

§12.4

1. Since the five points are not collinear, they must form either a convex pentagon or a convex quadrangle with one point inside, or a triangle with two points inside. In the first and second cases the result is obvious. In the third case the two inner points lie on a line meeting two distinct sides of the triangle. The ends of the third side form, with these inner points, the desired convex quadrangle.

4. The first two lines, BC and CA, decompose the plane into four angular regions. The line AB has no intersection with the region bounded by C/A and C/B, but it decomposes each of the remaining three angular regions into two parts. The region bounded by the triangle is the only finite part, since at least one side of each of the others is a ray.

5. Consider any $m - 1$ of the m lines, and the $f(2, m - 1)$ regions formed by them. They decompose the mth line into m parts (namely, two rays and $m - 2$ segments), lying respectively in m of the $f(2, m - 1)$ regions. These m regions are each decomposed into two, whereas the rest are unaffected. Hence

$$f(2, m) = f(2, m - 1) + m.$$

Combining this with the analogous equations

$$f(2, m - 1) = f(2, m - 2) + m - 1,$$
$$\cdots \qquad\qquad \cdots$$
$$f(2, 1) = f(2, 0) + 1,$$

we obtain

$$
\begin{aligned}
f(2, m) &= f(2, m - 1) + m \\
&= f(2, m - 2) + (m - 1) + m \\
&= \cdots \\
&= f(2, 0) + 1 + 2 + \cdots + m \\
&= 1 + \binom{m + 1}{2} = \binom{m}{0} + \binom{m}{1} + \binom{m}{2}.
\end{aligned}
$$

6. The first $m - 1$ planes decompose the mth into $f(2, m - 1)$ plane regions lying respectively in $f(2, m - 1)$ of the $f(3, m - 1)$ solid regions. These $f(2, m - 1)$ solid regions are each decomposed into two, whereas the rest are unaffected. Hence

$$
\begin{aligned}
f(3, m) &= f(3, m - 1) + f(2, m - 1) \\
&= f(3, m - 2) + f(2, m - 2) + f(2, m - 1) \\
&= \cdots \\
&= f(3, 0) + f(2, 0) + f(2, 1) + \cdots + f(2, m - 1) \\
&= 1 + \sum_{r=1}^{m}\left\{1 + \binom{r}{2}\right\} = 1 + m + \binom{m + 1}{3} \\
&= \binom{m + 1}{1} + \binom{m + 1}{3} \\
&= \binom{m}{0} + \binom{m}{1} + \binom{m}{2} + \binom{m}{3}.
\end{aligned}
$$

7. $f(n, m) = f(n, m - 1) + f(n - 1, m - 1)$, with $f(n, 0) = 1$. Therefore

$$f(n, m) = f(n, 0) + \sum_{r=0}^{m-1} f(n - 1, r).$$

To prove

$$f(n, m) = \binom{m}{0} + \binom{m}{1} + \cdots + \binom{m}{n}$$

by induction, assume the same formula with n replaced by $n - 1$.

$$f(n, m) = f(n\ 0) + \sum_{r=0}^{m-1}\left\{\binom{r}{0} + \binom{r}{1} + \cdots + \binom{r}{n-1}\right\}$$

$$= 1 + \sum_{0}^{m-1}\binom{r}{0} + \sum_{1}^{m-1}\binom{r}{1} + \cdots + \sum_{n-1}^{m-1}\binom{r}{n-1}$$

$$= \binom{m}{0} + \binom{m}{1} + \binom{m}{2} + \cdots + \binom{m}{n}.$$

The final step makes use of the familiar series

$$\sum_{r=n-1}^{m-1}\binom{r}{n-1} = \sum_{n-1}^{m-1}\left\{\binom{r+1}{n} - \binom{r}{n}\right\} = \binom{m}{n} - \binom{n-1}{n} = \binom{m}{n}.$$

§12.6

1. The relation $[prs]$ tells us that the three lines do not meet one another and that they contain points A, C, B (respectively) such that $[ACB]$. Suppose that p_1 is parallel to s. Then any ray from A within the angle between AB and p_1 meets s, and therefore also r. Hence p_1 is parallel to r.

2. The two rays through a given point parallel to a given line appear as the segments joining the point to the ends of the chord.

§13.1

1. Let q and r be the two parallel lines. If p met q without meeting r, then p and q would be two lines through the point $p \cdot q$, both parallel to r, contradicting Axiom 13.11.

2. Yes, provided the three lines are coplanar.

§13.2

1. This is a one-dimensional version of the principle that every direct isometry (including the identity) is the product of an even number of opposite isometries. Translations preserve directions, whereas half-turns reverse directions.

2. By 13.25, $(A \to D) = (C \to B)$ implies $(A \leftrightarrow B) = (C \leftrightarrow D)$, that is, $(A \leftrightarrow B) = (D \leftrightarrow C)$, which, in turn, implies $(A \to C) = (D \to B)$.

3. This is the half-turn about C.

4. Any two opposite sides are interchanged by the half-turn about the common midpoint of the diagonals.

5. Using the symbol $\equiv$ to relate congruent segments, we have

$$BA_1 \equiv C_1B_1 \equiv A_2C, \qquad BA_2 \equiv C_2B_2 \equiv A_3C.$$

By ex. 2, $BA_1 \equiv A_2C$ implies $BA_2 \equiv A_1C$. Hence $A_3 = A_1$.

6. Dissect the quadrangle into two triangles by a diagonal, and use 13.26.

7. The three bimedians all have the same midpoint.

§13.3

1. Each point on AB is transformed into a point dividing the segment AB in the same ratio, that is, into itself. For any other point P, we can draw PM parallel to CA, and PN parallel to CB, with M and N on AB. The corresponding point P' is obtained by drawing MP' parallel to AC', and NP' parallel to BC'.

This is a typical *affine construction*. Instead of Euclid's "ruler and compasses" we are using a simpler instrument, the *parallel-ruler*, which consists of four rulers (or two rulers and two auxiliary bars) forming a parallelogram with pivots at the four vertices. This enables us to draw lines parallel to a given line (and, of course, also to join two given points).

2. If an affinity interchanges two points C and C', it leaves invariant the midpoint O of CC'. If O is the only invariant point, it is also the midpoint of PP' for any P, and the affinity is the half-turn $C \leftrightarrow C'$. But if there is another invariant point M, the affinity, transforming the triangle MCC' into $MC'C$, is the reflection $M(CC')$.

3. If the affinity is not a dilatation, it must transform at least one line a into a line a' not parallel to a. Another line b, parallel to a, will be transformed into another line b', parallel to a'. The point of intersection $A = a \cdot a'$ is invariant; for, if not, it would lie on an invariant line m and could be called either $a \cdot m$ or $a' \cdot m$, contradicting its noninvariance. Similarly, there is another invariant point $B = b \cdot b'$. Therefore, the affinity is either a shear or a strain: $ABC \rightarrow ABC'$.

4. Since lines have linear equations, any transformation of the form 13.33 preserves collinearity. Conversely, we can express any affinity $1XY \rightarrow 1'X'Y'$ in the form 13.33 with suitable values for a, b, c, d, l, m, For, the triangle $(0, 0)(1, 0)(0, 1)$ is transformed into

$$(l, m)(a + l, c + m)(b + l, d + m),$$

which can be identified with any given triangle. The non-collinearity of these three points is ensured by the condition $ad \neq bc$.

5. (i) Translation, (ii) Central dilatation,
    (iii) Shear, (iv) Strain

(including an affine reflection as the special case when $a = -1$).

§13.4

1. This follows from the remark after 13.41.

2. If the diagonals P_0P_3, P_1P_4, P_2P_0, P_3P_1 are parallel to the sides P_1P_2, P_2P_3, P_3P_4 P_4P_0, respectively, the following triangles all have the same area:

$$P_0P_1P_2, \; P_1P_2P_3, \; P_2P_3P_4, \; P_3P_4P_0, \; P_4P_0P_1.$$

Therefore, P_2P_4 is parallel to P_0P_1.

3. When it is a translation or a half-turn. In fact, a central dilatation $O(\lambda)$ has $ad - bc = \lambda^2$.

4. Always.

5. Never.

6. Each affine reflection reverses area.

7. A translation is the product of two affine reflections in the direction of the translation, the mirrors being parallel lines in any other direction. More precisely, in the notation of Figure 13.2d, the translation $A \rightarrow D$ is the product of reflections $A(BC)$ and $B(AD)$.

A half-turn is the product of reflections in two intersecting lines, the direction of each reflection being along the mirror for the other: the half-turn $A \leftrightarrow B$ is the product of $A(CD)$ and $C(AB)$.

A shear is the product of reflections in one mirror in two different directions. Alternatively, it is the product of reflections in one direction in two intersecting mirrors.

8. For any geometric transformation, the successive transforms of a noninvariant point P_0 comprise a set of points $P_0P_1P_2 \cdots$ called the *orbit* of P_0; the transformation takes P_0 to P_1, P_1 to P_2, and so on. Exercise 3 (on page 203), describing a situation in which the orbit of every point consists of a set of collinear points, shows that the only affinities of this kind are the "trivial" ones: the dilatations, shears and strains. For every other kind of affinity there is at least one noninvariant point P_0 lying on no invariant line; the line P_0P_1 is transformed into a different line P_1P_2, the orbit begins with three points forming a triangle, and the affinity can be expressed as $P_0P_1P_2 \rightarrow P_1P_2P_3$. In the case of an equiaffinity, the "trivial" kinds are those considered in ex. 7. For any other kind, $P_0P_1P_2$ and $P_1P_2P_3$ are two triangles of equal area. Since they have a common side P_1P_2, ex. 1 shows that P_0P_3 must be parallel to P_1P_2.

9. Since the translation, half-turn and shear have already been covered in ex. 7, we may restrict consideration to $P_0P_1P_2 \rightarrow P_1P_2P_3$ with P_0P_3 parallel to P_1P_2. Letting M denote the midpoint of P_0P_3, as in Figure 13.4*b*, we see that this equiaffinity is the product of the two affine reflections

$$R_1 = M(P_1P_2) \qquad \text{and} \qquad R_2 = P_2(P_1P_3)$$

(compare §2.7 on page 34). For, these reflections have the effect

$$P_0P_1P_2 \rightarrow P_3P_2P_1 \rightarrow P_1P_2P_3.$$

10. Since R_1 transforms the points $\cdots P_0P_1P_2P_3 \cdots$ into $\cdots P_3P_2P_1P_0 \cdots$, while R_2 transforms $\cdots P_0P_1P_2P_3P_4 \cdots$ into $\cdots P_4P_3P_2P_1P_0 \cdots$, the stated parallelism certainly occurs when $i + j = h + k = 3$ or 4. For other values, we simply have to transform by a suitable power of the equiaffinity R_1R_2.

11. Because affine geometry cannot distinguish between a circle and any other ellipse. The elliptic shadow cast by a coin illustrates the fact that an ellipse is a "strained circle." We are free to use a strain as a coordinate transformation, writing ϵx for x, so that the ellipse becomes

$$\epsilon^2 x^2 + y^2 = 1$$

and the elliptic rotation 13.49 becomes

$$x' = x \cos \theta - \frac{y}{\epsilon} \sin \theta, \qquad y' = \epsilon x \sin \theta + y \cos \theta.$$

This equiaffinity reduces to a half-turn when we set $\theta = \pi$. If instead we set $\theta = \pi - \epsilon$, $\epsilon = \pi/(2d + 1)$, and make d tend to infinity, we obtain a new equiaffinity: *the focal rotation*

$$x' = -x - y, \qquad y' = -y,$$

which leaves invariant the pair of parallel lines $y^2 = 1$. In other words, the affinely regular star polygons of type $\{2 + 1/d\}$ $(d = 2, 3, 4, \ldots)$ may be regarded as approximating either a digon (page 37) or a *focal polygon*, whose vertices

$$(0, 1), (-1, -1), (2, 1), (-3, -1), \ldots$$

lie alternately on these two parallel lines while its sides pass alternately through the two "foci" $(\mp \frac{1}{2}, 0)$.

12. Every triangle is affinely regular, but the only affinely regular quadrangles are the parallelograms.

13. Given P_0, P_1, P_2, complete the parallelogram $P_0P_1P_2O$. Then draw P_2P_3 parallel to P_1O (with P_3 on O/P_0), P_3P_4 parallel to P_2O (with P_4 on O/P_1), and P_4P_5 parallel to OP_0 (with P_5 on O/P_2).

14. It is important to remember that affine geometry admits no measure of angles. The symbol θ occurring in 13.49 must not be interpreted as an angle but simply as a number. The use of sines and cosines does not force us to work in Euclidean geometry: they are employed because of their convenient properties, such as $\cos^2 \theta + \sin^2 \theta = 1$. These functions can, of course, be defined by analytical means without any reference to geometry. After these words of caution, we take the typical vertex P_j to have affine coordinates

$$(\cos j\theta, \ \sin j\theta),$$

where $\theta = 2\pi/n$, and conclude that

$$\frac{P_0P_3}{P_1P_2} = \frac{\sin 3\theta}{\sin 2\theta - \sin \theta} = \frac{3 - 4 \sin^2 \theta}{2 \cos \theta - 1} = 2 \cos \theta + 1.$$

15. The values are 2, 3, 4, 6 (as in §4.5 on page 60). This conclusion can be justified as follows. We see from §13.3 that the parallel-ruler enables us to multiply the length of a given segment by any *rational* number. In fact, given $(0, 0)$, $(1, 0)$, $(0, 1)$, the points that can be constructed by means of the parallel-ruler are the points (x, y) whose affine coordinates are rational, and no others. We see from ex. 8 that the nature of an affinely regular polygon $P_0P_1P_2 \cdots$ is determined by the position of P_3 on the line through P_0 parallel to P_1P_2. It is clear from ex. 10 that we can then construct P_4 and all the other vertices in turn. Exercise 14 shows that, for a polygon of type $\{n\}$, P_3 can be constructed if and only if $\cos \theta$ *and* θ/π *are both rational.*

The following trick for determining the admissible values of θ was devised by the same H. W. Richmond who geometrized Gauss's solution of the cyclotomic equation $z^{17} - 1 = 0$ (Figure 2.1b on page 27).

Since $\cos 2\theta = 2 \cos^2 \theta - 1$, every rational $\cos \theta$ yields a rational $\cos 2\theta$. Since θ/π is rational, the expressions

$$\cos \theta, \ \cos 2\theta, \ \cos 4\theta, \ldots, \ \cos 2^k\theta, \ldots$$

comprise a *finite* set of rational numbers. When these rational numbers are expressed as fractions in their "lowest terms," let b be the greatest denominator that occurs, and let

$$\cos \phi = \frac{a}{b} \qquad (\phi = 2^k\theta)$$

be one of the numbers having this denominator. Since a and b are relatively prime, the denominator of

$$\cos 2\phi = \frac{2a^2 - b^2}{b^2}$$

is either b^2 or (if b is even) $\frac{1}{2}b^2$. But this denominator must not be greater than b. Therefore,

$$b \geqslant \tfrac{1}{2}b^2 > 0, \ b \leqslant 2, \ b = 1 \text{ or } 2, \ \cos \phi = 0 \text{ or } \pm 1 \text{ or } \pm\tfrac{1}{2},$$

and the only admissible values for ϕ are $j\pi/2$ and $j\pi/3$ for integers j. Since $\cos (\pi/4)$ and $\cos (\pi/6)$ are irrational, it follows that the only admissible values for θ, with $0 < \theta \leqslant \pi$, are

$$\pi, \ \frac{2\pi}{3}, \ \frac{\pi}{2}, \ \frac{\pi}{3}.$$

In our geometric application of this result, $\theta = 2\pi/n \leqslant \pi$. Hence $n = 2, 3, 4$ or 6. In other words, the only finite affinely regular polygons constructible with the parallel-ruler are the digon, triangle, parallelogram, and affinely regular hexagon. (For instance, the pentagram and pentagon shown in Figure 13.4e are *not* constructible. It is impossible to assign rational coordinates to all the five vertices simultaneously.)

The connection with §4.5 may be explained by observing that, in the formation of a crystal, Nature is, in effect, using a parallel-ruler to line up certain atoms in the straight rows of a lattice.

§13.5

1. Any common factor of x and x_1 would divide $xy_1 - yx_1$. By 13.52, this is impossible.

2. By 13.52, $x_0 y - y_0 x = 1 = xy_1 - yx_1$, and therefore

$$(x_0 + x_1)y = (y_0 + y_1)x.$$

3. We can systematically assign the symbols $0, 1, \ldots, 6$ in cyclic order to the points of the basic lattice, as follows: each of A, B, C gets the label 0, and then we proceed with $1, 2, 3, 4, 5, 6, 0, 1, \ldots$ to the right; from A toward L we have $0, 3, 6, 2, 5, 1, 4, 0, \ldots$; and from B toward M we have $0, 5, 3, 1, 6, 4, 2, 0, \ldots$. All the points numbered alike form a "sublattice," and since there are seven such sublattices, each has a unit cell seven times as big as that of the basic lattice.

Alternatively, let the basic lattice consist of the points whose affine coordinates are integers. Take B at $(0, 0)$, C at $(2, 1)$, A at $(-1, 3)$. Then the only lattice points inside the triangle ABC are $(1, 1), (0, 2), (0, 1)$, forming a triangle of area $\frac{1}{2}$ (that is, half a unit cell). By Pick's theorem, the area of ABC is $\frac{3}{2} + 3 - 1 = \frac{7}{2}$.

3. (a) The triangle $(0, 0)$ $(3, 1)$ $(-1, 4)$ has area $\frac{3}{2} + 6 - 1 = \frac{13}{2}$, whereas the Cevians form an inner triangle of area $\frac{6}{2} + 0 - 1 = 2$. The ratio is $\frac{4}{13}$. (b) The triangle $(0, 0)$ $(3, 2)$ $(-2, 5)$ has area $\frac{3}{2} + 9 - 1 = \frac{19}{2}$, whereas the Cevians form an inner triangle of area $\frac{1}{2}$.

5. The parallelogram $(0, 0)$ $(2, -1)$ $(3, 1)$ $(1, 2)$ has area $\frac{4}{2} + 4 - 1 = 5$, whereas the small parallelogram in the middle has area 1.

The parallelogram $(\pm 6, \pm 6)$ has area 144, whereas the small octagon in the middle, namely $(\pm 3, 0)$ $(\pm 2, \pm 2)$ $(0, \pm 3)$, with 21 interior points, has area 24.

6.
$$\frac{1}{\lambda + 1}\frac{\mu}{\mu + 1} + \frac{1}{\mu + 1}\frac{\nu}{\nu + 1} + \frac{1}{\nu + 1}\frac{\lambda}{\lambda + 1}$$

$$= \frac{\mu(\nu + 1) + \nu(\lambda + 1) + \lambda(\mu + 1)}{(\lambda + 1)(\mu + 1)(\nu + 1)}$$

$$= 1 - \frac{\lambda\mu\nu + 1}{(\lambda + 1)(\mu + 1)(\nu + 1)}.$$

7. This is obvious unless λ, μ, ν are either all $\geqslant 1$ or all $\leqslant 1$. Assuming one of these eventualities, suppose, if possible, that LMN is the smallest of the four triangles. Then $\lambda\mu\nu + 1$ must be less than or equal to each of

$$(\lambda + 1)\nu, \qquad (\mu + 1)\lambda, \qquad (\nu + 1)\mu.$$

By addition,

$$3(\lambda\mu\nu + 1) \leqslant (\lambda + 1)\nu + (\nu + 1)\lambda + (\nu + 1)\mu$$
$$= (\mu\nu + \lambda) + (\nu\lambda + \mu) + (\lambda\mu + \nu),$$

that is,

$$(\lambda - 1)(\mu v - 1) + (\mu - 1)(v\lambda - 1) + (v - 1)(\lambda\mu - 1) \leqslant 0.$$

Since λ, μ, v are all $\geqslant 1$ or all $\leqslant 1$, each of these three terms must be zero, so that

$$\lambda = \mu = v = 1.$$

§13.6

1. (i), (ii), and (iv) lack the additive identity (zero); (iii) contains 1 and i but not $1 + i$. The remaining four sets include zero, which has no multiplicative inverse.

2. If B is the centroid of masses a at A and c at C, B' is the centroid of masses a at A' and c at C'. Points dividing AA', BB', CC' in the ratio $\mu : 1$ are the centroids of masses 1 at A and μ at A', 1 at B and μ at B', 1 at C and μ at C'. Of these three points, the middle one is the centroid of masses a at the first and c at the last.

In the concise notation of Möbius' *barycentric calculus*, we have

$$B = aA + cC, \qquad B' = aA' + cC',$$
$$B + \mu B' = a(A + \mu A') + c(C + \mu C').$$

3. In Möbius' notation, the centroid of equal masses at the four vertices of the quadrangle $ABCD$ is $A + B + C + D$. The vertices of the Varignon parallelogram are $A + B$, $B + C$, $C + D$, $D + A$, and its center is $(A + B) + (C + D) = (B + C) + (D + A)$.

4. Cutting the quadrangle along either diagonal, we obtain two triangles whose centroids are the midpoints of two of the broken line segments in Figure 13.6b.

5. A centrally symmetrical quadrangle, that is, a parallelogram.

§13.7

1. Inside the triangle $A_1 A_2 A_3$ we have $+ + +$; beyond the side $A_2 A_3$, $- + +$; and beyond the vertex A_1, $+ - -$.

2. $\begin{vmatrix} 1 & 0 & 1 \\ 0 & 1 & 1 \\ x & y & 1 \end{vmatrix}$.

3. $\frac{1}{2}(\overrightarrow{OS} + \overrightarrow{OT}) = \frac{1}{2}(\Sigma s_i \, \overrightarrow{OA_i} + \Sigma t_i \, \overrightarrow{OA_i})$

$$= \Sigma \tfrac{1}{2}(s_i + t_i)\overrightarrow{OA_i}.$$

4. $\sigma \Sigma s_i \overrightarrow{OA_i} + \tau \Sigma t_i \overrightarrow{OA_i} = \Sigma(\sigma s_i + \tau t_i)\overrightarrow{OA_i}.$

5. In this formulation it is no longer necessary to assume $\Sigma s_i = \Sigma t_i$.

6. $\begin{vmatrix} 0 & 1 & \lambda \\ \mu & 0 & 1 \\ 1 & v & 0 \end{vmatrix} = \lambda\mu v + 1.$

This has to be divided by $(\lambda + 1)(\mu + 1)(v + 1)$. When L, M, N are collinear, it becomes zero, in agreement with Menelaus's theorem.

7. When the line is entirely outside the triangle, the signs are all alike (say all plus). When the line penetrates the sides $A_2 A_3$ and $A_3 A_1$, T_3 differs in sign from T_1 and T_2.

§13.8

1. Any common point of a and b would be a common point of a and α. Apply 13.82 to b, c, and a.

2. Since a is parallel to b, it is parallel to the plane α through b, and we can use ex. 1.

3. Our proof of 13.81 shows that all the lines through A in the plane $q'r'$ are parallel to lines in the plane qr.

4. The centroid of equal masses at A_1, A_2, A_3, A_4 is the centroid of masses 1 at A_1, 1 at A_2, and 2 at the midpoint of A_3A_4.

5. (t_1, t_2, t_3, t_4) is the centroid of masses t_i at A_i ($i = 1, 2, 3, 4$).

§13.9

1. In the notation of Figure 13.8b, if A', B', C', O' are the vertices opposite to A, B, C, O, the six sides of the skew hexagon $AB'CA'BC'$ can be joined to the diagonal OO' to form a cycle of six tetrahedra, consecutive pairs of which are related by affine reflections; e.g., the tetrahedra $OO'AB'$ and $OO'B'C$ are related by the affine reflection $OO'(AC)$, which leaves invariant the plane $OO'B'$ while interchanging A and C.

2. An affine reflection interchanges pairs of points, P and P', in such a way that all the joining lines PP' are parallel and all the segments PP' are bisected by the mirror (which, in the three-dimensional case, is a plane).

3. The points (a, b, c) and (a', b', c') are interchanged by the central inversion $(x, y, z) \rightarrow (a + a' - x, b + b' - y, c + c' - z)$.

4. If (x, y, z) is a lattice point lying in a first rational plane $Xx + Yy + Zz = \pm 1$, any common divisor of x, y, z would have to divide ± 1.

5. No. For instance, $(1, 1, 0)$ is a visible point in the "second" rational plane $x + y = 2$.

6. When $x = 1$, we have $2y + 3z = -1$. Two obvious solutions are $y = 1, z = -1$, and $y = -2, z = 1$. When $x = -4$, we have $2y + 3z = 5$, with the obvious solution $y = z = 1$. We thus obtain the triangle $(1, 1, -1)(1, -2, 1)(-4, 1, 1)$.

7. The triangle $(1, 1, -1)(1, -2, 1)(-4, 1, 1)$, whose determinant is -1, is half a unit cell for the lattice in the plane $6x + 10y + 15z = 1$. Hence the general lattice point in this plane is

$$(1, 1, -1) + m(0, -3, 2) + n(-5, 0, 2) = (1 - 5n, 1 - 3m, -1 + 2m + 2n),$$

where m and n run through all the integers.

8. The given equation implies $x^2 + 2\sqrt{2}xy + 2y^2 = 3z^2$. Since $\sqrt{2}$ is irrational, any solution in integers woud require $xy = 0$ and $x^2 + 2y^2 = 3z^2$, which is impossible by the usual argument for establishing the irrationality of $\sqrt{3}$.

§14.1

1. By 14.11 the four points described in 14.13 are joined in pairs by six lines which, by 14.12, meet any other line in at least three points. Also each of the six lines meets the others in three points.

2. The m points, with c lines through each, apparently make a total of cm lines; but in this estimate each of the n lines is counted d times, once for each of the d points on it. Therefore $cm = dn$. The Pappus configuration 9_3 may be regarded as a cycle of three "Graves triangles" in six ways. (Coxeter, *Projective Geometry*, 1964, p. 39.)

3. Here is the table:

12	11	10	9	8	7	6	5	4	3	2	1	0
1	2	3	4	5	6	7	8	9	10	11	12	0
2	3	4	5	6	7	8	9	10	11	12	0	1
4	5	6	7	8	9	10	11	12	0	1	2	3
10	11	12	0	1	2	3	4	5	6	7	8	9

The last column indicates that the points on p_0 are P_0, P_1, P_3, P_9, and that the lines through P_0 are p_0, p_1, p_3, p_9. The other columns have an analogous interpretation. The columns of the table

P_0	P_{10}	P_9	P_6	P_5	P_2	P_8	P_3
P_{10}	P_9	P_6	P_5	P_2	P_8	P_3	P_0
P_6	P_5	P_2	P_8	P_3	P_0	P_{10}	P_9

indicate the eight lines of the configuration 8_3 formed by the cycle of eight points $P_0 P_{10} P_9 P_6 P_5 P_2 P_8 P_3$. The two mutually inscribed quadrangles are obtained by taking alternate points of this cycle.

4. Through any one of the points we have $p + 1$ lines, each containing p further points. This makes $1 + (p + 1)p$ points altogether.

5. The whole finite geometry provides a counterexample to refute Sylvester's theorem. Every line joining two of the points belongs to the geometry and thus contains not only two but $p + 1$ of the points.

§14.2

1. Let A, B, C, D be the points 14.23, and continue as follows:

$$E = AD \cdot BC = (0, 1, 1), \qquad F = BD \cdot CA = (1, 0, 1),$$
$$G = AB \cdot EF = (-1, 1, 0), \qquad H = BC \cdot DG = (0, 2, 1),$$
$$I = AD \cdot FH = (1, 2, 2), \qquad J = EF \cdot CI = (1, 2, 3).$$

2. The three pairs of opposite sides are

$$x_2 \pm x_3 = 0, \qquad x_3 \pm x_1 = 0, \qquad x_1 \pm x_2 = 0.$$

3. We see that $P_5 = P_2 P_3 \cdot P_4 P_7$, $P_6 = P_1 P_7 \cdot P_3 P_4$. The collinearity of $P_0 P_5 P_6$ makes $x^2 + x + 1 = 0$.

4.
$$P_4 = P_0 P_5 \cdot P_1 P_2 = (0, 1, 1), \qquad P_8 = P_1 P_5 \cdot P_2 P_0 = (1, 0, 1),$$
$$P_3 = P_2 P_5 \cdot P_0 P_1 = (1, 1, 0), \qquad P_6 = P_1 P_5 \cdot P_3 P_4 = (1, 2, 1)$$
$$P_7 = P_0 P_5 \cdot P_3 P_8 = (2, 1, 1), \qquad P_9 = P_0 P_1 \cdot P_4 P_8 = (1, 2, 0),$$
$$P_{10} = P_1 P_2 \cdot P_0 P_6 = (0, 1, 2), \qquad P_{11} = P_2 P_5 \cdot P_0 P_6 = (1, 1, 2),$$
$$P_{12} = P_0 P_2 \cdot P_1 P_7 = (2, 0, 1).$$

The lines are as follows:

$p_0 : x_3 = 0.$ $p_1 : x_2 = 0.$ $p_2 : x_1 + x_3 = 0.$

$p_3 : x_2 + x_3 = 0.$ $p_4 : x_1 + x_2 + x_3 = 0.$ $p_5 : x_1 + x_2 - x_3 = 0.$

$p_6 : -x_1 + x_2 + x_3 = 0.$ $p_7 : x_1 + x_2 = 0.$ $p_8 : x_1 - x_2 = 0.$

$p_9 : x_2 - x_3 = 0.$ $p_{10} : x_1 - x_2 + x_3 = 0.$ $p_{11} : x_1 - x_2 = 0.$

$p_{12} : x_1 = 0.$

5. The points $P_0 P_1 P_2 P_3 P_4 P_5 P_6$ may be taken to be

$$(1, 0, 0) \quad (0, 1, 0) \quad (0, 0, 1) \quad (1, 1, 0) \quad (0, 1, 1) \quad (1, 1, 1) \quad (1, 0, 1).$$

§14.3

1. The points $(1, 0, 0)$, $(1, 1, 1)$, $(p, 1, 1)$ all lie on the line $x_2 = x_3$. We obtain $(0, q - 1, 1 - r)$ by subtracting $(1, 1, r)$ from $(1, q, 1)$.

3. $S = P_{11}$, $T = P_5$, $U = V = F = P_{10}$. (The point P_6 is not used.)

§14.4

1. The definition for a harmonic set involves A and B symmetrically, also C and F in the same way.

2. Draw any triangle RSP whose sides SP, PR, RS pass respectively through A, B, C. Let AR meet BS in Q. Then PQ meets AB in the desired harmonic conjugate.

3. Taking RAB as triangle of reference, let C and S be $(0, 1, \lambda)$ and $(1, 1, \lambda)$. Then Q is $(1, 1, 0)$, P is $(1, 0, \lambda)$, and F is $(1, 1, 0) - (1, 0, \lambda) = (0, 1, -\lambda)$.

4. On any line in $PG(2, 3)$, there are exactly four points. The harmonic conjugate of any one of these with respect to any two others is a fourth point on the line and therefore can only be *the* fourth point on the line!

5. The same harmonic set is determined projectively by the quadrangle, and affinely by dividing the segment AA' internally at A_1 and externally at A_2, in the same ratio.

§14.5

1. If x and x' are corresponding lines of two perspective pencils, their point of intersection $x \cdot x'$ continually lies on the axis o.

2. Any section of a harmonic set of lines is a harmonic set of points, and any harmonic set of points is projected by a harmonic set of lines.

5. Whenever a projectivity on a line g is the product of two perspectivities, the join of the two centers meets g in an invariant point.

7. If the given projectivity is an involution, say $(AA')(BB')$, it is expressible as the product of the two involutions $(AB)(A'B')$ and $(AB')(BA')$. If the given projectivity is not an involution, and A is any noninvariant point, the projectivity may be expressed as $AA'A'' \barwedge A'A''A'''$ (where possibly A''' coincides with A); it is then seen to be the product of the two involutions

$$(AA'')(A'A') \qquad \text{and} \qquad (AA''')(A'A'').$$

8. (i) $(c_{11} - c_{22})^2 + 4c_{12}c_{21} = 0$.
  (ii) $c_{11} + c_{22} = 0$.

§14.6

1. Let AA' be the given pair of corresponding points, collinear with the center O. Let AX meet the axis in C. Then the collineation takes AC to $A'C$ and leaves invariant the line OX. Therefore X' is the point of intersection of $A'C$ and OX.

2. In the notation of Figure 14.3a, consider the perspective collineation with center O and axis DE that transforms P into P'. When the construction in ex. 1 is applied to Q it yields Q', and when it is applied to R it yields R'.

3. Let two points A and X, outside the line o of invariant points, be transformed into A' and X'. Since AA' and XX' are invariant lines, their common point O is an invariant point and therefore lies on o. Hence all joins of pairs of corresponding points meet o in the same point O.

4. In the notation of ex. 3, let O_1 be the harmonic conjugate of O with respect to A and A'. Then the harmonic homologies with centers A and O_1 will have the desired effect, since the former leaves A invariant and the latter takes A to A'.

5. Yes. For the unique projectivity $P_0P_1P_3 \barwedge P_1P_2P_4$ must transform the remaining point on P_0P_1 into the remaining point on P_1P_2. It is not necessary to give actual perspectivities, but in case they are desired, one possibility is

$$P_0P_1P_3P_9 \overset{P_{10}}{\underset{\wedge}{=}} P_0P_2P_8P_{12} \overset{P_9}{\underset{\wedge}{=}} P_1P_2P_4P_{10}.$$

$P_i \to P_{3i}$ is a projective collineation of period 3.

6. (i) A homology with center $(0, 0, 1)$ and axis $x_3 = 0$.
  (ii) An elation with center $(c_1, c_2, 0)$ and axis $x_3 = 0$.

7. Consider a quadrilateral $APXA_1P_1X_1$, as in Figure 14.6b, with A conjugate to A_1 and P to P_1. The polars a and p pass through A_1 and P_1, respectively. By 14.64, the polar triangles APX and apx are perspective from the line A_1P_1. Therefore x passes through X_1, and X is conjugate to X_1.

8. The condition for the two points $(0, 1, \pm1)$ to be conjugate is $c_{22} - c_{33} = 0$; for $(\pm1, 0, 1)$, $c_{33} - c_{11} = 0$. These two conditions imply $c_{11} - c_{22} = 0$, which is the conjugacy condition for $(1, \pm1, 0)$.

9. The given bilinear relation makes $x_1 = 0$ the polar of $(1, 0, 0)$, and $x_1 + x_2 + x_3 = 0$ the polar of $(1, 1, 1)$. Any self-conjugate point (x) must satisfy $x_1{}^2 + x_2{}^2 + x_3{}^2 = 0$. This is impossible in the real field but happens for all the four points $(1, \pm1, \pm1)$ in $PG(2, 3)$.

§14.7

2. When $B = D$, we have $x = p$, $y = PQ = d$, and $x \cdot y = P$.
When $A = D$, we have $y = q$, $x = d$, and $x \cdot y = Q$.

3. See Coxeter **2**, pp. 88–89.

7. The hint shows that the correlation $P_i \to p_i$ is projective. Being obviously of period 2, it is a polarity. The triangle $P_4P_{10}P_{12}$, whose sides are p_4, p_{10}, p_{12}, is self-polar. Finally, since the residues 0, 7, 8, 11 are the halves (mod 13) of 0, 1, 3, 9, the points P_0, P_7, P_8, P_{11} (and no others) lie on their polars.

Thus the four lines p_0, p_7, p_8, p_{11} are tangents, the six lines $p_1, p_2, p_3, p_5, p_6, p_9$ are secants, and the three lines p_4, p_{10}, p_{12} are non-secants. The three non-secants are the sides of the self-polar triangle $P_4P_{10}P_{12}$ which was used in describing the polarity. Since each non-secant is a common side of two self-polar triangles, there are three further self-polar triangles

$$P_4P_5P_9, \qquad P_{10}P_3P_6, \qquad P_{12}P_1P_2,$$

each having for its sides one non-secant and two secants [like the triangle EHH of Coxeter **2**, p. 82, Fig. 6.2C]. Each secant, containing only one pair of distinct conjugate points, is a side of only one self-polar triangle. Hence the only self-polar triangles are the four already mentioned.

Other geometries, such as $PG(2, 5)$, admit self-polar triangles formed by three secants or by one secant and two non-secants.

8. The sides of this hexagon are

$$x_1 = 0, \qquad x_1 = x_2 + x_3, \qquad x_2 = 0,$$
$$x_1 + x_3 = 2x_2, \qquad x_1 + x_2 = \tfrac{7}{5}x_3, \qquad \tfrac{1}{2}x_1 + x_2 = x_3.$$

Opposite sides meet in the three points

$$(0, 1, 2), \qquad (6, 1, 5), \qquad (2, 0, 1),$$

which all lie on the line $x_1 + 4x_2 = 2x_3$.

§14.8

1. Two distinct transversals from R would determine a plane containing both a and b.

2. Let a, b, c be three skew generators. Let an arbitrary plane through a meet c in R. This plane contains also the generator $Ra \cdot Rb$.

3. The four lines A_iB_i all intersect one another, and since they are not all coplanar they must be concurrent.

4. Calling the centers of perspective C_1, C_2, C_3, C_4, we see that $C_1C_2C_3C_4$ is perspective with $B_2B_3B_4B_1$ from A_1, with $B_1B_4B_3B_2$ from A_2, with $B_4B_1B_2B_3$ from A_3, with $B_3B_2B_1B_4$ from A_4, with $A_2A_3A_4A_1$ from B_1, and so on (with A's and B's consistently interchanged).

5. Each point of $PG(3, p)$ lies on $p^2 + p + 1$ lines, each containing p further points. Hence there are altogether $1 + p(p^2 + p + 1) = p^3 + p^2 + p + 1$ points, and, by duality, the same number of planes. Each of the $p^3 + p^2 + p + 1$ planes contains $p^2 + p + 1$ lines, but each line lies in $p + 1$ planes; therefore the total number of lines is

$$\frac{(p^3 + p^2 + p + 1)(p^2 + p + 1)}{p + 1} = (p^2 + 1)(p^2 + p + 1).$$

This expression was obtained by Von Staudt in 1856. (See the footnote on p. 237. See also P. H. Schoute, *Mehrdimensionale Geometrie*, vol. 1, p. 5, Leipzig, 1902.) When $p = 3$, the number is 130.

§14.9

1. Because Euclidean geometry does not admit self-perpendicular lines.

3. This is Clifford's first theorem in its original form, which can be derived from the form given in our text by inversion in a circle with center S.

§15.1

1. By 15.11 there is, on the ray AB, a point B' such that $CD \equiv AB'$. Thus we have $AB \equiv CD$ and $CD \equiv AB'$. By 15.12, $AB \equiv AB'$. But $AB \equiv AB$, and both B and B' are on the ray AB. Hence, by 15.11, $B' = B$.

2. Any triangle has an incircle, and the lengths of the tangents to it from A, B, C are $s - a$, $s - b$, $s - c$, as in §1.5, ex. 3. We have to abandon all the formulas involving trigonometrical functions, but ex. 1 remains valid. Even an acute-angled triangle may fail to have a circumcircle.

§15.2

1. See Coxeter 3, p. 189.

2. If two lines have a common perpendicular m, they are symmetrical by reflection in m. Any point of intersection on one side of m would yield another on the opposite side, contradicting 12.2511.

§15.3

1. The plane through l perpendicular to the plane ABC meets the latter in a line m which may intersect l or be parallel or ultraparallel to l. In the first case, all the planes Al, Bl, Cl pass through the point of intersection. In the second case, they pass through the common end of l and m. In the third case, by 15.26, l and m have a common perpendicular EF, and the planes Al, Bl, Cl are all perpendicular to the plane through EF perpendicular to l.

2. By 15.23, p and r are parallel. Therefore the product of reflections in them is a parallel displacement. The first reflection leaves J invariant; the second transforms J into L.

§15.4

1. A tetragonal rotation about the front vertex on the left, a trigonal rotation about the center of the face d, the half-turn about the line joining the midpoints of two opposite edges, and the half-turn about the line joining two opposite vertices: $(ab)(cd) = (acbd)^2$. Not counting the identity, we have $6 + 8 + 6 + 3 = 23$.

2. The cyclic group C_p has two "sets of poles," each consisting of just one pole. By adding an ineffective p_3-gonal pole with $p_3 = 1$, we are able to include this group as a "trivial" solution of 15.42.

§15.5

1. (a) C_2.

(b) $D_{2n}D_n$ (n even), $D_n \times \{I\}$ (n odd).

2. C_6C_3.

§15.6

(a) $D_2 \times \{I\}$.　　(b) $D_3 \times \{I\}$.　　(c) $A_4 \times \{I\}$.

§15.7

1. The vertices of $\{p, 2\}$ are p points evenly spaced along a great circle (say, the equator), and its edges are the p arcs that join neighboring vertices. The vertices of $\{2, p\}$ are two antipodal points (say, the north and south poles), and its edges are evenly spaced great semicircles (meridians).

2. The tetrahedron has six planes of symmetry, each joining an edge to the midpoint of the opposite edge. The cube and octahedron have nine, one parallel to each pair of opposite faces of the cube and one joining each pair of opposite edges. The dodecahedron and icosahedron have fifteen, one joining each pair of opposite edges.

3. $4\pi \Big/ \left(\dfrac{\pi}{p} + \dfrac{\pi}{q} + \dfrac{\pi}{2} - \pi \right)$. Two corresponding edges of the blown-up $\{p, q\}$ and $\{q, p\}$ perpendicularly bisect each other at the common right angle of four specimens of the fundamental region. Accordingly, it is natural that the order of the group should turn out to be $4E$.

§15.8

1. If the nth radius is k_n, we have $k_{n-1}k_{n+1} = k_n^2$.

3. Using the abbreviation

$$k^2 = \cos^2 \frac{\pi}{p_1} - \sin^2 \frac{\pi}{p_2} = \cos^2 \frac{\pi}{p_2} - \sin^2 \frac{\pi}{p_1}$$

(cf. 10.42), we find that the radius and distance are

$$\frac{1}{k} \sin \frac{\pi}{p_1} \quad \text{and} \quad \frac{1}{k} \cos \frac{\pi}{p_2}.$$

§16.1

If AB and AM, respectively, are perpendicular and parallel to r, as in Figure 16.3a, we have an acute angle at A and a right angle at B, and yet the rays do not meet.

§16.2

1. In the projective model, points and lines are represented by points and lines. Therefore isometries are represented by collineations. Since parallel lines are transformed into parallel lines, these collineations must transform ω into itself. Since a reflection leaves invariant every point on the mirror, the corresponding collineation must be an elation or a homology (§14.6). Since it is of period 2, it can only be a harmonic homology. Since it preserves ω, its center is the pole of the mirror (i.e., the point of intersection of the tangents at the ends of the chord).

When the mirror is represented by one of the vertical circles, the reflection appears as the inversion in the sphere, through this circle, orthogonal to Klein's sphere.

2. A pencil of concentric circles may be described as the orthogonal trajectories of a pencil of lines. In the conformal model these lines appear as circles through a pair of inverse points (with respect to Ω). Therefore the circles belong to the orthogonal pencil of coaxal circles (including Ω and having the pair of inverse points for limiting points).

§16.3

1. In the projective model, the common perpendicular to two ultraparallel lines joins their poles with respect to ω, and the common parallel to two rays joins their ends.

2. For any point G on A/B, we have $\angle MAG > \angle MBA$. But $\angle DAM = \angle EBM$. Therefore $\angle DAG > \angle EBA = \angle BAD$.

3. By considering congruent right-angled triangles, we see that $AD = CF = BE$. Since the angle C of the triangle ABC is equal to $\angle CAD + \angle EBC$, the sum of all three angles of the triangle is

$$\angle BAD + \angle EBA,$$

that is, the sum of two (equal) acute angles.

4. This is a generalization of the theorem that the altitude lines of a triangle are concurrent (which is a corollary of Fagnano's problem). The simplest proof uses the projective model and refers to Chasles's theorem (14.64): *Any two polar triangles are perspective triangles.*

5. Either a translation or a glide reflection, according as the triangles are, or are not, on the same side of their common side.

6. To a circle of the indicated radius, draw tangents at the ends of three radii making angles $120°$ with one another. These tangents form a trebly asymptotic triangle.

7. Draw the Cevian through the given point and compare Figure 16.3*a*.

§16.5

3. Consider how successive translations along CA, AB, BC will affect the side CA of the triangle ABC. The first translation slides this segment CA along itself to a position AX. The second (along AB) takes this to BY, where $\angle ABY = A$. The third (along BC) takes this to CZ, where $\angle ZCB = \pi - \angle CBY = \pi - (A + B)$, so that $\angle ZCA = \pi - A - B - C$. (This result can evidently be extended from triangles to higher polygons.)

4. Consider how successive half-turns about the midpoints DA, AB, BC, CD will affect the side DA of the quadrangle $ABCD$. The first half-turn reverses this side, yielding AD. The second (about the midpoint of AB) takes this to BX, where $\angle ABX = A$. The third (about the midpoint of BC) takes this to CY, where $\angle BCY = A + B$. The fourth (about the midpoint of CD) takes this to DZ, where $\angle CDZ = A + B + C$, so that $\angle ZDA = 2\pi - A - B - C - D$.

5. At any vertex we find one specimen of each angle of the polygon, in natural order. The cycle may be repeated any number of times (if the polygon has a sufficiently large area).

§16.6

1. Compare §16.3, ex. 3. The perpendicular bisectors of the sides of the triangle may be either intersecting or parallel or ultraparallel.

2. The external bisectors of two angles of the triangle may be either intersecting or parallel or ultraparallel.

3. The horocycle is symmetrical by reflection in any diameter. The diameter r reflects J into L.

4. Two. Their centers are the two ends of the perpendicular bisector of the segment joining the two points.

5. Remember that an equidistant curve has two branches.

7. Use §16.3, ex. 2.

§16.7

1. An equidistant curve.

2. The additive property, described in §6.6, ex. 5, shows that hyperbolic distance is *proportional* to inversive distance. The factor of proportionality is a matter of convention, like the value $\mu = 1$ that led to 16.53.

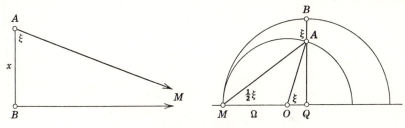

Figure 16.7a

3. The following proof of Lobachevsky's famous formula is credited to Paul Szász [see Coxeter, *Annali di Matematica, pura ed applicata*, (4), **71** (1966), p. 82]. In Figure 16.7a, the segment AB of length x is represented by the part AB of the line perpendicular to Ω at Q. The circle through B with center Q represents the line through B perpendicular to AB, and the tangent circle through A with center O (also on Ω) represents a parallel line having the same end M. The angle of parallelism

$$\xi = \Pi(x) = \angle BAM$$

appears as the angle at A in the "curvilinear triangle" BAM, and again as $\angle QOA$. Since x is the inversive distance between concentric circles with radii QB and QA (the latter not drawn), we have

$$e^x = \frac{QB}{QA} = \frac{QM}{QA} = \cot \tfrac{1}{2}\xi,$$

whence

$$\xi = 2 \arctan e^{-x}.$$

Any reader who dislikes using Euclidean trigonometry to obtain a hyperbolic result may turn to page 377 for a purely hyperbolic proof.

§16.8

By all the circles through one point (the point of contact).

§17.1

1. The vectors **a** and **c** must be either parallel to each other or both perpendicular to **b**.

2. $(\mathbf{a} \times \mathbf{b}) \times (\mathbf{c} \times \mathbf{d}) = [\mathbf{acd}]\mathbf{b} - [\mathbf{bcd}]\mathbf{a} = [\mathbf{abd}]\mathbf{c} - [\mathbf{abc}]\mathbf{d}$

3. $(\mathbf{a} \times \mathbf{b}) \cdot (\mathbf{a} \times \mathbf{b}) = (\mathbf{a} \cdot \mathbf{a})(\mathbf{b} \cdot \mathbf{b}) - (\mathbf{b} \cdot \mathbf{a})(\mathbf{a} \cdot \mathbf{b})$
$$= |\mathbf{a}|^2|\mathbf{b}|^2 - |\mathbf{a}|^2|\mathbf{b}|^2 \cos^2 \theta$$
$$= |\mathbf{a}|^2|\mathbf{b}|^2 \sin^2 \theta$$
$$= |\mathbf{a} \times \mathbf{b}|^2.$$

§17.2

The velocity is in the direction of the tangent. The acceleration is directed towards the center along the radius.

§17.3

1. $x = u - \sin u, y = -1 - \cos u$. This is, of course, a congruent cycloid.

2. $x = \cos u + u \sin u, y = \sin u - u \cos u$. This is a kind of spiral (but, of course, not an equiangular spiral).

3. Since $r = s \cos \varphi$.

§17.4

1. At the origin.

4. (a) $s = 4 \sin \psi$.

(b) $s = \frac{1}{2}l(\csc \psi \cot \psi + \log \tan \frac{1}{2}\psi)$.

5. Since $u = \log (\cosh u + \sinh u) = \log (\sec \psi + \tan \psi)$,
$$\int \sec \psi \, d\psi = \log (\sec \psi + \tan \psi) + C.$$

§17.5

$\rho = a \sinh u$. At the cusp it is zero: the curvature is infinite.

§17.6

1. Differentiating $\mathbf{r}^2 = 1$ twice, we obtain $\mathbf{r} \cdot \mathbf{t} = 0$ and $\mathbf{r} \cdot \mathbf{p} + \rho = 0$, whence
$$(\mathbf{r} + \rho\mathbf{p}) \cdot \mathbf{t} = (\mathbf{r} + \rho\mathbf{p}) \cdot \mathbf{p} = 0.$$

2. Because $\mathbf{t} \cdot \dfrac{d}{ds} (\mathbf{r} + \rho\mathbf{p}) = 0$.

§17.7

2. The helicoid $\dfrac{y}{x} = \tan \dfrac{z}{c}$.

3. $\dfrac{c}{a} = 1$.

§17.8

2. $\kappa = \tau = \dfrac{1}{3(1 + u^2)^2}$.

§17.9

1. $x = a\mu^u \cos u, \quad y = a\mu^u \sin u, \quad z = c\mu^u.$

3. The angle is arccot $(\sqrt{1 + c^2/a^2} \log \mu)$.

5. The cylinder based on an equiangular spiral.

§18.1

$$J[\mathbf{r}^1\mathbf{r}^2\mathbf{r}^3] = J\mathbf{r}^1 \cdot (\mathbf{r}^2 \times \mathbf{r}^3) = (\mathbf{r}_2 \times \mathbf{r}_3) \cdot (\mathbf{r}^2 \times \mathbf{r}^3)$$
$$= (\mathbf{r}_2 \cdot \mathbf{r}^2)(\mathbf{r}_3 \cdot \mathbf{r}^3) - (\mathbf{r}_3 \cdot \mathbf{r}^2)(\mathbf{r}_2 \cdot \mathbf{r}^3) = 1.$$

§18.2

2. $\mathbf{u} \cdot \mathbf{v} = g_{11}u^1v^1 + g_{22}u^2v^2 + g_{33}u^3v^3$
$$+ g_{12}(u^1v^2 + u^2v^1) + g_{23}(u^2v^3 + u^3v^2) + g_{31}(u^3v^1 + u^1v^3).$$

5. $\det g^{\alpha\beta} = G^{-1}.$

§18.3

2. $g_{\alpha\beta} = \delta_{\alpha\beta} + 1, \quad g^{\alpha\beta} = 4\delta^{\alpha\beta} - 1, \quad$ where $\quad \delta_{\alpha\beta} = \delta^{\alpha\beta} = \delta^\alpha_\beta.$

§18.5

3. $\Sigma x_\alpha{}^2 = 1, \quad \Sigma y_\alpha z_\alpha = 0.$

4. $g_{11} = g_{33} = 1, \quad g_{22} = (u^1)^2, \quad g_{\alpha\beta} = 0 (\alpha \neq \beta).$

5. $g_{11} = (u^3)^2, \quad g_{22} = (u^3 \sin u^1)^2, \quad g_{33} = 1, \quad g_{\alpha\beta} = 0 \ (\alpha \neq \beta).$

6. $g_{\alpha\beta} = \dfrac{1}{4}\left(\dfrac{x^2}{(A - u^\alpha)(A - u^\beta)} + \dfrac{y^2}{(B - u^\alpha)(B - u^\beta)} + \dfrac{z^2}{(C - u^\alpha)(C - u^\beta)}\right)$

§18.6

1. $\Sigma\Sigma\epsilon^{\alpha\beta\gamma}y_\beta z_\gamma.$

§19.1

2. $\Sigma\mathbf{r}^j \times \mathbf{r}_j = \Sigma\Sigma g^{ij}\mathbf{r}_i \times \mathbf{r}_j = \sqrt{g}\ \Sigma\Sigma g^{ij}\epsilon_{ij}\mathbf{n} = \mathbf{0}.$

The triangle formed by $\mathbf{r}^1$ and $\mathbf{r}_1$ has the same area, apart from sign, as the triangle formed by $\mathbf{r}^2$ and $\mathbf{r}_2$.

3. $\mathbf{r}^1 = \mathbf{r}_1/(1 + z_1{}^2), \quad \mathbf{r}^2 = \mathbf{r}_2/(u^1)^2.$

4. $g_{11} = g^{11} = 1, \quad g_{22} = \sin^2 u^1,$
$g^{22} = \csc^2 u^1, \quad g_{ij} = g^{ij} = 0 \ (i \neq j).$

§19.2

2. $\tan \phi = \sqrt{g}/g_{12}.$

4. $\cos \phi = \cos \theta \cos (\phi - \theta) - \sin \theta \sin (\phi - \theta)$

$$= \frac{a_1 a_2}{g_1 g_2} - \frac{a^2 a^1}{g^2 g^1}$$

$$= \frac{a_1}{g_1 g_2}(g_{12}a^1 + g_{22}a^2) - \frac{a^2}{g^2 g^1}(g^{11}a_1 + g^{12}a_2)$$

$$= \frac{g_{12}}{g_1 g_2}(a_1 a^1 + a^2 a_2) + \left(\frac{g_2}{g_1} - \frac{g^1}{g^2}\right)a_1 a^2$$

$$= \frac{g_{12}}{g_1 g_2}.$$

5. $g_1 a^1 = g_2 a^2$.

6. The net consists of the parametric curves, which are orthogonal if $g^{12} = 0$. The identity $g^{11}g_{11} - g^{22}g_{22} = 0$ is in agreement with the fact that the internal and external bisectors of an angle are perpendicular.

7. $S = \int_0^\pi \int_0^{2\pi} \sin u^1 \, du^1 \, du^2$.

§19.3

1. $\mathbf{r}^1 \times \mathbf{r}^2 = \mathbf{n}/\sqrt{g}$.

2. $b_{11} = -1$, $b_{22} = -\sin^2 u^1$, $b_{ij} = 0 \; (i \neq j)$.

§19.4

1. $H = 0$.

2. $H = \{z_1(1 + z_1{}^2) + u^1 z_{11}\}/2u^1(1 + z_1{}^2)^{\frac{3}{2}}$.

3. $u^1 = \frac{1}{2}\pi$, $u^1 = \frac{3}{2}\pi$.

5. At an umbilic.

§19.5

3. $u^1 \pm u^2 = k$.

4. $u^1 = \pm c \sinh (u^2 - k)$.

§19.6

1. At an umbilic, 19.52 is an identity.

2. The expression is a perfect square.

3. At an umbilic, $K = \kappa^2 > 0$.

4. The conditions $b_{11}:b_{12}:b_{22} = g_{11}:g_{12}:g_{22}$ become

$$-\sin^2 u^2 : \sin u^1 \cos u^1 \sin u^2 \cos u^2 : -\sin^2 u^1$$

$$= 2 \sin^2 u^1 + \cos^2 u^1 \sin^2 u^2 : \sin u^1 \cos u^1 \sin u^2 \cos u^2 : 2 \sin^2 u^2 + \sin^2 u^1 \cos^2 u^2.$$

5. No. When there is a curve of umbilics, this curve is itself a line of curvature, and the only lines of curvature that cross it do so at right angles.

§19.7

1. $b_{33} = 0$.

3. The lines of curvature are the intersections of the ellipsoid with the other quadrics of the system.

§19.8

1. $\theta = \frac{1}{4}\pi$, $\frac{3}{4}\pi$.

§20.1

1. $\sqrt{g}\, K = \dfrac{\partial}{\partial u^1}\left(\dfrac{\sqrt{g}}{g_{22}}\, \Gamma_{22}^1\right) - \dfrac{\partial}{\partial u^2}\left(\dfrac{\sqrt{g}}{g_{22}}\, \Gamma_{12}^1\right)$.

$\Gamma_{ij,i} = \frac{1}{2}(g_{ii})_j$, $\Gamma_{ij}^k = \Gamma_{ij,k}/g_{kk}$.

2. $\Gamma_{ii,i} = \frac{1}{2}(g_{ii})_i$, $\Gamma_{ii,j} = -\frac{1}{2}(g_{ii})_j$, $\Gamma_{ij,i} = \frac{1}{2}(g_{ii})_j$, $\Gamma_{ij}^k = \Gamma_{ij,k}/g_{kk}$.

3. $\Gamma_{12,2} = u^1$, $\Gamma_{22,1} = -u^1$, $\Gamma_{12}^2 = 1/u^1$, $\Gamma_{22}^1 = -u^1$; all others are 0.

5. By the equations just before 19.33,

$$\sqrt{g}\,\mathbf{r}^i = \Sigma \epsilon^{hi}\mathbf{n} \times \mathbf{r}_h.$$

Hence
$$\sqrt{g}\Sigma\Gamma^i_{ij} = \sqrt{g}\Sigma\mathbf{r}^i \cdot \mathbf{r}_{ij} = \Sigma\Sigma\epsilon^{hi}[\mathbf{n}\mathbf{r}_h\mathbf{r}_{ij}]$$
$$= \mathbf{n} \cdot \Sigma\Sigma\epsilon^{hi}\mathbf{r}_h \times \mathbf{r}_{ij} = \mathbf{n} \cdot (\mathbf{r}_1 \times \mathbf{r}_2)_j$$
$$= \mathbf{n} \cdot (\sqrt{g}\,\mathbf{n})_j = (\sqrt{g})_j.$$

6. $K = 1.$

§20.2

3. No; the tangents do not make a constant angle with the z-axis.

§20.3

2. $(\sqrt{g})_{11} = -\sqrt{g}.$

§20.4

2. Another circle of radius b and one of radius $a - b$.

§20.5

(i) $2\pi \sin r$,   (ii) $2\pi \sinh r$.

§20.7

$\cosh^2 u^1 + (u^2 + c)^2 = k^2.$

§21.1

Yes; it forms a map of three hexagons on the torus.

§21.2

2. Let $P_1P_2P_3P_4P_5P_6$ be a regular hexagon concentric with and interior to the disk. Join P_1P_4, P_2P_5, P_3P_6 through the boundary of the disk rather than through the center.
3. They form the Thomsen graph.
4. Yes, if q > 0. Both are nonorientable with $\chi = 2 - 2p - q$.

§21.3

1. $\{2, 1\}$ may be drawn as a great semicircle joining 2 antipodal vertices. $\{1, 2\}$ may be drawn as a great circle with one point on it specified as a vertex.
2. $\{3, 5\}/2$ has 6 vertices, each joined to every other. The vertices and edges of $\{5, 3\}/2$ form the Petersen graph [Ball **1**, p. 225].
5. $\{4, 4\}_{1,1}$ has 2 quadrangular faces, 4 edges, and 2 vertices; each vertex belongs to all 4 edges.
$\{4, 4\}_{2,0}$ has 4 quadrangular faces, 8 edges, and 4 vertices; each vertex belongs to all 4 faces.
$\{3, 6\}_{1,1}$ has 6 triangular faces, 8 edges, and 3 vertices; each vertex belongs to all 6 faces.
$\{6, 3\}_{1,1}$ has 3 hexagonal faces, 9 edges, and 6 vertices; each vertex belongs to all 3 faces.
$\{6, 3\}_{2,0}$ has 4 hexagonal faces, 12 edges, and 8 vertices.
All 5 maps are of genus 1.
7. $V = 3, 4, 4, 5, 6, 7.$

9. The positive integers p and q are not quite arbitrary. If one of them is 1, the other can only be 2. For instance, $q = 1$ imples $E = pr$, $F = 2r$, whence

$$E + F = (p + 2)r = (2p + 2q - pq)r = \chi \leqslant 2, \quad E = F = 1, \quad p = 2.$$

§21.4

1. The cube in only one way; the dodecahedron two ways.

§21.6

$$\chi = 2, \quad 1, \quad 0, \quad -1, \quad -2, \quad -3, \quad -4, \quad -5, \quad -6, \quad -7, \quad -8, \quad -9;$$
$$[N] = 4, \quad 6, \quad 7, \quad 7, \quad 8, \quad 9, \quad 9, \quad 10, \quad 10, \quad 10, \quad 11, \quad 11.$$

§22.1

1. $rN_1 = E'N_0$, where, by ex. 1 at the end of §10.3,

$$\frac{1}{E'} = \frac{1}{q} + \frac{1}{r} - \frac{1}{2}.$$

Similarly, $pN_2 = EN_3$.

2. This is derived from the analogous cube in the space $x_4 = 0$ by translating it through distance 1 along the fourth dimension.

3. The cube $(\pm 1, \pm 1, \pm 1, -1)$ is translated through distance 2 along the fourth dimension.

4. $(1, 1, 1, 1)$.

§22.2

1. Since $\cos \dfrac{\pi}{q} < \sin \dfrac{\pi}{p}$, $\dfrac{\pi}{p} + \dfrac{\pi}{q} > \dfrac{\pi}{2}$. Similarly, $\cos \dfrac{\pi}{q} < \sin \dfrac{\pi}{r}$.

2. $\{3, 3, 4\}$.

§22.3

1. $(\pm 1, 0, 0, 0)$, $(0, \pm 1, 0, 0)$, $(0, 0, \pm 1, 0)$, $(0, 0, 0, \pm 1)$.

2. $(\pm 1, \pm 1, 0, 0)$, permuted.

3. $(\tau, 1, \tau^{-1}, 0) = \tau^{-2}(\tau, 0, \tau, 0) + \tau^{-1}(\tau, \tau, 0, 0)$.

4. The extra points correspond to the centers of the 24 icosahedra.

§22.4

1. No.

2. (a) No. (b) Yes.

3. Yes. The twelve can have their centers at the vertices of a regular icosahedron.

4. $1^2 + 2^2 + \cdots + n^2 = n(n + 1)(2n + 1)/6$.

$$\binom{2}{2} + \binom{3}{2} + \cdots + \binom{n + 1}{2} = \binom{n + 2}{3}.$$

§22.5

No.

Index